T0214684

Lecture Notes in Computer Science 11349

Commenced Publication in 1973
Founding and Former Series Editors:
Gerhard Goos, Juris Hartmanis, and Jan van Leeuwen

Editorial Board

More information about this series at http://www.springer.com/series/7410

Carlos Cid · Michael J. Jacobson, Jr. (Eds.)

Selected Areas in Cryptography – SAC 2018

25th International Conference
Calgary, AB, Canada, August 15–17, 2018
Revised Selected Papers

 Springer

Editors
Carlos Cid (iD)
Royal Holloway, University of London
Egham, UK

Michael J. Jacobson, Jr. (iD)
University of Calgary
Calgary, AB, Canada

ISSN 0302-9743   ISSN 1611-3349   (electronic)
Lecture Notes in Computer Science
ISBN 978-3-030-10969-1   ISBN 978-3-030-10970-7   (eBook)
https://doi.org/10.1007/978-3-030-10970-7

Library of Congress Control Number: 2018965923

LNCS Sublibrary: SL4 – Security and Cryptology

This Springer imprint is published by the registered company Springer Nature Switzerland AG
The registered company address is: Gewerbestrasse 11, 6330 Cham, Switzerland

Preface

The Conference on Selected Areas in Cryptography (SAC) is the leading Canadian venue for the presentation and publication of cryptographic research, and has been held annually since 1994. SAC celebrated its 25th anniversary in 2018, taking place for the second time at the University of Calgary in Calgary, Alberta. In keeping with its tradition, SAC 2018 offered a relaxed and collegial atmosphere for researchers to present and discuss new results.

There are four areas covered at each SAC conference. Three of them are permanent:

- Design and analysis of symmetric key primitives and cryptosystems, including block and stream ciphers, hash functions, MAC algorithms, and authenticated encryption schemes
- Efficient implementations of symmetric and public key algorithms
- Mathematical and algorithmic aspects of applied cryptology

A fourth area varies from year to year, and the special selected topic for SAC 2018 was "Cryptography for the Internet of Things."

SAC 2018 received a total of 57 submissions, out of which the Program Committee (PC) selected 22 papers for presentation. The review process was thorough, with each submission receiving the attention of at least three reviewers (at least four for submissions involving a PC member). We would like to thank all authors for their submissions, and are very grateful to the PC members and reviewers for their effort and contribution to the selection of a high-quality program for SAC 2018.

There were three invited talks. The Stafford Tavares Lecture was given by Adi Shamir, who presented "Machine Learning in Security: Applications and Implications." The second invited talk was given by Andrey Bogdanov, who spoke about "Whitebox Cryptography." This year, in honor of its 25th anniversary, SAC had a special third invited talk by Carlisle Adams, who presented "SAC^{25}: A Retrospective." Stafford Tavares, one of the co-founders of SAC, was also a special invited guest, and gave a retrospective presentation of SAC at the conference banquet.

This year SAC also hosted what is now the fourth iteration of the SAC Summer School (S3). S3 is intended to be a place where early-career researchers can increase their knowledge of cryptography through instruction by, and interaction with, leading researchers in the field. We were fortunate to have Daniel J. Bernstein (Cryptographic Software Engineering), Andrey Bogdanov (Design of Lightweight Symmetric-Key Algorithms), Francesco Regazzoni (Cryptographic Hardware Engineering), and Meltem Sonmez Turan (Applications and Standardization of Lightweight Cryptography). We would like to express our sincere gratitude to these four presenters for dedicating their time and effort to what has become a highly anticipated and highly beneficial event for all participants.

A special thanks also goes to the team at the University of Calgary Conference Services, our technical and administrative support (Coral Burns, Mitra Mottaghi, and

Humaira Waqar), and our local student volunteers (Sepideh Avizheh, Shuai Li, Simpy Parveen, and Randy Yee) for their tireless support to the organisation of SAC 2018, both before and during the conference. Finally, we are very grateful to our sponsors, the Communications Security Establishment, Alberta Innovates, the Institute for Security, Privacy and Information Assurance, the Pacific Institute for the Mathematical Sciences, Springer, and the University of Calgary's Department of Computer Science, Faculty of Science, and Office of the Vice-President (Research), whose enthusiastic support (both financial and otherwise) greatly contributed to the success of SAC 2018.

November 2018   Carlos Cid
                                                    Michael J. Jacobson, Jr.

Organization

General and Program Chairs

Carlos Cid   Royal Holloway University of London, UK
Michael J. Jacobson, Jr. University of Calgary, Canada

Program Committee

Carlisle Adams   University of Ottawa, Canada
Diego Aranha   University of Campinas, Brazil
Frederik Armknecht   Universität Mannheim, Germany
Roberto Avanzi   ARM, Germany
Steve Babbage   Vodafone, UK
Paulo Barreto   University of Washington Tacoma, USA
Daniel J. Bernstein   University of Illinois at Chicago, USA
Alex Biryukov   University of Luxembourg, Luxembourg
Andrey Bogdanov   DTU, Denmark
Vassil Dimitrov   University of Calgary, Canada
Itai Dinur   Ben-Gurion University, Israel
Maria Eichlseder   TU Graz, Austria
Pierre-Alain Fouque   Université Rennes and Institut Universitaire de France,
                          France
Guang Gong   University of Waterloo, Canada
Johann Groszschaedl   University of Luxembourg, Luxembourg
M. Anwar Hasan   University of Waterloo, Canada
Howard Heys   Memorial University of Newfoundland, Canada
Jérémy Jean   ANSSI, France
Elif Bilge Kavun   Infineon Technologies, Germany
Stefan Kölbl   DTU, Denmark
Gaëtan Leurent   Inria, France
Subhamoy Maitra   Indian Statistical Institute, India
Brice Minaud   Royal Holloway University of London, UK
Nicky Mouha   NIST, USA
Michael Naehrig   Microsoft Research, USA
Svetla Nikova   KU Leuven, Belgium
Ludovic Perret   Sorbonne University/Inria/CNRS, France
Josef Pieprzyk   Data61, CSIRO, Australia
Francesco Regazzoni   Università della Svizzera Italiana, Switzerland
Matt Robshaw   Impinj, USA
Sondre Rønjom   University of Bergen, Norway
Fabrizio De Santis   Siemens AG, Germany
Sujoy Sinha Roy   KU Leuven, Belgium

Jörn-Marc Schmidt	secunet Security Networks, Germany
Peter Schwabe	Radboud University, The Netherlands
Kyoji Shibutani	Sony Corporation, Japan
Paul Stankovski	Lund University, Sweden
Frederik Vercauteren	KU Leuven, Belgium
Meiqin Wang	Shandong University, China
Hongjun Wu	Nanyang Technological University, Singapore
Huapeng Wu	University of Windsor, Canada
Bo-Yin Yang	Academia Sinica, Taiwan
Kan Yasuda	NTT, Japan
Amr Youssef	Concordia University, Canada

Additional Reviewers

Josep Balasch
Ward Beullens
Wouter Castryck
Morten Dahl
Jan-Pieter D'Anvers
Lauren De Meyer
Sébastien Duval
Wieland Fischer
Benedikt Gierlichs

Florian Goepfert
Angela Jäschke
Tanja Lange
Erik Mårtensson
Rachel Player
Vincent Rijmen
Hermann Seuschek
Alan Szepieniec
Zhenfei Zhang

Contents

Post-Quantum Cryptography

Lattice-Based Cryptography

Classical Public Key Cryptography

Machine Learning and Cryptography

Design of Symmetric Key Primitives

Targeted Ciphers for Format-Preserving Encryption

Sarah Miracle[✉] and Scott Yilek

University of St. Thomas, St. Paul, USA
{sarah.miracle,syilek}@stthomas.edu

Abstract. We introduce Targeted Ciphers, which typically encipher points on domain $\mathcal{X}$, but can be easily modified to instead encipher points on some subset $\mathcal{S} \subseteq \mathcal{X}$. Ciphers that can directly support this domain targeting are useful in Format-Preserving Encryption, where one wishes to encipher points on a potentially complex domain $\mathcal{S}$. We propose two targeted ciphers and analyze their security. The first, Targeted Swap-or-Not, is a modification of the Swap-or-Not cipher proposed by Hoang, Morris, and Rogaway (CRYPTO 2012). The second, a new cipher we call Mix-Swap-Unmix, achieves the stronger notion of full security. Our targeted ciphers perform domain targeting more efficiently than the recently proposed Cycle Slicer algorithm of Miracle and Yilek (ASIACRYPT 2017).

Keywords: Format-preserving encryption
Small-domain block ciphers · Markov chains · Matchings

1 Introduction

In this era of "big data," where organizations regularly harvest and store large amounts of customer data, the need to secure personal information in the face of data breaches has become essential. Encrypting sensitive personal and financial data like credit card numbers, social security numbers, and birth dates is an obvious way to defend against data breaches, but *how* to encrypt these diverse types of data is not always obvious. Practitioners are faced with the challenge of introducing encryption into large databases that interact with a potentially complex system of hardware and legacy software, while trying not to break anything. Given these challenges and constraints, it is easy to see the appeal of *Format-Preserving Encryption* (FPE) schemes, in which ciphertexts have the same format as plaintexts. For example, if one encrypts a 9 decimal digit US social security number with an FPE scheme, the resulting ciphertext would also be a 9 digit number. Such FPE schemes can often be "dropped in" to existing systems with little disruption.

Early attempts at constructing and analyzing FPE schemes were conducted by Brightwell and Smith [8] and later Spies [25]. The increasing practical interest in the problem, especially related to credit card encryption, has led to a recent

© Springer Nature Switzerland AG 2019
C. Cid and M. J. Jacobson, Jr. (Eds.): SAC 2018, LNCS 11349, pp. 3–26, 2019.
https://doi.org/10.1007/978-3-030-10970-7_1

surge in academic research on FPE and related problems [1–3,6,11,14–17,19–22,24]. There are even FPE standards NIST SP 800-38G [12] and ANSI ASC X9.124 that include Feistel-based FPE schemes like FF1 [4] and FF3 [7].

There are a variety of techniques known for constructing a format-preserving encryption scheme to encipher points in domain $\mathcal{S}$. Since block ciphers have traditionally been designed for bitstring domains, we cannot use an existing cipher (e.g., AES) without modification. Instead, there are generally three main strategies for constructing the desired encryption scheme. First, we could try to construct a cipher that is customized to work directly on domain $\mathcal{S}$. This works best when $\mathcal{S}$ has a relatively simple structure, like integers in the range $\{0, \ldots, N-1\}$, and many ciphers used in FPE are designed to work on this domain. (For example, such a cipher would work well for our social security number example; we would just need a cipher on $\{0, \ldots, N-1\}$ with $N = 10^9$).

If the domain $\mathcal{S}$ is more complicated, then a second option for building an FPE scheme is to try to find a way to *rank* the elements of the domain, then employ a cipher that works on $\{0, \ldots, N-1\}$ with $N = |\mathcal{S}|$, and then *unrank*. Ranking the elements of $\mathcal{S}$ means finding an efficient way to map (and unmap) each element $m \in \mathcal{S}$ to a unique element $x \in \{0, \ldots, |\mathcal{S}|-1\}$. The FPE scheme just described is called rank-encipher-unrank [3]. Rank-encipher-unrank only works on domains for which efficient ranking and unranking algorithms are known. Thus, practitioners, when faced with the task of enciphering points in some domain $\mathcal{S}$, must either invent a custom ranking and unranking procedure,[1] or, if $\mathcal{S}$ can be specified with a DFA or regular expression, apply known algorithms to rank regular languages [3]. In this latter case, there are toolkits written to aid practitioners [16], though there can still be some subtle efficiency issues depending on whether one starts with a regular expression or a DFA.

Finally, a third option that only assumes the ability to test membership in $\mathcal{S}$ is to find a larger domain $\mathcal{X}$ for which an efficient cipher already exists, and then try to somehow use or modify this cipher to get a new cipher on the target domain $\mathcal{S}$. For example, if we need a cipher on *valid* social security numbers (e.g., do not start with 000), we could try to take a cipher on $\{0, \ldots, 10^9\}$ and somehow cleverly use it to get a cipher on our desired domain. Black and Rogaway [6] were the first to analyze a folklore technique for doing this, called Cycle Walking, in which the cipher on the larger set is applied repeatedly to a point $m \in \mathcal{S}$ until the resulting ciphertext also is an element of $\mathcal{S}$. If the size of $\mathcal{X}$ is not too large relative to the size of $\mathcal{S}$, then we can expect this procedure to terminate quickly, though the running time can vary across different inputs. Recent work [19,20] by Miracle and Yilek has explored ways to make this task of transforming a cipher on $\mathcal{X}$ into a cipher on $\mathcal{S} \subseteq \mathcal{X}$, which they refer to as *domain targeting*, possible in constant time, meaning the running time does not depend on the input. Looking ahead, our results can be seen as bringing the very theoretical results of [19,20] closer to practice.

---

[1] For example, if one must encipher dates in the form mm/dd, then a custom ranking might map each date to a day numbered 0–365 in the obvious way.

We emphasize that while ranking/unranking and domain targeting might seem like two distinct ways to build FPE schemes on domains S, they can actually be complementary techniques. For example, a practitioner might have a very complicated domain S for which they wish to do FPE. Perhaps the domain is specified by a complex regular expression, and so the general techniques for ranking/unranking are impractical. An alternative option may be to find a larger, simpler set $X \supseteq S$ that is easier to rank. Then, the rank-encipher-unrank algorithm would need to apply something like domain targeting before unranking, since applying rank-encipher-unrank might yield an element in $X - S$.

CONSTANT-TIME DOMAIN TARGETING. Our main goal in this paper is to make constant-time domain targeting more efficient. Before getting to our new results, we first give an overview of previous techniques.

The constructions provided by Miracle and Yilek for domain targeting from set X to set S, called Reverse Cycle Walking in [19] and Cycle Slicer in [20], are both based on the same underlying idea: take a cipher on X and use it to construct a random matching (i.e., permutation with only 2-cycles or transpositions) on $S \subseteq X$; then swap some of the points that are paired together based on bit flips. Said another way, both the Reverse Cycle Walking (RCW) and Cycle Slicer (CS) constructions give a way to build matchings on the target set S out of arbitrary permutations on the larger set X. Once a matching on S is formed, pairs of points in the matchings are swapped based on additional bit flips. This procedure, called a matching exchange process, is repeated over many rounds, and Miracle and Yilek use a result of Czumaj and Kutolowski [9] to argue the resulting ciphers are secure. Further, the number of rounds needed for security does not depend on the specific inputs, so constant-time implementations that do not leak timing information are possible.

Unfortunately, RCW and CS are both rather inefficient, requiring many rounds for security. For example, the Cycle Slicer paper uses social security numbers as an example, with $X = \{0, 1\}^{30}$ and $S = \{0, \ldots, 10^9 - 1\}$, and claims that about 12,000 rounds of Cycle Slicer are needed for security. If we plug an existing, provably-secure cipher like Swap-or-Not (SN) [15] into the construction, we would end up with hundreds of rounds of Swap-or-Not times 12,000 rounds of Cycle Slicer, meaning overall we need millions of Swap-or-Not rounds. If full security [13, 24] is desired, in which ciphers are required to be indistinguishable from random permutations even when adversaries can query all domain points, the situation is even worse. The key idea in this paper is that instead of applying a general transformation to convert any cipher into one that supports domain targeting, perhaps we can instead specifically design ciphers (or slightly modify existing ones) to directly support domain targeting.

OUR RESULTS. We take a step toward bringing constant-time domain targeting closer to practice. We propose using what we refer to as *targeted ciphers* for the task. The idea is to design new ciphers (or find existing ones) that already can support domain targeting with only small modification. Informally, a targeted cipher will proceed in rounds to encipher points in some domain X, yet can be slightly modified to have the property that after every round, every point

$x \in \mathcal{S} \subseteq \mathcal{X}$ is still mapped to another point in $\mathcal{S}$. In other words, over the entire course of the algorithm, elements of the target set $\mathcal{S}$ never "leave" the target set, and every additional round of the cipher further mixes up these elements.

With this informal idea in mind, we present two targeted ciphers and formally analyze their security. Our first targeted cipher, *Targeted Swap-or-Not* (TSN), is a modification of the Swap-or-Not cipher, proposed by Hoang, Morris, and Rogaway. The second, which achieves the stronger notion of full security, is a new cipher we design and analyze called *Mix-Swap-Unmix* (MSU). With both ciphers, we achieve a substantial increase in efficiency when compared to constructions that achieve a similar level of security by using a general transformation like Cycle Slicer, bringing domain targeting closer to practicality.

TECHNIQUES. Like previous work on domain targeting, both of our targeted ciphers are matching-based, or swap-based, meaning that every round pairs up points and then swaps some of them. To construct a cipher on $\mathcal{S} \subseteq \mathcal{X}$, previous work, specifically Cycle Slicer, in each round builds a *random* matching on the larger set $\mathcal{X}$ and then, for each pair of points x, x' paired together in the matching, only swaps x and x' if both points are in the target set $\mathcal{S}$ and an additional bit flip is 1. The security analysis heavily relies on the fact that the matchings are random, which allows [20] to apply an existing result of Czumaj and Kutolowski [9].

Our first targeted cipher, Targeted Swap-or-Not, stems from the observation that the Swap-or-Not cipher is already matching-based: focusing on the version of SN for domain $\mathcal{X} = [N]$, in round i of SN, point x is paired up with point $x' = K_i - x \mod N$, where K_i is the random round key. The points x and x' are then swapped if a random function applied to them is 1. This operation clearly results in a matching on $[N]$, so our targeted version adds the constraint that points should only be swapped if they are both in the target set $\mathcal{S} \subseteq [N]$.

Since the high-level idea in TSN is the same as in Cycle Slicer, it might appear that the same security analysis should follow. But, there is a key difference: in Cycle Slicer, each round is a random matching, while in TSN we get a very non-random matching completely determined by the round key (which can be computed from any known pair x, x'). Thus, for TSN's analysis we cannot rely on the matching exchange process results. Instead, we modify the original Swap-or-Not security proof of [15], using a recent refinement by Dai, Hoang, and Tessaro [10]. Our final security bounds show that TSN needs only a modest increase in rounds over SN to support targeting. As an example, our bounds show that if TSN is applied to domain $[N]$ for $N = 2^{30}$ and targeted to a target set of size $|\mathcal{S}| = 10^9$, and if we allow a CCA adversary $q = |\mathcal{S}|/2$ queries, then we need just under 600 rounds of Swap-or-Not to get advantage less than 10^{-9}. Using Cycle Slicer and Swap-or-Not for the same parameters would require hundreds of thousands of rounds.

For our second targeted cipher, Mix-Swap-Unmix (MSU), we aim to build a targeted cipher that can achieve *full security*. A fully secure cipher is one that indistinguishable from a random permutation by an adversary who can query all N domain points. Only a few fully-secure ciphers are known, and they tend to be inefficient; for example, the Mix-and-Cut cipher of [24] uses about 10,000 rounds of Swap-or-Not to encipher 30-bit inputs. If one wishes to do domain targeting and still maintain full security, the efficiency problem gets even worse. Combining a fully-secure cipher like Mix-and-Cut with a general domain targeting transformation like Cycle Slicer can result in 100s of millions of rounds of Swap-or-Not. Thus, we aim to build a new fully-secure cipher that directly supports domain targeting.

Like previous fully-secure ciphers [21,24], our new cipher MSU is built from Swap-or-Not. At the same time, since we want to support targeting, we need each round of MSU to give a matching on the larger domain $\mathcal{X} = [N]$, and then we can only swap elements that are both in the target set $\mathcal{S}$. To build this matching, we use an idea from Naor and Reingold [23]. They used the fact that for permutations π and σ, the cycle structure of $\pi \circ \sigma \circ \pi^{-1}$ is the same as the cycle structure of the inner permutation σ, to build permutations with particular cycle structures. Since we want a matching, or a permutation made up of just 2-cycles, we let π (the outer permutation) be Swap-or-Not, and then σ (the inner permutation) simply be the permutation that swaps adjacent elements. This is one round of MSU. While this gives us a targeted cipher, we still need to argue full security. In Sect. 4, we show that this construction boosts the security of Swap-or-Not and gives us full security. The final construction is also much more efficient than using an existing fully-secure cipher with Cycle Slicer, requiring about 100 times fewer rounds of Swap-or-Not.

EXTENSIONS AND FUTURE WORK. We mention a few other related results we have included in the paper. First, the MSU construction described above uses an additional bit flip for each pair of points in each round. This bit flip seems unnecessary and leads to an increase in the number of rounds in the case where $\mathcal{S}$ is much smaller than $\mathcal{X}$. In Appendix A, we show that in this setting, the bit flip can in fact be eliminated. The proof involves finding an equivalent underlying matching exchange process that mimics MSU without the bit flips, and the techniques may be of independent interest. We also show in Appendix A that if domain targeting is not needed and one simply wants to use the MSU cipher on domain $[N]$, then we can prove that significantly less rounds are needed by applying a recent result of Bernstein [5]. In short, MSU without targeting results in something called an *involution walk*, and techniques from representation theory can be applied.

One last extension of our results is that our targeted ciphers can be used in a straightforward way to solve the *domain completion* problem, recently introduced in [14] and further studied in [20], in which we wish to construct a cipher that stays consistent with a table of existing input-output mappings that were manually chosen. Specifically, our constructions can take the place of Cycle Slicer in the CSDC algorithm of [20], resulting in efficiency gains in that setting.

Looking forward, an obvious question is whether other well-known cipher design techniques can be modified to directly support targeting. For example, Feistel-based ciphers are widely used and, in fact, the standardized FPE schemes are Feistel-based, so it would be convenient if they could be made to support targeting with simple modifications. Unfortunately, this seems unlikely. A card-shuffling view of Feistel is that the input points are cut into many piles, and then the bottom cards are dropped from the piles in different orderings depending on the internal random round function. Imagine some of the cards at the bottom of the cut piles are initially in positions in the target set S. These cards will end up near the bottom of the deck after one round of Feistel, but the positions near the bottom of the deck might not correspond to positions in S. Thus, we immediately lose our desired property of targeted ciphers that points in S always stay in S after each round.

Finally, though we used the MSU construction to build a (rather slow) fully-secure cipher by applying in each round Swap-or-Not, a swap, and then Swap-or-Not inverse, we believe Swap-or-Not could be replaced by something much faster (e.g., a few rounds of Feistel) in the MSU construction, and the resulting (targeted) cipher could provide strong security with a modest number of rounds.

2   Preliminaries

NOTATION. If x is a bitstring with length n, then we denote by $x \oplus 1$ the bitwise exclusive-OR of the n bits of x with the bitstring $0^{n-1}1$ ($n-1$ zeroes followed by a single one). If S is a set, then $x \leftarrow_{\$} S$ means we choose an element of S uniformly at random and assign it to x. If S is instead an algorithm, then the same notation represents running S with uniformly random coins and assigning the output to x. For permutations $\pi, \sigma : \mathcal{M} \to \mathcal{M}$ with π having inverse π^{-1}, we denote by $\pi \circ \sigma \circ \pi^{-1}$ the permutation that computes $\pi^{-1}(\sigma(\pi(x)))$ on $x \in \mathcal{M}$. We let $[N]$ denote the set $\{0, \ldots, N-1\}$. For $X \in [N]$, then we let $X \oplus 1$ denote the result of taking the binary representation of X and applying a bitwise-XOR with the binary representation of 1; in other words, if X is even (resp. odd), then $X \oplus 1$ will be the next (resp. previous) number. Let $\mathsf{odd}(N)$ denote the odd elements of $[N]$.

BLOCK CIPHERS. We say that $E : \mathcal{K} \times \mathcal{M} \to \mathcal{M}$ for finite sets $\mathcal{K}$ and $\mathcal{M}$ (sometimes referred to as the key space and domain, respectively) is a block cipher if $E_K(\cdot) = E(K, \cdot)$ is a permutation on $\mathcal{M}$ for every $K \in \mathcal{K}$. Let E^{-1} be the inverse block cipher of E.

The standard notion of security for block ciphers is security against adaptive chosen-ciphertext attack (CCA), sometimes called Strong PRP Security. To define this security notion, we describe the security games SPRP1 and SPRP0. In SPRP1, the game starts with a **main** procedure that chooses a random key for the cipher and then runs the adversary with oracles for procedures Enc and Dec, which answer queries using the cipher and the chosen key. The final output of the game is the bit the adversary outputs. The game SPRP0 works the same, but with **main** choosing a random permutation from $\mathsf{Perm}(\mathcal{M})$, defined as the set of all permutations $\pi : \mathcal{M} \to \mathcal{M}$, and using that to answer oracle queries to Enc and Dec. We can then define the CCA advantage of an adversary A against E by $\mathbf{Adv}_E^{\mathrm{cca}}(A) = |\mathrm{Pr}\left[\mathsf{SPRP1}_E^A \Rightarrow 1 \right] - \mathrm{Pr}\left[\mathsf{SPRP0}_E^A \Rightarrow 1 \right]|$ where the probabilities are over the random coins used in the security games. If the adversary A

is non-adaptive (meaning it makes the same queries every run) and only makes queries to Enc in the SPRP security games above, then we say it is a NCPA (short for non-adaptive chosen-plaintext attack) adversary and we refer to its advantage in the games against block cipher E as $\mathbf{Adv}_E^{\text{ncpa}}(A)$.

As has become standard, we overload notation and denote by $\mathbf{Adv}_E^{\text{cca}}(q)$ the maximum CCA advantage over all adversaries making at most q adaptive oracle queries. Similarly, the maximum advantage over all adversaries making at most q non-adaptive oracle queries to only the forward direction subroutine Enc we denote by $\mathbf{Adv}_E^{\text{ncpa}}(q)$. We will be interested in *full security* or *fully-secure ciphers*, meaning $\mathbf{Adv}_E^{\text{cca}}(N)$ is low, where $N = |\mathcal{M}|$. Said another way, a fully-secure block cipher will be one for which the CCA advantage is low despite the adversary being able to query every domain point. As explained in the introduction, such fully-secure ciphers have been the target of a number of recent papers [13, 21, 24].

CHERNOFF BOUND. Later in the paper, we will need to upper bound the probability that among t independent coin flips there are more than $(3/4)t$ heads.

Proposition 1. *Let $X_1, \ldots, X_t$ be independent random variables such that each $X_i = 1$ with prob. $1/2$ and $X_i = 0$ with prob. $1/2$. Let $X = \sum_{i=1}^t X_i$. Then, $\Pr[X \geq (3/4)t] \leq e^{-t/20}$.*

MATCHINGS. In this paper we use the term *matching* on $\mathcal{M}$ to refer to a permutation $\tau : \mathcal{M} \to \mathcal{M}$ made up of only transpositions, also called 2-cycles or swaps. A matching is an involution, so $\tau(\tau(x)) = x$ for all $x \in \mathcal{M}$. Let $\mathsf{Match}(\mathcal{M}, k)$ be the set of all matchings on $\mathcal{M}$ that are made up of exactly k transpositions. For a set $\mathcal{M}$ with an even number N of elements, we use the term *perfect matching* to refer to a matching on $\mathcal{M}$ with exactly $N/2$ transpositions, meaning every point is swapped with another distinct point. Thus, $\mathsf{Match}(\mathcal{M}, |\mathcal{M}|/2)$ is the set of such perfect matchings when $|\mathcal{M}|$ is even.

A *matching exchange process* on $\mathcal{M}$ proceeds in rounds. In each round, k is sampled from some probability distribution on $\{0, \ldots, |\mathcal{M}|/2\}$, then τ is chosen randomly from $\mathsf{Match}(\mathcal{M}, k)$. Finally, for each pair of points $x, \tau(x) \in \mathcal{M}$ such that $x \neq \tau(x)$, we flip a random bit $b_{\{x,\tau(x)\}}$ and define a new matching $\bar{\tau}$ by $\bar{\tau}(x) = \tau(x)$ if $b_{\{x,\tau(x)\}} = 1$, and $\bar{\tau}(x) = x$ otherwise. We then apply this new matching $\bar{\tau}$ to each point in $\mathcal{M}$. This process may repeat for many rounds (with independently chosen k and matchings). We will also consider a special case of a matching exchange process called an *involution walk*. Here the matching τ generating at each step is always a perfect matching (i.e. k is always $|\mathcal{M}|/2$). In Appendix A we bound the number of rounds needed for MSU in part by relying on previous bounds for matching exchange processes and involution walks.

TOTAL VARIATION DISTANCE. In order to determine how many rounds of MSU are needed, we will bound the total variation distance for the underlying matching exchange process. Let $x, y \in \Omega$, $P^r(x, y)$ be the probably of going from x to y in r steps and μ be another distribution on Ω. In our case Ω will be the set of all permutations of a given size, $P^r(x, y)$ will be the probably of going between particular permutations x and y with r rounds of MSU and μ will be the uniform distribution on permutations. Specifically, for a permutation y, $\mu(y) = 1/|\Omega|$. Then the *total variation distance* is defined as

$$||P^r(x, y) - \mu|| = \max_{x \in \Omega} \frac{1}{2} \sum_{y \in \Omega} |P^r(x, y) - \mu(y)|.$$

COMPOSITION AND CYCLE STRUCTURE. We will use the following well-known fact from group theory, which was used in the cryptographic realm by Naor and Reingold [23].

Proposition 2. *For any permutations π and σ, the cycle structures of permutations σ and $\pi \circ \sigma \circ \pi^{-1}$ are the same. Thus, if τ is a matching, then $\pi \circ \tau \circ \pi^{-1}$ is also a matching with the same number of transpositions.*

3   Targeted Swap-or-Not

We begin by describing the Swap-or-Not cipher introduced by Hoang, Morris, and Rogaway [15] and then present our new Targeted Swap-or-Not cipher (TSN).

SWAP-OR-NOT. Hoang, Morris, and Rogaway [15] showed that the Swap-or-Not (SN) cipher provides CCA security against adversaries who only make $q = (1 - \epsilon)N$ queries, where N is the size of the domain. In words, for domain $\mathcal{M} = [N]$, the r-round SN cipher has key KF specifying round keys $K_i \in \mathcal{M}$ and round functions $F_i : \mathcal{M} \to \mathcal{M}$. In round i, point X is paired with a "buddy" point $K_i - X \mod N$ (which could be the same point, i.e., $K_i - X \mod N = X$), and the result of F_i determines if X should *swap* positions with its buddy point *or not*.

Hoang, Morris, and Rogaway analyzed the security of Swap-or-Not and provided bounds on both the NCPA and CCA advantages of adversaries attacking the scheme. Recently, Dai, Hoang and Tessaro [10] improved these bounds using a technique they named the chi-squared method. We will need their bound

$$\mathbf{Adv}_{\mathsf{SN}}^{\mathsf{cca}}(A) \leq \frac{2N}{\sqrt{r/2 + 1}} \left(\frac{N + q}{2N} \right)^{(r/2+1)/2}, \tag{1}$$

where again N is the size of the domain, r is the number of SN rounds, and q is the number of adversarial queries.

OUR ALGORITHM. In each round i of TSN, point X is again paired with a "buddy" point $K_i - X \mod N$. However, regardless of the result of the round function F_i if either X or X's "buddy" point are not in the target set $\mathcal{S}$ then the points do not swap positions. If both points are in $\mathcal{S}$ then whether they swap (or not) is again determined by the round function F_i. The detailed description of how to encipher a single point using TSN can be found in Fig. 1. Note that if we

```
procedure TSN_KF(x)
    for i in 1 ... r do
        x' ← (K_i − x) mod N ;  X̂ ← max(x, x')
        if F_i(X̂) = 1 | and x ∈ S and x' ∈ S | then
            x ← x'
    return x
```

Fig. 1. The Targeted Swap-or-Not Cipher for target set $\mathcal{S}$. The addition of the boxed code is the only change from the original Swap-or-Not algorithm.

let $\mathcal{S} = [N]$ then TSN becomes the original Swap-or-Not cipher for the domain $\{0, \ldots, N-1\}$.

SECURITY ANALYSIS. Our analysis of TSN relies heavily on the original analysis done by Hoang, Morris and Rogaway to bound the NCPA security of the Swap-or-Not algorithm [15] and then improved by Dai, Hoang and Tessaro [10] using the χ^2 method. Our main contribution here lies in the application of this algorithm to the targeting setting and the analysis while quite technical is a generalization of the ideas and techniques used in the previous analysis.

Our goal is to bound the CCA security of TSN but, as in [10], we will begin by bounding the weaker NCPA security using the χ^2 method and then use a result of Maurer, Pietrzak, and Renner [18] to derive a bound on the CCA security. Specifically we adapt Lemma 3 from [10] to the TSN algorithm. Combining this lemma with the techniques from the proof of Lemma 5 from [10] and applying to TSN immediately gives the following lemma which shows that in order to bound the NCPA security of TSN it suffices to bound the χ^2-divergence.

Lemma 1 (adapted from Dai, Hoang, Tessaro [10]). *Let* TSN *represent the permutation generated by r rounds of Targeted Swap-Or-Not and* UN *represent a random permutation. Additionally, let $p_{TSN,r}(\cdot|Q_i)$ be the distribution on the $i+1$ query by a non-adaptive NCPA adversary A to* TSN *with r rounds conditioned on the output of the previous i queries represented by $Q_i = \{q_1, q_2, \ldots q_i\}$ and similarly $p_{UN}(\cdot|Q_i)$ is the distribution on the $i+1$ query to the uniformly random permutation (i.e. the uniform distribution on the remaining $|\mathcal{S}| - i$ elements). Given this the NCPA advantage of an NCPA adversary A making at most q non-adaptive queries is*

$$\mathbf{Adv}_{TSN}^{ncpa}(A) \leq \|p_{TSN,r}(\cdot) - p_{UN}(\cdot)\| \leq \sqrt{\frac{1}{2} \sum_{i=0}^{q-1} \mathbf{E}\left[\chi^2(p_{TSN,r}(\cdot|Q_i), p_{UN}(\cdot|Q_i))\right]}.$$

Where the expectation is taken over a vector $Q_i = \{q_1, q_2, \ldots q_i\}$ sampled according to the interaction with TSN *and the χ^2 divergence between $p_{TSN,r}(\cdot|Q_i)$ and $p_{UN}(\cdot|Q_i)$ is defined as*

$$\sum_{q_{i+1} \in \mathcal{S} \setminus Q_i} \frac{(p_{TSN,r}(q_{i+1}|Q_i) - p_{UN}(q_{i+1}|Q_i))^2}{p_{UN}(q_{i+1}|Q_i)}.$$

In order to bound $\mathbf{E}\left[\chi^2(Q_i)\right]$, we prove the following lemma which generalizes Eq. 5 from [15].

Lemma 2. *Let $|\mathcal{S}|$ be the number of elements in the target set $\mathcal{S}$ and $|\mathcal{X}|$ be the number of elements in the larger domain set $\mathcal{X}$. Then we have,*

$$\mathbf{E}\left[\sum_{q_{i+1}\in\mathcal{S}\setminus\{q_1,\dots,q_i\}}(p_{\mathsf{TSN},r}(q_{i+1}|Q_i)-p_{\mathsf{UN}}(q_{i+1}|Q_i))^2\right] \leq \left(\frac{2|\mathcal{X}|-|\mathcal{S}|+i}{2|\mathcal{X}|}\right)^r,$$

where the expectation is taken over a vector $Q_i = \{q_1, q_2, \dots q_i\}$ sampled according to the interaction with Targeted Swap-Or-Not.

Proof. Again we point out that the following proof uses the same techniques and is a relatively straightforward generalization of the proof of Eq. 5 from [15]. Our proof proceeds by induction on r. We let $r = 0$ be our base case (the proof here follows directly from [15]). When $r = 0$ the elements are in their initial deterministic location and

$$\mathbf{E}\left[\sum_{q_{i+1}\in\mathcal{S}}(p_{\mathsf{TSN},0}(q_{i+1})-p_{\mathsf{UN}}(q_{i+1}|Q_i))^2\right] = \mathbf{E}\left[\sum_{q_{i+1}\in\mathcal{S}}(p_{\mathsf{TSN},0}(q_{i+1})-1/|\mathcal{S}|)^2\right]$$

$$= (1-1/|\mathcal{S}|)^2 + (|\mathcal{S}|-1)(-1/|\mathcal{S}|)^2$$

$$= 1-1/|\mathcal{S}| < \left(\frac{2|\mathcal{X}|-|\mathcal{S}|+i}{2|\mathcal{X}|}\right)^0.$$

Next we assume inductively that the lemma holds for r and prove that it holds for $r+1$. In order to analyze this case we will need to use some additional terminology. For clarity we will use the same terminology as in [15] and [10] and redefine it here for readability. Let $K_1, \dots, K_{r+1}$ be the random keys for the first $r+1$ rounds. Let $S_r = \mathcal{S} - Q_{i,r}$ be the set of available positions for the $i+1$ query where $Q_{i,r}$ is set of positions for the first i queries given r rounds of TSN. We will abbreviate $p_r(x)$ to mean $p_{\mathsf{TSN},r}(x|Q_i)$ (i.e. the probability the $i+1$ query is x given r rounds of TSN) and define $s_r = \sum_{x \in S_r}(p_r(x)-1/(|\mathcal{S}|-i))^2$.

Definition 1 (Hoang, Morris, Rogaway [15]). *Let f be a bijection from S_r to S_{r+1} given by*

$$f(x) = \begin{cases} x & x \in S_{r+1}, \\ K_{r+1} - x & otherwise. \end{cases}$$

Given this, Hoang, Morris and Rogaway [10] point out the following.

$$p_{r+1}(f(x)) = \begin{cases} p_r(x) & \text{if } K_{r+1} - x \notin S_r, \\ \frac{1}{2}p_r(x) + \frac{1}{2}p_r(K_{r+1} - x) & \text{otherwise.} \end{cases}$$

The size of S_r is $|\mathcal{S}| - i$ and thus in our targeted setting, the probability that $K_{r+1} - x \notin S_r$ is $(|\mathcal{X}| - (|\mathcal{S}| - i))/|\mathcal{X}|$. Combining these and letting $Q = \mathbf{E}\left[(p_{r+1}(f(x)) - (1/(|\mathcal{S}| - i)))^2 | s_r\right]$ gives the following.

$$
Q = \frac{|\mathcal{X}| - |\mathcal{S}| + i}{|\mathcal{X}|}\left(p_r(x) - \frac{1}{|\mathcal{S}| - i}\right)^2 + \frac{1}{|\mathcal{X}|}\sum_{y \in S_r}\left[\frac{p_r(x) + p_r(y)}{2} - \frac{1}{|\mathcal{S}| - i}\right]^2
$$

$$
= \frac{|\mathcal{X}| - |\mathcal{S}| + i}{|\mathcal{X}|}\left(p_r(x) - \frac{1}{|\mathcal{S}| - i}\right)^2 + \frac{1}{|\mathcal{X}|}\left(\frac{s_r}{4} + \frac{|\mathcal{S}| - i}{4}\left(p_r(x) - \frac{1}{|\mathcal{S}| - i}\right)^2\right)
$$

$$
= \frac{s_r}{4|\mathcal{X}|} + \frac{3i + 4|\mathcal{X}| - 3|\mathcal{S}|}{4|\mathcal{X}|}\left(p_r(x) - \frac{1}{|\mathcal{S}| - i}\right)^2
$$

Note that the expansion of the first sum uses the definition of s_r and the fact that $\sum_{y \in S_r}(p_r(y) - \frac{1}{|\mathcal{S}| - i}) = 0$. Details can be found in [10]. Using the fact that f gives a bijection from S_r to S_{r+1} and the equation above, we have the following,

$$
\mathbf{E}\left[s_{r+1} | s_r\right] = \sum_{x \in S_{r+1}} \mathbf{E}\left[[p_{r+1}(x) - 1/(|\mathcal{S}| - i)]^2 | s_r\right]
$$

$$
= \sum_{y \in S_r} \mathbf{E}\left[[p_{r+1}(f(y)) - 1/(|\mathcal{S}| - i)]^2 | s_r\right]
$$

$$
= \sum_{y \in S_r}\left(\frac{s_r}{4|\mathcal{X}|} + \frac{3i + 4|\mathcal{X}| - 3|\mathcal{S}|}{4|\mathcal{X}|}\left(p_r(y) - \frac{1}{(|\mathcal{S}| - i)}\right)^2\right)
$$

$$
= \frac{s_r(|\mathcal{S}| - i)}{4|\mathcal{X}|} + \frac{3i + 4|\mathcal{X}| - 3|\mathcal{S}|}{4|\mathcal{X}|}\sum_{y \in S_r}\left(p_r(y) - \frac{1}{(|\mathcal{S}| - i)}\right)^2
$$

$$
= s_r\left(\frac{2|\mathcal{X}| - |\mathcal{S}| + i}{2|\mathcal{X}|}\right).
$$

Using the law of iterated expectations and our inductive hypothesis we have,

$$
\mathbf{E}\left[s_{r+1}\right] = \mathbf{E}\left[\sum_{q_{i+1} \in \mathcal{S}\backslash\{q_1,\ldots,q_i\}}\left(p_{\mathsf{TSN},r}(q_{i+1}|Q_i) - \frac{1}{|\mathcal{S}| - i}\right)^2\right]
$$

$$
\leq \left(\frac{2|\mathcal{X}| - |\mathcal{S}| + i}{2|\mathcal{X}|}\right)^r.
$$

□

Next, we use Lemma 2 to bound the χ^2 divergence and subsequently to bound the NCPA security of our Targeted Swap-or-Not Cipher.

Theorem 1. *Let TSN_r represent the permutation generated by r rounds of Targeted Swap-Or-Not. The NCPA advantage of an NCPA adversary A making at most q non-adaptive queries is*

$$\mathbf{Adv}_{TSN}^{ncpa}(A) \leq ||p_{TSN,r}(\cdot) - p_{UN}(\cdot)|| \leq \left(\frac{|\mathcal{S}| \cdot |\mathcal{X}|}{r+1}\right)^{\frac{1}{2}} \left(\frac{2|\mathcal{X}| - |\mathcal{S}| + q + 1}{2|\mathcal{X}|}\right)^{\frac{r+1}{2}}.$$

Proof. Using the definition of the χ^2 divergence given in Lemmas 1 and 2 we have the following.

$$\chi^2(p_{TSN,r}(\cdot|Q_i), p_{UN}(\cdot|Q_i)) = \sum_{q_{i+1} \in \mathcal{S} \setminus Q_i} \frac{(p_{TSN,r}(q_{i+1}|Q_i) - p_{UN}(q_{i+1}|Q_i))^2}{p_{UN}(q_{i+1}|Q_i)}$$

$$= (|\mathcal{S}| - i) \sum_{q_{i+1} \in \mathcal{S} \setminus Q_i} (p_{TSN,r}(q_{i+1}|Q_i) - (1/|\mathcal{S}| - i))^2$$

$$\leq (|\mathcal{S}| - i) \left(\frac{2|\mathcal{X}| - |\mathcal{S}| + i}{2|\mathcal{X}|}\right)^r$$

Next we substitute this result into Lemma 1 and bound the subsequent summation with an integral (similar to what was done in [10]) to get the following, which implies our theorem.

$$(||p_{TSN,r}(\cdot) - p_{UN}(\cdot)||)^2 \leq \frac{1}{2}\sum_{i=1}^{q} \mathbf{E}\left[\chi^2(Q_i)\right] \leq \frac{1}{2}\sum_{i=1}^{q}(|\mathcal{S}| - i)\left(\frac{2|\mathcal{X}| - |\mathcal{S}| + i}{2|\mathcal{X}|}\right)^r$$

$$\leq \frac{|\mathcal{S}|}{2}\int_0^{q+1}\left(\frac{2|\mathcal{X}| - |\mathcal{S}| + i}{2|\mathcal{X}|}\right)^r di$$

$$\leq \frac{|\mathcal{S}| \cdot |\mathcal{X}|}{r+1}\left(\frac{2|\mathcal{X}| - |\mathcal{S}| + q + 1}{2|\mathcal{X}|}\right)^{r+1}.$$

Finally, to bound the CCA security of TSN we will use a well-known result of Maurer, Pietrzak, and Renner [18]. As in the analysis by Dai, Hoang and Tessaro [10] we note that the inverse of r rounds of TSN is also r rounds of TSN and thus applying [18] allows us to amplify our NCPA security bound to CCA security and gives the following corollary.

Corollary 1. *Let TSN represent the permutation generated by r rounds of Targeted Swap-Or-Not. The CCA advantage of a CCA adversary A making at most q queries is*

$$\mathbf{Adv}_{TSN}^{cca}(A) \leq 2\left(\frac{|\mathcal{S}| \cdot |\mathcal{X}|}{r/2+1}\right)^{1/2}\left(\frac{2|\mathcal{X}| - |\mathcal{S}| + q + 1}{2|\mathcal{X}|}\right)^{(r/2+1)/2}.$$

4   Mix-Swap-Unmix

MOTIVATION. In the previous section, we saw a way to modify Swap-or-Not to get a targeted cipher, and the resulting cipher is indistinguishable from a random permutation when the adversary queries at most a constant fraction of the points. Recent papers [21,24] have introduced small-domain ciphers that provide *full security*, meaning the ciphers are indistinguishable from random permutations even to an adversary allowed to query all domain points. This leaves the question of whether we can build fully-secure small-domain ciphers that support targeting without too much loss in efficiency.

---

procedure $\mathrm{MSU}_{\mathbf{KF},\mathbf{G}}(X)$
    for j in $1 \ldots m$ **do**
        $Z \leftarrow \mathsf{SN}_{\mathrm{KF}_j}(X)\,;\ Z' \leftarrow Z \oplus 1$
        $X' \leftarrow \mathsf{SN}^{-1}_{\mathrm{KF}_j}(Z')\,;\ \hat{X} \leftarrow \max(X, X')$
        if $G_j(\hat{X}) = 1$ $\boxed{\text{and } X \in \mathcal{S} \text{ and } X' \in \mathcal{S}}$ **then**
            $X \leftarrow X'$
        else
            $X \leftarrow X$
    return X

Fig. 2. Mix-Swap-Unmix Cipher. The boxed code is for domain targeting, and can be excluded if a cipher on $[N]$ is desired.

We could certainly take an existing fully-secure cipher and apply a general transformation like Reverse Cycle Walking or Cycle Slicer to get a matching, and then only swap points that both lie in the target set. Unfortunately, fully-secure ciphers are already significantly less efficient than partially-secure counterparts like Swap-or-Not (which itself is far less efficient than the Feistel-based, standardized schemes), so using many rounds of a general transformation like Cycle Slicer is simply too slow to ever be practical. To be more concrete, if we start with the fully-secure cipher Mix-and-Cut [24] on domain $\mathcal{X} = \{0,1\}^{30}$, that cipher internally needs 10,000 rounds of Swap-or-Not to achieve full security. If we then apply Cycle Slicer to target the set of bitstrings that represent 9-digit numbers, then [20] states we need 12,000 rounds of Cycle Slicer, with each of those 12,000 rounds applying the 10,000 rounds of Swap-or-Not inside of Mix-and-Cut. Thus, to get a targeted, fully-secure cipher with this method, we would need $10000 \times 12000 = 120$ million rounds of Swap-or-Not!

Clearly, there is a lot of efficiency loss in using a general transformation like Cycle Slicer on an existing fully-secure cipher. Thus, we instead turn to a different approach: directly constructing a fully-secure cipher that is matching-based and thus supports domain targeting. Like the existing fully-secure ciphers Mix-and-Cut and Sometimes-Recurse, we build our new fully-secure cipher from Swap-or-Not. We call our new algorithm Mix-Swap-Unmix (MSU). MSU, by default, enciphers points in the general domain $[N] = \{0, \ldots, N-1\}$ for even N, and can support targeting to any domain $\mathcal{S} \subseteq [N]$.

THE ALGORITHM. Let $\mathsf{SN}_{\mathrm{KF}}$ denote the Swap-or-Not cipher with domain $[N]$, with key KF consisting of round keys $K_1, \ldots, K_r$ and round functions

$F_i : [N] \rightarrow \{0, 1\}$. Our new cipher MSU will have domain $\mathcal{S} \subseteq [N]$ and keys $(\mathbf{KF}, \mathbf{G})$ consisting of m Swap-or-Not keys $\mathbf{KF} = \{KF_1, \ldots, KF_m\}$ and m round functions $\mathbf{G} = \{G_1, \ldots, G_m\}$ with each $G_j : [N] \rightarrow \{0, 1\}$. The code is shown in Fig. 2. The boxed statements are for domain targeting; if one's desired domain is simply $[N]$, the boxed portion can be excluded.

In words, to encipher a point $X \in \mathcal{S} \subseteq [N]$ with MSU, we first apply r rounds of the Swap-or-Not cipher to get a new point Z. If Z is even, it is swapped with $Z+1$, otherwise it is swapped with $Z-1$. We then apply the inverse of the Swap-or-Not cipher applied earlier in the round to get a new point X'. If X and X' are both in $\mathcal{S}$ and an additional bit flip is 1, then the swap of X and X' becomes official; otherwise if either the bit flip is 0 or one of both of the points is not in $\mathcal{S}$, then X and X' are simply mapped to themselves for this round of MSU. Thus, in one round of MSU, a point X is either mapped to $\mathsf{SN}_{KF_j}^{-1}(\mathsf{SN}_{KF_j}(X) \oplus 1)$ or it is simply mapped back to itself.

Each round of MSU gives a matching, a permutation on $\mathcal{X}$ made up of only transpositions. This follows from Proposition 2 that states that if π and σ are permutations, then the permutation $\pi \circ \sigma \circ \pi^{-1}$ has the same cycle structure as σ. Since in MSU the "inner" permutation σ simply consists of swaps of points Z with $Z \oplus 1$, the overall cycle structure of MSU will also be made up of just swaps/transpositions.

SECURITY. We now formally show this construction gives a fully-secure cipher on $\mathcal{S}$, meaning it is indistinguishable from a random permutation even to an adversary that can see all $|\mathcal{S}|$ input-output mappings.

Theorem 2. *Let MSU be described as above, with m rounds, each of which uses r rounds of Swap-or-Not. Then,*

$$\mathbf{Adv}_{MSU}^{cca}(A) \leq m \cdot \Delta_1 + \Delta_2$$

where $\Delta_1 = \dfrac{2N}{\sqrt{r/2+1}} \left(\dfrac{7}{8}\right)^{(r/2+1)/2} + e^{\frac{-N}{40}}$ *and* $\Delta_2 = |\mathcal{S}|^{1-(2m/T)}$, *where*

$$T = \max\left(40\ln(2|\mathcal{S}|^2), \frac{10\ln(|\mathcal{S}|/9)}{\ln(1 + (7/36N^2)((7/9)|\mathcal{S}|^2 - |\mathcal{S}|))}\right) + \frac{72N\ln(2|\mathcal{S}|^2)}{|\mathcal{S}|}.$$

Before proving the theorem, we note that the presence of the $e^{\frac{-N}{40}}$ term means that MSU does not provide good security for very small domains. Yet, this term is not problematic for domains like those discussed in the Introduction where N is, say, 2^{30}.

Proof. Let $\mathcal{S} \subseteq [N]$ and let $\mathsf{MSU} : \mathcal{K} \times \mathcal{S} \rightarrow \mathcal{S}$ be the m-round Mix-Swap-Unmix algorithm as defined in Sect. 4 with randomly chosen round keys $\mathbf{KF}$, randomly chosen round functions $\mathbf{G}$, and using the r-round Swap-or-Not cipher on domain $[N]$. Let A be a CCA adversary against MSU that queries every point in $\mathcal{S}$. We wish to bound the following advantage

$$\mathbf{Adv}_{MSU}^{cca}(A) = \Pr\left[\mathsf{SPRP1}_{MSU}^A \Rightarrow 1\right] - \Pr\left[\mathsf{SPRP0}_{MSU}^A \Rightarrow 1\right].$$

To do so, we will use a sequence of game transitions, starting with $\mathsf{Gm}_0 = \mathsf{SPRP1}$ and making small changes to the games until we have $\mathsf{SPRP0}$.

For the rest of the proof, we will write $\Pr[\mathsf{Gm}]$ instead of $\Pr\left[\mathsf{Gm}^A \Rightarrow 1\right]$ for brevity. For our first game transition, we will modify the Enc procedure to apply the round functions $\mathbf{G}$ to the maximum of Z and Z', instead of to the max of X and X'. Let the resulting game be Gm_1. Regardless of this change, the round function still just associates a random bit flip with the pairs of points that are matched by this round, so $\Pr[\mathsf{Gm}_0] = \Pr[\mathsf{Gm}_1]$.

Our next game, Gm_2, is the same as Gm_1 but with the random round functions $\mathbf{G}$ replaced by bit flips that take place in the **main** function and are associated to every possible odd Z value in $[N]$; there are separate sets of bit flips for each round of MSU (placed into a table B), just as there are separate round functions G_j for each round. The Enc procedure then uses the table B with these bit flips in place of G_j in each round to determine if swaps take place. Detailed code for game Gm_2 is given in Appendix B. Since random round functions with $0/1$ outputs have just been replaced by random bit flips, $\Pr[\mathsf{Gm}_2] = \Pr[\mathsf{Gm}_1]$.

Notice that if, for any round j, too many bit flips are 1, then a bad flag bad_j is set. This will be needed later in the proof, but we point out here that the bad events only depend on the sum of independent bit flips in **main**, so we will be able to easily bound the probability of these events with a Chernoff bound.

Our next sequence of game transitions will replace Swap-or-Not in each round of MSU with a randomly chosen permutation on $[N]$. But, care must be taken, since our adversary A against MSU may query all domain points, yet Swap-or-Not is only proven secure against adversaries that query a constant fraction of the domain points. Intuitively, we will be able to overcome this "gap" by only making queries to Swap-or-Not when the round bits in the table B are 1.

More formally, we define a sequence of hybrid games $\mathsf{H}_0, \ldots, \mathsf{H}_m$. The first hybrid game, H_0, is identical to Gm_2, meaning it uses the bit table B in place of random round functions. In game H_ℓ, the first ℓ rounds of MSU use a completely random permutation, while the remaining rounds use Swap-or-Not. This means that the last hybrid game, H_m, is identical to Gm_2 but with every round of MSU using a random permutation on $[N]$ in place of Swap-or-Not. We now claim that for every $i \in \{1, \ldots, m\}$, $\Pr[\mathsf{H}_{i-1}] - \Pr[\mathsf{H}_i] \le \mathbf{Adv}_{\mathsf{SN}}^{\mathsf{cca}}(3N/4) + e^{\frac{-N}{40}}$.

To prove this, we provide a CCA adversary B against Swap-or-Not that makes at most $3N/4$ oracle queries. These queries will all be to the decryption oracle on the elements of $\mathsf{odd}(N)$ for which a bit flip is 1. Adversary B will run adversary A, answering its queries using its own oracles. If adversary B has a SN oracle, then it will end up simulating H_{i-1} for A, while if it has a random permutation oracle, it will end up simulating H_i.

Before we get to the exact details of this adversary B, we expand the equation in the above claim to take into account the event that bad_i is set to true. In the following equations, let bad_i denote the event that the flag bad_i (which, recall, means in the part of the B used in round i of MSU) is set to true during the execution of the game. Note that the probability of bad_i being set to true is the same in any hybrid game, since they all have identical **main** procedures. Now,

$\Pr[H_{i-1}] - \Pr[H_i]$

$= \left(\Pr[H_{i-1} \wedge \mathsf{bad}_i] + \Pr[H_{i-1} \wedge \overline{\mathsf{bad}_i}]\right) - \left(\Pr[H_i \wedge \mathsf{bad}_i] + \Pr[H_i \wedge \overline{\mathsf{bad}_i}]\right)$

$= \Pr[\mathsf{bad}_i] \cdot (\Pr[H_{i-1} \mid \mathsf{bad}_i] - \Pr[H_i \mid \mathsf{bad}_i])$

$\quad + \Pr[\overline{\mathsf{bad}_i}] \left(\Pr[H_{i-1} \mid \overline{\mathsf{bad}_i}] - \Pr[H_i \mid \overline{\mathsf{bad}_i}]\right)$

$\leq \Pr[\mathsf{bad}_i] + \Pr[\overline{\mathsf{bad}_i}] \left(\Pr[H_{i-1} \mid \overline{\mathsf{bad}_i}] - \Pr[H_i \mid \overline{\mathsf{bad}_i}]\right).$

We are now ready to specify our adversary B against Swap-or-Not. Adversary B is given a Swap-or-Not oracle and will run adversary A and try to simulate its environment to match the hybrid games H_{i-1} and H_i. If B has a real Swap-or-Not oracle, then it will end up simulating H_{i-1}, while if it has a random permutation oracle it will end up simulating H_i. To simulate round i of the MSU algorithm, B first flips coins just like in the **main** procedure of the hybrid games to populate the B table. If the bad_i (the bad flag for round i) gets set to true, meaning too many coin flips ended up as 1 for that round of MSU, then adversary B needs to stop and simply output a random $0/1$ guess. If bad_i is not set, then B proceeds by querying its own SN oracle with all z and $z \oplus 1$ in which $B[i][z] = 1$. B now runs A and can properly complete round i of MSU for A on any query, since the only way round i can affect a point X is if the corresponding bit in B is 1. Because B queried every such point, it will know what to do with any given X or X'. Thus, as long as the bad_i flag is not set, B will perfectly simulate the hybrid game for A.

In the equations below, let S1 be short for $\mathsf{SPRP1}^B \Rightarrow 1$ and S0 be short for $\mathsf{SPRP0}^B \Rightarrow 1$. We can now see adversary B's advantage

$\mathbf{Adv}_{\mathsf{SN}}^{\mathsf{cca}}(B) = \Pr[S1] - \Pr[S0]$

$= \left(\Pr[S1 \wedge \mathsf{bad}_i] + \Pr[S1 \wedge \overline{\mathsf{bad}_i}]\right) - \left(\Pr[S0 \wedge \mathsf{bad}_i] + \Pr[S0 \wedge \overline{\mathsf{bad}_i}]\right)$

$= \Pr[\mathsf{bad}_i] \cdot (\Pr[S1 \mid \mathsf{bad}_i] - \Pr[S0 \mid \mathsf{bad}_i])$

$\quad + \Pr[\overline{\mathsf{bad}_i}] \cdot \left(\Pr[S1 \mid \overline{\mathsf{bad}_i}] - \Pr[S0 \mid \overline{\mathsf{bad}_i}]\right)$

$= \Pr[\mathsf{bad}_i] \cdot (1/2 - 1/2) + \Pr[\overline{\mathsf{bad}_i}] \cdot \left(\Pr[S1 \mid \overline{\mathsf{bad}_i}] - \Pr[S0 \mid \overline{\mathsf{bad}_i}]\right)$

$= \Pr[\overline{\mathsf{bad}_i}] \cdot \left(\Pr[S1 \mid \overline{\mathsf{bad}_i}] - \Pr[S0 \mid \overline{\mathsf{bad}_i}]\right)$

When the bad_i flag is not set, adversary B running in the SPRP1 game is perfectly simulating the hybrid game H_{i-1} and B running in SPRP0 is perfectly simulating the hybrid game H_i. Thus, combining the equations above gives

$\Pr[H_{i-1}] - \Pr[H_i] \leq \Pr[\mathsf{bad}_i] + \Pr[\overline{\mathsf{bad}_i}] \left(\Pr[H_{i-1} \mid \overline{\mathsf{bad}_i}] - \Pr[H_i \mid \overline{\mathsf{bad}_i}]\right)$

$\leq \Pr[\mathsf{bad}_i] + \mathbf{Adv}_{\mathsf{SN}}^{\mathsf{cca}}(B)$

where adversary B makes at most $q = (3/4)N$ queries to its oracle. Applying the bound from Eq. (1) in Sect. 3 and Proposition 1 to our hybrid argument over m rounds gives us the Δ_1 bound in our theorem statement.

Now, continuing with our game transitions, let Gm_3 be the same as H_m, but with the bit flips moved into the Enc procedure and taking place at the time they are needed (in the **if**). This syntactic change has no effect on the output of the game. Next, we will transition from Gm_3 to a game Gm_4 in which each round of MSU now applies a randomly chosen perfect matching to X to get X' instead of computing $X' \leftarrow \pi^{-1}(\pi(X) \oplus 1)$.

We now claim that the new version of MSU in Gm_4 is actually a matching exchange process. This specific matching exchange process, where a perfect matching on $[N]$ is then restricted to a subset $\mathcal{S}$ (i.e., matchings that do not pair up points in $\mathcal{S}$ are thrown out), is analyzed in Appendix A. We can apply Theorem 5 in that appendix to show that $\Pr\left[\mathsf{Gm}_4^A \Rightarrow 1\right] - \Pr\left[\mathsf{SPRPO}_{\mathsf{MSU}}^A \Rightarrow 1\right] \le \Delta_2$ where Δ_2 is the bound from Theorem 5. Combining all of our bounds on the above game transitions completes the proof of Theorem 2. □

DISCUSSION AND EXTENSIONS. Using the Δ_1 and Δ_2 bounds above, we can see that we need a few hundred rounds of Swap-or-Not within each of about 5000 rounds of MSU, to get low adversarial advantage. While this is still a lot of rounds, it is substantially less than the 100 s of millions of rounds needed in previous work.

Additionally, we mention two extensions of this result. First, we have presented MSU as a cipher on $[N]$ that can be targeted to a domain $\mathcal{S} \subseteq [N]$. If we are only interested in a cipher on $[N]$ and do not need targeting, then we can improve the full security bound in Theorem 2 by applying a recent result of Bernstein [5] on the mixing time of involution walks, which are especially one type of matching exchange process. More details can be found in Appendix A, but our Δ_2 term in the above theorem will become the value in Corollary 2. Then, for the case where $N = 10^9$, we will only need about 220 rounds of MSU to get the Δ_2 term less than 10^{-9}.

Second, our MSU algorithm as described and analyzed above works best when $|\mathcal{S}| \ge |\mathcal{X}|/2$. If the target set is smaller than that, we can show the round function $\mathbf{G}$ (which essentially does bit flips that determine if a swap should take place) can be removed, which speeds up mixing. Showing this is non-trivial, since the resulting algorithm no longer appears to be a matching exchange process. We analyze the resulting process in more detail in Appendix A, Corollary 3.

Acknowledgements. We thank the SAC 2018 anonymous reviewers for their detailed and helpful comments.

A   Analyzing a Matching Exchange Process

In order to bound the number of rounds of MSU that are needed we analyze the underlying matching exchange process. We obtain three different bounds depending on how the size of the domain $\mathcal{X}$ relates the size of the target set $\mathcal{S}$. Our best bound is when the $|\mathcal{S}| = |\mathcal{X}|$ and we show that the process is an involution walk and rely on a recent result of Bernstein [5]. When $|\mathcal{S}| < |\mathcal{X}|$

we rely on previous work [9, 20] to bound the variation distance of a general matching exchange process. In the case where $|\mathcal{S}| \geq |\mathcal{X}|/2$ in order for MSU to be a matching exchange process we have added an additional bit flip to each pair selected (the round function G). When $|\mathcal{S}| < |\mathcal{X}|/2$ we prove that there exists a matching exchange process that results in the identical distribution on matchings generated by MSU and thus we do not need to add the additional bit flip. By eliminating this extra bit flip we improve the parameters of the matching exchange process and provide a tighter bound on the variation distance.

Recall that at each step of a matching exchange process a parameter $\kappa \leq |S|/2$ is selected according to some distribution. Next a matching of size κ on the set S is selected uniformly at random. Finally for each pair in the matching a bit is flipped independently to determine whether that particular pair is kept in the matching. For the purposes of this section, we will view MSU as generating a perfect matching on $\mathcal{X}$ and then ignoring all pairs in the matching except for those where both points are in our target set $\mathcal{S} \subseteq \mathcal{X}$. We consider the ideal scenario where each round of MSU generates a uniformly random perfect matching on $\mathcal{X}$.

AN INVOLUTION WALK. An involution walk is defined as a random walk on the symmetric group S_n for n even where at each step a uniformly random perfect matching on S_n is generated and then each pair in the matching is applied with probability $1 - p$ and discarded with probability p. It is straightforward to see that as intended, when $|\mathcal{X}| = |\mathcal{S}|$, MSU is indeed an involution walk on the set $\mathcal{X}$. Bernstein proves the following theorem for any involution walk.

Theorem 3 (Bernstein [5]). *For* $t = \log_{\frac{2}{1+p}}(n) + \frac{c}{\ln(\frac{2}{1+p})}$, n *such that* $\frac{10 \ln(n+2)}{\sqrt{(n+2)/2-1}} \leq \ln\left(\frac{2}{1+p}\right)$ *and* $n-1 > \sqrt{n/2}(1+\ln(n))$, *then* $||P^{*t} - U||_{TV} \leq e^{-c/2}$.

In order to apply this theorem to MSU we will require $n \geq 2^{19}$ and let $p = 1/2$ which gives the following corollary.

Corollary 2. *For* $n \geq 2^{19}$ *the involution walk with parameter* $p = 1/2$ *satisfies*

$$||P^{*t} - U||_{TV} \leq n^{1/2}e^{-t \ln(4/3)/2}.$$

Proof. Solving for c in the expression $t = \log_{\frac{2}{1+p}}(n) + \frac{c}{\ln(\frac{2}{1+p})}$ and then simplifying gives $c = t\ln(\frac{2}{1+p}) - \ln n$. Substituting this into the equation for variation distance and simplifying gives $||P^{*t} - U||_{TV} \leq n^{1/2}e^{-t \ln(\frac{2}{1+p})/2}$. Fixing $p = 1/2$ gives the desired result $||P^{*t} - U||_{TV} \leq n^{1/2}e^{-t \ln(4/3)/2}$. Requiring that $n \geq 2^{19}$ satisfies the requirements $\frac{10 \ln(n+2)}{\sqrt{(n+2)/2-1}} \leq \ln(4/3)$ and $n - 1 > \sqrt{n/2}(1 + \ln(n))$, and completes the proof. $\qquad\square$

GENERAL MATCHING EXCHANGE PROCESSES. When $|\mathcal{S}| < |\mathcal{X}|$, we will use the following result of Miracle and Yilek [20] which bounds the variation distance of a matching exchange process.

Theorem 4 (Miracle, Yilek [20]).
Let $T = \max\left(40\ln(2n^2), \frac{10\ln(n/9)}{\ln(1+p_1p_2(7/36)((7/9)n^2-n))}\right) + \frac{72\ln(2n^2)}{p_1 n}$, then

$$\|\nu_{ME^r} - \mu_{UN}\| \leq n^{1-(2r/T)},$$

where ν_{ME^r} is the distribution after r rounds of a matching exchange process on n elements and μ_{UN} is the uniform distribution on permutations of n elements.

In order to apply the theorem we need to bound two parameters p_1 and p_2 of the associated matching exchange process which are defined below.

Definition 2 (Miracle, Yilek [20]).

1. For any points x, y the probability that a pair (x, y) is part of a matching is at least p_1.
2. For any points x, y, z, and w conditioned on (x, y) being a pair in the matching, the probability that (z, w) is also in the matching is at least p_2.

We begin by consider the MSU process as defined in Fig. 2 and prove the following. We will use this bound for the case when $|\mathcal{X}| > |\mathcal{S}| \geq |\mathcal{X}|/2$.

Theorem 5.
Let $T = \max\left(40\ln(2|\mathcal{S}|^2), \frac{10\ln(|\mathcal{S}|/9)}{\ln(1+(7/36|\mathcal{X}|^2)((7/9)|\mathcal{S}|^2-|\mathcal{S}|))}\right) + \frac{72|\mathcal{X}|\ln(2|\mathcal{S}|^2)}{|\mathcal{S}|}$, then

$$\|\nu_{MSU^r} - \mu_{UN}\| \leq |\mathcal{S}|^{1-(2r/T)},$$

where ν_{MSU^r} is the distribution after r rounds of MSU, $|\mathcal{S}|$ is the size of the target set $\mathcal{S}$, $|\mathcal{X}|$ is the size of the larger domain set $\mathcal{X}$, and μ_s is the uniform distribution on permutations of $|\mathcal{S}|$ elements.

Proof. In order to apply Theorem 4 we first bound the parameters p_1 and p_2. In MSU the probability that we select a pair (x, y) with $x, y \in \mathcal{S}$ is $1/(|\mathcal{X}| - 1)$ since there are $|\mathcal{X}| - 1$ choices for a particular point to get mapped to and each are equally likely. Thus $p_1 = 1/(|\mathcal{X}| - 1) > 1/|\mathcal{X}|$. Given that a pair (x, y) is already included in the matching, the probability that a second pair (z, w) is also included is $1/(|\mathcal{X}| - 3)$ since there are $|\mathcal{X}| - 3$ remaining choices for z to get mapped to and each are equally likely. Thus $p_2 = 1/(|\mathcal{X}| - 3) > 1/|\mathcal{X}|$. Directly substituting these parameters into Theorem 4 completes the proof.   □

ELIMINATING THE BIT FLIP. When $|\mathcal{S}| < |\mathcal{X}|/2$ we are able to show that MSU is a matching exchange process without adding an additional bit flip for each pair in the matching and thus we can remove the round function G from Fig. 2. We prove the following.

Theorem 6. *The distribution on matchings on the target set $\mathcal{S}$ generated by MSU without the round function G is identical to the final distribution generated by a matching exchange process on $\mathcal{S}$ with parameters $p_1 = 2/|\mathcal{X}|$ and $p_2 = 2/|\mathcal{X}|$, where $|\mathcal{X}|$ is the size of the domain set $\mathcal{X}$ and $|\mathcal{S}| < |\mathcal{X}|/2$.*

Proof. Our proof begins by giving a particular matching exchange process $\mathcal{P}$ and associated distribution on κ and then proving that the distribution on matching that results from this process is identical to the distribution that results from the MSU process. We then bound the matching exchange process parameters p_1 and p_2 for our given process.

Let $\mathcal{P}$ be a matching exchange process where the probability that $\kappa = i$ is given by p_i. Let G_i be the probability that a particular matching of size i on $\mathcal{S}$ (i.e. a matching with $2i$ points) is selected by MSU. It is straightforward to see from the definition of MSU that G_i is the same for each matching of size i. Let $m = |\mathcal{S}|/2$ be the size of a perfect matching on $\mathcal{S}$ and $\mathcal{M}_i$ be the number of perfect matchings on a set with $2i$ points. We now define p_i as follows,

$$p_i = \begin{cases} G_m \cdot 2^m \cdot \mathcal{M}_m & \text{if } i = m \\ (G_i - G_{i+1}) \cdot \frac{2^m \mathcal{M}_i^m}{\mathcal{M}_{m-i}} & \text{if } 0 \le i < m \end{cases}$$

Consider a particular matching m_i of size i on $\mathcal{S}$. By definition, it is selected with probability G_i in MSU. We will show that the probability it is selected by $\mathcal{P}$ is also G_i. In $\mathcal{P}$ this matching is selected if we select any matching that contains m_i as a sub-matching and then flip the bits appropriately to just select the edges in m_i. Thus in $\mathcal{P}$ the probability that m_i is selected is the sum over matching from size i to m of the number of matchings that contain m_i times the probability a matching of that size is selected times the probability we select the exact edges in m_i which is $(2^{-1})^m$. This gives us the following

$$\Pr[m_i] = \sum_{x=i}^{m} p_i(\mathcal{M}_i^m)^{-1} \mathcal{M}_{m-i} 2^{-m}.$$

We will prove by induction on i that $G_i = \Pr[m_i]$ for $0 \le i \le m$. For our base case let $i = m$. Then we have,

$$\Pr[m_m] = p_m(\mathcal{M}_m)^{-1} 2^{-m} = G_m \cdot 2^m \mathcal{M}_m (\mathcal{M}_m)^{-1} 2^{-m} = G_m.$$

Next we assume inductively that $G_{i+1} = \Pr[m_{i+1}]$ and then show that this holds for i as follows.

$$\Pr[m_i] = \sum_{x=i}^{m} p_x(\mathcal{M}_x^m)^{-1} \mathcal{M}_{m-x} 2^{-m}$$

$$= p_i(\mathcal{M}_i^m)^{-1} \mathcal{M}_{m-i} 2^{-m} + \left(\sum_{x=i+1}^{m} p_x(\mathcal{M}_x^m)^{-1} \mathcal{M}_{m-x} 2^{-m} \right)$$

$$= \left(p_i(\mathcal{M}_i^m)^{-1} \mathcal{M}_{m-i} 2^{-m} \right) + G_{i+1}$$

$$= \left((G_i - G_{i+1}) \cdot \frac{2^m \mathcal{M}_i^m}{\mathcal{M}_{S-i}} \right) ((\mathcal{M}_i^m)^{-1} \mathcal{M}_{m-i} 2^{-m}) + G_{i+1}$$

$$= (G_i - G_{i+1}) + G_{i+1} = G_i.$$

It remains to show that these choices of p_i form a probability distribution. To show this we need to show that for all $0 \leq i \leq m$, $p_i \geq 0$ and that $\sum_{i=0}^{m} p_i = 1$. Given the above definition of the $\{p_i\}$'s to show that for all $0 \leq i \leq m$, $p_i \geq 0$ it suffices to show that $G_i - G_{i+1} > 0$ for all $0 \leq i < m$. Recall that G_i is the probability that the MSU algorithm results in a particular matching on $\mathcal{S}$ of size i. Additionally recall that the MSU process is equivalent to first generating a uniformly random perfect matching on $\mathcal{X}$ and then removing all edges except those where both points are in $\mathcal{S}$. Thus G_i is the number of matchings consistent with a particular matching of size i divided by the total number of matchings. If we fix a particular matching of size i on $\mathcal{S}$ then there are $2(m-i)$ remaining points in $\mathcal{S}$ that are unmatched. In all consistent matchings these are matched with points in $\mathcal{X} - \mathcal{S}$ of which there are $|\mathcal{X}| - |\mathcal{S}|$ remaining. There are $\binom{|\mathcal{X}|-|\mathcal{S}|}{|\mathcal{S}|-2i}$ ways to choose these points and $(|\mathcal{S}| - 2i)!$ ways to match them with the remaining points in $\mathcal{S}$. Finally there are $\mathcal{M}_{|\mathcal{X}|/2-|\mathcal{S}|+i}$ ways to match up the remaining points in $\mathcal{S}$. Combining these observations gives the following.

$$G_i = \binom{|\mathcal{X}| - |\mathcal{S}|}{|\mathcal{S}| - 2i} \cdot (|\mathcal{S}| - 2i)! \cdot \mathcal{M}_{|\mathcal{X}|/2-|\mathcal{S}|+i} \cdot (\mathcal{M}_{|\mathcal{X}|/2})^{-1}$$

$$= \frac{(|\mathcal{X}| - |\mathcal{S}|)!}{(|\mathcal{X}| - 2|\mathcal{S}| + 2i)!} \cdot (|\mathcal{X}| - 2|\mathcal{S}| + 2i - 1)!! \cdot (\mathcal{M}_{|\mathcal{X}|/2})^{-1}.$$

Since our goal is to show that $G_i - G_{i+1} > 0$ for $0 \leq i < m$, it suffices to show

$$\frac{(|\mathcal{X}| - 2|\mathcal{S}| + 2i - 1)!!}{(|\mathcal{X}| - 2|\mathcal{S}| + 2i)!} > \frac{(|\mathcal{X}| - 2|\mathcal{S}| + 2(i+1) - 1)!!}{(|\mathcal{X}| - 2|\mathcal{S}| + 2(i+1) - 1)!}.$$

This simplifies to the following which holds as long as $|\mathcal{S}| < |\mathcal{X}|/2$,

$$(|\mathcal{X}| - 2|\mathcal{S}| + 2i + 2) \cdot (|\mathcal{X}| - 2|\mathcal{S}| + 2i + 1) > (|\mathcal{X}| - 2|\mathcal{S}| + 2i + 1).$$

We know that the distribution on matchings given by MSU is a valid probability distribution. Above we proved that the probability of any particular matching of size i is the same under both MSU and $\mathcal{P}$. This implies that the $\sum_{i=0}^{m} G_i \times \mathcal{M}_i^m = 1$. Similarly this implies that $\sum_{i=0}^{m} p_i = \sum_{i=0}^{m} p_i (\mathcal{M}_i^m)^{-1} \cdot \mathcal{M}_i^m = \sum_{i=0}^{m} G_i \times \mathcal{M}_i^m = 1$. Thus the $\{p_i\}$'s form a valid probability distribution as long as $|\mathcal{S}| < |\mathcal{X}|/2$.

It remains to bound the two parameters p_1 and p_2 for the matching exchange process $\mathcal{P}$. Recall from Definition 2 that p_1 is a lower bound on the probably that for any two points x and y the pair (x, y) is included in the matching. Note that this is the probability in the matching exchange process before a bit is flipped for each pair in the matching. Recall that in MSU the probability that we select a pair (x, y) with $x, y \in \mathcal{S}$ is $1/(|\mathcal{X}| - 1)$ since there are $|\mathcal{X}| - 1$ choices for a particular point to get mapped to and each are equally likely. Let p_1 be the probability that a particular pair (x, y) is select to be part of the matching in the corresponding matching exchange process $\mathcal{P}$ that we analyzed above. This implies that $p_1 \cdot (1/2) = 1/(|\mathcal{X}| - 1)$ and thus $p_1 = 2/(|\mathcal{X}| - 1) > 2/|\mathcal{X}|$.

Next, the parameter p_2 is a lower bound on the probability that for any four points x, y, z, and w in $\mathcal{S}$ that conditioned on the pair (x, y) being part of the matching, the probability that the pair (z, w) is also part of the matching. Again these are the probabilities for the underlying matching exchange process $\mathcal{P}$. Let P_1 be the event that the pair (x, y) is part the original matching (before the bit flip) and P_2 be the event that pair (z, w) is part of the original matching. Similarly let F_1 be the event that the pair (x, y) is part of the final matching and F_2 be the event that the pair (z, w) is part of the final matching. We are interested in $p_2 = \Pr[P_2 \mid P_1]$. Note that $\Pr[P_1 \cap P_2] = 4\Pr[F_1 \cap F_2]$. By the laws of conditional probability we have,

$$p_2 = \frac{\Pr[P_2 \cap P_1]}{\Pr[P_1]} = \frac{4\Pr[F_1 \cap F_2]}{p_1} = \frac{4\Pr[F_2 \mid F1]\Pr[F_1]}{p_1} = 2\Pr[F_2 \mid F_1].$$

Recall that for MSU given that a pair (x, y) is already included in the matching, the probability that a second pair (z, w) is also included is $1/(|\mathcal{X}|-3)$ since there are $|\mathcal{X}| - 3$ remaining choices for z to get mapped to and each are equally likely. Thus $p_2 = 2\Pr[F_2 \mid F_1] = 2/(|\mathcal{X}| - 3) > 2/|\mathcal{X}|$.   □

Directly substituting the parameters on the matching exchange process given by Theorem 6 into Theorem 4 gives the following corollary.

Corollary 3.
Let $T = \max\left(40\ln(2|\mathcal{S}|^2), \frac{10\ln(|\mathcal{S}|/9)}{\ln(1+(7/9|\mathcal{X}|^2)((7/9)|\mathcal{S}|^2-|\mathcal{S}|))}\right) + \frac{36|\mathcal{X}|\ln(2|\mathcal{S}|^2)}{|\mathcal{S}|}$, then

$$||\nu_{MSU^r} - \mu_{UN}|| \leq |\mathcal{S}|^{1-(2r/T)},$$

where ν_{MSU^r} is the distribution after r rounds of MSU without the round function G, $|\mathcal{S}|$ is the size of the target set $\mathcal{S}$, $|\mathcal{X}|$ is the size of the larger domain set $\mathcal{X}$, $|\mathcal{S}| < |\mathcal{X}|/2$, and μ_s is the uniform distribution on permutations of $|\mathcal{S}|$ points.

B   Game for Proof of Theorem 2

main Gm_2^A	proc. $\mathsf{Enc}(X)$
$(\mathbf{KF}, \mathbf{G}) \leftarrow_\$ \mathcal{K}$	for j in $1 \ldots m$ do
for j in $1 \ldots m$ do	$\quad Z \leftarrow \mathsf{SN}_{\mathrm{KF}_j}(X)\,;\; Z' \leftarrow Z \oplus 1$
$\quad t \leftarrow 0$	$\quad X' \leftarrow \mathsf{SN}_{\mathrm{KF}_j}^{-1}(Z')\,;\; \hat{Z} \leftarrow \max(Z, Z')$
$\quad$ for z in $\mathrm{odd}(N)$ do	$\quad$ if $\mathsf{B}[j][\hat{Z}] = 1 \wedge X \in \mathcal{S} \wedge X' \in \mathcal{S}$
$\quad\quad \mathsf{B}[j][z] \leftarrow_\$ \{0,1\}$	$\quad$ then
$\quad\quad t \leftarrow t + \mathsf{B}[j][z]$	$\quad\quad X \leftarrow X'$
$\quad$ if $t \geq (3/4) \cdot (N/2)$ then	$\quad$ else
$\quad\quad \mathrm{bad}_j \leftarrow \mathrm{true}$	$\quad\quad X \leftarrow X$
$b' \leftarrow A^{\mathsf{Enc},\mathsf{Dec}}$	return X
return b'	

References

1. Bellare, M., Hoang, V.T.: Identity-based format-preserving encryption. In: Thuraisingham, B.M., Evans, D., Malkin, T., Xu, D. (eds.) ACM CCS 2017, pp. 1515–1532. ACM Press, October/November 2017

2. Bellare, M., Hoang, V.T., Tessaro, S.: Message-recovery attacks on Feistel-based format preserving encryption. In: Weippl, E.R., Katzenbeisser, S., Kruegel, C., Myers, A.C., Halevi, S. (eds.) ACM CCS 2016, pp. 444–455. ACM Press, October 2016

3. Bellare, M., Ristenpart, T., Rogaway, P., Stegers, T.: Format-preserving encryption. In: Jacobson, M.J., Rijmen, V., Safavi-Naini, R. (eds.) SAC 2009. LNCS, vol. 5867, pp. 295–312. Springer, Heidelberg (2009). https://doi.org/10.1007/978-3-642-05445-7_19

4. Bellare, M., Rogaway, P., Spies, T.: The FFX mode of operation for format-preserving encryption, February 2010. http://csrc.nist.gov/groups/ST/toolkit/BCM/documents/proposedmodes/ffx/ffx-spec.pdf

5. Bernstein, M.: The mixing time for a random walk on the symmetric group generated by random involutions. In: Proceedings of the 28th International Conference on Formal Power Series and Algebraic Combinatorics (FPSAC) (2016)

6. Black, J., Rogaway, P.: Ciphers with arbitrary finite domains. In: Preneel, B. (ed.) CT-RSA 2002. LNCS, vol. 2271, pp. 114–130. Springer, Heidelberg (2002). https://doi.org/10.1007/3-540-45760-7_9

7. Brier, E., Peyrin, T., Stern, J.: BPS: a format-preserving encryption proposal. http://csrc.nist.gov/groups/ST/toolkit/BCM/documents/proposedmodes/bps/bps-spec.pdf

8. Brightwell, M., Smith, H.: Using datatype-preserving encryption to enhance data warehouse security. In: National Information Systems Security Conference (NISSC) (1997)

9. Czumaj, A., Kutylowski, M.: Delayed path coupling and generating random permutations. Random Struct. Algorithms 17, 238–259 (2000)

10. Dai, W., Hoang, V.T., Tessaro, S.: Information-theoretic indistinguishability via the chi-squared method. In: Katz, J., Shacham, H. (eds.) CRYPTO 2017. LNCS, vol. 10403, pp. 497–523. Springer, Cham (2017). https://doi.org/10.1007/978-3-319-63697-9_17

11. Durak, F.B., Vaudenay, S.: Breaking the FF3 format-preserving encryption standard over small domains. In: Katz, J., Shacham, H. (eds.) CRYPTO 2017. LNCS, vol. 10402, pp. 679–707. Springer, Cham (2017). https://doi.org/10.1007/978-3-319-63715-0_23

12. Dworkin, M.: Recommendation for block cipher modes of operation: methods for format preserving-encryption. NIST Special Publication 800–38G (2016). http://dx.doi.org/10.6028/NIST.SP.800-38G

13. Granboulan, L., Pornin, T.: Perfect block ciphers with small blocks. In: Biryukov, A. (ed.) FSE 2007. LNCS, vol. 4593, pp. 452–465. Springer, Heidelberg (2007). https://doi.org/10.1007/978-3-540-74619-5_28

14. Grubbs, P., Ristenpart, T., Yarom, Y.: Modifying an enciphering scheme after deployment. In: Coron, J.-S., Nielsen, J.B. (eds.) EUROCRYPT 2017. LNCS, vol. 10211, pp. 499–527. Springer, Cham (2017). https://doi.org/10.1007/978-3-319-56614-6_17

15. Hoang, V.T., Morris, B., Rogaway, P.: An enciphering scheme based on a card shuffle. In: Safavi-Naini, R., Canetti, R. (eds.) CRYPTO 2012. LNCS, vol. 7417, pp. 1–13. Springer, Heidelberg (2012). https://doi.org/10.1007/978-3-642-32009-5_1

16. Luchaup, D., Dyer, K.P., Jha, S., Ristenpart, T., Shrimpton, T.: LibFTE: a toolkit for constructing practical, format-abiding encryption schemes. In: Proceedings of the 23rd USENIX Security Symposium, pp. 877–891 (2014)

17. Luchaup, D., Shrimpton, T., Ristenpart, T., Jha, S.: Formatted encryption beyond regular languages. In: Ahn, G.J., Yung, M., Li, N. (eds.) ACM CCS 2014, pp. 1292–1303. ACM Press, November 2014

18. Maurer, U., Pietrzak, K., Renner, R.: Indistinguishability amplification. In: Menezes, A. (ed.) CRYPTO 2007. LNCS, vol. 4622, pp. 130–149. Springer, Heidelberg (2007). https://doi.org/10.1007/978-3-540-74143-5_8

19. Miracle, S., Yilek, S.: Reverse cycle walking and its applications. In: Cheon, J.H., Takagi, T. (eds.) ASIACRYPT 2016. LNCS, vol. 10031, pp. 679–700. Springer, Heidelberg (2016). https://doi.org/10.1007/978-3-662-53887-6_25

20. Miracle, S., Yilek, S.: Cycle slicer: an algorithm for building permutations on special domains. In: Takagi, T., Peyrin, T. (eds.) ASIACRYPT 2017. LNCS, vol. 10626, pp. 392–416. Springer, Cham (2017). https://doi.org/10.1007/978-3-319-70700-6_14

21. Morris, B., Rogaway, P.: Sometimes-recurse shuffle - almost-random permutations in logarithmic expected time. In: Nguyen, P.Q., Oswald, E. (eds.) EUROCRYPT 2014. LNCS, vol. 8441, pp. 311–326. Springer, Heidelberg (2014). https://doi.org/10.1007/978-3-642-55220-5_18

22. Morris, B., Rogaway, P., Stegers, T.: How to encipher messages on a small domain. In: Halevi, S. (ed.) CRYPTO 2009. LNCS, vol. 5677, pp. 286–302. Springer, Heidelberg (2009). https://doi.org/10.1007/978-3-642-03356-8_17

23. Naor, M., Reingold, O.: Constructing pseudo-random permutations with a prescribed structure. J. Cryptol. 15(2), 97–102 (2002)

24. Ristenpart, T., Yilek, S.: The mix-and-cut shuffle: small-domain encryption secure against N queries. In: Canetti, R., Garay, J.A. (eds.) CRYPTO 2013. LNCS, vol. 8042, pp. 392–409. Springer, Heidelberg (2013). https://doi.org/10.1007/978-3-642-40041-4_22

25. Spies, T.: Format-preserving encryption. Unpublished whitepaper (2008). https://www.voltage.com/wp-content/uploads/Voltage-Security-WhitePaper-Format-Preserving-Encryption.pdf

Variants of the AES Key Schedule for Better Truncated Differential Bounds

Patrick Derbez[1], Pierre-Alain Fouque[1], Jérémy Jean[2],
and Baptiste Lambin[1(✉)]

[1] Univ Rennes, CNRS, IRISA, Rennes, France
{patrick.derbez,baptiste.lambin}@irisa.fr,
pierre-alain.fouque@univ-rennes1.fr
[2] ANSSI, Paris, France
Jeremy.Jean@ssi.gouv.fr

Abstract. Differential attacks are one of the main ways to attack block ciphers. Hence, we need to evaluate the security of a given block cipher against these attacks. One way to do so is to determine the minimal number of active S-boxes, and use this number along with the maximal differential probability of the S-box to determine the minimal probability of any differential characteristic. Thus, if one wants to build a new block cipher, one should try to maximize the minimal number of active S-boxes. On the other hand, the related-key security model is now quite important, hence, we also need to study the security of block ciphers in this model.

In this work, we search how one could design a key schedule to maximize the number of active S-boxes in the related-key model. However, we also want this key schedule to be efficient, and therefore choose to only consider permutations. Our target is AES, and along with a few generic results about the best reachable bounds, we found a permutation to replace the original key schedule that reaches a minimal number of active S-boxes of 20 over 6 rounds, while no differential characteristic with a probability larger than 2^{-128} exists. We also describe an algorithm which helped us to show that there is no permutation that can reach 18 or more active S-boxes in 5 rounds. Finally, we give several pairs (P_s, P_k), replacing respectively the ShiftRows operation and the key schedule of the AES, reaching a minimum of 21 active S-boxes over 6 rounds, while again, there is no differential characteristic with a probability larger than 2^{-128}.

Keywords: AES · Key schedule · Related-key · Truncated differential

Patrick Derbez was supported by the French Agence Nationale de la Recherche through the CryptAudit project under Contract ANR-17-CE39-0003.

Pierre-Alain was supported by the French Agence Nationale de la Recherche through the BRUTUS project under Contract ANR-14-CE28-0015.

Baptiste Lambin was supported by the Direction Générale de l'Armement (Pôle de Recherche CYBER).

C. Cid and M. J. Jacobson, Jr. (Eds.): SAC 2018, LNCS 11349, pp. 27–49, 2019.
https://doi.org/10.1007/978-3-030-10970-7_2

1 Introduction

First introduced in 1991 by Biham and Shamir [2], differential cryptanalysis is one of the main tool to analyze and attack symmetric primitives. The main idea is to introduce some differences in the plaintext, and see how these differences propagate through the different steps of the algorithm, independently from the key. For example, given an encryption function $\mathcal{E}(p, k)$ encrypting the plaintext $p \in \mathbb{F}_2^{n_b}$ using a key $k \in \mathbb{F}_2^{n_k}$, if one is able to prove that there exists a pair of differences $\Delta_{in}, \Delta_{out} \in \mathbb{F}_2^{n_b}$ such that $\mathcal{E}(p \oplus \Delta_{in}, k) = \mathcal{E}(p, k) \oplus \Delta_{out}$ for all keys, then it gives a strong distinguisher for the encryption function $\mathcal{E}$. Moreover, due to the non-linearity of $\mathcal{E}$, such a differential relation could only hold with a certain probability. Consequently, a lot of work has been put into designing algorithms that search for the best possible differential characteristics of a given cipher. For instance, Matsui's algorithms [18] were the first designed. Most of modern ciphers are now built as *iterated ciphers*, i.e., a round function f is built and repeated several times, XOR-ing a round key between each application of f, see Fig. 1. Thus, to search for such a pair $(\Delta_{in}, \Delta_{out})$, one often studies the propagation of the input difference through each round of the cipher, thus leading to a *differential characteristic* consisting of all differences in each state s_i.

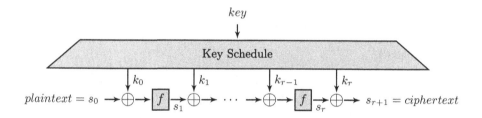

Fig. 1. Generic iterated cipher construction [11]

One can also choose to consider only *truncated differences*, that is, only look at whether or not the difference in one byte is zero. While this can also directly lead to various attacks, e.g., impossible differential attacks [1,16], it can also be used to get some results in differential cryptanalysis. Indeed, in most cipher designs, the non-linear component consists of an S-box, a small non-linear function applied several times over all iterations. This S-box is the reason that some differential characteristic only holds with a certain probability. Given an S-box S acting on a small number of s bits, and for each pair $(\Delta_{in}, \Delta_{out}) \in \mathbb{F}_2^{2s}$, one can easily compute how many $x \in \mathbb{F}_2^s$ verifies the relation $S(x \oplus \Delta_{in}) = S(x) \oplus \Delta_{out}$. This allow to compute the Difference Distribution Table (DDT) of the S-box, which gives the probability that the above relation holds for each $(\Delta_{in}, \Delta_{out})$. Thus, given a differential characteristic, one can easily compute the probability that it holds, simply by multiplying all differential probabilities of each S-box together.[1] Hence, given a *truncated* differential characteristic, while we cannot

---

[1] Using the fair assumption that each round is independent, which while obviously not true, is admitted as a reasonable assumption.

determine the exact probability that this characteristic holds, we can deduce its minimal probability. Indeed, if the S-box has a maximal differential probability of p, and there are n S-boxes with a non-zero difference (called *active S-boxes*), then the truncated differential characteristic holds with a probability at most p^n. Thus, given the maximal differential probability of the S-box used and the bit-length n_k of the key, one can easily deduce the minimal number of active S-boxes n_{min} that leads to $p^{n_{min}} < 2^{-n_k}$. So, if for a given number of round, we can prove that there is at least n_{min} active S-boxes, we know that there would be no differential characteristic with a probability better than 2^{-n_k}, which would mean that finding a pair of plaintexts satisfying this characteristic would *a priori* costs more than an exhaustive search for the key.

Such differentials and truncated differentials can also be considered in the *related-key model*. First introduced in 2009 to attack AES-192 and AES-256 [3,4], this model allows the attacker to inject differences in the plaintext, but also in the key. Another worth-mentioning model is the more recent *related-tweak model* for tweakable block ciphers, where the attacker fully controls an additional input for the block cipher called a *tweak* [17,21]. While this model is closer to chosen-plaintext attacks, the tweak is often (but not necessarily) used alongside the key and thus involved in the key schedule, such as in the TWEAKEY framework [13]. Since the attacker can now inject some differences in both the plaintext and the key, this causes a large increase in the complexity to search differential and truncated differential characteristics. Nonetheless, several tools have been designed to tackle this problem [5,9,10]. Hence, a few proposals were made to give another, more secure, key schedule for some primitives, such as [7,19] for AES and [20] for SKINNY and AES-based constructions from FSE 2016 [12]. However, their main concern was mostly to design a more secure key schedule, without considering the possible loss in efficiency. To that regard, Khoo et al. [14] proposed a new key schedule for AES which consists in only a permutation at the byte level, based on their proof on the number of active S-boxes in the related-key model for AES. Using a permutation thus leads to a very efficient key schedule, both in software and hardware, and can also make the analysis easier.

Our Contributions. In this paper, we go further and study how we can design a *good* permutation to use as the key schedule in AES-128. More precisely, we first start by giving some bounds on the reachable minimal number of active S-boxes for up to 7 rounds of AES if we use a simple permutation as its key schedule. Especially, we show that there is no permutation that can reach a minimal number of active S-boxes of 18 or more over 5 rounds. These bounds allow us to know the results that a "perfect" permutation could reach. Then, we provide a method to search for such a permutation. To do so, we reused the meta-heuristic approach given by Nikolić in [20], combined with a Constraint Programming model inspired from the work of Gerault et al. in [10]. Especially, we give a way to model the underlying equations of a truncated differential characteristic, leading to a more precise model than the original one from [10].

Namely, the truncated differential characteristics found are always valid unless we consider the DDT of the S-box.

We also went further and modified both the key schedule and one step of the AES round function (namely, ShiftRows) to see whether we can achieve better bounds. As a result, we exhibit a permutation P_k which, when used as the AES key schedule, lead to a minimal number of active S-boxes of 20 over 6 rounds, while no characteristic has a probability larger than 2^{-128}. When changing both the key schedule and the ShiftRows step, we give several pairs of permutations (P_k^i, P_s^i) that have a minimal number of active S-boxes of 21 over 6 rounds, while again, no characteristic has a probability larger than 2^{-128}. While we applied this method to AES, it is quite generic and could also be used on any block cipher, as long as one have an efficient enough way to compute the minimal number of active S-boxes. Our implementation is available at https://github.com/TweakAESKS/TweakAESKS.

2 Background

Differential cryptanalysis was first introduced by Biham and Shamir in 1991 [2] and mainly consists in studying the propagation of differences between two plaintexts through the cipher. Here, we only consider *truncated differences*, that is, we are only interested in whether a byte does have a non-zero difference (*active* byte) or not (*inactive* byte). Our work is centered around AES, for which we make a few remainders. AES is the NIST block cipher standard, derived from Rinjdael [8]. It uses an internal state of 128 bits, and several key sizes are available, namely 128, 192 and 256. Here, when mentioning AES, we refer to the 128-bit version.

It is an SPN block cipher, iterating a round function $R = \mathsf{MC} \circ \mathsf{SR} \circ \mathsf{SB} \circ \mathsf{ARK}$ 10 times, where each component of the round function is quickly described in the following. The state can be viewed as a 4×4 byte array, and thus we will often talk about *columns* of the state. The round function consists in four operations: AddRoundKey (ARK), SubBytes (SB), ShiftRows (SR) and MixColumns (MC). ARK XORs the round key into the internal state. This round key is derived from the master key using a key schedule KS, for which we do not give details, our ultimate goal being to change it. We refer the interested reader to [8] for the original descriptions. SB applies a non-linear operation (called S-box) on each byte of the state, then SR performs a cyclic shift of each row, where Row j is shifted by $j - 1$ bytes to the left, $j \in \{1, 2, 3, 4\}$. Finally, MC is a linear operation that multiplies each column of the internal state by an MDS matrix with coefficients in $\mathbb{F}_{2^8}$.

We first recall several well known properties of the MC operation, which will be used is the rest of the article. Here, $w(x)$ correspond to the number of active bytes in x, which is either a state or a column of the state.

Proposition 1 (MixColumns MDS property). *Let z and y be two state columns such that $\mathsf{MC}(z) = y$. Then, either $w(z) + w(y) = 0$ or $w(z) + w(y) \geq 5$.*

Moreover, for any five bytes in y and z, there exists one linear equation between those five bytes.

Proof. This comes directly from the fact that the matrix used in the MC operation is MDS.

Proposition 2 (MixColumns linear property). *Let z, z', y, y' be four state columns such that $MC(z) = y$ and $MC(z') = y'$. Then, the MixColumns MDS property also holds for $(z \oplus z')$ and $(y \oplus y')$, that is: either $w(z \oplus z') + w(y \oplus y') = 0$ or $w(z \oplus z') + w(y \oplus y') >= 5$.*

Proof. This comes directly from the previous proposition and the fact that MC is linear.

Lemma 1. *Let k, x, y, z be four state columns such that $MC(z) = y$, z contains at least one active byte and $x = y \oplus k$. Denote by $i_{y,z}$ the number of inactive bytes in y and z (i.e., $i_{y,z} = 8 - w(y) - w(z)$) and $c_{z,k,x}$ the number of bytes from z that are cancelled by k in x. If $i_{y,z} + c_{y,z,k} \geq 5$, then there is at least one linear equation on some bytes of k. Moreover, this can only happens if $c_{y,z,k} \geq 2$.*

Proof. If $i_{y,z} + c_{y,z,k} \geq 5$, then from the MixColumns MDS property, it follows that there is an equation between any five bytes chosen from the inactive ones in y and z, and the bytes from z which are cancelled by k. If we denote such a cancelled byte by z_i, that is, $z_i \oplus k_i = 0$, then we have $k_i = z_i$, hence the equation involves some bytes of k and some inactive bytes from y and z, which are zeros.

Since z contains at least one active byte, we have $w(z) + w(y) \geq 5$, hence $i_{y,z} \leq 3$. Therefore, if $c_{y,z,k} = 1$ (i.e., only one byte if cancelled), we have $i_{y,z} + c_{y,z,k} \leq 4$, and thus no equation is implied.

When considering truncated differentials, we are often interested in the number of active S-boxes, that is, the number of active bytes going through an S-box (i.e., active bytes at the beginning of the round). We will often refer to the (minimal) number of active S-boxes in a characteristic as the *length* of the characteristic, and to a *minimal characteristic* to refer to a characteristic which reaches the minimal number of active S-boxes. Given a truncated differential characteristic of length n, one can deduce the maximal probability that this characteristic can have once being instantiated. Indeed, if the S-box has a maximal non-zero differential probability of p, then the maximal probability of this characteristic is p^n. If one studies a block cipher with a key of length n_k bits, then the goal is to prove that no characteristics can be instantiated with a probability larger than 2^{-n_k}. Hence, for AES, since the maximal differential probability of the S-box is 2^{-6}, we know that if for a given number of rounds the minimal number of active S-boxes is greater or equal than 22, then no differential characteristic with a differential probability larger than 2^{-128} exists.

Searching whether a characteristic reaching a given length or maximal probability exists has been a major focus in academic research. One way to find the best probability is to proceed in two steps. First, one try to find a truncated differential characteristic with a minimal number of active S-boxes, and then try to instantiate this characteristic. When searching for such a truncated differential characteristic, one can choose to consider additional information about the cipher along with "basic" propagation rules coming from the round function, to avoid trying to instantiate characteristics that would not be instantiable anyway. Hence for AES, we give the following definitions.

Definition 1. *A characteristic is said to be valid in the "truncated differential setting" if and only if the* MixColumns *linear property is always verified and there is at least one non-trivial solution to the system of equations (if any) induced by Lemma 1.*

A characteristic that remains valid even when one does not consider the MixColumns *linear property nor the equations is said to be valid in the* pure truncated differential setting.

The point of these definitions is twofold. On the one hand, since the pure truncated differential setting contains significantly less constraints, the minimal characteristic could be a lot easier to find. However, it may result in an invalid characteristic when one tries to instantiate it, which could have been detected in the truncated differential setting. Conversely, finding the minimal characteristic in the truncated differential setting could be harder, but the only thing that could invalidate this characteristic is the S-box DDT.

We chose to use the same approach as Gerault et al. [10], who proposed to use two Constraint Programming models. The first one was used to find the minimal characteristics for AES, considering only the MixColumns linear property. The second one takes a list of truncated characteristic and tries to find the best instantiation (if any) of each characteristic with respect to its probability. As we aim at changing the key schedule, we changed these models, detailed in the following.

Model 1. *This model takes as input a permutation P_k to use as the key schedule and a number of rounds, and output the minimal number of active S-boxes with these parameters in the truncated differential setting. Compared to the first model of [10], we directly model the equations coming from the* MixColumns *operation (see Lemma 1), resulting in a more reliable result, albeit being slower. We refer the reader to Appendix A for the method used to model these equations.*

Model 2. *This model also takes as input a permutation P_k for the key schedule and a number of rounds, along with a list of truncated differential characteristics. It then goes through each of these truncated characteristics, and tries to find a instantiation with a probability larger than 2^{-128}. If such an instantiation is found, it gives its probability and the differential characteristic, otherwise it just stops without trying to find an instantiation with a probability smaller than 2^{-128}.*

3   Generic Bounds

Before trying to find a permutation that reaches a certain number of active S-boxes, we need to study which number of S-boxes we can reach. From the fact that using a permutation as the key schedule implies that the number of active bytes in the key is constant, we can deduce several bounds on the number of active S-boxes. To demonstrate these bounds, we show that there is always a differential characteristic of a certain length, independently from the permutation used in the key schedule.

Proposition 3. *Using a permutation as the key schedule, there is always a differential characteristic of length 1 (resp. 5). for 2 (resp. 3) rounds. For 4 rounds, there is always a characteristic of length either 8, 9 or 10. Moreover, these differential characteristics always remain valid in the truncated differential setting.*

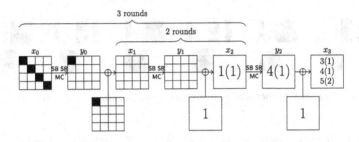

Fig. 2. Characteristic always valid for 2,3 and 4 rounds. $x(y)$ means that there are x active Sboxes somewhere in the state, with y columns containing at least one active bytes. Multiple $x(y)$ in a state means that one of them must be true

Proof. Such a characteristic is depicted in Fig. 2. For 2 rounds, there is only one active byte in the second state, which is cancelled by the active byte in the key. For 3 rounds, the previous characteristic is extended by adding one more round before it, and comes directly from the MixColumns MDS property.

For 4 rounds, we add one more round after the 3-round differential characteristic. Since y_2 has four active bytes on the same column, and since the key has one active byte anywhere in the key state, x_3 can have either 3, 4 or 5 active bytes, which results in a differential characteristic of length either 8, 9 or 10.

No equation is implied since there is always at most one active key byte that is cancelled with the ARK operation for each round (Lemma 1). Finally, there are only two MixColumn transitions with active bytes, one of the form $MC(z) = y$ where z and y are one column of the state with $w(z) = 4$ and $w(y) = 1$ and another of the form $MC(z') = y'$, where $w(z') = 1$ and $w(y') = 4$. Hence, $w(z \oplus z') \geq 3$ and $w(y \oplus y') \geq 3$, and thus the MixColumns linear property is always valid.

Corollary 1. *Using a permutation as the key schedule, the optimal bounds on the number of active S-boxes that can be proven for 2, 3 and 4 rounds is respectively 1, 5 and 10 in the truncated differential setting.*

The proof of this corollary comes directly from the previous proposition. If we try to extend the previous characteristic with one more round, we obtain that there is always a characteristic of length either 19, 20, 21, 24 or 25 in the truncated differential setting. However, if we only consider the pure truncated differential setting, then we have the following proposition.

Proposition 4. *For 5, 6 and 7 rounds, there is always a characteristic of length respectively 14, 18 and 21 in the pure truncated differential setting.*

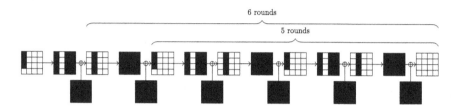

Fig. 3. Characteristic always valid for 5, 6 and 7 rounds.

Proof. Such a characteristic is depicted in Fig. 3. Note that considering how this kind of characteristic is built, there are a lot of underlying equations in the truncated differential setting, which is very likely to make this characteristic invalid. However, in the pure differential setting, these characteristics always remains valid as they come directly from the propagation rules of the AES round function.

Corollary 2. *Using a permutation as the key schedule, the optimal bounds on the number of active S-boxes that can be proven for 5, 6 and 7 rounds is respectively 14, 18 and 21 in the pure truncated differential setting.*

Now the first question that we may ask is whether or not there exists a permutation which reaches all those bounds. Fortunately, such a permutation was already found by Khoo et al. in [14], which is

$$P_{KLPS} = (5\ \ 2\ \ 3\ \ 8\ \ 9\ \ 6\ \ 7\ \ 12\ \ 13\ \ 10\ \ 11\ \ 0\ \ 1\ \ 14\ \ 15\ \ 4).$$

However, if we study this permutation in the truncated differential setting for 7 rounds using Model 1, then we have that the minimum number of active S-boxes becomes 22, proving that no differential characteristic with a probability larger than 2^{-128} can be found, hence the following theorem.

Theorem 1. *We can find a permutation for the key schedule which guarantees that no differential characteristic with a probability larger than 2^{-128} exists for 7 or more rounds of AES. Moreover, this does not depend on the S-box DDT.*

Obviously, now the main question is: How far can we go? Can we find a permutation that reach 22 S-boxes for 6 rounds or lower, or at least a permutation such that no differential characteristic with probability larger than 2^{-128} exists? This would allow us to show that even with an extremely simple and efficient key schedule, we can still have a rather good security against differential attacks in the related-key model. We study this in the next section.

4   Searching for a Permutation

4.1   Bound on 5 Rounds

In this section, we show that there is no permutation that can reach a minimal number of active S-boxes of 18 over 5 rounds. While this does not imply that we cannot find a permutation such that there is no differential characteristic with a probability better than 2^{-128}, this still gives us a good idea of what we can reach for 5 rounds.

To achieve this, we proceed in two steps. First, we search for a set of cycles such that using a given cycle of this set, one cannot build a truncated differential characteristic of length strictly lower than 18, which induces equations (according to Lemma 1) on at most 1 round. Since all permutations can be decomposed into a composition of cycles, this would not only speed up the search (since we do not need to check every permutation one at a time), but also gives a way to build all permutations that *could* reach 18 S-boxes on 5 rounds. To build such a set of cycles, we used a quite straightforward algorithm.

First, we suppose that the cycle starts with 0. Then, we guess the image of 0, and for each of those guesses, we have two cases: either the cycle is not complete, and thus we need to make another guess on the next element of the cycle, or the cycle is closed. Whenever we make a new guess or decide that the cycle is closed, we can build several truncated key characteristics $k_0 \rightarrow k_1 \rightarrow \ldots \rightarrow k_4$ according to the current (partial) cycle examined: each active byte in this truncated key characteristic must be a byte that belongs to the current (partial) cycle. Then, for each of those truncated key characteristics, we search the minimal number of active S-boxes that we can reach using this characteristic. To speed up the search, we only consider truncated characteristics that induces equations on at most 1 rounds, such that these characteristics are always valid in the truncated differential model. If, for a given (partial) cycle, one can find a corresponding truncated characteristic with strictly less than 18 S-boxes, then we know that this (partial) cycle cannot be part of the permutation we are looking for. If we were in the case where the cycle was not complete, then we know that we do not need any more guesses, and if the cycle was closed, we can dismiss it. Thus in the end, we will have a set of closed cycles which start with 0, and for which all truncated characteristics that induces equations on at most 1 rounds have at least 18 active S-boxes. We then need to apply the same algorithm, but this time with cycles beginning by 1 and not containing 0 (to avoid repetitions) and so on.

In the end, we have a set of permutations for which we know that, if a permutation reaches a minimal number of active S-boxes of 18 (or higher), then it must be built from this set of cycles. Thus, we just need to built all possible permutations from these cycles, and plug them into Model 1 to see if the actual minimal number of S-boxes is indeed 18 or higher. The number of cycles which can be used to build a permutation reaching 18 S-boxes is given in Appendix B, and by testing all possible combinations, we found out that there is no such permutation, hence the following theorem.

Theorem 2. *There is no permutation that, when used as key schedule, can reach a minimal number of active S-boxes of 18 or higher over 5 rounds. Using the same method, we were also able to find at least one permutation which have a minimal number of active S-boxes of 16 over 5 rounds, namely:*

$$(15\ 0\ 2\ 3\ 4\ 11\ 5\ 7\ 6\ 12\ 8\ 10\ 9\ 1\ 13\ 14).$$

However, the possibility of reaching 17 S-boxes over 5 rounds is still unknown, and the complexity of the algorithm for 6 rounds is too high. Hence, we focused our search for a permutation reaching 22 active S-boxes over 6 rounds, using another approach we detail in the next section.

4.2   Finding a Permutation over 6 Rounds

First of all, let us take a quick look at how we could naively search for such a permutation. This is rather straightforward: for each possible permutation, we check whether the minimal number of S-boxes is at least 22. Since we are looking for a permutation over 16 bytes, we have $16! > 2^{44}$ possible permutations. While 2^{44} basic operations could be achievable in a reasonable amount of time, the computation of the minimal number of S-boxes is actually quite costly. For example, if one would use the algorithm from [9] which has an approximate complexity of 2^{34} operations, this would raise the total cost to 2^{78} operations, which is clearly impractical. While we do not have a complexity estimation for our constraint programming tool, the average time to solve Model 1 is about 40 min for 6 rounds, which would lead to way too much time to try each permutation, so exhausting all permutations is clearly not a viable way to proceed.

On the other hand, one could try to pick a random permutation, evaluate its minimal number of S-boxes, and try again if this number is lower than 22. While the cost of computing the minimal number of S-boxes remains, this approach could be successful if the density of the set of permutation reaching 22 S-boxes overall permutations is high enough. Indeed, if we do this for 7 rounds, we are able to find a permutation reaching the same number of S-boxes for 7 rounds and lower as the permutation from [14] in about 200 tries. However, this approach was not able to find a permutation reaching 22 S-boxes over 6 rounds.

Hence, we need something more efficient for 6 rounds. Inspired by the work of Nikolić [20], we choose to use a meta-heuristic called *simulated annealing*. Meta-heuristics are a class of search algorithms which aim to find an (almost) optimal solution to an optimization problem, often inspired by some real-life phenomenon. To be more precise, unlike *Constraint Programming* or *Integer Linear Programming* which aims at recovering an *optimal* solution, meta-heuristics only look for a *good enough* solution: it may not be optimal, but it should be rather close to an optimal solution. In our case, we could define our optimization problem as: Which permutation maximize the minimal number of active S-boxes over 6 rounds? However, we are not really interested in maximizing the minimal number of S-boxes, we only need to find a permutation which reaches 22 S-boxes. Moreover, our problem is of the form *"Maximize the minimum value of a given function"*, which is not something easily handled by classical techniques like Constraint or Linear Programming. Finally, meta-heuristics are designed to be both relatively easy to implement and rather efficient, hence they seem quite appropriate to tackle this problem.

We give a generic algorithm for simulated annealing in Appendix C, also given in [20]. The main idea of this algorithm is to try to maximize a function $f(x)$ (called *objective function*) by progressively improving a solution, starting from a random one, while allowing degradation. To be more precise, starting from a random x_0, the algorithm builds another solution x_i from x_{i-1} using the function ϵ. Then, if $f(x_i) > f(x_{i-1})$, then x_i is accepted and the algorithm continues. However, if $f(x_i) \leq f(x_{i-1})$, which would mean that x_i is worse than the previous solution x_{i-1}, x_i is only accepted with some probability depending on a value T, and if it is rejected, another x_i is generated from x_{i-1}. Then, the value T is updated with a function $\alpha(T)$. For more details about this algorithm and the choice of its parameters, we refer the reader to [6, 15, 20].

Now, we need to see how we implement this algorithm in practice. As in [20], we did not observe major differences between different parameters for the initial temperature T_0 and the cooling schedule $\alpha(T)$. Hence, we only give one set of parameters, from which all our following results come from. For the initial temperature, we used $T_0 = 2$. For the cooling schedule, we used the same one as in [20], i.e., $\alpha(T) = \frac{T}{1+\beta T}$ with $\beta = 0.001$ Finally, the neighbor function ϵ generates a new permutation from the one that has been tested. This new permutation should be "close" to the previous one, hence we use a random transposition to generate a new permutation, namely, $\epsilon(x) = \tau \circ x$ where τ is a random transposition.

The only thing missing to implement the algorithm is a way to evaluate $f(x)$. Recall that in our case, $f(x)$ is the minimal number of active S-boxes for a given permutation x. A naive way to compute $f(x)$ would be to solve Model 1 with the permutation x. However, as mentioned before, solving this model is quite costly, which would results in a very slow meta-heuristic. Instead, we make the following observation. Let n be the number of active S-boxes we want to prove, that is, we want to find a permutation for which the minimal number of active S-boxes is at least n. Then, given a certain permutation, we are only interested in one fact: does this permutation have a characteristic with a length strictly less than n? If so, then even if this characteristic is not a minimal one, we still know that this permutation will not reach our goal of a minimum of n active S-boxes. This allows to slightly modify the original algorithm for a much quicker execution, which lead to more permutation being evaluated and thus better chances to find a good one. The complete algorithm is given as Algorithm 1, with a more detailed explanation below.

Algorithm 1. Tweaked Simulated Annealing

Input: Target length n
1: $x \leftarrow$ *random permutation*, $T \leftarrow 2$, $l \leftarrow 0$
2: **while** $l < n$ **do**
3:    $\tau \leftarrow$ *random transposition*, $x' \leftarrow \tau \circ x$
4:    $l' \leftarrow$ quicksearch(x', n)
5:    **if** $l' \geq n$ **then**
6:       $x \leftarrow x'$, $l \leftarrow$ fullsearch(x)
7:    **else if** $l' > l$ **then**
8:       $x \leftarrow x'$, $l \leftarrow l'$
9:    **else**
10:       $r \leftarrow U[0, 1]$ *Generate a uniformly random real number in [0,1]*
11:       **if** $r < e^{\frac{l'-l}{T}}$ **then**
12:          $x \leftarrow x'$, $l \leftarrow l'$
13:       **end if**
14:    **end if**
15:    $T \leftarrow \frac{T}{1+0.001T}$
16: **end while**
Output: x

So instead of directly computing the minimal number of active S-boxes for a given permutation, we do the following. We first use the algorithm quicksearch, which is a classical dynamic programming algorithm which, given a permutation x and a target number of S-boxes n, search for a *relatively* short characteristic of length $\leq n$. As mentioned before, the idea is to use the fact that we are mostly interested in whether or not a characteristic of length strictly less than n exists. This algorithm performs this relatively quickly, without having to find *the* minimal number of S-boxes. Once we get such a characteristic of length l', three cases can happen.

- If $l' \geq n$, then the permutation might be a good one. However, since the quicksearch algorithm does not return the length of the shortest characteristic, we need to call the fullsearch algorithm, which basically solves Model 1 using the provided permutation, and returns the real minimal number of S-boxes. If the output of fullsearch is greater or equal than n, then we found a permutation and the algorithm terminates. If not, we still choose to update x to x', because the fact that quicksearch returned a value greater or equal than n means that the permutation looked quite good at first glance. We also update l to the real minimal number of active S-boxes of x, since otherwise the algorithm would terminate while it did not found a permutation reaching n S-boxes.

- Otherwise if $l' > l$, that is, the permutation x' seems to have a minimal number of S-boxes greater than the previous one, then we update x to x' too. This corresponds to the case $f(x') > f(x)$ in the original Simulated Annealing algorithm.
- Finally, if $l' \leq l$, this is the same as the original algorithm. We accept the solution x' and update x to it only with a certain probability depending on the current temperature T and the respective number of S-boxes found for x and x'.

We first launched this algorithm using $n = 20$, and were able to find the permutation P_k (given below) reaching this minimal number of S-boxes in about 2^{16} tries:

$$P_k = (8 \ 1 \ 7 \ 15 \ 10 \ 4 \ 2 \ 3 \ 6 \ 9 \ 11 \ 0 \ 5 \ 12 \ 14 \ 13).$$

Reaching 21 S-boxes is still an open question and for reference, we were able to test about 2^{24} permutations in several days. However, we were able to show that using P_k as the key schedule, while only reaching a minimum amount of 20 S-boxes in the truncated setting, still guarantee that no characteristic with a probability better then 2^{-128} can be found when one use the DDT of the AES S-box. To do that, we used Model 2, which allows to check if there is a characteristic with a better probability than 2^{-128} and to exhibit one if that is the case. To make this model work, we need to give it a list of truncated differential characteristics, and it will check if such a characteristic can be instantiated with a probability better than 2^{-128}. Hence, to prove that P_k has no such characteristic, we need a list of all valid truncated characteristics of 20 and 21 S-boxes (since 22 S-boxes already guarantees that no characteristic will be instantiable with a probability better than 2^{-128}). This can be computed rather quickly using Model 1 and asking the solver to find all characteristics of length 20 and 21. There are 253 characteristics of length 20 and 3284 of length 21. After about nine hours on a standard desktop to loop through all these characteristics, it turns out that none of them can be instantiated[2] with a probability better than 2^{-128}. In conclusion, we were able to find a permutation P_k such that using this permutation as the key schedule of AES-128 guarantees that no differential characteristic with a probability better than 2^{-128} exists over 6 or more rounds. For reference, we also ran Model 1 on this permutation to get the minimal number of active S-boxes for a lower amount of rounds, summarized in Table 1.

Now, even if we were able to find a permutation leading to no differential characteristic of probability better than 2^{-128} for 6 rounds or more, it still only reaches 20 S-boxes in the truncated setting. Hence, we would like to see if by

---

[2] For reference, the best probability we could reach among all the characteristics of length 20 was 2^{-134}.

Table 1. Minimal number of S-boxes that our permutation P_k reaches on a given number of rounds compared to the one from [14]. [a]No instantiation with a better probability than 2^{-128}.

Number of rounds	2	3	4	5	6	7
Original key schedule	1	3	9	11	13[a]	15
P_{KLPS}		1	5	10	14	18[a]	22
P_k		1	5	10	15	20[a]	23

modifying further the AES round function, we could reach more active S-boxes. This is treated in the next section.

5   Tweaking both ShiftRows and the Key Schedule

Using the approach given in the previous section allowed to find a permutation for the key schedule, which induces a minimal number of S-boxes of 20 for 6 rounds. Here, we would like to see if by changing the ShiftRows operation in the AES-128, we could reach a better number of active S-boxes, namely 21 or 22. Obviously, we cannot try all possible permutations for ShiftRows, as again, there are 2^{44} permutations over 16 elements. Hence, we show here how we restricted ourselves to only a few thousand candidates for ShiftRows, which are the most likely to lead to a good minimal number of active S-boxes, and give a few examples of pairs (P_s, P_k) that reach 21 S-boxes for 6 rounds, where P_s is used instead of the ShiftRows operation, and P_k instead of the original key schedule KS of AES.

First, we can see that we can drastically reduce the number of candidates for P_s using the following two propositions. We denote $\mathcal{P}_i$ the set of all permutations P_i acting insides the columns of the state, i.e., there exists four permutations $P_i^0, P_i^1, P_i^2, P_i^3$ over four elements such that P_i^j acts on the j-th column and $P_i = P_i^0 \circ P_i^1 \circ P_i^2 \circ P_i^3$, and $\mathcal{P}_c$ the set of all permutations which permutes the columns of the state.

Proposition 5. *Let P_s and P_s' be two permutations over 16 elements such that $P_s' = P_i' \circ P_s \circ P_i$, where $P_i, P_i' \in \mathcal{P}_i$, and let $P_k' = P_i^{-1} \circ P_k \circ P_i$. Then using (P_s', P_k') instead of (SR, KS) will lead to the same minimal number of active S-boxes that using (P_s, P_k) instead of (SR, KS). Hence, we can build equivalence classes $\mathcal{E}_i(P_s) = \{P_s' \mid \exists\, P_i, P_i' \text{ s.t. } P_s' = P_i \circ P_s \circ P_i'\}$, and there are 10147 such equivalence classes.*

Proof. We need to show that, for each characteristic we can build using (P_s, P_k), one can find a characteristic with the same number of active S-boxes using (P_s', P_k'), where $P_s' = P_i' \circ P_s \circ P_i$ and $P_k' = P_i^{-1} \circ P_k \circ P_i$.

Given a characteristic $(X_0, \ldots, X_r)$ such that the length of the characteristic is given by $\sum_{i=0}^{r} X_i$, and denote Y_i the state after the MC operation such that

$X_{i+1} = Y_i \oplus K_i$. We have $Y_{i+1} = \mathsf{MC} \circ P_s \circ \mathsf{SB}(Y_i \oplus K_i)$ and $K_{i+1} = P_k(K_i)$, where P_k is a bytewise permutation. For all i, let $K'_i = P_i^{-1}(K_i)$ and $Y'_i = P_i^{-1}(Y_i)$, hence we have

$$
\begin{aligned}
K'_{i+1} = P_i^{-1}(K_{i+1}) &= P_i^{-1} \circ P_k(K_i) \\
&= P_i^{-1} \circ P_k \circ P_i \circ P_i^{-1}(K_i) \\
&= P'_k \circ P_i^{'-1}(K_i) \\
&= P'_k(K'_i).
\end{aligned}
$$

So P'_k is a valid key schedule. Furthermore, note that when considering the propagation of active bytes through MC, one only need to consider the number of active bytes before MC in one given columns to know the number of active byte after MC in that same column. Hence, since $P_i \in \mathcal{P}_i$ only permutes bytes inside each column, the number of active bytes does not change in each column and thus for any $P_i \in \mathcal{P}_i$, MC and $\mathsf{MC}' = \mathsf{MC} \circ P_i$ behave similarly when searching for truncated differential characteristics, i.e., replacing MC by MC' has no effect. In the same way, one can replace MC by $P_i \circ \mathsf{MC}$ with $P_i \in \mathcal{P}_i$. Moreover, SB acts on each byte separately, hence $P_i \circ \mathsf{SB} = \mathsf{SB} \circ P_i$. Thus, we have:

$$
\begin{aligned}
Y'_{i+1} = P_i^{-1}(Y_i) &= P_i^{-1} \circ \mathsf{MC} \circ P_s \circ \mathsf{SB}(Y_i \oplus K_i) \\
&= P_i^{-1} \circ \mathsf{MC} \circ P_s \circ \mathsf{SB}(P_i \circ P_i^{-1}(Y_i) \oplus P_i \circ P_i^{-1}(K_i)) \\
&= \mathsf{MC} \circ P_s \circ \mathsf{SB}(P_i(Y'_i) \oplus P_i(K'_i)) \quad \text{\textit{replacing} } P_i^{-1} \circ \textit{MC by MC has no effect} \\
&= \mathsf{MC} \circ P'_s \circ P_s \circ P_i \circ SB(Y'_i \oplus K'_i) \quad \text{\textit{replacing MC by MC} } \circ P'_i \textit{ has no effect} \\
&= \mathsf{MC} \circ P'_s \circ \mathsf{SB}(Y'_i \oplus K'_i).
\end{aligned}
$$

So (P'_s, P'_k) correctly defines a round function and we have $X'_{i+1} = Y'_i \oplus K'_i = P_i^{-1}(Y_i \oplus K_i) = P_i^{-1}(X_{i+1})$ for all i. Hence, each X'_i is a permutation of X_i, and thus the corresponding characteristic $(X'_0, \ldots, X'_r)$ has the same number of active S-boxes as $(X_0, \ldots, X_r)$.

Proposition 6. *Let P_s and P'_s be two permutations over 16 elements such that $P'_s = P_c^{-1} \circ P_s \circ P_c$ where $P_c \in \mathcal{P}_c$, and let $P'_k = P_c^{-1} \circ P_k \circ P_c$. Then, using (P'_s, P'_k) instead of (SR, KS) will lead to the same minimal number of active S-boxes that using (P_s, P_k) instead of (SR, KS). Hence we can combine this with the previous proposition, and for each class representative P_s of some class $\mathcal{E}_i(P_s)$ defined previously, we can build equivalence classes $\mathcal{E}(P_s) = \{P'_s \mid \exists\, P_c \in \mathcal{P}_c \text{ s.t. } P'_s = P_c^{-1} \circ P_s \circ P_c\}$, and there are 9186 such equivalence classes.*

The proof of the previous theorem is very similar to the proof of Proposition 5 and is given in Appendix D. Hence, we only need to consider 9186 possible candidates P_s to replace SR, instead of 2^{44}. Moreover, we would like to avoid weakening AES in the single-key model. In that model, the original ShiftRows allows to reach full diffusion after 3 rounds. So we only considered the permutations that also reached full diffusion in at most 3 rounds, and there are 4381 of them. Finally, recall that in the pure truncated differential setting, using the original ShiftRows implies that there is always a characteristic of length 18 which is built using a fully active key. While this characteristic has high chances of being invalidated once we consider the equations it implies on the key, we still would like to avoid it. To do that, we used the following proposition.

Proposition 7. *If one uses a permutation P_s instead of ShiftRows such that P_s send the bytes from any one column to at most three columns, then the characteristic from Proposition 4 cannot happen.*

Proof. The characteristic from Proposition 4 can be built because a state containing a single fully active column lead to a fully active state after MC ∘ SR. However, if one uses a permutation P_s which send the bytes from any one column to at most three columns, then the state after MC ∘ P_s will contain at most 3 fully active column. Thus, when XOR-ing the key afterwards, the resulting state would have at least 4 active bytes, instead of 3 in the characteristic from Proposition 4, thus this characteristic cannot happen.

Hence, we only want to try some permutations P_s instead of ShiftRows which verify the previous propositions and achieve a full diffusion in at most 3 rounds in the single-key model, which lead to 3288 possible candidates for P_s. Now everything is quite straightforward. We reuse Algorithm 1 to search for a permutation leading to 21 S-boxes, except that we use a different permutation than ShiftRows in the quicksearch algorithm and modified Model 1 to use that permutation instead of ShiftRows for the fullsearch algorithm. We also added the additional condition that it should stop after 24 hours if no permutation reaching the objective was found. Surprisingly, the quicksearch algorithm ran faster with those permutations than with the original SR, which allowed us to test about 2^{25} permutations P_k on average in 24 hours for a specific candidate P_s. After a few more than 100 possible P_s tried, we were able to find several pairs (P_s, P_k) that reach 21 S-boxes (see Appendix E). After testing about 1100 candidates for P_s, finding a pair (P_s, P_k) that reaches 22 S-boxes is still an open problem.

We also used Model 2, tweaked to use a different permutation instead of SR, to check if there is a differential characteristic with a probability better than 2^{-128} over 6 rounds with these pairs (P_s, P_k), and again, none of these permutation allows such a characteristic.

6   Conclusion

In this paper, we studied how AES would behave in the related-key model if we change its key schedule to a much simpler and efficient one, namely a permutation. We first gave a few generic bounds about the best number of active S-boxes reachable for a given number of round, and especially, we showed that no permutation can reach a minimal number of 18 or more active S-boxes over 5 rounds. However we were able to exhibit a permutation reaching 16 S-boxes over 5 rounds, hence closing the gap a bit further. We showed that we can find a permutation which allows to have at least 20 active S-boxes over 6 rounds, while guaranteeing that no characteristic with a probability larger than 2^{-128} exists. This allows us to reach the same amount round than with the original AES-128 key schedule (see [9]), but with a more efficient key schedule which is also easier to analyze and has a higher minimal number of active S-boxes. We also took a look at how modifying the SR operation could improve the minimal number of S-boxes over 6 rounds. It turns that we can find several pairs (P_s, P_k) to use instead of SR and the key schedule (respectively) which allows to have at least 21 S-boxes over 6 rounds, and again, no characteristic with a probability better than 2^{-128}. We also provided a Constraint Programming model which can handle directly the equations coming from MixColumns, thus allowing to find the exact minimal number of active S-boxes considering everything but the S-box DDT in a reasonable amount of time and memory. Our implementation is available at https://github.com/TweakAESKS/TweakAESKS.

A few open questions remain. First, could we reach a minimal number of 22 active S-boxes changing only the key schedule (and possibly SR) for 6 rounds? In the same idea, could we close the gap for 5 rounds? We know that we cannot get 18 or more active S-boxes, but 16 S-boxes is reachable, thus the possibility of reaching 17 S-boxes is still unknown. Finally, we chose to change the SR operation, but how about changing either MC or the S-box? While changing everything would lead to a cipher that does not have much in common with AES, it could answer the following generic question: Can we build an AES-like SPN (with a round function structured as $MC \circ P_s \circ SB$ where P_s is a permutation and MC uses an MDS matrix) using a permutation as the key schedule, which could reach either 22 S-boxes over 6 rounds, or guarantee that no characteristic with probability better than 2^{-128} exists over 5 rounds?

A   Modelizing the MC Equations in Constraint Programming

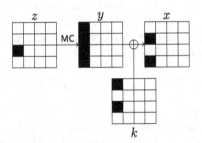

Fig. 4. A partial round that implies one equation

We will give here an example as how we generate constraints to modelize the equations coming from the MC operation. From the MDS property of MC, we know that there is en equation between any set of five bytes taken from the same column of z and y. Specifically, we have the following equation, where coefficient are in $\mathbb{F}_{256}$:

$$5.z[0] + 7.z[1] + z[3] = 2.y[0] + y[2].$$

Now we take the situation given in Fig. 4. First, all bytes 0,1 and 3 of z are inactive, hence we can replace $z[0], z[1]$ and $z[3]$ in the previous equation by zeros. Moreover, we can see that both $y[0]$ and $y[2]$ are cancelled by some bytes in k, i.e. $y[i] \oplus k[i] = 0, i \in \{0,2\}$. Hence, our equation becomes $2.k[0] + k[2] = 0$.

So, if this situation occurs, we know that we have a specific equation involving bytes of k. However, this equation has coefficient in $\mathbb{F}_{256}$, which are not handled by Constraint Programming solvers. Hence, we modelize this equation at a bit-level, using the fact that the scalar multiplication in $\mathbb{F}_{256}$ corresponds to a linear operation in $\mathbb{F}_2^8$. By denoting $k_j^i, j \in [0,7], i \in 0, 2$ the j-th bit of $k[i]$, we have

$$
\begin{pmatrix}
0\,0\,0\,0\,0\,0\,0\,1 \\
1\,0\,0\,0\,0\,0\,0\,1 \\
0\,1\,0\,0\,0\,0\,0\,0 \\
0\,0\,1\,0\,0\,0\,0\,1 \\
0\,0\,0\,1\,0\,0\,0\,1 \\
0\,0\,0\,0\,1\,0\,0\,0 \\
0\,0\,0\,0\,0\,1\,0\,0 \\
0\,0\,0\,0\,0\,0\,1\,0
\end{pmatrix}
\cdot
\begin{pmatrix}
k_0^0 \\ k_1^0 \\ k_2^0 \\ k_3^0 \\ k_4^0 \\ k_5^0 \\ k_6^0 \\ k_7^0
\end{pmatrix}
+
\begin{pmatrix}
k_0^1 \\ k_1^1 \\ k_2^1 \\ k_3^1 \\ k_4^1 \\ k_5^1 \\ k_6^1 \\ k_7^1
\end{pmatrix}
= 0.
$$

We now have everything to modelize this case using an *if-constraint*. In our model, we have a binary variable for each byte of the state which is set to 0 if the corresponding byte is inactive, and 1 otherwise. Since all the equations only involves some key-bits, we also have binary variables for each bit of each subkey. Restricting this in the situation given in Fig. 4, we would have binary variables $z[i], y[i], x[i], k[i], i \in [0,15]$ modelizing whether or not bytes are active, and binary variables $k_j^i, i \in [0,15], j \in [0,7]$ for each bit of the key. Obviously, we need to modelize the fact that if a key byte is inactive, then its bits are all zeros, which is easily modelized with

$$k[i] = 0 \iff k_j^i = 0 \; \forall j \in [0,7].$$

Hence, the above equation only holds when $z[0] = z[1] = z[2] = 0, y[0] = y[2] = 1$ and $x[0] = x[2] = 0$. Note that we do not need to check that $k[0] = 1$ since the fact that $y[0] = 1$ and $x[0] = 0$ necessarily implies that $k[0] = 1$ (and the same argument goes for $k[2]$). So, to modelize this case, we use an *if-constraint*. Such a constraint is of the form $E \Rightarrow C$, and means that if the expression E is true, then the constraint C must hold. Thus, we modelize the above situation with the constraint

$$z[0] = 0 \wedge z[1] = 0 \wedge z[2] = 0 \wedge y[0] = 1$$
$$\wedge\, y[1] = 1 \wedge x[0] = 0 \wedge x[2] = 0$$

$$\Rightarrow$$

$$
\begin{aligned}
k_7^0 + k_0^1 &= 0 \bmod 2 \wedge \\
k_0^0 + k_7^0 + k_1^1 &= 0 \bmod 2 \wedge \\
k_1^0 + k_2^1 &= 0 \bmod 2 \wedge \\
k_2^0 + k_7^0 + k_3^1 &= 0 \bmod 2 \wedge \\
k_3^0 + k_7^0 + k_4^1 &= 0 \bmod 2 \wedge \\
k_4^0 + k_5^1 &= 0 \bmod 2 \wedge \\
k_5^0 + k_6^1 &= 0 \bmod 2 \wedge \\
k_6^0 + k_7^1 &= 0 \bmod 2
\end{aligned}
$$

Hence in our model, we need to do this for all rounds and for each column of the state. The number of constraints coming from this is easy to compute. For a fixed round and column, denote i the number of inactive bytes taken in z, hence $\binom{4}{i}$ possibilities, with $1 \leq i \leq 3$. Denote j the number of inactive bytes taken in y hence $\binom{4}{j}$ possibilities. Hence, we have $5 - i - j$ active bytes (that are cancelled) in y, taken in the remaining $4 - j$ bytes, thus $\binom{4-j}{5-i-j}$ possibilities. Moreover, we know from Lemma 1 that we must have $5 - i - j \geq 2$. So the number of constraints for a fixed round and a fixed column is

$$\sum_{i=1}^{3} \sum_{j=0}^{3-i} \binom{4}{i}\binom{4}{j}\binom{4-j}{5-i-j} = 164,$$

hence $656r$ constraints for r rounds.

B   Number of Cycles to Build a Permutation Reaching 18 S-Boxes over 5 Rounds

If one would want to build a permutation reaching 18 active S-boxes over 5 rounds, then Table 2 gives the number of possible cycle which can be used to build such a permutation. For example, this table means that if the permutation contains a cycle of length 11, then there are only 48 cycles of this length which can be used to build the permutation. This table also implies that the permutation should not contain a cycle of length ≥ 12. As mentioned in Sect. 4.1, none of the possible combinations of those cycles allows to build a permutation reaching 18 active S-boxes over 5 rounds.

Table 2. Number of cycles which must be use to build a permutation reaching 18
A-boxes over 5 rounds

Length of the cycle	Number of cycles
1	16
2	120
3	796
4	6576
5	25656
6	78448
7	112608
8	74904
9	15576
10	1344
11	48

C   Generic Simulated Annealing Algorithm

See Algorithm 2

Algorithm 2. Simulated Annealing [20]

  Input: initial temperature T_0, cooling schedule $\alpha(T)$, neighbor function $\epsilon(x)$
1:   $x \leftarrow random, \quad T \leftarrow T_0$
2: **while** termination criteria not met **do**
3:   $x' \leftarrow \epsilon(x)$
4:   **if** $f(x') > f(x)$ **then**
5:    $x \leftarrow x'$
6:   **else**
7:    $r \leftarrow U[0,1]$ *Generate a uniformly random real number in [0,1]*
8:    **if** $r < e^{\frac{f(x')-f(x)}{T}}$ **then**
9:     $x \leftarrow x'$
10:    **end if**
11:   **end if**
12:   $T \leftarrow \alpha(T)$
13: **end while**
  Output: x

D   Proof of Proposition 6

As in the proof of Proposition 5, we need to show that, for each characteristic
we can build using (P_s, P_k), one can find a characteristic with the same number

of active S-boxes using (P'_s, P'_k), with $P'_s = P_c^{-1} \circ P_s \circ P_c$ and $P'_k = P_c^{-1} \circ P_k \circ P_c$, $P_c \in \mathcal{P}_c$.

Given a characteristic $(X_0, \ldots, X_r)$, and using the same notation as in the proof of Proposition 5, for all i let $K'_i = P_c^{-1}(K_i)$ and $Y'_i = P_c^{-1}(Y_i)$. Showing that P'_k is a valid key-schedule is done in the same way as for Proposition 5. Furthermore, note that since MC acts on each column separately, we have MC $\circ$ $P_c^{-1} = P_c^{-1} \circ$ MC. In the same way, SB acts on each byte separately, hence $P_c \circ$ SB = SB $\circ P_c$. Thus we have

$$
\begin{aligned}
Y'_{i+1} = P_c^{-1}(X_{i+1}) &= P_c^{-1} \circ \mathsf{MC} \circ P_s \circ \mathsf{SB}(Y_i \oplus K_i) \\
&= P_c^{-1} \circ \mathsf{MC} \circ P_s \circ \mathsf{SB}(P_c(Y'_i) \oplus P_c(K'_i)) \\
&= P_c^{-1} \circ \mathsf{MC} \circ P_s \circ \mathsf{SB} \circ P_c(Y'_i \oplus K'_i) \\
&= \mathsf{MC} \circ P_c^{-1} \circ P_s \circ P_c \circ \mathsf{SB}(Y'_i \oplus K'_i) \\
&= \mathsf{MC} \circ P'_s \circ \mathsf{SB}(Y'_i \oplus K'_i)
\end{aligned}
$$

So again, (P'_s, P'_k) correctly defines a round function and $X'_{i+1} = P_c^{-1}(X_{i+1})$ for all i. Thus each X'_i is a permutation of X_i, hence the corresponding characteristic $(X'_0, \ldots, X'_r)$ has the same number of active S-boxes as the characteristic $(X_0, \ldots, X_r)$.

E Pairs (P_s, P_k) Reaching 21 Sboxes over 6 Rounds

Table 3. Pairs (P_s, P_k) which reach 21 S-boxes, along with the number of P_k tried before founding it

(P_s, P_k)	# iterations
$P_s^1 = (0\ 1\ 2\ 4\ 3\ 8\ 9\ 12\ 5\ 13\ 14\ 15\ 6\ 7\ 10\ 11)$ $P_k^1 = (10\ 4\ 12\ 11\ 6\ 2\ 5\ 1\ 8\ 0\ 9\ 7\ 13\ 14\ 15\ 3)$	$3151253 \sim 2^{21.6}$
$P_s^2 = (0\ 1\ 2\ 4\ 3\ 8\ 9\ 12\ 5\ 6\ 13\ 14\ 7\ 10\ 11\ 15)$ $P_k^2 = (15\ 14\ 11\ 10\ 6\ 12\ 4\ 0\ 3\ 8\ 1\ 9\ 2\ 5\ 13\ 7)$	$42414349 \sim 2^{25.3}$
$P_s^3 = (0\ 1\ 4\ 8\ 9\ 10\ 12\ 13\ 5\ 6\ 14\ 15\ 2\ 3\ 7\ 11)$ $P_k^3 = (14\ 12\ 8\ 6\ 7\ 4\ 0\ 1\ 3\ 11\ 10\ 2\ 9\ 5\ 13\ 15)$	$8588115 \sim 2^{23}$
$P_s^4 = (0\ 1\ 2\ 8\ 4\ 9\ 12\ 13\ 5\ 6\ 7\ 14\ 3\ 10\ 11\ 15)$ $P_k^4 = (12\ 14\ 11\ 4\ 8\ 0\ 3\ 7\ 10\ 15\ 2\ 9\ 6\ 13\ 5\ 1)$	$15016901 \sim 2^{23.8}$
$P_s^5 = (0\ 1\ 2\ 8\ 4\ 9\ 12\ 13\ 3\ 5\ 14\ 15\ 6\ 7\ 10\ 11)$ $P_k^5 = (5\ 9\ 15\ 13\ 3\ 4\ 6\ 2\ 11\ 7\ 10\ 0\ 8\ 14\ 1\ 12)$	$51700477 \sim 2^{25.6}$

For a given P_s^i, we also took a look at the permutations P'_k that are rather "close" to the ones we found, that is, permutations P'_k which are one or two transpositions away from each P_k^i. It turns out that, except for (P_s^4, P_k^4), none of these permutations also reach 21 S-boxes. Oddly, there are 3 permutations that are 1 transposition away from P_k^4 which also reach 21 S-boxes when using P_s^4 instead of SR, and again, none of them has a differential characteristic with a probability better than 2^{-128} over 6 rounds. Those three permutations are

$$P_k^{4'} = \begin{pmatrix} 14 & 12 & 11 & 4 & 8 & 0 & 3 & 7 & 10 & 15 & 2 & 9 & 6 & 13 & 5 & 1 \end{pmatrix},$$

$$P_k^{4''} = \begin{pmatrix} 12 & 14 & 11 & 4 & 10 & 0 & 3 & 7 & 8 & 15 & 2 & 9 & 6 & 13 & 5 & 1 \end{pmatrix},$$

$$P_k^{4'''} = \begin{pmatrix} 12 & 14 & 11 & 4 & 8 & 0 & 3 & 7 & 2 & 15 & 10 & 9 & 6 & 13 & 5 & 1 \end{pmatrix}.$$

References

1. Biham, E., Biryukov, A., Shamir, A.: Cryptanalysis of Skipjack reduced to 31 rounds using impossible differentials. In: Stern, J. (ed.) EUROCRYPT 1999. LNCS, vol. 1592, pp. 12–23. Springer, Heidelberg (1999). https://doi.org/10.1007/3-540-48910-X_2
2. Biham, E., Shamir, A.: Differential cryptanalysis of DES-like cryptosystems. J. Cryptol. 4(1), 3–72 (1991)
3. Biryukov, A., Khovratovich, D.: Related-key cryptanalysis of the full AES-192 and AES-256. In: Matsui, M. (ed.) ASIACRYPT 2009. LNCS, vol. 5912, pp. 1–18. Springer, Heidelberg (2009). https://doi.org/10.1007/978-3-642-10366-7_1
4. Biryukov, A., Khovratovich, D., Nikolić, I.: Distinguisher and related-key attack on the full AES-256. In: Halevi, S. (ed.) CRYPTO 2009. LNCS, vol. 5677, pp. 231–249. Springer, Heidelberg (2009). https://doi.org/10.1007/978-3-642-03356-8_14
5. Biryukov, A., Nikolić, I.: Automatic search for related-key differential characteristics in byte-oriented block ciphers: application to AES, Camellia, Khazad and others. In: Gilbert, H. (ed.) EUROCRYPT 2010. LNCS, vol. 6110, pp. 322–344. Springer, Heidelberg (2010). https://doi.org/10.1007/978-3-642-13190-5_17
6. Černỳ, V.: Thermodynamical approach to the traveling salesman problem: an efficient simulation algorithm. J. Optim. Theory Appl. 45(1), 41–51 (1985)
7. Choy, J., Zhang, A., Khoo, K., Henricksen, M., Poschmann, A.: AES variants secure against related-key differential and boomerang attacks. In: Ardagna, C.A., Zhou, J. (eds.) WISTP 2011. LNCS, vol. 6633, pp. 191–207. Springer, Heidelberg (2011). https://doi.org/10.1007/978-3-642-21040-2_13
8. Daemen, J., Rijmen, V.: AES Proposal: Rijndael (1999)
9. Fouque, P.-A., Jean, J., Peyrin, T.: Structural evaluation of AES, and chosen-key distinguisher of 9-round AES-128. In: Canetti, R., Garay, J.A. (eds.) CRYPTO 2013. LNCS, vol. 8042, pp. 183–203. Springer, Heidelberg (2013). https://doi.org/10.1007/978-3-642-40041-4_11
10. Gérault, D., Lafourcade, P., Minier, M., Solnon, C.: Revisiting AES Related-Key Differential Attacks with Constraint Programming. IACR Cryptology ePrint Archive 2017/139 (2017)
11. Jean, J.: TikZ for Cryptographers (2016). https://www.iacr.org/authors/tikz/
12. Jean, J., Nikolić, I.: Efficient design strategies based on the AES round function. In: Peyrin, T. (ed.) FSE 2016. LNCS, vol. 9783, pp. 334–353. Springer, Heidelberg (2016). https://doi.org/10.1007/978-3-662-52993-5_17

13. Jean, J., Nikolić, I., Peyrin, T.: Tweaks and keys for block ciphers: the TWEAKEY framework. In: Sarkar, P., Iwata, T. (eds.) ASIACRYPT 2014. LNCS, vol. 8874, pp. 274–288. Springer, Heidelberg (2014). https://doi.org/10.1007/978-3-662-45608-8_15

14. Khoo, K., Lee, E., Peyrin, T., Sim, S.M.: Human-readable proof of the related-key security of AES-128. IACR Trans. Symmetric Cryptol. **2017**(2), 59–83 (2017)

15. Kirkpatrick, S., Gelatt, C.D., Vecchi, M.P.: Optimization by simulated annealing. Science **220**(4598), 671–680 (1983)

16. Knudsen, L.: DEAL-a 128-bit block cipher (1998)

17. Liu, G., Ghosh, M., Song, L.: Security analysis of SKINNY under related-tweakey settings. IACR Trans. Symmetric Cryptol. **2017**(3), 37–72 (2017)

18. Matsui, M.: On correlation between the order of S-boxes and the strength of DES. In: De Santis, A. (ed.) EUROCRYPT 1994. LNCS, vol. 950, pp. 366–375. Springer, Heidelberg (1995). https://doi.org/10.1007/BFb0053451

19. Nikolić, I.: Tweaking AES. In: Biryukov, A., Gong, G., Stinson, D.R. (eds.) SAC 2010. LNCS, vol. 6544, pp. 198–210. Springer, Heidelberg (2011). https://doi.org/10.1007/978-3-642-19574-7_14

20. Nikolić, I.: How to use metaheuristics for design of symmetric-key primitives. In: Takagi, T., Peyrin, T. (eds.) ASIACRYPT 2017. LNCS, vol. 10626, pp. 369–391. Springer, Cham (2017). https://doi.org/10.1007/978-3-319-70700-6_13

21. Zong, R., Dong, X., Wang, X.: MILP-Aided Related-Tweak/Key Impossible Differential Attack and Its applications to QARMA, Joltik-BC. Cryptology ePrint Archive, Report 2018/142 (2018). https://eprint.iacr.org/2018/142

Analysis and Improvement of an Authentication Scheme in Incremental Cryptography

Louiza Khati[1,2]($\boxtimes$) and Damien Vergnaud[3,4]

[1] Département d'informatique de l'ENS, École normale supérieure, CNRS,
PSL Research University, 75005 Paris, France
`louiza.khati@ens.fr`
[2] ANSSI, Paris, France
[3] Sorbonne Université, CNRS, Laboratoire d'Informatique de Paris 6, LIP6,
75005 Paris, France
[4] Institut Universitaire de France, Paris, France

Abstract. Introduced in cryptography by Bellare, Goldreich and Goldwasser in 1994, *incrementality* is an attractive feature that enables to update efficiently a cryptographic output like a ciphertext, a signature or an authentication tag after modifying the corresponding input. This property is very valuable in large scale systems where gigabytes of data are continuously processed (e.g. in cloud storage). Adding cryptographic operations on such systems can decrease dramatically their performance and incrementality is an interesting solution to have security at a reduced cost.

We focus on the so-called *XOR-scheme*, the first incremental authentication construction proposed by Bellare, Goldreich and Goldwasser, and the only *strongly* incremental scheme (*i.e.* incremental regarding insert *and* delete update operations at *any* position in a document). Surprisingly, we found a simple attack on this construction that breaks the *basic security* claimed by the authors in 1994 with only one authentication query (not necessarily chosen). Our analysis gives different ways to fix the scheme; some of these patches are discussed in this paper and we provide a security proof for one of them.

1 Introduction

Bellare, Goldreich and Goldwasser initiate the study on *incremental cryptography* in [3] and then refined it in [4]. Cryptographic incremental constructions are meant to provide efficient updates compared to classical algorithms. Usually, the result of a cryptographic algorithm (such as encryption or authentication) over a document has to be re-computed entirely if any change is applied to the document (and this regardless of the modification size). Incremental cryptography enables to update a signature, a message authentication code (MAC) or a ciphertext in time proportional to the number of modifications applied to the

© Springer Nature Switzerland AG 2019
C. Cid and M. J. Jacobson, Jr. (Eds.): SAC 2018, LNCS 11349, pp. 50–70, 2019.
https://doi.org/10.1007/978-3-030-10970-7_3

corresponding document. This attractive feature leads to build many incremental cryptographic primitives such as encryption schemes [1,2,4], signature [3,9,16], MACs [4,9,14], hash functions [6,11] and authenticated encryption constructions [2,8,18].

An algorithm is incremental regarding specific *update* operations such as inserting, deleting or replacing a data block inside a document. A desirable incremental algorithm should support all these operations for *any* positions: it should be possible to insert, delete or replace a data block of the document for all positions without breaking the security of the cryptographic algorithm. Most known algorithms only support replacement of data blocks and the algorithms that support insertion, deletion and replacement[1] are deemed *strongly incremental*.

Virus protection is the first application of incremental cryptography quoted in the seminal paper [3]. They consider the usage scenario where processor accesses files on a remote host and a virus can alter these files. A simple idea is to compute authentication tags for all files with a key stored securely by the processor and any modification by a virus will be detected by verifying the corresponding tag. Knowing that these files will be updated often enough, using an incremental authentication algorithm preserves the processor by performing a lighter computation.

Bellare *et al.* also introduced in [3] the corresponding security notions. In the *basic security* model, the adversary can obtain a valid authentication tag for any message it wanted (as in classical MAC security) and it can also update (with the supported update operations) valid pairs message/tag. This is a first security level but it is reasonable to consider a stronger adversary that can alter files *and* tags before applying update operations; it corresponds to the *tamper-proof security* notion introduced in [3].

Nowadays this use case can be extended to the "digital world". Large amount of data [10,15] are processed every day by different services like cloud services, distributed networks and distributed storage. It is clear that all these data require integrity and/or privacy at a low computational cost otherwise going through gigabytes of data for minor changes without incremental primitives is really demanding in term of time and energy. A concrete example is the Cloud Bigtable by Google [7] that stores petabytes of data across thousands of commodity servers. This Bigtable has a particular data structure that links a unique index number to each block. In this case an incremental hashing that supports replacement and insertion operations is suitable as mentioned in [15]. A more critical usage is storage services in mobile cloud computing where a mobile client device is in addition limited in term of energy consumption. To solve this issue, Itani, Kayssi and Chehab provide an energy-efficient protocol in [13] that guarantee data integrity based on incremental MACs. Another use case is sensor networks and more specifically environmental sensors [12,15]: several sensors are deployed at different physical positions and they record continuously data.

---

[1] Actually supporting insertion and deletion is sufficient as replacement can be obtain by combining these two update operations.

At some point, all the data ends up in a big public database that has to be publicly checkable. The database is updated (insertion operation mainly) at a high frequency and if the hash value over all the database is entirely re-computed for each insertion it will be very consuming. All these use cases are examples among many others. Incremental cryptography is clearly an area to explore to solve practical issues.

For now incrementality is mainly investigated for hashing and signing even if it was also considered for encryption in [2,3]. It is not surprising regarding all the practical use cases that need incremental authenticated constructions. Recently, the CAESAR[2] competition stimulates research on authenticated encryption algorithm. Sasaki and Yasuda analysed several candidates and found that none of them performs incrementality. That is why they designed their own authenticated encryption mode with associated data [18] based on existing constructions. This new mode is incremental for the replace, insert and delete operations, but the insert and delete operations of this mode concern only the last block of the authenticated data or the last block of the message (and it remains open to design a strongly incremental authenticated encryption algorithm).

Actually, as far as we know, the only authentication scheme that is strongly incremental is the *XOR-scheme* designed by Bellare, Goldreich and Goldwasser in [4] (*cf.* Fig. 2). This strong property comes with a cost: only basic security is claimed in [3] and this algorithm needs to generate and store a lot of randomness. The MAC operation generates a random value for each data block and these random values are necessary for the verification and the update operations. The XOR-scheme is based on a pseudo-random function (PRF) and a pseudo-random permutation (PRP) and the incremental algorithms for (single block) insert and delete operations require only two applications of the underlying PRF and two applications of the underlying PRP. The XOR-scheme relies on the concept of *pair block chaining* (which was later used in [11] which involves taking each pair of two consecutive blocks of a message and feeding them into a pseudo-random function before chaining all the outputs of the PRF into the final hash. This scheme extends another scheme, called the *randomized XOR-scheme*, from [5] which is incremental only for replacement. Even if they share a similar name, these two algorithms are different: the randomized XOR-scheme is not based on a pair block chaining structure and requires actually much less randomness. To distinguish the two schemes, in this paper, we will call this second scheme the *unchained XOR-scheme*.

An analysis on some incremental hash functions was provided by Phan and Wagner [17]. They give, inter alia, patterns that could give collisions on a hash function based on pair block chaining. Two cases are of interest for the XOR-scheme: *non-distinct blocks* and *cycling chaining*. The first one considers repeated blocks messages like $A||B||C||B||A$ and $B||C||B||A||B$ that would have the same sum value if no randomness was used (*cf.* Fig. 2) but as underlined by the authors the random values appended to each message block prevents these repetitions.

---

[2] Competition for Authenticated Encryption: Security, Applicability, and Robustness.

The second one considers a variant of the XOR-scheme [11] where the first and the last block are chained then some repeated patters like $A||B||A$ and $B||A||B$ would have the same sum value but it not the case in the original version from [4]. Therefore, in the present state-of-the-art, no attacks are known against the original strongly incremental *XOR-scheme* proposed in [4].

1.1 Contributions of the Paper

In this paper, we analyse the security of the original XOR-scheme construction proposed by Bellare, Goldreich and Goldwasser in [4] and based on a chained structure as defined in [3].

ATTACKS. We provide an attack that breaks the XOR-scheme *basic security* claimed by the authors[3]. It succeeds with probability 1 using only one MAC query. It takes advantage of the chaining structure of this scheme and some xor function properties. This attack is very simple and it is surprising that it remained unnoticed until now (especially since the paper [4] appeared in a major computer science conference and was extensively cited since 1994).

ANALYSIS AND PATCHED CONSTRUCTIONS. We analyse our attack and the original XOR-scheme to find where its security breaks down. We show that the main flaw is that the XOR-scheme does not explicitly take into account the document length and we noticed that adding the number of data block to the construction prevents this kind of attacks. We analyse different ways to patch the scheme by introducing the document block length in the construction and found that the scheme can still be weak for some options.

We propose a modified version of the XOR-scheme and prove its basic security. Our security proof for the patched XOR-scheme uses tool from the unchained XOR-scheme security proof [5].

ORGANIZATION OF THE PAPER. We introduce some mathematical backgrounds, recall the security models for incremental MAC constructions and we give a detailed description of the XOR-scheme construction in Sect. 2. Then we present a general forgery attack and its analysis in Sect. 3. In Sect. 4, we discuss different solutions to patch efficiently the scheme without making it more complicated neither breaking the structure of the algorithm. We choose one construction and give its detailed description. Its security proof is given in Sect. 5 before the conclusion in Sect. 6.

2 Preliminaries

2.1 Notations

For any integer n, $\{0,1\}^n$ denotes the set of bit strings of length n and we let $\{0,1\}^*$ denote the set of all finite-length bit strings. For two bit strings X and

---

[3] In [4, Theorem 3.1], Bellare, Goldreich and Goldwasser stated a security result for their scheme but no proofs are provided in their paper.

Y, $X||Y$ denotes their concatenation. For a finite set S, we use $x \xleftarrow{\$} S$ to denote sampling x uniformly at random from S. For $X \in \{0,1\}^*$, we use $|X|$ to denote the bit length of X and $|X|_\ell$ denotes the number of ℓ-bit block in the bit-string X (and in particular $|X|_1 = |X|$).

RANDOM FUNCTIONS/RANDOM PERMUTATIONS. The set of all functions $\{0,1\}^\ell \rightarrow \{0,1\}^L$ is denoted $\mathcal{F}_{\ell,L}$. A *random function* F is a randomly chosen function in $\mathcal{F}_{\ell,L}$. The set of all permutations $\{0,1\}^\ell \rightarrow \{0,1\}^\ell$ is denoted $\mathcal{P}_\ell$. A *random permutation* P is a randomly chosen permutation in $\mathcal{P}_\ell$.

PSEUDO-RANDOM FUNCTIONS/PSEUDO-RANDOM PERMUTATIONS. Given a non-empty subset $\mathcal{K}$, a function family $F_k \colon \mathcal{K} \times \{0,1\}^\ell \rightarrow \{0,1\}^L$, where $k \in \mathcal{K}$, is a (t, ϵ)-pseudo-random function (PRF) if for any algorithm $\mathcal{A}$ running in a time at most t, the following holds:

$$|Pr[k \xleftarrow{\$} \mathcal{K} : \mathcal{A}^{F_k(.)} = 1] - Pr[F \xleftarrow{\$} \mathcal{F}_{\ell,L} : \mathcal{A}^{F(.)} = 1]| \leq \epsilon.$$

Given a non-empty subset $\mathcal{K}$, a permutation family $P_k \colon \mathcal{K} \times \{0,1\}^L \rightarrow \{0,1\}^L$, where $k \in \mathcal{K}$, is a (t, ϵ)-pseudo-random permutation (PRP) if for any algorithm $\mathcal{A}$ running in a time at most t, the following holds:

$$|Pr[k \xleftarrow{\$} \mathcal{K} : \mathcal{A}^{P_k(.)} = 1] - Pr[F \xleftarrow{\$} \mathcal{F}_{L,L} : \mathcal{A}^{F(.)} = 1]| \leq \epsilon.$$

2.2 Definitions

SYNTACTIC DEFINITION. We begin by describing the syntactic definition of strongly incremental MAC algorithms. In the following, we consider authentication of messages whose length is a multiple of an integer b (which is usually smaller than the block length of the underlying PRF or PRP) but is obviously possible to handle messages of arbitrary finite length using padding. A *document* $D \in \mathcal{D}$ with $\mathcal{D} = \bigcup_{i=1}^{\infty} \{0,1\}^{ib}$ is a sequence of n b-bit blocks, for some integer $n \geq 1$ denoted $D = (D_1, D_2, \ldots, D_n)$ where D_i is the i-th b-bit block of D.

Definition 1. *A strongly incremental MAC scheme is a 5-tuple*

$$\Pi = (\mathcal{K}, \text{MAC}, \text{V}, \text{I}, \text{D})$$

in which:

$\mathcal{K}$ *is the key space. A key k is randomly chosen in the key space $\mathcal{K}$. The key k is an input for the* MAC, V, I *and* D *algorithms.*
MAC, *the MAC algorithm, is a probabilistic algorithm that takes as input the key k, a document D and returns an authentication tag T.*
V, *the verification algorithm, is a deterministic algorithm that takes as input the key k, a document D and a tag T and returns 1 if the tag is valid and 0 otherwise.*
I, *the incremental insert algorithm, is a probabilistic algorithm that takes as input the key k, the insertion position j, the message block to add D'_j, the document D and a tag T to update.*

D, *the incremental delete algorithm, is a probabilistic algorithm that takes as input the key k, the deletion position j, the document D and a tag T to update.*

with the three following correctness properties:

- $(\forall k \in \mathcal{K})(\forall n \in \mathbb{N})(\forall D \in \{0,1\}^{nb})(\forall T \in \{\texttt{MAC}(k,D)\})(\{\texttt{V}(k,D,T)\} = \{1\})$
- $(\forall k \in \mathcal{K})(\forall n \in \mathbb{N})(\forall D = (D_1, D_2, \ldots, D_n) \in \{0,1\}^{nb})(\forall T \in \{\texttt{MAC}(k,D)\})$
  $(\forall j \in \{1, \ldots, n+1\})(\forall D^* \in \{0,1\}^b)(\forall T' \in \{\texttt{I}(k,j,D^*,D,T)\})$
  $(\{\texttt{V}(k,(D_1, \ldots, D_{j-1}, D^*, D_j, \ldots D_n), T')\} = \{1\})$
- $(\forall k \in \mathcal{K})(\forall n \in \mathbb{N})(\forall D = (D_1, D_2, \ldots, D_n) \in \{0,1\}^{nb})(\forall T \in \{\texttt{MAC}(k,D)\})$
  $(\forall j \in \{1, \ldots, n\})(\forall T' \in \{\texttt{D}(k,j,D,T)\})$
  $(\{\texttt{V}(k,(D_1, \ldots, D_{j-1}, D_{j+1}, \ldots D_n), T')\} = \{1\})$

REMARK 1. All these algorithms take as input a key k and to lighten the notation, in the following, the key k is put as subscript. For example, the MAC algorithm is simply denoted $\texttt{MAC}_k(D)$.

REMARK 2. In practice, the incremental algorithms I and D have to be more efficient than re-computing an entire authentication tag T and cryptographers are looking for scheme where these algorithms are constant-time (*i.e.* independent of the number of b-bit blocks of the document D).

SECURITY MODEL. The adversary $\mathcal{A}$ is an algorithm (*i.e.* an oracle probabilistic Turing machine) playing in a computational security game denoted $G^{BS}_{\texttt{MAC},\texttt{V},\texttt{I},\texttt{D}}$ (*cf.* Fig. 1). A key k is picked uniformly at random in the key space of the strongly incremental MAC and the adversary has access to all the following oracles

- **a MAC oracle:** the adversary can ask to compute a MAC (for k) on any document of its choice;
- **a verifying oracle:** the adversary can ask to verify (for k) the validity of any pair document/authentication tag;
- **an update oracle(s):** the adversary can use the incremental operations (for k) on chosen document/authentication tag pairs (in a way depending on the security models as defined below).

At the end of each oracle query (except verification queries), the corresponding authenticated document/authentication tag (D, T) is added to a list $\mathcal{L}$ and the adversary wins the game if it outputs eventually a pair $(D^*, T^*) \notin \mathcal{L}$ that is accepted by the verification algorithm.

BASIC SECURITY. As defined in [3], in basic security settings the adversary $\mathcal{A}$ is **not allowed** to do incremental operations on a couple (D, T) where the verification algorithm $\texttt{V}_k(D, T)$ *will* fail. It can only apply incremental operations on couples (D, T) that belong to the list $\mathcal{L}$. As mentioned above, to win the security game, $\mathcal{A}$ must provide a forgery that is to say a document D^* and a tag T^* such that $\texttt{V}_k(D^*, T^*)$ returns 1 and the couple (D^*, T^*) is not in the list $\mathcal{L}$.

Remark 1. The verification $V_k(D, T)$ is not applied before incremental operation to check authenticity otherwise the low computational cost is lost. It is simply **assumed** that $\mathcal{A}$ does not query incremental operations on altered couples. In this paper, we focus on this basic security notion only.

Definition 2. *Let* $\Pi = (\mathcal{K}, \text{MAC}, \text{V}, \text{I}, \text{D})$ *be a strongly incremental MAC scheme and let* $\mathcal{A}$ *be an adversary. Let* $\mathbf{Adv}_{A,\Pi}^{BS} :=$

$$\Pr[k \xleftarrow{\$} \mathcal{K}; \mathcal{L} \leftarrow \{\}; (D^*, T^*) \leftarrow \mathcal{A}^{\text{MAC}_k^{\mathcal{L}}, \text{V}_k^{\mathcal{L}}, \text{I}_k^{\mathcal{L}}, \text{D}_k^{\mathcal{L}}} : 1 \leftarrow V_k^{\mathcal{L}}(D^*, T^*) \wedge (D^*, T^*) \notin \mathcal{L}].$$

Π is $(\lambda, q_m, q_v, q_{inc}; \epsilon)$*-BS-secure in the basic sense if, for any adversary* $\mathcal{A}$ *which runs in time* λ*, making* q_m *queries to the* MAC *oracle,* q_v *to the* V *oracle and* q_{inc} ***valid*** *queries to the incremental oracles* (I, D) *we have* $\mathbf{Adv}_{A,\Pi}^{BS} < \epsilon$.

Game $G_{\text{MAC},\text{V},\text{I},\text{D}}^{BS}$

$k \xleftarrow{\$} \mathcal{K}$
$\mathcal{L} \leftarrow \{\}$
If $\mathcal{A}^{\text{MAC}_k^{\mathcal{L}}, \text{V}_k^{\mathcal{L}}, \text{I}_k^{\mathcal{L}}, \text{D}_k^{\mathcal{L}}}$ makes a query (D^*, T^*) such that
  - $V_k^{\mathcal{L}}(D^*, T^*)$ returns 1 and
  - $(D^*, T^*) \notin \mathcal{L}$
Then Return 1 **Else** return 0.

Fig. 1. Game defining basic security (BS) for an incremental authentication scheme.

TAMPER-PROOF SECURITY. As defined in [4] *tamper-proof security* is a stronger security notion since the adversary $\mathcal{A}$ is allowed to query incremental operation on **any** couple (D, T) even new couples (couples that do not belong to $\mathcal{L}$). Then $\mathcal{A}$ wins the security game if it provides a new couple (D^*, T^*) such that $V_k(D^*, T^*)$ returns 1. It was already mentioned in [4], that the XOR-scheme does not achieve tamper-proof security and this is also the case of our modified XOR-scheme.

2.3   Description of the XOR-Scheme

The XOR-scheme $(\mathcal{XS})$ as defined in [3] is an incremental authenticated algorithm based on pair-wise chaining as shown in Fig. 2. Let ℓ and L be two positive integers and let $b < \ell$ be some positive integer. The $\mathcal{XS}$ scheme is based on a pseudo-random function $F : \mathcal{K}_F \times \{0, 1\}^{2\ell} \longrightarrow \{0, 1\}^L$ and a pseudo-random permutation $P : \mathcal{K}_P \times \{0, 1\}^L \longrightarrow \{0, 1\}^L$. The incremental algorithms for (single block) insert and delete operations require only two applications of the underlying PRF and two applications of the underlying PRP. The $\mathcal{XS}$ scheme generates an authentication tag for a document D by repeatedly applying the PRF to pairs of blocks – each made of a b-bit data block from the document D and an $\ell - b$ random block (pick uniformly at random and independently for each block). In the following, for simplicity, we consider only documents whose binary length is a multiple of b and we denote $\mathcal{D} = (\{0, 1\}^b)^*$.

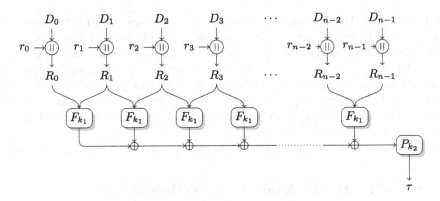

Fig. 2. Description of the XOR-scheme

- The **key space** $\mathcal{XS.K}$ is $\mathcal{K}_F \times \mathcal{K}_P$ the Cartesian product of the key space of the underlying PRF F and PRP P.
- The **MAC algorithm** $\mathcal{XS}$.MAC takes as input a document $D \in \mathcal{D}$ and outputs a tag $T := (r, \tau)$. For each document block D_i, an $(\ell - b)$-bits block r_i is randomly generated. The concatenation of these values is denoted $R_i := D_i \| r_i$. Each couple (R_{i-1}, R_i) is processed by the function F_{k_1} and outputs a value denoted h_i then the bitwise XOR (eXclusive OR) of all the values (denoted Σ) is processed by the permutation P_{k_2} to give the value τ.
- The **Verification algorithm** $\mathcal{XS}$.V takes as inputs the document D and a tag $T := (r, \tau)$. It re-computes the value τ from the inputs r and D. It returns 1 if this value is equal to the input τ and 0 otherwise.
- The **Insert operation** $\mathcal{XS}$.I enables to insert a block value in a document. It takes as inputs the position j where the block value has to be inserted, the previous block value D_{j-1}[4], the block value D_j[5], the new block value D'_j and the tag T. It outputs the new tag.
- The **Delete operation** $\mathcal{XS}$.D enables to delete a block from the document. It takes as inputs the position j where the block as to be deleted, the block value to delete D_j, the previous and next block values D_{j-1} and D_{j+1} and the tag T.

The update algorithms are intuitive and given in Fig. 8 (in Appendix) for update operations at a position different from the first block position. They can be adapted to be applied to the first block. In the original version, it is specified that a prefix and postfix are added to the document. For a document $D = D_1 \ldots D_n$, the authentication tag is computed on $D_0 \| D_1 \ldots D_n \| D_{n+1}$ where D_0 and D_{n+1} are specific prefix and postfix values. In this paper, this specification is not taken into account: it does not prevent our attack and the repaired scheme is proven secure without it.

---

[4] For the first position, there is no previous block.
[5] For the last position, there is no next block.

XOR-SCHEME LIMITS. Supporting insert, delete and consequently replace operations should make the XOR-scheme very efficient in term of update time running. The fresh random values r_i generated by this scheme for each new document block are necessary for security. But generating so much randomness is *time consuming*: for an n-block document D, a $n(\ell - b)$-random bits value r is generated. Random generation also slows down the insertion operation. Another drawback is the tag *expansion*: the random value r is part of the tag and needs to be stored. For a n block document, random storage costs $n(\ell - b)$ bits. Even if today storage is not an issue, having short tags is desirable.

3   Forgery Attacks Against the XOR-Scheme

According to the basic security game described in Fig. 1, the adversary $\mathcal{A}$ wins the game if it finds a new pair (D^*, T^*) such that the verification operation returns 1. If an adversary has access to any tag T (such that $T = (r, \tau)$) returned by the MAC algorithm on a document D (for example $D_0 \| D_1 \| D_2$), it can forge a different document D^* having the same value τ. The value τ is computed as follows:

$$\tau = P_{k_2}[F_{k_1}(D_0 \| r_0, D_1 \| r_1) \oplus F_{k_1}(D_1 \| r_1, D_2 \| r_2)] \tag{1}$$
$$\Sigma = F_{k_1}(R_0, R_1) \oplus F_{k_1}(R_1, R_2) = h_1 \oplus h_2 \tag{2}$$

$\mathcal{A}$ can build a document $D^* \neq D$ and a value r^* such that the corresponding Σ^* value collides with Σ even if there is no weakness on F. A way to do so is inserting a specific block chain in document D in order to cancel all the new values h_i' introduced by these repetitions as shown in Fig. 3. It seems that the chaining structure of the XOR-scheme should prevent this behavior because changing or inserting a block value will affect two values h_i then the tag τ will be different. These modifications have to be compensated: the values h_i' introduced has to be canceled by xoring the same value and all the original values h_i that are deleted has to be re-introduced. We use this trick to break the claimed *basic security*.

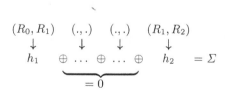

Fig. 3. Xor cancellation strategy in the XOR-scheme

FORGERY ATTACK. Applying this strategy gives us an adversary $\mathcal{A}^{\text{MAC}_{k_1, k_2}}$ wining the game $G_{\text{MAC,V,I,D}}^{BS}$ with probability 1 and that requires only 1 MAC_{k_1, k_2} query (Fig. 4).

Adversary $\mathcal{A}^{\mathrm{MAC}^{\mathcal{L}}_{k_1,k_2}}$

1 $\mathcal{A}$ asks the MAC of a short document D where $D = D_0||D_1||D_2$ and receives the corresponding authentication-tag $T = (r, \tau)$.

$$
\begin{array}{ccc}
(R_0, R_1) & (R_1, R_2) & \\
\downarrow & \downarrow & \\
h_1 & \oplus \quad h_2 & = \Sigma
\end{array}
$$

Fig. 4. Σ computation for 3 block document

2 $\mathcal{A}$ builds a document D^* from D such that $D^* = D_0||D_1||D_2||D_1||D_2||D_1||D_2$ and a value r^* from r such that $r^* = r_0||r_1||r_2||r_1||r_2||r_1||r_2$.

$$
\begin{array}{ccccccccccccc}
\substack{(D_0||r_0, D_1||r_1) \\ (R_0, R_1)} & & \substack{(D_1||r_1, D_2||r_2) \\ (R_1, R_2)} & & \substack{(D_2||r_2, D_1||r_1) \\ (R_2, R_1)} & & \substack{(D_1||r_1, D_2||r_2) \\ (R_1, R_2)} & & \substack{(D_2||r_2, D_1||r_1) \\ (R_2, R_1)} & & \substack{(D_1||r_1, D_2||r_2) \\ (R_1, R_2)} & & \\
\downarrow & & \downarrow & & \downarrow & & \downarrow & & \downarrow & & \downarrow & & \\
h_1 & \oplus & h_2 & \oplus & \not{h_2} & \oplus & \not{h_2} & \oplus & \not{h_2} & \oplus & \not{h_2} & = \Sigma
\end{array}
$$

Fig. 5. Attack on the XOR-scheme

The document D^* is different from D but it has the same value τ. The document D^* given in Fig. 5 is an example of a forgery and many other examples can be given. To be more general, for any $x \in \{0,1\}^b$, any $x' \in \{0,1\}^{(\ell-b)}$ and for any valid pair (D, T) such that $D = D_0 \ldots D_i||D_{i+1} \ldots D_n$, many forgeries $(D^*, (r^*, \tau))$ can be built by inserting the specific block chain $D_i||x||D_i||x$ in D (and the corresponding random value chain $r_i||x'||r_i||x'$ in r for any x') such that:

$$
D^* = D_0 \ldots D_{i-1}|| \underbrace{D_i||x||D_i||x||D_i} ||D_{i+1} \ldots D_n
$$

$$
r^* = r_0 \ldots r_{i-1}|| \underbrace{r_i||x'||r_i||x'||r_i} ||r_{i+1} \ldots r_n.
$$

A variant of this forgery is to insert only a repeated document block D_i (and r_i) is the following:

$$
D^* = D_0 \ldots D_{i-1}|| \underbrace{D_i||D_i||D_i} ||D_{i+1} \ldots D_n.
$$

$$
r^* = r_0 \ldots r_{i-1}|| \underbrace{r_i||r_i||r_i} ||r_{i+1} \ldots r_n.
$$

A more powerful forgery can be built from (D, T) by inserting any values x and y in D (and any values x' and y' in r) such that:

$$
D^* = D_0 \ldots D_{i-1}|| \underbrace{D_i||x||y||x||y||x||D_i||x||D_i} ||D_{i+1} \ldots D_n.
$$

$$
r^* = r_0 \ldots r_{i-1}|| \underbrace{r_i||x'||y'||x'||y'||x'r_i||x'||r_i} ||r_{i+1} \ldots r_n.
$$

For all these attacks, the underbraced chains can be repeated many times. These three attacks are some of the possible attacks, following this *canceling* strategy, some exotic chains can be inserted in order to ends with a value τ that corresponds to a legitimate tag. A first observation is that all these attacks are performed by inserting blocks and providing a forgery D^* that has the same length that the original one looks impossible or at least harder.

4   Modification of the XOR-Scheme

The previous section described an attack that breaks the basic security of the XOR-scheme by producing a document D^* using a MAC query (D, T) where $\tau = \tau^*$ and $|D|_b \neq |D^*|_b$. All the forgeries D^* produced are longer that the original document D. One can notice that if the adversary $\mathcal{A}$ is only allowed to MAC and verify documents that have the same length n then the attack presented in Sect. 3 will fail. A first naive idea is to force all documents to have the same length (documents that are too small can be padded in the MAC algorithm) but this solution is not realistic and the incremental property will be lost. A natural way to fix this flaw is to use the document length n for the computation of the value τ in order to make it *size dependent*. The size can be expressed according to any units: number of bits, bytes, blocks. Choosing the number of blocks n is sufficient. A postfix block containing the number of blocks n can be added at the end of the document then the computation of the value Σ will become:

$$\Sigma = F_{k_1}(D_0||r_0, D_1||r_1) \oplus F_{k_1}(D_1||r_1, D_2||r_2) \oplus \cdots \oplus F_{k_1}(D_{n-1}||r_{n-1}, r_n||n)$$

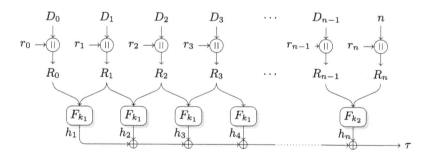

Fig. 6. Description of the fixed XOR-scheme

The last block works as a mask for each value τ: incremental operation will refresh the last random value r_n in order to have a different mask value for any modification. As a consequence, the pseudo-random permutation P is not necessary anymore ($\tau = \Sigma$), it is removed in the modified scheme (Fig. 6). This version of the XOR-scheme is proven in Sect. 5.

The last random block value r_n (concatenated with the document length n) is necessary otherwise the corresponding h_n value can be canceled. If it is omitted, the following attack is indeed possible:

1. $\mathcal{A}$ asks the MAC of a document $D = D_0||D_1$ and receives the tag $T_1 = (r_0||r_1, \tau_1)$.
2. $\mathcal{A}$ asks to delete the first block of D by with the delete query $\mathtt{D}(0, ., D_0, D_1, \tau_1)$ and receives $T_2 = (r_1, \tau_2)$.
3. $\mathcal{A}$ asks to insert the block D_0' at the first position of the resulting document with the query $\mathtt{I}_k(0, D_1, ., D_0', \tau_2)$ and receives $T_3 = (r_0'||r_1, \tau_3)$.
4. $\mathcal{A}$ asks to insert the block D_2 at the position 2 of the original document D with the query $\mathtt{I}_k(2, ., D_1, D_2, \tau_1)$ and receives $T_4 = (r_0||r_1||r_2, \tau_4)$.
5. $\mathcal{A}$ builds the document $D^* = D_0'||D_1||D_2$ and the tag $T^* = (r_0'||r_1||r_2, \tau_1 \oplus \tau_3 \oplus \tau_4)$.

The couple (D^*, r^*) is a forgery: it is not in the list $\mathcal{L}$ of tagged document and it has a valid tag. To avoid such attacks, for incremental operations the last random block (concatenated with the document length) needs to be always refreshed. To be sure that none of the previous attack will not be practical, an independent key is used to process the last couple (R_{n-1}, R_n). That way, it would be hard for an adversary to make a forgery from a linear combination of tagged documents.

COMPLEXITY. The modified XOR-scheme is slightly slower than the original one. For the MAC and the incremental algorithms the P call is removed but a call to the function F_{k_2} is added as shown in Fig. 7. The delete D and insert I operations are slightly slower because of the last block update: the last value R_i depending on the document length has to be removed and the a new value R_i' with the new document length n' is added.

Functions	scheme	$(\ell - b)$-bits Generation	F	P	xor
MAC	XS	n	$n - 1$	1	$n - 1$
	M-XS	$n + 1$	$(n - 1) + 1$	0	n
V	XS	0	$n - 1$	1	$n - 1$
	M-XS	0	$(n - 1) + 1$	0	n
D	XS	0	3	2	3
	M-XS	1	5	0	5
I	XS	1	3	2	3
	M-XS	2	5	0	5

Fig. 7. Complexity: XOR-scheme (XS) and modified XOR-scheme (M-XS)

OTHER SOLUTIONS. In the original XOR-scheme (Fig. 2), the document length can be added differently in the algorithm (but still with a random value r_n):

1 Before the last operation $P_{k_2}(\Sigma)$, an intermediate operation $F_{k_3}(r_n||n, \Sigma)$ can be added such that $\tau = P_{k_2}[F_{k_3}(r_n||n, \Sigma)]$.
2 The block length can be processed individually as a last block such that $\tau = P_{k_2}[F_{k_3}(r_n||n) \oplus \Sigma]$.

5   Security Proof

The security proof follows the proof strategy used in [5] for proving the unchained XOR-scheme.

INFORMATION THEORETIC CASE. As in [5], we first consider the case where the two underlying PRFs F_{k_1} and F_{k_2} are replaced by two truly random functions F_1 and F_2 from $\{0,1\}^{2\ell}$ to $\{0,1\}^L$. We consider an unbounded adversary and the following theorem claims the security of this modified scheme in the information theoretic case. More precisely, it provides an absolute bound on the success of the adversary in terms of the number of oracle queries it makes.

Theorem 1. *Let $\mathcal{F}_{2\ell,L}$ be the family of random functions with input length 2ℓ and output length L. Let $\mathcal{A}$ be any (computationally unbounded) adversary, in the basic security settings, making a (q_m, q_v, q_{inc})-attack against the modified XOR-scheme with two functions picked uniformly at random from $\mathcal{F}_{2\ell,L}$. The probability that $\mathcal{A}$ is successful is at most*

$$q^2 \cdot 2^{b-\ell} + q_v \cdot (t^2 \cdot 2^{b-\ell} + \cdot 2^{-L}).$$

where $q = q_m + q_{inc}$ and t denotes the maximal block-length of the documents authenticated in the security game.

Proof (Theorem 1 (Sketch)). The proof follows closely the proof from [5]. The main difference is that we use two different random functions in the modified scheme and that we need the following simple lemma to prove that some specific matrix (close to the one used in [5]) is full-rank. For the reader familiar with [5], we use similar notations in the following.

Lemma 1. *Let X be some finite set and let $n \in \mathbb{N}$. Let $(R_0, R_1, \ldots, R_n) \in X^{n+1}$ with $R_i \neq R_j$ for all $i \neq j$ then if there exists $(R_0^*, R_1^*, \ldots, R_n^*) \in X^{n+1}$ such that*

$$\{(R_0, R_1), (R_1, R_2), \ldots (R_{n-1}, R_n)\} = \{(R_0^*, R_1^*), (R_1^*, R_2^*), \ldots (R_{n-1}^*, R_n^*)\}$$

then for all $i \in \{0, \ldots, n\}$, $R_i = R_i^*$.

Proof (Lemma 1). This lemma can be easily proved by induction over n. Let us denote S_n the first set $\{(R_0, R_1), (R_1, R_2), \ldots (R_{n-1}, R_n)\}$ where all R_i are distinct. In particular, the set S_n contains exactly n different couples. One can notice that the first member of each couple is the second member of the previous couple except the first and the last couples. In others words a value R_i appears in two couples: once as a first member and once as a second member except the first one R_0 and the last one R_n.

The case $n = 1$ is trivial. We consider the case $n = 2$ that provides greater clarity. Let assume that it exists $(R_0^*, R_1^*, R_2^*) \in X^3$ such that

$$\{(R_0, R_1), (R_1, R_2)\} = \{(R_0^*, R_1^*), (R_1^*, R_2^*)\}$$

and $\#\{R_0, R_1, R_2\} = 3$. As there are exactly two couples in each set, we have the following two cases:

- **case 1:** $(R_0, R_1) = (R_0^*, R_1^*)$ and $(R_1, R_2) = (R_1^*, R_2^*)$ then in this case, we get $R_0 = R_0^*, R_1 = R_1^*, R_2 = R_2^*$;
- **case 2:** $(R_0, R_1) = (R_1^*, R_2^*)$ and $(R_1, R_2) = (R_0^*, R_1^*)$. The first equality implies $R_1^* = R_0$ and the second equality implies $R_1^* = R_2$ and thus $R_0 = R_2$ which contradicts the statement $R_0 \neq R_2$.

Suppose now that Lemma 1 holds for all integers $k \leq n - 1$ for some $n \in \mathbb{N}^*$. We will show that it holds for n.

Let us suppose that there exists $(R_0^*, R_1^*, \ldots, R_n^*) \in X^{n+1}$ such that $S_n = S_n^*$ where S_n^* is the set $\{(R_0^*, R_1^*), (R_1^*, R_2^*), \ldots (R_{n-1}^*, R_n^*)\}$. Again, as all the values R_i are different in S_n then the n couples are different. The equality of these two sets S_n and S_n^* implies that they contain exactly the same n couples and that in each set a couple appears only once. We have the following two cases:

- **case 1:** $(R_{n-1}, R_n) = (R_{n-1}^*, R_n^*)$ and $S_{n-1} = S_{n-1}^*$. From the induction hypothesis, for all $i \in \{0, \ldots, n-1\}$, $R_i = R_i^*$.
- **case 2:** $(R_{n-1}, R_n) \neq (R_{n-1}^*, R_n^*)$ Then there exists $i \in \{0, \ldots, n-1\}$ such that $(R_{n-1}, R_n) = (R_{i-1}^*, R_i^*)$. It implies $R_i^* = R_n$ and according to the structure of these sets, there is a couple in S_n^* that has a first member equal to $R_i^* = R_n$) and it has to be the case in S_n. But as mentioned above, R_n is a value that appears only in one couple of S_n and we get a contradiction. $\square$

We will use this lemma with $X = \{0, 1\}^{\ell}$ at the end of the proof to show that different messages of the same block-length involve different input pairs for the underlying PRF F_1.

Since the adversary $\mathcal{A}$ is computationally unbounded we may assume without loss of generality that it is deterministic. The probabilistic choices in $\mathcal{A}$'s attack on the scheme are thus the initial choice of F_1 and F_2 of the random functions in $\mathcal{F}_{2\ell, L}$ and the choices of random coins made by the authentication oracles in the security game. We assume (again without loss of generality) that $\mathcal{A}$ makes exactly $q = q_s + q_{inc}$ authentication queries (either as a direct MAC query or as an update query using the insert or the delete oracle). As in [5], there is no loss of generality to assume that $\mathcal{A}$ makes all its authentication queries and then makes exactly one verify query (for its purported forgery). We prove that in this case the probability of the event (denoted Succ) $\mathcal{A}$'s forgery is valid is upper-bounded by

$$q^2 \cdot 2^{b-\ell} + t^2 \cdot 2^{b-\ell} + 2^{-L}.$$

and using a classical argument (see e.g. [5]) we get the claimed bound for general adversaries.

We consider the simple case where all the random coins used in the last block of each authenticated document are different. Note that in all authentication queries (from a fresh MAC query or an update query), this random block is picked uniformly at random and independently of the previous blocks. To analyze the probability of this event (denoted Distinct), we can therefore use the following simple lemma:

Lemma 2 ([5, Fact A.1]). *Let $P(m, t)$ denote the probability of at least one collision in the experiment of throwing t balls, independently at random, into m buckets. Then $P(m, t) \leq t^2/m$.*

We thus have

$$\Pr[\mathsf{Succ}] = \Pr[\mathsf{Succ}|\mathsf{Distinct}] \cdot \Pr[\mathsf{Distinct}] + \Pr[\mathsf{Succ}|\overline{\mathsf{Distinct}}] \cdot \Pr[\overline{\mathsf{Distinct}}]$$
$$\leq \Pr[\mathsf{Succ}|\mathsf{Distinct}] + \Pr[\overline{\mathsf{Distinct}}]$$
$$\leq \Pr[\mathsf{Succ}|\mathsf{Distinct}] + P(2^{b-\ell}, q)$$
$$\leq \Pr[\mathsf{Succ}|\mathsf{Distinct}] + q^2 \cdot 2^{b-\ell}.$$

and it remains to upper-bound $\Pr[\mathsf{Succ}|\mathsf{Distinct}]$.

Let us fix a particular sequence of q documents $D^1, \ldots, D^q$ (each made of at most t blocks of b bits) corresponding to all documents authenticated in the security game by some authentication queries (either as a direct MAC query or as an update query using the insert or the delete oracle). We also fix $r^1, \ldots, r^q$ some bit-strings possibly used as random values in the modified XOR-scheme for these documents (*i.e.* r^i consists in $1 \leq t_i \leq t$ blocks of $\ell - b$ bits if D^i is made of t_i blocks of b bits) and we assume that the last blocks of all of them are all different. Finally we fix $\tau^1, \ldots, \tau^q$ some possible corresponding tags in $\{0, 1\}^L$ for these documents. We consider only bit-strings $(D^1, \ldots, D^q)$, $(r^1, \ldots, r^q)$ and $(\tau^1, \ldots, \tau^q)$ for which the probability that there exists two functions F_1 and F_2 such that $T^i = (r^i, \tau^i)$ is a valid MAC for D^i (for all $i \in \{1, \ldots, q\}$) for F_1 and F_2 is non-zero.

We will compute the probability of the event that $\mathcal{A}$'s forgery is valid conditioned on the event that the authentication queries made by $\mathcal{A}$ are on the documents $D^1, \ldots, D^q$, use the random coins $(r^1, \ldots, r^q)$ and result in the tags $(\tau^1, \ldots, \tau^q)$. More precisely, we will show that this probability is upper-bounded by $t^2 \cdot 2^{b-\ell} + 2^{-L}$ (and since the bit-strings $(D^1, \ldots, D^q)$, $(r^1, \ldots, r^q)$ and $(\tau^1, \ldots, \tau^q)$ are arbitrary, we will get the result by standard conditioning arguments).

We consider a possible forgery output by $\mathcal{A}$ and we denote D^{q+1} the corresponding document, r^{q+1} the used randomness and τ^{q+1} the tag. It is worth noting that the pair (D^{q+1}, r^{q+1}) is different from all pairs (D^i, r^i) for all $i \in \{1, \ldots, q\}$ (since otherwise, this is not an actual forgery) but we cannot assume that the last block of r^{q+1} is different from the last blocks of all previous random values r^i for all $i \in \{1, \ldots, q\}$ (since $\mathcal{A}$ may choose it arbitrarily and it can reuse a value obtained in a previous authentication query).

For $i \in \{1, \ldots, q+1\}$, we denote D^i_j for all $j \in \{1, \ldots, t_i\}$, the j-th block of the document D^i and similarly r^i_j for all $j \in \{1, \ldots, t_i+1\}$, the j-th block of the randomness r^i. As in [5], we consider the matrix B with $q+1$ rows and $2^{2\ell+1}$ columns over $\mathbb{F}_2 = \{0, 1\}$ where the entry in row $i \in \{1, \ldots, q+1\}$ and column $j \in \{1, \ldots, 2^{2\ell+1}\}$ is defined as follows:

- for $j \in \{1, \ldots, 2^{2\ell}\}$, the entry is equal to 1 if j is the index of the 2ℓ-bit string $(D^i_{t_i} \| r^i_{t_i} \| t_i \| r^i_{t_i+1})$ in lexicographic order (and 0 otherwise).

– for $j \in \{2^{2\ell} + 1, \ldots, 2^{2\ell+1}\}$, the entry is equal to 1 if $j - 2^{2\ell}$ is the index of the 2ℓ-bit string $(D_k^i \| r_k^i \| D_{k+1}^i \| r_{k+1}^i)$ in lexicographic order for some $k \in \{1, \ldots, t_i - 1\}$ (and 0 otherwise).

In other words, the matrix B contains a 1 on the row i for $i \in \{1, \ldots, q+1\}$ only at positions corresponding to bit-strings of length 2ℓ used as inputs to the random functions F_1 and F_2 in the modified XOR-scheme (where the left part consisting of the first $2^{2\ell}$ columns of the matrix corresponds to the unique input of F_2 and the right part corresponds to all inputs to F_1).

We have the following lemma:

Lemma 3. *The matrix B has full rank with probability at least $1 - t^2 \cdot 2^{b-\ell}$.*

Proof (Lemma 3). The proof is similar to the proof of [5, Lemma A.3]. If the pair $(t_{q+1}, r_{t_{q+1}+1}^{q+1})$ is different from all $(t_i, r_{t_i+1}^i)$ for $i \in \{1, \ldots, q\}$, then the matrix B is in echelon form (in its left part) and is thus trivially of full rank.

Otherwise, we assume that $r_{t_{q+1}+1}^{q+1}$ is equal to some $r_{t_i+1}^i$ and if $t_{q+1} = t_i$ (the last block of randomness of $\mathcal{A}$'s forgery is equal to the last block of randomness of the i-th authenticated message and the block-length of these two messages are equal). It is worth noting that there exists only one index $i \in \{1, \ldots, q\}$ such that this is the case (since we assume that these last blocks of randomness are all different). For this i-th document, the random blocks r_j^i for $j \in \{1, \ldots, t_i\}$ are all different with probability at least $1 - t_i^2 \cdot 2^{b-\ell} \geq 1 - t^2 \cdot 2^{b-\ell}$ by Lemma 2. Since the pair (D^{q+1}, r^{q+1}) is different from (D^i, r^i) and since the pairs (D_k^i, r_k^i) are all different for $k \in \{1, \ldots, t_i\}$ (with probability at least $1 - t^2 \cdot 2^{b-\ell}$), we can apply Lemma 1 to the sets (of the same length $t_i = t_{q+1}$):

$$\left\{ D_1^i \| r_1^i \| D_2^i \| r_2^i, D_2^i \| r_2^i \| D_3^i \| r_3^i, \ldots, D_{t_i-1}^i \| r_{t_i-1}^i \| D_{t_i}^i \| r_{t_i}^i \right\}$$

and

$$\left\{ D_1^{q+1} \| r_1^{q+1} \| D_2^{q+1} \| r_2^{q+1}, D_2^{q+1} \| r_2^{q+1} \| D_3^{q+1} \| r_3^{q+1}, \ldots, D_{t_i-1}^{q+1} \| r_{t_i-1}^{q+1} \| D_{t_i}^{q+1} \| r_{t_i}^{q+1} \right\}.$$

We thus obtain that there exist an index $k \in \{1, \ldots, t_i - 1\}$ such that

$$(D_k^{q+1} \| r_k^{q+1} \| D_{k+1}^{q+1} \| r_{k+1}^{q+1}) \neq (D_k^i \| r_k^i \| D_{k+1}^i \| r_{k+1}^i).$$

Therefore in this case the left part of the last row (consisting of the first $2^{2\ell}$ columns) is identical to the left part of the i-th row but these rows differ in at least one position in the right part of the matrix B. By elementary operations on the rows, one can easily transform the matrix B in echelon form and it is therefore of full rank (with probability at least $1 - t^2 \cdot 2^{\ell-b}$). $\qquad\square$

To conclude the proof, one can identify the functions F_1 and F_2 to their vector of values in $(\{0,1\}^{2\ell})^L$ by denoting $F_i(x) = (\varphi_{i,1}^{(x)}, \ldots, \varphi_{i,L}^{(x)})$ for $x \in \{0,1\}^{2\ell}$ and $i \in \{1,2\}$, where $\varphi_{i,j} \in \{0,1\}^{2\ell}$ for $i \in \{1,2\}$ and $j \in \{1, \ldots, L\}$. In this case by construction, τ^i is the authentication tag of the document D^i with randomness r^i for all $i \in \{1, \ldots, q+1\}$ if and only if for all $j \in \{1, \ldots, L\}$, the j-th bit $\tau_i^{(j)}$

of τ_i is equal to the dot product of the i-th row of the matrix B and the vector $\varphi_{2,i}\|\varphi_{1,i}$. Using the same argument as in [5], since B is of full rank, the number of vectors satisfying this $q+1$ equations is 2^L times smaller than the number of vectors satisfying only the first q equations (corresponding to the first q rows of B), and therefore we obtained that the forgery τ^{q+1} output by the adversary is valid with probability 2^{-L} if the matrix B is full rank.

We have thus proved that, in the simplified case, the probability that $\mathcal{A}$'s forgery is valid is upper-bounded by

$$q^2 \cdot 2^{b-\ell} + t^2 \cdot 2^{b-\ell} + 2^{-L}.$$

and thus the claimed bound for general adversaries. □

COMPUTATIONAL CASE. If we replace the (truly) random functions by pseudo-random functions in the previous result, we obtain readily the following computational security result:

Theorem 2. *Let F be the family of pseudo-random functions with input length $2 \cdot \ell$ and output length L. Let $\mathcal{A}$ be any adversary making a (q_m, q_v, q_{inc})-attack against the modified XOR-scheme with two functions picked uniformly at random from F and running in time λ.*

There exist an adversary $\mathcal{B}$ against the pseudo-randomness property of F that makes $q' = q \cdot t$ queries to F, runs in time $\lambda' = \lambda + O(q'(\ell + L))$ such that

$$\mathbf{Adv}_{\mathcal{B},F}^{PRF} \geq \mathbf{Adv}_{\mathcal{A},\mathcal{XS}}^{BS} - [(q^2 + t^2) \cdot 2^{b-\ell} + 2^{-L}].$$

where $q = q_m + q_{inc}$ and t denotes the maximal block-length of the documents authenticated in the security game.

Proof (Theorem 2). The proof is identical to the proof of [5, Theorem 4.2] and is left to the reader. □

6   Conclusion

We showed that the XOR-scheme as described in [3] does not provide the claimed basic security: a forgery can be easily built from any tag by inserting specific document block chains to a legitimate document and the corresponding random value chains to the legitimate random value. We proposed a modified XOR-scheme that is not vulnerable to these attacks and we proved its security in the *basic sense*. Our modified XOR-scheme is the only secure strongly incremental algorithm but unfortunately it still has some drawbacks: the randomness generation slows down the algorithm and the tag length makes it unpractical because the random values have to be stored. But it is definitely worth analyzing its structure in order to improve it or to build another strongly incremental authenticated scheme (or prove a lower bound on the length of strongly incremental MAC algorithms). Another interesting open problem is to design a strongly incremental authenticated scheme that achieves tamper-proof security.

Acknowledgments. The authors are supported in part by the French ANR ALAM-BIC Project (ANR-16-CE39-0006). The authors thank Mihir Bellare for helpful discussions and for pointing out references.

A   Appendix

See Fig. 9.

Original XOR-scheme Algorithms

Algo. $\text{MAC}_{k_1,k_2}(D)$
$\Sigma \leftarrow 0^L$
$n \leftarrow |D|_b$
$r_0 \xleftarrow{\$} \{0,1\}^{(\ell-b)}$
$r \leftarrow r_0$
$R_0 \leftarrow (D_0 || r_0)$
For $i = 1$ **to** $n-1$ **do**
$\quad r_i \xleftarrow{\$} \{0,1\}^{(\ell-b)}$
$\quad r \leftarrow r || r_i$
$\quad R_i \leftarrow r_i || D_i$
$\quad h_i \leftarrow F_{k_1}(R_{i-1}, R_i)$
$\quad \Sigma \leftarrow \Sigma \oplus h_i$
$\tau \leftarrow P_{k_2}(\Sigma)$
$T \leftarrow (r, \tau)$
Return (T)

Algo. $\text{V}_{k_1,k_2}(D,T)$
$\Sigma \leftarrow 0^L$
$n \leftarrow |D|_b$
$r \leftarrow T[0 : (n-1)(\ell - b) - 1]$
$\tau \leftarrow T[(n-1)(\ell - b) :]$
$R_0 \leftarrow r_0 || D_0$
For $i = 1$ **to** $n-1$ **do**
$\quad R_i \leftarrow r_i || D_i$
$\quad h_i \leftarrow F_{k_1}(R_{i-1}, R_i)$
$\quad \Sigma \leftarrow \Sigma \oplus h_i$
$\tau' \leftarrow P_{k_2}(\Sigma)$
If $\tau' = \tau$
$\quad$ **Return** 1
Else
$\quad$ **Return** 0

Algo. $\text{D}_{k_1,k_2}(j, D, T)$
$n \leftarrow |D|_b$
$r \leftarrow T[0 : (n-1)(\ell - b) - 1]$
$\tau \leftarrow T[(n-1)(\ell - b) :]$
$R_{j-1} \leftarrow r_{j-1} || D_{j-1}$
$R_j \leftarrow r_j || D_j$
$R_{j+1} \leftarrow r_{j+1} || D_{j+1}$
$h_j \leftarrow F_{k_1}(R_{j-1}, R_j)$
$h_{j+1} \leftarrow F_{k_1}(R_j, R_{j+1})$
$h'_j \leftarrow F_{k_1}(R_{j-1}, R_{j+1})$
$\Sigma \leftarrow h_j \oplus h_{j+1} \oplus h'_j$
$\Sigma \leftarrow \Sigma \oplus P^-_{k_2}(\tau)$
$\tau \leftarrow P_{k_2}(\Sigma)$
$r \leftarrow r_0 \ldots r_{j-1} || r_{j+1} \ldots r_n$
$T \leftarrow (r, \tau)$

Algo. $\text{I}_{k_1,k_2}(j, D'_j, D, T)$
$n \leftarrow |D|_b$
$r \leftarrow T[0 : (n-1)(\ell - b) - 1]$
$\tau \leftarrow T[(n-1)(\ell - b) :]$
$r'_j \xleftarrow{\$} \{0,1\}^{(\ell-b)}$
$R'_j \leftarrow r'_j || D'_j$
$R_j \leftarrow r_j || D_j$
$R_{j-1} \leftarrow r_j || D_{j-1}$
$h'_j \leftarrow F_{k_1}(R_{j-1}, R'_j)$
$h'_{j+1} \leftarrow F_{k_1}(R_{j'}, R_j)$
$h_j \leftarrow F_{k_1}(R_{j-1}, R_j)$
$\Sigma \leftarrow h'_j \oplus h'_{j+1} \oplus h_j$
$\Sigma \leftarrow \Sigma \oplus P^-_{k_2}(\tau)$
$\tau \leftarrow P_{k_2}(\Sigma)$
$r \leftarrow r_0 \ldots r_{j-1} || r'_j || r_j \ldots r_n$
$T \leftarrow (r, \tau)$

Fig. 8. Original XOR-scheme algorithm

Algo. $\mathtt{D}_{k_1,k_2}(j, D, T)$

$n \leftarrow |D|_b$

$r \leftarrow T[0 : n(\ell - b) - 1]$

$\tau \leftarrow T[n(\ell - b) :]$

$R_{j-1} \leftarrow r_{j-1}\|D_{j-1}$

$R_j \leftarrow r_j\|D_j$

$R_{j+1} \leftarrow r_{j+1}\|D_{j+1}$

$h_j \leftarrow F_{k_1}(R_{j-1}, R_j)$

$h_{j+1} \leftarrow F_{k_1}(R_j, R_{j+1})$

$h'_j \leftarrow F_{k_1}(R_{j-1}, R_{j+1})$

$R_{n-1} \leftarrow r_{n-1}\|D_{n-1}$

$R_n \leftarrow r_n\|n$

$h_n \leftarrow F_{k_2}(R_{n-1}, R_n)$

$r'_n \xleftarrow{\$} \{0,1\}^{(\ell-b)}$

$R'_n \leftarrow r'_n\|(n-1)$

$h'_n \leftarrow F_{k_2}(R_{n-1}, R'_n)$

$\Sigma \leftarrow h_j \oplus h_{j+1} \oplus h'_j \oplus h_n \oplus h'_n$

$\tau \leftarrow \tau \oplus \Sigma$

$r \leftarrow r_0 \ldots r_{j-1}\|r_{j+1} \ldots r_{n-1}\|r'_n$

$T \leftarrow (r, \tau)$

Modified XOR-scheme Algorithms

Algo. $\mathtt{MAC}_{k_1,k_2}(D)$

$\Sigma \leftarrow 0^L$

$n \leftarrow |D|_b$

$r_0 \xleftarrow{\$} \{0,1\}^{(\ell-b)}$

$r \leftarrow r_0$

$R_0 \leftarrow r_0\|D_0$

For $i = 1$ **to** $n - 1$ **do**

$\quad r_i \xleftarrow{\$} \{0,1\}^{(\ell-b)}$

$\quad r \leftarrow r\|r_i$

$\quad R_i \leftarrow r_i\|D_i$

$\quad h_i \leftarrow F_{k_1}(R_{i-1}, R_i)$

$\quad \Sigma \leftarrow \Sigma \oplus h_i$

$r_n \xleftarrow{\$} \{0,1\}^{(\ell-b)}$

$R_n \leftarrow r_n\|n$

$h_n \leftarrow F_{k_2}(R_{n-1}, R_n)$

$\tau \leftarrow \Sigma \oplus h_n$

$T \leftarrow (r, \tau)$

Algo. $\mathtt{I}_{k_1,k_2}(j, D_j, D_{j-1}, D'_j, D_n, T)$

$n \leftarrow |D|_b$

$r \leftarrow T[0 : n(\ell - b) - 1]$

$\tau \leftarrow T[n(\ell - b) :]$

$r'_j \xleftarrow{\$} \{0,1\}^{(\ell-b)}$

$R'_j \leftarrow r'_j\|D'_j$

$R_j \leftarrow r_j\|D_j$

$R_{j-1} \leftarrow r_j\|D_{j-1}$

$R_{n-1} \leftarrow r_{n-1}\|D_{n-1}$

$R_n \leftarrow r_n\|n$

$h_n \leftarrow F_{k_2}(R_{n-1}, R_n)$

$r'_{n+1} \xleftarrow{\$} \{0,1\}^{(\ell-b)}$

$R'_{n+1} \leftarrow r'_{n+1}\|(n+1)$

$h'_{n+1} \leftarrow F_{k_2}(R_{n-1}, R'_{n+1})$

$h'_j \leftarrow F_{k_1}(R_{j-1}, R'_j)$

$h'_{j+1} \leftarrow F_{k_1}(R_{j'}, R_{j+1})$

$h_j \leftarrow F_{k_1}(R_{j-1}, R_j)$

$\Sigma \leftarrow h'_j \oplus h'_{j+1} \oplus h_j \oplus h_n \oplus h'_{n+1}$

$\tau \leftarrow \tau \oplus \Sigma$

$r \leftarrow r_0 \ldots r_{j-1}\|r'_j\|r_j \ldots r_{n-1}\|r'_{n+1}$

$T \leftarrow (r, \tau)$

Algo. $\mathtt{V}_{k_1,k_2}(D, T)$

$\Sigma \leftarrow 0^L$

$n \leftarrow |D|_b$

$r \leftarrow T[0 : n(\ell - b) - 1]$

$\tau \leftarrow T[n(\ell - b) :]$

$R_0 \leftarrow D_0\|r_0$

For $i = 1$ **to** $n - 1$ **do**

$\quad R_i \leftarrow r_i\|D_i$

$\quad h_i \leftarrow F_{k_1}(R_{i-1}, R_i)$

$\quad \Sigma \leftarrow \Sigma \oplus h_i$

$R_n \leftarrow (n\|r_n)$

$h_n \leftarrow F_{k_2}(R_{n-1}, R_n)$

$\tau \leftarrow \Sigma \oplus h_n$

If $\tau' = \tau$

$\quad$ **Return** 1

Else

$\quad$ **Return** 0

Fig. 9. Modified XOR-scheme algorithm.

References

1. Atighehchi, K.: Space-efficient, byte-wise incremental and perfectly private encryption schemes. Cryptology ePrint Archive, Report 2014/104 (2014). http://eprint.iacr.org/2014/104

2. Atighehchi, K., Muntean, T.: Towards fully incremental cryptographic schemes. In: Chen, K., Xie, Q., Qiu, W., Li, N., Tzeng, W.G. (eds.) ASIACCS 2013, 8–10 May 2013, pp. 505–510. ACM Press, Hangzhou (2013)

3. Bellare, M., Goldreich, O., Goldwasser, S.: Incremental cryptography: the case of hashing and Signing. In: Desmedt, Y.G. (ed.) CRYPTO 1994. LNCS, vol. 839, pp. 216–233. Springer, Heidelberg (1994). https://doi.org/10.1007/3-540-48658-5_22

4. Bellare, M., Goldreich, O., Goldwasser, S.: Incremental cryptography and application to virus protection. In: 27th ACM STOC, 29 May– 1 June 1995, pp. 45–56. ACM Press, Las Vegas (1995)

5. Bellare, M., Guérin, R., Rogaway, P.: XOR MACs: new methods for message authentication using finite pseudorandom functions. In: Coppersmith, D. (ed.) CRYPTO 1995. LNCS, vol. 963, pp. 15–28. Springer, Heidelberg (1995). https://doi.org/10.1007/3-540-44750-4_2

6. Bellare, M., Micciancio, D.: A new paradigm for collision-free hashing: incrementality at reduced cost. Cryptology ePrint Archive, Report 1997/001 (1997). http://eprint.iacr.org/1997/001

7. Bershad, B.N., Mogul, J.C. (eds.): 7th Symposium on Operating Systems Design and Implementation (OSDI 2006), 6–8 November, Seattle, WA, USA. USENIX Association (2006). https://www.usenix.org/publications/proceedings/?f[0]=im$_$group$_$audience3A137

8. Buonanno, E., Katz, J., Yung, M.: Incremental unforgeable encryption. In: Matsui, M. (ed.) FSE 2001. LNCS, vol. 2355, pp. 109–124. Springer, Heidelberg (2002). https://doi.org/10.1007/3-540-45473-X_9

9. Fischlin, M.: Lower bounds for the signature size of incremental schemes. In: 38th FOCS, 19–22 October 1997, pp. 438–447. IEEE Computer Society Press, Miami Beach (1997)

10. Gantz, J., Reinsel, D.: The digital universe in 2010: big data, bigger digital shadows, and biggest growth in the far east. EMC report (2013). https://www.emc.com/collateral/analyst-reports/idc-the-digital-universe-in-2020.pdf

11. Goi, B.M., Siddiqi, M.U., Chuah, H.T.: Incremental hash function based on pair chaining & modular arithmetic combining. In: Rangan, C.P., Ding, C. (eds.) Progress in Cryptology – INDOCRYPT 2001, vol. 2247, pp. 50–61. Springer, Heidelberg (2001). https://doi.org/10.1007/3-540-45311-3_5

12. Hart, J.K., Martinez, K.: Environmental sensor networks: a revolution in the earth system science? Earth-Sci. Rev. **78**(3), 177–191 (2006). http://www.sciencedirect.com/science/article/pii/S0012825206000511

13. Itani, W., Kayssi, A.I., Chehab, A.: Energy-efficient incremental integrity for securing storage in mobile cloud computing. In: 2010 International Conference on Energy Aware Computing, pp. 1–2 (2010)

14. Micciancio, D.: Oblivious data structures: applications to cryptography. In: 29th ACM STOC, 4–6 May 1997, pp. 456–464. ACM Press, El Paso (1997)

15. Mihajloska, H., Gligoroski, D., Samardjiska, S.: Reviving the idea of incremental cryptography for the zettabyte era use case: incremental hash functions based on SHA-3. Cryptology ePrint Archive, Report 2015/1028 (2015). http://eprint.iacr.org/2015/1028

16. Mironov, I., Pandey, O., Reingold, O., Segev, G.: Incremental deterministic public-key encryption. In: Pointcheval, D., Johansson, T. (eds.) EUROCRYPT 2012. LNCS, vol. 7237, pp. 628–644. Springer, Heidelberg (2012). https://doi.org/10.1007/978-3-642-29011-4_37

17. Phan, R.C., Wagner, D.A.: Security considerations for incremental hash functions based on pair block chaining. Comput. Secur. **25**(2), 131–136 (2006). https://doi.org/10.1016/j.cose.2005.12.006

18. Sasaki, Y., Yasuda, K.: A new mode of operation for incremental authenticated encryption with associated data. In: Dunkelman, O., Keliher, L. (eds.) SAC 2015. LNCS, vol. 9566, pp. 397–416. Springer, Cham (2016). https://doi.org/10.1007/978-3-319-31301-6_23

Cryptanalysis of Symmetric Key Primitives

Integral Attacks on Round-Reduced Bel-T-256

Muhammad ElSheikh, Mohamed Tolba, and Amr M. Youssef[(⊠)]

Concordia Institute for Information Systems Engineering, Concordia University,
Montréal, QC, Canada
youssef@ciise.concordia.ca

Abstract. Bel-T is the national block cipher encryption standard of the Republic of Belarus. It has a 128-bit block size and a variable key length of 128, 192 or 256 bits. Bel-T combines a Feistel network with a Lai-Massey scheme to build a complex round function with 7 S-box layers per round then iterate this round function 8 times to construct the whole cipher. In this paper, we present integral attacks against Bel-T-256 using the propagation of the bit-based division property. Firstly, we propose two 2-round integral characteristics by employing a Mixed Integer Linear Programming (MILP) (Our open source code to generate the MILP model can be downloaded from https://github.com/mhgharieb/Bel-T-256) approach to propagate the division property through the round function. Then, we utilize these integral characteristics to attack $3\frac{2}{7}$ rounds (out of 8) Bel-T-256 with data and time complexities of 2^{13} chosen plaintexts and $2^{199.33}$ encryption operations, respectively. We also present an attack against $3\frac{6}{7}$ rounds with data and time complexities of 2^{33} chosen plaintexts and $2^{254.61}$ encryption operations, respectively. To the best of our knowledge, these attacks are the first published theoretical attacks against the cipher in the single-key model.

Keywords: Bel-T · Integral attacks · Bit-based division property
MILP

1 Introduction

In 2011, the Republic of Belarus, formerly known by its Russian name Byelorussia, has approved the Bel-T block cipher family as the state standard cryptographic encryption algorithm [1]. The Bel-T family consists of three block ciphers, denoted as Bel-T-k, with the same block size of 128 bits and key length $k = 128$, 192 or 256 bits. Bel-T merges a Lai-Massey scheme [8] with a Feistel network [5]. To the authors' knowledge, there are only two published cryptanalysis results on Bel-T's; fault-based attacks are considered in [6], and related-key differential attack on round-reduced Bel-T-256 are presented in [2]. In this paper, we present the first published single-key attack against Bel-T-256. Table 1 contrasts the result of our attacks with the related-key differential attack in [2].

C. Cid and M. J. Jacobson, Jr. (Eds.): SAC 2018, LNCS 11349, pp. 73–91, 2019.
https://doi.org/10.1007/978-3-030-10970-7_4

Table 1. Attack results on Bel-T-256

Model	Attack	#Round	Data	Time	Reference
Related key	Differential	$5\frac{6}{7}$	$2^{123.28}$	$2^{228.4}$	[2]
Single key	Integral	$3\frac{2}{7}$	2^{13}	$2^{199.33}$	Sect. 3.3
		$3\frac{6}{7}$	2^{33}	$2^{254.61}$	Sect. 3.4

Integral Attacks. In [4], Daemen *et al.* proposed a new cryptanalysis technique to analyze the security of the block cipher SQUARE. Subsequently, Knudsen and Wagner [7] formalized this technique and called it *integral attack*. The integral attack is a chosen-plaintext attack where the set of plaintext used in the attack is chosen to have XOR sum of 0. Firstly, the cryptanalyst constructs a multiset of plaintext such that it has a constant value at some bits while the other bits vary through all possible values. After that, the cryptanalyst calculates the XOR sum of all bits (or some of them) on the corresponding ciphertext after r rounds. If it is always 0 irrespective of the used secret key, we conclude that the cipher under test has an integral distinguisher.

The major techniques used to construct an integral characteristic include estimating the algebraic degree of the nonlinear parts of the cipher, and evaluating the propagation characteristic of the following integral properties [7]: ALL ($\mathcal{A}$) where every member appears the same number in the multiset; BALANCE ($\mathcal{B}$) where the XOR sum of all members in the multiset is 0; CONSTANT ($\mathcal{C}$) where the value is fixed to a constant for all members in the multiset; and UNKNOWN ($\mathcal{U}$) where the multiset is indistinguishable from one of n-bit random values.

Recently, Todo and Morii [16] proposed a generalization of the integral property called *bit-based integral property*. Unfortunately, the searching algorithm which they proposed to construct the integral distinguisher is restricted to ciphers whose block size is less than 32 bits due to its exponential time and memory complexities. To overcome this problem, Xiang *et al.* [17] proposed systematic rules to easily search for such integral distinguishers by employing a Mixed Integer Linear Programming (MILP) approach.

The rest of this paper is organized as follows. In Sect. 2, we briefly revisit the bit-based division property and summarize how to present its propagation through the basic cipher operations with MILP models. We also describe our approach to model the modular subtraction operation. In Sect. 3, we investigate the security of Bel-T block cipher against the integral attacks utilizing this MILP approach, finally, the conclusion is presented in Sect. 4.

2   Bit-Based Division Property

The division property, introduced by Todo [14], is a generalization of the integral property to utilize the hidden relations between the traditional $\mathcal{A}$ and $\mathcal{B}$ properties by exploiting the algebraic degree of the nonlinear components of the block cipher. Later, Todo in [15] proposed the first theoretical attack against the full round MISTY1 based on a 6-round integral distinguisher. To construct this distinguisher, Todo utilized an improved version of the division property after analyzing the Algebraic Normal Form (ANF) of the S-boxes.

Recently, Todo and Morii [16] proposed a special case of the division property, called *bit-based division property*, in which each bit is traced independently. The bit-based division property allows us to exploit both of the algebraic degree and the details of the round function's structure. The bit-based division property is defined as follows:

Definition 1 (Bit-based Division Property [14]). *Let* $\mathbb{X}$ *be a multiset whose elements take a value of* $\mathbb{F}_2^n$. *When the multiset* $\mathbb{X}$ *has the division property* $\mathcal{D}_{\mathbb{K}}^{1^n}$, *where* $\mathbb{K}$ *denotes a set of n-dimensional vectors whose i-th element takes 0 or 1, it fulfills the following conditions:*

$$\bigoplus_{x \in \mathbb{X}} x^u = \begin{cases} unknown & if\ there\ exists\ k \in \mathbb{K}\ s.t.\ u \succeq k, \\ 0 & otherwise. \end{cases}$$

where $x^u = \prod_{i=1}^n x[i]^{u[i]}$, $u \succeq k$ *if* $u[i] \geq k[i]\ \forall i$, *and* $x[i]$, $u[i]$ *are the i-th bits of* x *and* u, *respectively.*

In the following, we present some propagation rules of the division property and show how to utilize MILP for automating the search for integral distinguishers based on the bit-based division property.

2.1 MILP Modeling for Propagation Rules of the Bit-Based Division Property

The advantage of the bit-based division property, over the traditional one, is its ability to exploit both the algebraic degree and the details of the round function structure by tracing each bit independently. The technique presented in [16] to find such distinguishers, however, is restricted to primitives whose block sizes are less than 32 bits due to its time and memory complexities. As mentioned above, to overcome this limitation, Xiang *et al.* [17] defined a new notation called *Division Trail*. With the division trail, it becomes easy to employ MILP for constructing the integral distinguisher. Later, Sun *et al.* complemented this work by handling ARX-based ciphers (`modulo` operations) [10] and ciphers with non-bit-permutation linear layers [11].

In the following subsection, we briefly describe how to model the division trail through several operations using MILP constraints. We firstly start by introducing the notation of a division trail.

Definition 2 (Division Trail [17]). *Let* f_r *denote the round function of an iterated block cipher. Assume that the input multiset to the block cipher has the initial division property* $\mathcal{D}_{\{k\}}^{1^n}$, *and denote the division property after i-round propagation through* f_r *by* $\mathcal{D}_{\mathbb{K}_i}^{1^n}$. *Thus, we have the following chain of division property propagations:*

$$\{k\} \overset{\text{def}}{=} \mathbb{K}_0 \xrightarrow{f_r} \mathbb{K}_1 \xrightarrow{f_r} \mathbb{K}_2 \xrightarrow{f_r} \cdots \xrightarrow{f_r} \mathbb{K}_r.$$

Moreover, for any vector $\boldsymbol{k}_i^* \in \mathbb{K}_i (i \geq 1)$, there must exist a vector $\boldsymbol{k}_{i-1}^* \in \mathbb{K}_{i-1}$ such that $\boldsymbol{k}_{i-1}^*$ can propagate to $\boldsymbol{k}_i^*$ by the division property propagation rules. Furthermore, for $(\boldsymbol{k}_0, \boldsymbol{k}_1, \ldots, \boldsymbol{k}_r) \in \mathbb{K}_0 \times \mathbb{K}_1 \times \cdots \times \mathbb{K}_r$, if $\boldsymbol{k}_{i-1}$ can propagate to $\boldsymbol{k}_i$ for all $i \in \{1, 2, \ldots, r\}$, we call $(\boldsymbol{k}_0, \boldsymbol{k}_1, \ldots, \boldsymbol{k}_r)$ an r-round division trail. Thus, the set of the last vectors of all r-round division trails which start with $\{\boldsymbol{k}\}$ is equal to $\mathbb{K}_r$. Then, the i-th bit of r-round ciphertext is balanced if e_i (a unit vector whose i-th element is 1) does not exist in $\mathbb{K}_r$.

The propagation rules of the bit-based division property through basic operations in block ciphers can be found in [15]. In here, we only summarize the MILP models associated with such rules.

Model for COPY [11]. Let $(a) \xrightarrow{COPY} (b_1, b_2, \ldots, b_m)$ denote the division trail through COPY function, where one bit is copied to m bits. Then, it can be described using the following MILP constraints:

$$\begin{cases} a - b_1 - b_2 - \cdots - b_m = 0, \\ a, b_1, b_2, \ldots, b_m \text{ are binary variables} \end{cases}$$

Model for XOR [11]. Let $(a_1, a_2, \ldots, a_m) \xrightarrow{XOR} (b)$ denote the division trail through an XOR function, where m bits are compressed to one bit using an XOR operation. Then, it can be described using the following MILP constraints:

$$\begin{cases} a_1 + a_2 + \cdots + a_m - b = 0, \\ a_1, a_2, \ldots, a_m, b \text{ are binary variables} \end{cases}$$

Model for AND [17]. Let $(a_0, a_1) \xrightarrow{AND} (b)$ denote the division trail though an AND function, where two bits are compressed using an AND operation. Then, it can be described using the following MILP constraints:

$$\begin{cases} b - a_0 \geq 0, \\ b - a_1 \geq 0, \\ a_0, a_1, b \text{ are binary variables} \end{cases}$$

MILP Model for S-Boxes. The original version of the bit-based division introduced in [16] is limited to bit-oriented ciphers and cannot be applied to ciphers with S-boxes. Xiang et al. overcome this problem by representing the S-Box using its algebraic normal form (ANF) (Algorithm 2 in [17]), also see [9].

The division trail though an n-bit S-box can be represented as a set of $2n$-dimensional binary vectors $\in \{0, 1\}^{2n}$ which has a convex hull. The H-Representation of this convex hull can be computed using readily available functions such as `inequality_generator()` function in Sage[1] which returns a set of linear inequalities that describe these vectors. We use this set of inequalities as MILP constraints to present the division trail though the S-box.

---

[1] http://www.sagemath.org/.

MILP Model for Modular Addition. In [10], Sun *et al.* proposed a systematic method to deduce an MILP model for the modular addition operation of 4-bit variables by expressing the operation at the bit-level. Then this method is generalized for n-bit variables in [12].

Let $\boldsymbol{x} = (x_0, x_1, \ldots, x_{n-1})$, $\boldsymbol{y} = (y_0, y_1, \ldots, y_{n-1})$, and $\boldsymbol{z} = (z_0, z_1, \ldots, z_{n-1})^2$ be n-bit vectors where $\boldsymbol{z} = \boldsymbol{x} \boxplus \boldsymbol{y}$. Then, z_i can be iteratively expressed as follows:

$$z_{n-1} = x_{n-1} \oplus y_{n-1} \oplus c_{n-1}, \ c_{n-1} = 0,$$

$$z_i = x_i \oplus y_i \oplus c_i, \ c_i = x_{i+1}y_{i+1} \oplus (x_{i+1} \oplus y_{i+1})c_{i+1}, \ i = n-2, n-3, \ldots, 0.$$

Consequently, the division trail through the modular addition can be deduced in terms of COPY, AND, and XOR operations [12].

MILP Model for Modular Addition with a Constant. In [10], Sun *et al.* explain how to deduce an MILP model for the modular addition of a 4-bit variable with a constant. The authors expressed the operation at the bit-level and exploited that the operations of XOR/AND with a constant do not influence the division property [10]. We can generalize this method for n-bit variables as follows. Let $(a_0, a_1, \ldots, a_{n-1}) \to (d_0, d_1, \ldots, d_{n-1})$ denote the division trail through n-bit modular addition with a constant, the division property propagation can be decomposed as COPY, AND, and XOR operations as follows:

$$\begin{cases} (a_{n-1}) \xrightarrow{COPY} (d_{n-1}, f_0, g_0) \\ (a_{n-2}) \xrightarrow{COPY} (a_{n-2,0}, a_{n-2,1}, a_{n-2,2}) \\ (a_{n-2,0}, f_0) \xrightarrow{XOR} (d_{n-2}) \\ (a_{n-2,1}, g_0) \xrightarrow{AND} (e_0) \\ (a_{n-2,2}, e_0) \xrightarrow{XOR} (v_0) \\ (v_{i-1}) \xrightarrow{COPY} (f_i, g_i) \\ (a_{n-2-i}) \xrightarrow{COPY} (a_{n-2-i,0}, a_{n-2-i,1}, a_{n-2-i,2}) \\ (a_{n-2-i,0}, f_i) \xrightarrow{XOR} (d_{n-2-i}) \\ (a_{n-2-i,1}, g_i) \xrightarrow{AND} (e_i) \\ (a_{n-2-i,2}, e_i) \xrightarrow{XOR} (v_i) \\ (a_0, v_{n-3}) \xrightarrow{XOR} (d_0) \end{cases} \left. \begin{array}{c} \\ \\ \\ \\ \end{array} \right\} iterated \ for \ i = 1, \ldots, n-3$$

where the intermediate variables $a_{i,0}$, $a_{i,1}$, $a_{i,2}$, f_i, g_i, e_i, and v_i are as shown in Table 2.

MILP Model for Modular Subtraction. In this section, we present an approach to deduce an MILP model for the modular subtraction operation using the same methodology used for Modular Addition. For consistency, we use the same notation as in [10].

---

[2] Big-endian representation.

Table 2. The intermediate variables for modular addition with a constant

$\underbrace{z_{n-1}}_{d_{n-1}}$	$\underbrace{x_{n-1}}_{a_{n-1}}$		
$\underbrace{z_{n-2}}_{d_{n-2}}$	$\underbrace{x_{n-2}}_{a_{n-2,0}} \oplus \underbrace{c_{n-2}}_{f_0}$	c_{n-2}	x_{n-1}
$\underbrace{z_{n-3}}_{d_{n-3}}$	$\underbrace{x_{n-3}}_{a_{n-3,0}} \oplus \underbrace{c_{n-3}}_{f_1}$	$\underbrace{c_{n-3}}_{v_0}$	$\underbrace{x_{n-2}}_{a_{n-2,2}} \oplus \overbrace{\underbrace{x_{n-2}}_{a_{n-2,1}}\ \underbrace{c_{n-2}}_{g_0}}^{e_0}$
$\underbrace{z_{n-4}}_{d_{n-4}}$	$\underbrace{x_{n-4}}_{a_{n-4,0}} \oplus \underbrace{c_{n-4}}_{f_2}$	$\underbrace{c_{n-4}}_{v_1}$	$\underbrace{x_{n-3}}_{a_{n-3,2}} \oplus \overbrace{\underbrace{x_{n-3}}_{a_{n-3,1}}\ \underbrace{c_{n-3}}_{g_1}}^{e_1}$
...		...	
$\underbrace{z_1}_{d_1}$	$\underbrace{x_1}_{a_{1,1}} \oplus \underbrace{c_1}_{f_{n-3}}$	$\underbrace{c_1}_{v_{n-4}}$	$\underbrace{x_2}_{a_{2,2}} \oplus \overbrace{\underbrace{x_2}_{a_{2,1}}\ \underbrace{c_2}_{g_{n-4}}}^{e_{n-4}}$
$\underbrace{z_0}_{d_0}$	$\underbrace{x_0}_{a_0} \oplus c_0$	$\underbrace{c_0}_{v_{n-3}}$	$\underbrace{x_1}_{a_{1,2}} \oplus \overbrace{\underbrace{x_1}_{a_{1,1}}\ \underbrace{c_1}_{g_{n-3}}}^{e_{n-3}}$

Let x, y and z be n-bit vectors where $z = x \boxminus y$. This relation can be rewritten as $z = x \boxplus$ (2's complement of y) $= x \boxplus (\bar{y} \boxplus 1)$, where $\bar{y}$ is the 1's complement of y. Therefore, the division trail through the modular subtraction can be modelled as a division trail through a modular addition followed by a modular addition with a constant. This representation has two issues. The first issue is that two operations are used to present one operation which requires the use of more MILP constraints and variables, and consequently slowing down the search process. The second issue is that the information about the value of the constant, which is 1, in the modular addition with a constant is not utilized. This may lead the search process to conclude that some bits are not balanced even that they are balanced, as we show in Appendix A. Instead, at the bit level implementation, the modular subtraction operation is handled as a modular addition operation with two modifications: the first carry to the modular addition will be 1 instead of 0 ($c_{n-1} = 1$), and the second input to the modular addition will be the 1's complement of the second operand ($\bar{y}$).

Let $x = (x_0, x_1, \dots, x_{n-1})$, $y = (y_0, y_1, \dots, y_{n-1})$, and $z = (z_0, z_1, \dots, z_{n-1})$. Then, z_i can be iteratively expressed as follows:

$$z_{n-1} = x_{n-1} \oplus \bar{y}_{n-1} \oplus c_{n-1}, \ c_{n-1} = 1,$$
$$z_i = x_i \oplus \bar{y}_i \oplus c_i, \ c_i = x_{i+1}\bar{y}_{i+1} \oplus (x_{i+1} \oplus \bar{y}_{i+1})c_{i+1}, \quad \forall i = n-2, n-3, \dots, 0.$$
$$where \ \bar{y}_i = y_i \oplus 1$$

The operation of XOR/AND with a constant does not influence the division property [10]. Therefore, the division property of $\bar{y}$ is the same of y. Consequently, we can generalize the modular subtraction operation for n-bit variables as follows:

Proposition 1. *Let* $((a_0, a_1, \ldots, a_{n-1}), (b_0, b_1, \ldots, b_{n-1})) \to (d_0, d_1, \ldots, d_{n-1})$
be a division trail through n-bit modular subtraction operation. The division property propagation can be decomposed as COPY, AND, and XOR operations as follows:

$$
\begin{cases}
(a_{n-1}) \xrightarrow{COPY} (a_{n-1,0}, a_{n-1,1}, a_{n-1,2}) \\
(b_{n-1}) \xrightarrow{COPY} (b_{n-1,0}, b_{n-1,1}, b_{n-1,2}) \\
(a_{n-1,0}, b_{n-1,0}) \xrightarrow{XOR} (d_{n-1}) \\
(a_{n-1,2}, b_{n-1,2}) \xrightarrow{XOR} (t_0) \\
(a_{n-1,1}, b_{n-1,1}) \xrightarrow{AND} (t_1) \\
(t_0, t_1) \xrightarrow{XOR} (v_0) \\
(v_0) \xrightarrow{COPY} (g_0, r_0) \\
(a_{n-2}) \xrightarrow{COPY} (a_{n-2,0}, a_{n-2,1}, a_{n-2,2}) \\
(b_{n-2}) \xrightarrow{COPY} (b_{n-2,0}, b_{n-2,1}, b_{n-2,2}) \\
\left.\begin{array}{l}
(a_{n-i,0}, b_{n-i,0}, g_{i-2}) \xrightarrow{XOR} (d_{n-i}) \\
(a_{n-i,1}, b_{n-i,1}) \xrightarrow{AND} (v_{i-1}) \\
(a_{n-i,2}, b_{n-i,2}) \xrightarrow{XOR} (m_{i-2}) \\
(m_{i-2}, r_{i-2}) \xrightarrow{AND} (q_{i-2}) \\
(v_{i-1}, q_{i-2}) \xrightarrow{XOR} (w_{i-2}) \\
(w_{i-2}) \xrightarrow{COPY} (g_{i-1}, r_{i-1}) \\
(a_{n-i-1}) \xrightarrow{COPY} (a_{n-i-1,0}, a_{n-i-1,1}, a_{n-i-1,2}) \\
(b_{n-i-1}) \xrightarrow{COPY} (b_{n-i-1,0}, b_{n-i-1,1}, b_{n-i-1,2})
\end{array}\right\} \text{iterated for } i = 2, \ldots, n-2 \\
(a_{1,0}, b_{1,0}, g_{n-3}) \xrightarrow{XOR} (d_1) \\
(a_{1,1}, b_{1,1}) \xrightarrow{AND} (v_{n-2}) \\
(a_{1,2}, b_{1,2}) \xrightarrow{XOR} (m_{n-3}) \\
(m_{n-3}, r_{n-3}) \xrightarrow{AND} (q_{n-3}) \\
(v_{n-2}, q_{n-3}) \xrightarrow{XOR} (w_{n-3}) \\
(a_0, b_0, w_{n-3}) \xrightarrow{XOR} (d_0)
\end{cases}
$$

where the intermediate variables $a_{i,0}$, $a_{i,1}$, $a_{i,2}$, t_0, t_1, v_i, g_i, r_i, m_i, q_i, and w_i *are as shown in Table 3.*

In Appendix A, we present the results of an experiment we performed on a toy cipher to validate the model of the modular subtraction and show the effect of the first carry.

Table 3. The intermediate variables for modular subtraction

$$\underbrace{z_{n-1}}_{d_{n-1}} = \underbrace{x_{n-1}}_{a_{n-1,0}} \oplus \underbrace{\bar{y}_{n-1}}_{b_{n-1,0}} \oplus 1$$

$$\underbrace{z_{n-2}}_{d_{n-2}} = \underbrace{x_{n-2}}_{a_{n-2,0}} \oplus \underbrace{\bar{y}_{n-2}}_{b_{n-2,0}} \oplus \overbrace{c_{n-2}}^{g_0} \quad\middle|\quad \overbrace{c_{n-2}}^{v_0} = \underbrace{\overbrace{x_{n-1}}^{t_1}\,\bar{y}_{n-1}}_{a_{n-1,1}\ b_{n-1,1}} \oplus \overbrace{(\underbrace{x_{n-1}}_{a_{n-1,2}} \oplus \underbrace{\bar{y}_{n-1}}_{b_{n-1,2}})}^{t_0,\ q_0}$$

$$\underbrace{z_{n-3}}_{d_{n-3}} = \underbrace{x_{n-3}}_{a_{n-3,0}} \oplus \underbrace{\bar{y}_{n-3}}_{b_{n-3,0}} \oplus \overbrace{c_{n-3}}^{g_1} \quad\middle|\quad \overbrace{c_{n-3}}^{w_0} = \underbrace{\overbrace{x_{n-2}}^{v_1}\,\bar{y}_{n-2}}_{a_{n-2,1}\ b_{n-2,1}} \oplus \overbrace{(\underbrace{x_{n-2}}_{a_{n-2,2}} \oplus \underbrace{\bar{y}_{n-2}}_{b_{n-2,2}})}^{m_0,\ q_1} \oplus \overbrace{c_{n-2}}^{r_0}$$

$$\underbrace{z_{n-4}}_{d_{n-4}} = \underbrace{x_{n-4}}_{a_{n-4,0}} \oplus \underbrace{\bar{y}_{n-4}}_{b_{n-4,0}} \oplus \overbrace{c_{n-4}}^{g_2} \quad\middle|\quad \overbrace{c_{n-4}}^{w_1} = \underbrace{\overbrace{x_{n-3}}^{v_2}\,\bar{y}_{n-3}}_{a_{n-3,1}\ b_{n-3,1}} \oplus \overbrace{(\underbrace{x_{n-3}}_{a_{n-3,2}} \oplus \underbrace{\bar{y}_{n-3}}_{b_{n-3,2}})}^{m_1,\ q_2} \oplus \overbrace{c_{n-3}}^{r_1}$$

$$\ldots \qquad\qquad \ldots$$

$$\underbrace{z_1}_{d_1} = \underbrace{x_1}_{a_{1,0}} \oplus \underbrace{\bar{y}_1}_{b_{1,0}} \oplus \overbrace{c_1}^{g_{n-3}} \quad\middle|\quad \overbrace{c_1}^{w_{n-4}} = \underbrace{\overbrace{x_2}^{v_{n-3}}\,\bar{y}_2}_{a_{2,1}\ b_{2,1}} \oplus \overbrace{(\underbrace{x_2}_{a_{2,2}} \oplus \underbrace{\bar{y}_2}_{b_{2,2}})}^{m_{m-4},\ q_{n-4}} \oplus \overbrace{c_2}^{r_{n-4}}$$

$$\underbrace{z_0}_{d_0} = \underbrace{x_0}_{a_0} \oplus \underbrace{\bar{y}_0}_{b_0} \oplus c_0 \quad\middle|\quad \overbrace{c_0}^{w_{n-3}} = \underbrace{\overbrace{x_1}^{v_{n-2}}\,\bar{y}_1}_{a_{1,1}\ b_{1,1}} \oplus \overbrace{(\underbrace{x_1}_{a_{1,2}} \oplus \underbrace{\bar{y}_1}_{b_{1,2}})}^{m_{m-3},\ q_{n-3}} \oplus \overbrace{c_1}^{r_{n-3}}$$

3   Integral Attack on Bel-T-256

In this Section, we investigate the security of the Bel-T block cipher against the integral attack based on the bit-based division property.

3.1   Bel-T Specification

The official Bel-T specification is available only in Russian and the only version of the specification available in English is the one provided in its fault-based attacks analysis [6]. Bel-T has a 128-bit block size and a variable key length of 128, 192 or 256 bits. The 128-bit plaintext is divided into 4 32-bit words, i.e., $P = A^0||B^0||C^0||D^0$. Then, the round function illustrated in Fig. 1, is repeated eight times for all versions of Bel-T. Three mappings G_5, G_{13} and G_{21}: $\{0,1\}^{32} \to \{0,1\}^{32}$ are used, where G_r maps a 32-bit word $u = u_1||u_2||u_3||u_4$, with $u_i \in \{0,1\}^8$, as follows: $G_r(u) = (H(u_1)||H(u_2)||H(u_3)||H(u_4)) \lll r$. Here, H is an 8-bit S-box and $\lll r$ stands for left shift rotation by r positions. The specification of the 8-bit S-box can be found in [6].

Key Schedule. In all versions of Bel-T, the 128-bit plaintext block P is encrypted using a 256-bit encryption key denoted as $K_1||\ldots||K_8$, where K_i is a

32-bit word for $1 \leq i \leq 8$. The encryption key is distributed among the round keys as shown in Table 4. The encryption key is extracted from the master key as follows:

- Bel-T-256: the encryption key is identical to the master key.
- Bel-T-192: the master key is formatted as $K_1 || \ldots || K_6$ and K_7, K_8 are set to $K_7 := K_1 \oplus K_2 \oplus K_3$ and $K_8 := K_4 \oplus K_5 \oplus K_6$.
- Bel-T-128: the master key is formatted as $K_1 || \ldots || K_4$ and K_5, K_6, K_7, K_8 are set to $K_5 := K_1$, $K_6 := K_2$, $K_7 := K_3$ and $K_8 := K_4$.

3.2 Integral Distinguishers of Bel-T

As shown in Fig. 1, the Bel-T round function includes 7 S-boxes, modular additions, modular additions with key and modular subtractions. We construct an MILP model for the bit-based division property through Bel-T as follows. Firstly, we generate the division trail of the S-box using Algorithm 2 in [17]. Then, we deduce the inequalities of the S-box using inequality_generator() function in Sage. In the case of the Bel-T S-box, the number of generated inequalities is 71736, which is very large set to be handled by any MILP optimizer. Therefore, we reduce this set using a Greedy Algorithm which is proposed by Sun *et al.* in [13]. The size of the reduced set of the S-box representation inequalities is 28 and can be found in Appendix B.

Then, we implement the MILP model for modular addition and deduce the model for subtraction. Finally, we use the Gurobi[3] optimizer to search for the longest integral distinguisher for Bel-T. Based on our implementation, we found several 2-round integral distinguishers. Our code that is used to generate the MILP model for Bel-T and to search for an integral distinguisher can be downloaded from github.[4]

In here, we present two such distinguishers which are chosen in order to minimize the attack data and time complexities.

$$IC1 : ((\mathcal{C}_{0-31}), (\mathcal{C}_{0-31}), (\mathcal{C}_{0-17}||\mathcal{A}_{18-18}||\mathcal{C}_{19-31}), (\mathcal{A}_{0-7}||\mathcal{C}_{8-31}))$$
$$\xrightarrow{2R} ((\mathcal{U}_{0-31}), (\mathcal{U}_{0-31}), (\mathcal{U}_{0-26}||\mathcal{B}_{27-31}), (\mathcal{U}_{0-31}))$$
$$IC2 : ((\mathcal{C}_{0-31}), (\mathcal{C}_{0-31}), (\mathcal{C}_{0-10}||\mathcal{A}_{11-26}||\mathcal{C}_{27-31}), (\mathcal{A}_{0-15}||\mathcal{C}_{16-31}))$$
$$\xrightarrow{2R} ((\mathcal{U}_{0-26}||\mathcal{B}_{27-31}), (\mathcal{U}_{0-31}), (\mathcal{B}_{0-31}), (\mathcal{U}_{0-31}))$$

where $\mathcal{C}_{i-j}/\mathcal{A}_{i-j}/\mathcal{B}_{i-j}/\mathcal{U}_{i-j}$ denote CONSTANT/ALL/BALANCE/UNKNOWN from bit number i to bit number j respectively counting from the most significant bit of the branch. Both of these integral distinguishers have been verified experimentally using a set of 256 randomly generated keys.

---

[3] http://www.gurobi.com/.
[4] https://github.com/mhgharieb/Bel-T-256.

Table 4. Encryption key schedule of Bel-T, where i and K_{7i-j} denote the round number and the round key, respectively.

i	K_{7i-6}	K_{7i-5}	K_{7i-4}	K_{7i-3}	K_{7i-2}	K_{7i-1}	K_{7i}
1	K_1	K_2	K_3	K_4	K_5	K_6	K_7
2	K_8	K_1	K_2	K_3	K_4	K_5	K_6
3	K_7	K_8	K_1	K_2	K_3	K_4	K_5
4	K_6	K_7	K_8	K_1	K_2	K_3	K_4
5	K_5	K_6	K_7	K_8	K_1	K_2	K_3
6	K_4	K_5	K_6	K_7	K_8	K_1	K_2
7	K_3	K_4	K_5	K_6	K_7	K_8	K_1
8	K_2	K_3	K_4	K_5	K_6	K_7	K_8

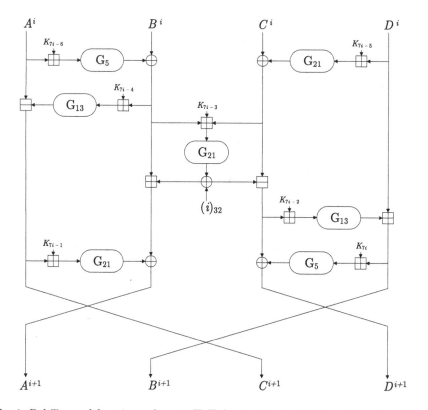

Fig. 1. Bel-T round function, where $\oplus, \boxplus, \boxminus$ denote bit-wise XOR, arithmetic addition and subtraction modulo 2^{32} respectively, and $(i)_{32}$ denotes the round number represented as 32-bit word.

3.3   Integral Cryptanalysis of $3\frac{2}{7}$-Round Bel-T-256

In this section, we present our Integral attack on $3\frac{2}{7}$-round Bel-T-256 by appending one round and two S-box layers on the above derived integral distinguisher $IC1$ as illustrated in Fig. 2.

Data Collection. We select m structures of plaintexts. In each structure, the 9 bits (bit number 18 in branch C^0 and bits 0-7 in branch D^0) vary through all 2^9 possible values and all other bits are fixed to an arbitrary constant value.

This ensures that each structure satisfies the required input division property of the integral distinguisher $IC1$. After that, we query the encryption oracle to obtain the corresponding ciphertexts. Subsequently, we apply the following key recovery procedure.

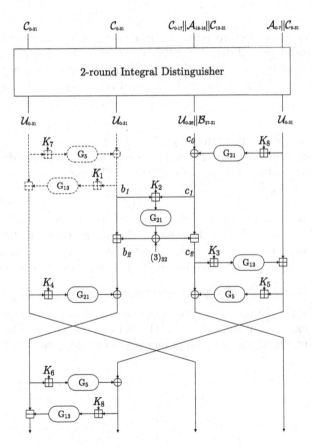

Fig. 2. $3\frac{2}{7}$-round attack on Bel-T-256

Key Recovery. For ciphertexts in each structure obtained in the data collection phase, we apply the following procedure:

1. Guess K_8 and K_4 and partially decrypt the ciphertext to obtain b_2.
2. Guess K_6 and K_5 and partially decrypt the ciphertext to obtain c_2.
3. Recall that $b_1 = b_2 - G_{21}(b_1 + c_1 + K_2) \oplus (3)_{32}$ and $c_1 = c_2 + G_{21}(b_1 + c_1 + K_2) \oplus (3)_{32}$. Hence $b_1 + c_1 = b_2 + c_2$. Therefore, by guessing K_2, we can deduce $G_{21}(b_1 + c_1 + K_2) = G_{21}(b_2 + c_2 + K_2)$ and then compute c_1 from b_2 and c_2.
4. Guess K_3 and use the previous guessed value of K_8 to compute c_0 from c_1 and c_2.
5. For each bit in the 5 least significant bits of the 32-bit word c_0, check that its XOR sum over the structure is zero. The probability that all these 5 bits are balanced is 2^{-5}. Therefore the probability that a key is survived after this test is also 2^{-5}. This means that the number of 192-bit key candidates passed this check is $2^{192} \times 2^{-5}$.

After repeating the above procedure for m structures, the number of surviving 192-bit key candidates will be $2^{192} \times (2^{-5})^m = 2^{192-5m}$. After that, we recover the 256-bit master key by testing the 2^{192-5m} 192-bit surviving key candidates along with the remaining 2^{64} values for K_1 and K_7 using 2 plaintext/ciphertext pairs.

Attack Complexity. The data complexity of the above attack is $m \times 2^9$ chosen plaintexts. The dominant part of time complexity is coming from deducing 192-bit key candidates after checking m structures. This part is equal to $\frac{7}{23} \times 2^9 \times 2^{192} \times [1 + 2^{-5} + (2^{-5})^2 + \cdots + (2^{-5})^{m-1}] = \frac{7}{23} \times 2^{201} \times \frac{1 - (2^{-5})^m}{1 - 2^{-5}}$. Additionally, the part due to exhaustively searching for the master key which is equal to $2 \times 2^{64} \times 2^{192-5m} = 2^{257-5m}$. To balance the attack between data and time complexities, we take $m = 16$. This means that the data complexity will be $16 \times 2^9 = 2^{13}$ chosen plaintexts and the time complexity will be $\frac{7}{23} \times 2^{201} \times \frac{1 - 2^{-80}}{1 - 2^{-5}} + 2^{177} \approx 2^{199.33}$ encryption operations.

   It should be noted that other choices of m can lead to possible data and time trade-off. For example, if we set $m = 1$, the data complexity will be reduced to 2^9 chosen plaintexts at the expense of increasing the time complexity to 2^{252}.

3.4   Integral Cryptanalysis of $3\frac{6}{7}$-Round Bel-T-256

In this section, we present our integral attack on $3\frac{6}{7}$-round Bel-T-256 by appending one round and six S-box layers on the above derived integral distinguisher $IC2$, which is the only distinguisher makes the attack feasible, as illustrated in Fig. 3.

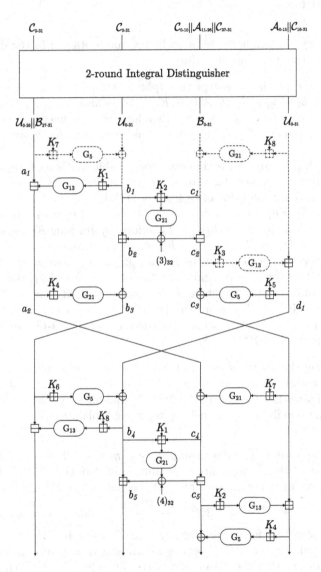

Fig. 3. $3\frac{6}{7}$-round attack on Bel-T-256

Data Collection. We select m structures of plaintexts. In each structure, the 32 bits (bits 11–26 in branch C^0 and bits 0–15 in branch D^0) vary through all 2^{32} possible values and all other bits are fixed to an arbitrary constant value. This ensures that each structure satisfies the required input division property of the integral distinguisher $IC2$. After that, we query the encryption oracle to obtain the corresponding ciphertexts. Subsequently, we apply the following key recovery procedure.

Key Recovery. For ciphertexts in each structure obtained in the data collection, we apply the following procedure:

1. Guess K_4 and partially decrypt the ciphertext to obtain c_5.
2. Recall that $b_4 = b_5 - G_{21}(b_4 + c_4 + K_1) \oplus (4)_{32}$ and $c_4 = c_5 + G_{21}(b_4 + c_4 + K_1) \oplus (4)_{32}$, hence $b_4 + c_4 = b_5 + c_5$. Therefore, by guessing K_1, we can deduce $G_{21}(b_4 + c_4 + K_1) = G_{21}(b_5 + c_5 + K_1)$ and then compute b_4 and c_4 from b_5 and c_5.
3. Guess K_2, K_6, K_7 and K_8 and deduce each 32-bit words a_2, b_3, c_3 and d_1.
4. Use the previous guessed value of K_4 to get the value of b_2 from a_2 and b_3.
5. Guess K_5 and get the value of c_2 from c_3 and d_1.
6. Recall that $b_1 = b_2 - G_{21}(b_1 + c_1 + K_2) \oplus (3)_{32}$ and $c_1 = c_2 + G_{21}(b_1 + c_1 + K_2) \oplus (3)_{32}$, hence $b_1 + c_1 = b_2 + c_2$. Therefore, by guessing K_2, we can deduce $G_{21}(b_1 + c_1 + K_2) = G_{21}(b_2 + c_2 + K_2)$ and then compute b_1 from b_2 and c_2.
7. Use the previous guessed value of K_1 to compute a_1 from a_2 and b_1.
8. For each bit in the 5 least significant bits of 32-bit word a_1, check that the XOR sum of it over the structure is zero. The probability that all these 5 bits are balanced is 2^{-5}. Therefore the probability that a key is survived after this test is also 2^{-5}. This means that the number of 224-bit key candidates passed this check is $2^{224} \times 2^{-5}$.

After repeating the above procedure for m structures, the number of surviving 224-bit key candidates will be $2^{224} \times (2^{-5})^m = 2^{224-5m}$. After that we recover the 256-bit master key by testing the 2^{224-5m} 192-bit surviving key candidates along with the remaining 2^{32} values for K_3 using 2 plaintext/ciphertext pairs.

Attack Complexity. The data complexity is $m \times 2^{32}$ chosen plaintexts. The dominant part of time complexity is coming from deducing 224-bit key candidates after checking m structure. This part is equal to $\frac{10}{27} \times 2^{32} \times 2^{224} \times [1 + 2^{-5} + (2^{-5})^2 + \cdots + (2^{-5})^{m-1}] = \frac{10}{27} \times 2^{256} \times \frac{1 - (2^{-5})^m}{1 - 2^{-5}}$. Additionally, the part due to exhaustively searching for the master key which is equal to $2 \times 2^{32} \times 2^{224-5m} = 2^{257-5m}$. To balance the attack between data and time complexities, we take $m = 2$. This means that the data complexity will be $2 \times 2^{32} = 2^{33}$ chosen plaintexts and the time complexity will be $\frac{10}{27} \times 2^{256} \times \frac{1 - 2^{-10}}{1 - 2^{-5}} + 2^{247} \approx 2^{254.61}$ encryption.

4   Conclusion

In this paper, we investigated the security of Bel-T-256 against integral attacks based on the bit-based division property. In particular, we have built a MILP model for the Bel-T round function to automate the search for integral distinguishers based on the bit-based division property. Using two of the obtained integral distinguishers, we presented attacks on $3\frac{2}{7}$ and $3\frac{6}{7}$ rounds of Bel-T-256 with data and time complexities of 2^{13}, 2^{33} chosen plaintexts and $2^{199.33}$, $2^{254.61}$ encryption operations, respectively.

A   Validation of the MILP Model for the Division Trail Through a Modular Subtraction Operation

In this appendix, we provide the result of our experiments on a toy cipher in order to validate the MILP model for the division trail through a modular subtraction operation. Moreover, we show that the proposed model of the division trail through the modular subtraction at the bit-level ($z = x \boxminus y$) gives better results than modelling it as a division trail through a modular addition followed by a modular addition with a constant ($z = x \boxplus \bar{y} \boxplus 1$).

The round function of the toy cipher used during the experiments is a small version of the SPECK round function [3] with modular subtraction instead of modular addition as shown in Fig. 4 where the block size is 8 bits, (X_L^i, X_R^i) is the input of the i-th round, and k_i is the subkey used in the i-th round.

We follow the same approach used in [10] to validate their MILP model for modular addition. The experimental procedure is as follows:

1. For an initial division property, use our MILP model for the modular subtraction at the bit-level ($z = x \boxminus y$) to find the set of balanced bits at the output of the toy cipher.
2. Use the other MILP model ($z = x \boxplus \bar{y} \boxplus 1$) to find the balanced bits corresponding the same initial division property.
3. Exhaustively search for the balanced bits as follows:
   (a) Divide the space of the plaintexts (2^8 plaintexts) to a group of multi-sets of plaintexts. Each one of these multi-sets satisfies the initial division property.
   (b) Encrypt each multi-set of the plaintexts using a randomly chosen key and find the bits with zero-sum over all the corresponding ciphertexts of that multi-set, and then find the common zero-sum bits over all the multi-sets.
   (c) Repeat the previous step 2^{10} iterations and find the common zero-sum bits at the output of the toy cipher over all the iterations.

4. Compare the results from the previous three steps for the same initial division property.
5. Repeat the previous steps for all possible values of the initial division property and for a toy cipher consists of up to 6 rounds similar to the one in the Fig. 4.

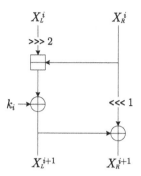

Fig. 4. The round function of the toy cipher.

Table 5. Comparison of zero-sum bits found by using three methods for the toy cipher, where #{Bits} is the number of balanced bits and 'Bits' is the position of these bits counted from the most significant bit.

Input Division property	Rounds	Exhaustive search		MILP-aided Bit-based Division property			
				$z = x \boxminus y$		$z = x \boxplus \bar{y} \boxplus 1$	
		#{Bits}	Bits	#{Bits}	Bits	#{Bits}	Bits
$\mathcal{D}^{1^8}_{\{[01111111]\}}$	1	8	$0 \sim 7$	8	$0 \sim 7$	8	$0 \sim 7$
	2	8	$0 \sim 7$	8	$0 \sim 7$	8	$0 \sim 7$
	3	6	$1 \sim 3, 5 \sim 7$	6	$1 \sim 3, 5 \sim 7$	4	$2 \sim 3, 6 \sim 7$
	4	1	3	1	3	0	-
	5	0	-	0	-	0	-
$\mathcal{D}^{1^8}_{\{[11111110]\}}$	1	8	$0 \sim 7$	8	$0 \sim 7$	8	$0 \sim 7$
	2	8	$0 \sim 7$	8	$0 \sim 7$	8	$0 \sim 7$
	3	6	$1 \sim 3, 5 \sim 7$	6	$1 \sim 3, 5 \sim 7$	6	$1 \sim 3, 5 \sim 7$
	4	3	$2 \sim 3, 6$	1	3	1	3
	5	0	-	0	-	0	-
$\mathcal{D}^{1^8}_{\{[00001111]\}}$	1	8	$0 \sim 7$	8	$0 \sim 7$	8	$0 \sim 7$
	2	4	$2 \sim 3, 6 \sim 7$	4	$2 \sim 3, 6 \sim 7$	4	$2 \sim 3, 6 \sim 7$
	3	0	-	0	-	0	-
$\mathcal{D}^{1^8}_{\{[11110000]\}}$	1	8	$0 \sim 7$	8	$0 \sim 7$	8	$0 \sim 7$
	2	8	$0 \sim 7$	8	$0 \sim 7$	6	$1 \sim 3, 5 \sim 7$
	3	2	$3, 7$	2	$3, 7$	1	3
	4	0	-	0	-	0	-

From the result of the experiments, we can conclude that the balanced bits found by the MILP-aided bit-based division property are indeed balanced. Moreover, the MILP model for the division trail through the modular subtraction at the bit-level ($z = x \boxminus y$) also uses less number of constraints and gives same or better results (in terms of number of the balanced bits) than modelling it as a division trail through a modular addition followed by a modular addition with a constant ($z = x \boxplus \bar{y} \boxplus 1$). A sample of our results can be found in Table 5 and the mismatch between the two approaches for modelling the division trail through a modular subtractions is summarized in Table 6.

Table 6. Mismatch between the two approaches for modelling the division trail through a modular subtraction.

Rounds	Inputs Division property	MILP-aided Bit-based Division property			
		$z = x \boxminus y$		$z = x \boxplus \bar{y} \boxplus 1$	
		#{Bits}	Bits	#{Bits}	Bits
1	{[10000011]}, {[11000010]}, {[11000011]}	8	$0 \sim 7$	6	$1 \sim 3, 5 \sim 7$
2	{[01101101]}, {[01111001]}, {[10100101]}, {[10101100]}, {[10110001]}, {[10111000]}, {[11100100]}, {[11110000]}	8	$0 \sim 7$	6	$1 \sim 3, 5 \sim 7$
	{[10001111]}, {[10011011]}, {[11001110]}, {[11001111]}, {[11011010]}, {[11011011]}	6	$1 \sim 3, 5 \sim 7$	4	$2 \sim 3, 6 \sim 7$
	{[10000011]}, {[11000010]}	2	3,7	1	3
3	{[11000011]}	4	$2 \sim 3, 6 \sim 7$	1	3
	{[01110111]}, {[01111111]}, {[10110110]}, {[10110111]}, {[10111101]}, {[10111110]}, {[11110110]}, {[11111100]}	6	$1 \sim 3, 5 \sim 7$	4	$2 \sim 3, 6 \sim 7$
	{[01101101]}, {[01111001]}, {[10100101]}, {[10101100]}, {[10110001]}, {[10111000]}, {[11100100]}, {[11110000]}	2	3,7	1	3
	{[10001111]}, {[10011011]}, {[11001110]}, {[11001111]}, {[11011010]}, {[11011011]}	1	3	0	-
4	{[11111011]}	6	$1 \sim 3, 5 \sim 7$	4	$2 \sim 3, 6 \sim 7$
	{[01110111]}, {[01111111]}, {[10110110]}, {[10110111]}, {[10111101]}, {[10111110]}, {[11110110]}, {[11111100]}	1	3	0	-
5	{[11111011]}	1	3	0	-

B   Division Trail Representation of Bel-T S-Box

$$
\begin{bmatrix}
1 & 1 & 1 & 36 & 1 & 1 & 1 & 1 & -6 & -6 & -6 & -6 & -6 & -6 & -6 & -6 & 5 \\
1 & 1 & 39 & 1 & 1 & 1 & 1 & 1 & -7 & -7 & -7 & -7 & -6 & -6 & -6 & -6 & 6 \\
2 & 68 & 2 & 2 & 2 & 2 & 2 & 2 & -11 & -11 & -12 & -12 & -12 & -11 & -12 & -11 & 10 \\
1 & 1 & 1 & 1 & 1 & 1 & 35 & 1 & -6 & -6 & -6 & -6 & -6 & -6 & -6 & -5 & 5 \\
14 & 0 & 0 & 0 & 0 & 0 & 0 & 0 & -1 & -1 & -3 & -3 & -3 & -2 & -3 & -1 & 3 \\
0 & 0 & 0 & 0 & 9 & 0 & 0 & 0 & -2 & -2 & -1 & 0 & -2 & -1 & -2 & -1 & 2 \\
-6 & -2 & 0 & 0 & -5 & 2 & 2 & 1 & -7 & -6 & -6 & -6 & 30 & -5 & -7 & -5 & 20 \\
-8 & -12 & -6 & -11 & -8 & -6 & -13 & -2 & -6 & -10 & -5 & -5 & -12 & -2 & 32 & 4 & 70 \\
-1 & -2 & -2 & -3 & -2 & -2 & -2 & -4 & 16 & 16 & 17 & 17 & 17 & 14 & 17 & 15 & 0 \\
0 & -1 & -2 & 0 & 0 & 1 & 0 & 0 & -2 & 2 & 5 & -3 & -3 & -2 & -3 & 2 & 6 \\
1 & 1 & -1 & 2 & 0 & 2 & 0 & 2 & -1 & -6 & -7 & 19 & 1 & -7 & -7 & -7 & 8 \\
-9 & -1 & 0 & -1 & -1 & 3 & 3 & -2 & -7 & -7 & -6 & -7 & -2 & 29 & -5 & -8 & 21 \\
0 & -3 & -1 & -1 & -3 & -1 & 0 & 2 & -4 & 10 & -4 & -3 & -1 & -1 & 2 & -5 & 13 \\
-2 & -1 & -8 & -2 & -5 & -5 & -8 & -12 & 10 & -15 & 3 & 3 & -11 & 3 & 4 & -7 & 53 \\
-23 & -24 & -23 & -21 & -21 & -23 & -22 & -26 & 4 & 5 & 2 & 3 & 4 & -1 & 4 & 4 & 158 \\
-2 & -4 & 0 & -5 & -2 & -1 & -6 & -1 & -4 & -4 & -4 & -5 & -6 & 3 & 14 & 2 & 25 \\
-10 & 1 & 3 & 2 & 2 & 0 & 3 & 3 & -13 & -12 & -13 & 47 & 0 & -12 & -12 & -12 & 23 \\
-1 & -2 & 0 & -2 & -2 & -1 & 0 & -2 & 3 & 3 & 3 & 3 & 2 & 3 & 3 & 1 & 7 \\
-6 & -3 & -5 & -7 & -6 & -6 & -6 & -4 & 4 & 3 & 4 & 5 & 5 & 4 & 2 & 1 & 35 \\
-1 & -2 & -3 & -2 & 0 & -1 & -3 & -3 & 15 & 15 & 14 & 14 & 15 & 12 & 15 & 13 & 0 \\
-9 & 3 & -9 & -9 & 0 & 1 & -9 & -9 & -2 & -2 & 8 & -3 & -2 & 8 & -2 & -12 & 48 \\
-1 & -1 & -3 & 0 & 0 & -1 & -1 & -1 & -2 & 1 & 3 & -4 & -3 & -2 & -2 & 6 & 11 \\
-2 & -2 & 0 & -2 & -2 & 3 & -2 & -2 & -1 & -1 & -1 & -1 & 2 & -3 & -1 & 2 & 13 \\
-1 & -2 & -1 & 0 & -1 & -1 & -1 & 1 & -2 & -3 & 0 & 0 & -2 & 0 & 3 & 1 & 9 \\
0 & 0 & -1 & -1 & -1 & -1 & -1 & -1 & -3 & 3 & 1 & -2 & -1 & 1 & 1 & -2 & 8 \\
-1 & -1 & 0 & -1 & 0 & -1 & -1 & -1 & 0 & 0 & 0 & -1 & 0 & 1 & 0 & 0 & 6 \\
-1 & -1 & 0 & -1 & -1 & -1 & -1 & 1 & 0 & 0 & 0 & 0 & 1 & 0 & -1 & 0 & 6 \\
-2 & -1 & -2 & -2 & -2 & -2 & 0 & -2 & 2 & 2 & 0 & 2 & 2 & 1 & 2 & 1 & 11
\end{bmatrix}
\begin{bmatrix}
a_0 \\ a_1 \\ a_2 \\ a_3 \\ a_4 \\ a_5 \\ a_6 \\ a_7 \\ b_0 \\ b_1 \\ b_2 \\ b_3 \\ b_4 \\ b_5 \\ b_6 \\ b_7 \\ 1
\end{bmatrix}
\geq
\begin{bmatrix}
0 \\ 0 \\ 0 \\ 0 \\ 0 \\ 0 \\ 0 \\ 0 \\ 0 \\ 0 \\ 0 \\ 0 \\ 0 \\ 0 \\ 0 \\ 0 \\ 0 \\ 0 \\ 0 \\ 0 \\ 0 \\ 0 \\ 0 \\ 0 \\ 0 \\ 0 \\ 0 \\ 0
\end{bmatrix}
$$

References

1. Preliminary state standard of republic of belarus (stbp 34.101.312011) (2011). http://apmi.bsu.by/assets/files/std/belt-spec27.pdf
2. Abdelkhalek, A., Tolba, M., Youssef, A.M.: Related-key differential attack on round-reduced Bel-T-256. IEICE Trans. Fundam. Electron. Commun. Comput. Sci. **101**(5), 859–862 (2018)
3. Beaulieu, R., Treatman-Clark, S., Shors, D., Weeks, B., Smith, J., Wingers, L.: The SIMON and SPECK lightweight block ciphers. In: 2015 52nd ACM/EDAC/IEEE Design Automation Conference (DAC), pp. 1–6. IEEE (2015)
4. Daemen, J., Knudsen, L., Rijmen, V.: The block cipher square. In: Biham, E. (ed.) FSE 1997. LNCS, vol. 1267, pp. 149–165. Springer, Heidelberg (1997). https://doi.org/10.1007/BFb0052343
5. Feistel, H., Notz, W.A., Smith, J.L.: Some cryptographic techniques for machine-to-machine data communications. Proc. IEEE **63**(11), 1545–1554 (1975)
6. Jovanovic, P., Polian, I.: Fault-based attacks on the Bel-T block cipher family. In: Proceedings of the 2015 Design, Automation & Test in Europe Conference & Exhibition, pp. 601–604. EDA Consortium (2015)

7. Knudsen, L., Wagner, D.: Integral cryptanalysis. In: Daemen, J., Rijmen, V. (eds.) FSE 2002. LNCS, vol. 2365, pp. 112–127. Springer, Heidelberg (2002). https://doi.org/10.1007/3-540-45661-9_9

8. Lai, X., Massey, J.L.: A proposal for a new block encryption standard. In: Damgård, I.B. (ed.) EUROCRYPT 1990. LNCS, vol. 473, pp. 389–404. Springer, Heidelberg (1991). https://doi.org/10.1007/3-540-46877-3_35

9. Sun, L., Wang, M.: Toward a further understanding of bit-based division property. Sci. China Inf. Sci. **60**(12), 128101 (2017)

10. Sun, L., Wang, W., Liu, R., Wang, M.: MILP-aided bit-based division property for ARX-based block cipher. Cryptology ePrint Archive, report 2016/1101 (2016). https://eprint.iacr.org/2016/1101

11. Sun, L., Wang, W., Wang, M.: MILP-aided bit-based division property for primitives with non-bit-permutation linear layers. Cryptology ePrint Archive, report 2016/811 (2016). https://eprint.iacr.org/2016/811

12. Sun, L., Wang, W., Wang, M.: Automatic search of bit-based division property for ARX ciphers and word-based division property. In: Takagi, T., Peyrin, T. (eds.) ASIACRYPT 2017. LNCS, vol. 10624, pp. 128–157. Springer, Cham (2017). https://doi.org/10.1007/978-3-319-70694-8_5

13. Sun, S., et al.: Towards finding the best characteristics of some bit-oriented block ciphers and automatic enumeration of (related-key) differential and linear characteristics with predefined properties (2014). https://eprint.iacr.org/2014/747

14. Todo, Y.: Structural evaluation by generalized integral property. In: Oswald, E., Fischlin, M. (eds.) EUROCRYPT 2015. LNCS, vol. 9056, pp. 287–314. Springer, Heidelberg (2015). https://doi.org/10.1007/978-3-662-46800-5_12

15. Todo, Y.: Integral cryptanalysis on full MISTY1. J. Cryptol. **30**(3), 920–959 (2017)

16. Todo, Y., Morii, M.: Bit-based division property and application to SIMON family. In: Peyrin, T. (ed.) FSE 2016. LNCS, vol. 9783, pp. 357–377. Springer, Heidelberg (2016). https://doi.org/10.1007/978-3-662-52993-5_18

17. Xiang, Z., Zhang, W., Bao, Z., Lin, D.: Applying MILP method to searching integral distinguishers based on division property for 6 lightweight block ciphers. In: Cheon, J.H., Takagi, T. (eds.) ASIACRYPT 2016. LNCS, vol. 10031, pp. 648–678. Springer, Heidelberg (2016). https://doi.org/10.1007/978-3-662-53887-6_24

Cryptanalysis of Reduced sLiSCP Permutation in Sponge-Hash and Duplex-AE Modes

Yunwen Liu[1,2], Yu Sasaki[3(✉)], Ling Song[4,5], and Gaoli Wang[6]

[1] imec-COSIC, KU Leuven, Leuven, Belgium
yunwen.liu@esat.kuleuven.be
[2] College of Liberal Arts and Sciences, National University of Defense Technology,
Changsha, China
[3] NTT Secure Platform Laboratories, 3-9-11, Midori-cho Musashino-shi,
Tokyo 180-8585, Japan
sasaki.yu@lab.ntt.co.jp
[4] Nanyang Technological University, Singapore, Singapore
[5] Institute of Information Engineering, Chinese Academy of Sciences, Beijing, China
songling@iie.ac.cn
[6] Department of Cryptography and Network Security,
East China Normal University, Shanghai 200062, China
glwang@sei.ecnu.edu.cn

Abstract. This paper studies security of a family of lightweight permutations sLiSCP that was proposed by AlTawy et al. at SAC 2017. sLiSCP also specifies an authenticated encryption (AE) mode and a hashing mode based on the sponge framework, however the designers' analysis focuses on the indistinguishability of the permutation, and there is no analysis for those modes. This paper presents the first analysis of reduced-step sLiSCP in the AE and hashing modes fully respecting the recommended parameters and usage by the designers. Forgery and collision attacks are presented against 6 (out of 18) steps of the AE and hashing modes. Moreover, rebound distinguishers are presented against 15 steps of the permutation. We believe that those results especially about the AE and hashing modes provide a better understanding of sLiSCP, and bring more confidence about the lightweight version sLiSCP-light.

Keywords: sLiSCP · Simeck · Permutation · Sponge · Collision
Forgery

1 Introduction

Ubiquitous computing and the Internet of Things (IoT) are developing rapidly as the new computing paradigm in information technology. The deployment of small computing devices such as Radio-Frequency Identification (RFID) tags, sensor nodes and smart cards increases fast and plays an important role in various applications. At the same time, it also brings a wide range of new security and privacy

© Springer Nature Switzerland AG 2019
C. Cid and M. J. Jacobson, Jr. (Eds.): SAC 2018, LNCS 11349, pp. 92–114, 2019.
https://doi.org/10.1007/978-3-030-10970-7_5

concerns. These small devices demand harsh cost constraints like low memory availability, low area requirements and power consumptions, which makes it difficult to employ conventional cryptographic algorithms. Lightweight cryptography is a field of cryptography that caters for security concerns of resource-constrained devices. Dozens of symmetric-key primitives have been proposed to address the issues, such as lightweight block ciphers (LED [18], PRESENT [12], SIMON & SPECK [6], Simeck [30] etc.), lightweight hash functions (Spongent [11], Photon [17], Quark [3] etc.), lightweight stream ciphers (Grain [19], Mickey [5], Trivium [14] etc.) and lightweight authenticated encryptions (Ascon [15], Ketje-Jr [10] and NORX [4]). Meanwhile, lightweight cryptographic algorithms including PRESENT, Photon, Grain and Trivium are adopted by ISO as new standards. Recently, the National Institute of Standards and Technology of the U.S. (NIST) has started a process for standardizing lightweight authenticated encryptions with associated data (AEAD) and hashing [26].

Among the existing lightweight cryptographic algorithms, permutation-based designs are of special interest. They have an outstanding advantage for devices that have limited resources to provide multiple cryptographic functions with low overhead. In fact, encryption, authentication, hashing, and possibly pseudorandom-bit generation which are the basic functionalities required by a security protocol can be achieved by applying a cryptographic permutation in certain modes, such as Sponge [9]. Ascon, NORX and Ketje-Jr are examples of permutation-based designs to provide both encryption and authentication.

sLiSCP is a family of cryptographic permutations designed by AlTawy et al. and proposed at SAC 2017 [1]. It has two instances, namely, sLiSCP-192 and sLiSCP-256 which adopt a 4-branch type-2 generalized Feistel network (GFN) where the functions in GFN are instantiated with reduced-round Simeck-48/64 [30] whose secret key is replaced with a public constant. Both sLiSCP-192 and sLiSCP-256 have 18 steps. Besides, the designers use sLiSCP in the sponge framework to construct authenticated encryption (AE) [8] and hash functions [7]. Considering that the coming standardization activity for lightweight cryptography by NIST takes into account the designs that support both AE and hash function, security analysis of sLiSCP is of great interest as an example case.

Cryptanalysis is crucial for any design. The existing security analysis of sLiSCP by the designers focus on the indistinguishability of the permutation, and there is no analysis in the hashing and AE modes. The designers showed that impossible differential (or zero-correlation) distinguishers reach 9 steps of the sLiSCP permutation and zero-sum distinguishers utilizing division property [28] can achieve 17 steps of sLiSCP-192/256 with complexity 2^{190} (resp. 2^{255}) for sLiSCP-192 (resp. sLiSCP-256). Without rigorous cryptanalysis, it is hard to determine the most suitable number of steps. Recently, a lightweight variant of sLiSCP, sLiSCP-light [2], was proposed by the same designers, which replaces the 4-branch generalized Feistel network with the 4-branch generalized Misty structure, and the number of steps in the permutation is reduced to only 12.

Table 1. Summary of attacks against sLiSCP

Target	Version	Attacks	Steps	Time	Data	Memory	Ref.
AE	sLiSCP-192	Forgery	6/18	$2^{104.0}$	$2^{104.0}$	negl.	Sect. 4
	sLiSCP-256	Forgery	6/18	$2^{112.2}$	$2^{112.2}$	negl.	Sect. 4
	sLiSCP-192	State Recovery	6/18	$2^{105.6}$	$2^{105.6}$	negl.	Sect. 4
Hash	sLiSCP-192	Collision	6/18	$2^{69.8}$	N/A	$2^{32.1}$	Sect. 5
	sLiSCP-256	Collision	6/18	$2^{74.8}$	N/A	$2^{46.3}$	Sect. 5
Permutation	both	Imp Diff	9/18	N/A	N/A	N/A	[1]
	both	Zero Cor	9/18	N/A	N/A	N/A	[1]
	sLiSCP-192	Zero-sum	17/18	2^{190}	N/A	negl.	[1]
	sLiSCP-256	Zero-sum	17/18	2^{255}	N/A	negl.	[1]
	sLiSCP-192	Rebound	15/18	$2^{122.7}$	N/A	$2^{37.7}$	Sect. 6
	sLiSCP-256	Rebound	15/18	$2^{168.3}$	N/A	$2^{47.7}$	Sect. 6

Our Contributions. In this paper, we provide security analysis of sLiSCP, in particular, the first results of sLiSCP in AE and hashing modes. The number of attacked steps, 6, is small compared to the full steps, 18. However, 18 steps of sLiSCP uses 216 and 288 rounds of Simeck-48 and Simeck-64 to permute 192-bit and 256-bit states, respectively, which looks conservative. Indeed, the number of steps was later reduced in sLiSCP-light. We believe that our analysis helps to understand the suitable choice of the number of steps.

Our first analysis is the 6 (out of 18) steps forgery attacks in the AE mode. The attacks fully respect the limitation by the designers, i.e. we use the size and position of the inner and outer parts (or capacity and rate) according to the designer's recommendation and the nonce is never repeated. There are two versions of the AE mode; sLiSCP-192/112 and sLiSCP-256/128 that use 112-bit and 128-bit key and claim 112-bit and 128-bit security, respectively. The attack complexities are $2^{103.96}$ and $2^{112.2}$ queries for sLiSCP-192/112 and sLiSCP-256/128 respectively. Moreover, the state recovery is applied to sLiSCP-192/112.

We then convert the above attacks to find collisions in the hashing mode. The claimed security is 80 bits and 96 bits for sLiSCP-192 and sLiSCP-256 respectively, thus naively applying the attacks on AE to hashing modes is worse than the birthday attack. In the hash setting, attackers have access to the internal state value and can choose message values to control the differential propagation. To exploit this property, we use the multi-block strategy and find collisions with $2^{69.8}$ and $2^{74.8}$ computations for sLiSCP-192 and sLiSCP-256, respectively.

Finally, we evaluate sLiSCP as a permutation by applying rebound attacks [23,25]. Although the zero-sum distinguisher by the designers [1] can break more steps, their complexities are very close to the permutation size. Our rebound attacks reach only 15 rounds but the computational complexities, $2^{122.7}$ for sLiSCP-192 and $2^{168.3}$ for sLiSCP-256, are significantly smaller than the permutation size. The differential based approach can be applied to AE or hash settings and our rebound attacks provide better understandings to decide the suitable number of steps in the lightweight design e.g. sLiSCP-light. Our results are summarized in Table 1 along with the attacks by the designers.

The core of our attacks is the discovery of efficient differential trails for the sLiSCP permutation. Because of the large state size and the complex underlying Simeck permutation, it is infeasible to find useful trails with existing automated search tools.[1] In this paper, we start with our differential trail search strategy.

Paper Outline. Section 2 describes the sLiSCP specification. Section 3 explains how to search for differential trails for large sLiSCP permutations. Section 4 describes forgery and state-recovery attacks in the AE mode. Section 5 describes collision attacks in the hashing mode. Section 6 presents rebound attacks against sLiSCP permutations. We conclude this paper in Sect. 7.

2 Specification of sLiSCP

2.1 sLiSCP Permutation

The sLiSCP permutation F is denoted as sLiSCP-b, where $b = 4m$ and $m \in \{48, 64\}$. As depicted in Fig. 1, F updates the input $(X_0^0, X_1^0, X_2^0, X_3^0)$ of four m-bit words in s steps and gets the output $(X_0^s, X_1^s, X_2^s, X_3^s)$. The permutation

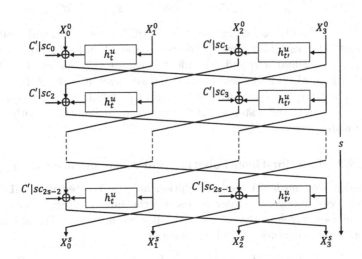

Fig. 1. sLiSCP permutation using Simeck^u-m as h_t^u

F can be described in terms of the step function f as

$$F(X_0^0, X_1^0, X_2^0, X_3^0) = f^s(X_0^0, X_1^0, X_2^0, X_3^0) = (X_0^s, X_1^s, X_2^s, X_3^s).$$

---

[1] We first tried to find the optimal 6-step differential trail for sLiSCP-192 with MILP. Even after 2,000,000 s (more than 23 days), we did not have any hope that the tool would finish. Searching for the optimized trail for sLiSCP-256 is even harder.

Table 2. Parameters for the permutation F in sLiSCP-192 and sLiSCP-256

Algorithm	Branch size m	Rounds u	Steps s	State size b	Total number of Simeck rounds $2us$
sLiSCP-192	48	6	18	192	216
sLiSCP-256	64	8	18	256	288

The step function f is built on a 4 branch Type-2 GFN and based on an u-round Simeck [1]. In step j ($0 \leq j \leq s - 1$), the step function $f(X_0^j, X_1^j, X_2^j, X_3^j)$ is defined as

$$\left(X_1^j, h_{t'}^u(X_3^j) \oplus X_2^j \oplus (C'|SC_{2j+1}), X_3^j, h_t^u(X_1^j) \oplus X_0^j \oplus (C'|SC_{2j})\right),$$

where C' and SC_j are a constant $2^m - 256$ and a step-dependent constant, respectively, "$|$" is a bitwise-OR. and $h_t^u(\cdot)$ is an u-round Simeck depending on the constant t. We sometimes omit t and denote the function by Simeck^u-$m(\cdot)$, which is further detailed as

$$h_t^u(x) = \text{Simeck}^u\text{-}m(x) = h_{u-1} \circ h_{u-2} \circ \ldots \circ h_0(x),$$

where $h_i(x) = h_i(x_0 \| x_1)$ is defined as follows (See also Fig. 5 in Appendix.):

$$h_i(x) = \left((x_0 \odot (x_0 \lll 5)) \oplus (x_0 \lll 1) \oplus x_1 \oplus (C|RC_i), x_0\right).$$

Here "$\oplus$," "$\odot$" and "$\lll$" denote bit-wise XOR, bitwise AND, and a left cyclic shift, respectively. x_0 and x_1 are $\frac{m}{2}$-bit words and C and RC_i are a constant defined as $2^{\frac{m}{2}} - 2$ and a round-dependent constant. The parameters for the permutation F in sLiSCP-192 and sLiSCP-256 are given in Table 2. Because the constants do not impact to our attacks, we omit the details of the constants. The schematic diagram of the s-step sLiSCP permutation instantiated with u-round Simeck-m is illustrated in Fig. 1.

2.2   sLiSCP Mode for Hash Function and Authenticated Encryption

Hash function and authenticated encryption are constructed using sLiSCP in the sponge-based modes. In order to specify the initialization, absorbing and squeezing phases conveniently, we use the following notations. For sLiSCP-192, the 192-bit state is denoted as 24-byte state as

$$(X_0, X_1, X_2, X_3) = (B_0, ..., B_5, B_6, ..., B_{11}, B_{12}, ..., B_{17}, B_{18}, ..., B_{23}),$$

where $X_i \in \mathbb{F}_2^{48}$ and $B_i \in \mathbb{F}_2^8$. For sLiSCP-256, the 256-bit state is denoted as

$$(B_0, ..., B_7, B_8, ..., B_{15}, B_{16}, ..., B_{23}, B_{24}, ..., B_{31}).$$

Initialization. In the hashing mode, the state is initialized to a constant value called IV. In the AE mode, the state is initialized to a mixture of nonce, key, and constant. Because we do not use those configurations in our attacks, we refer to [1] for the details of the initial set up.

Rate and Capacity. In the sponge-based construction, the b-bit state is divided into rate r and capacity c such that $r + c = b$. In both of the AE and hash modes, $r = 32$ and $r = 64$ are recommended when F is sLiSCP-192 and sLiSCP-256, respectively. (Accordingly, $c = 160$ and $c = 192$ for sLiSCP-192 and sLiSCP-256, respectively.) The byte positions of the rate are defined as B_i ($i = 6, 7, 18, 19$) for sLiSCP-192 and B_i ($i = 8, 9, 10, 11, 24, 25, 26, 27$) for sLiSCP-256.

Hash Mode. As depicted in Fig. 3 in Appendix, the message M is padded and split into blocks of r bits each. After the initialization, the message block is XORed with B_i ($i = 6, 7, 18, 19$) and ($i = 8, 9, 10, 11, 24, 25, 26, 27$) for sLiSCP-192-based and sLiSCP-256-based constructions, respectively, followed by the application of the permutation F. The absorbing phase finishes when all message blocks are processed. Then in the squeezing phase, extraction of the r' bits of the state and application of F is iterated until the entire digest is obtained. r' is recommended as $r' = 32$ for sLiSCP-192 and $r' \in \{32, 64\}$ for sLiSCP-256.

AE Mode. Firstly, the key K, the message M and the associated data A are padded. After the initialization, K and A are processed block-by-block with making appropriate separation by XORing constant in capacity. To convert M to C, for each block, r-bit M_i is XORed to the state and the result is output as C_i. Then, the state is updated by sLiSCP permutation F. After all the ciphertext blocks are generated, the key K is absorbed to the state again, and the tag T is extracted from the state. The AE mode is described in Fig. 4 in Appendix.

Recommended Parameters and Security. The recommended parameters and security claims of the hashing mode and the AE mode are presented in Tables 3 and 4, respectively.

Table 3. Recommended parameters and bit securities in hashing mode

Algorithm	IV	Digest	r	r'	c	Collision
sLiSCP-192	0x502020	160	32	32	160	80
sLiSCP-256	0x604040	192	64	64	192	96
sLiSCP-256	0x604020	192	64	32	192	96

Table 4. Recommended parameters and bit securities in AE mode

Algorithm	Key	Nonce	Tag	r	c	Confidentiality	Integrity
sLiSCP-192/80	80	80	80	32	160	80	80
sLiSCP-192/112	112	80	112	32	160	112	112
sLiSCP-256/128	128	128	128	64	192	128	128

3   Differential Trail Search on sLiSCP

The core of our attacks is to find good differential trails. While there are many existing results on automated differential trail search tool, it is infeasible to apply those to sLiSCP permutations owing to their large state size and complicated step function using Simeck. In this section, we introduce our strategy to reduce the search problem for the entire permutation to several iterations of Simeck.

The search strategy depends on which of the permutation or the sponge mode is attacked.

Permutation: The number of attacked steps is large (i.e. 15 for our attacks), thus we search for an iterative differential trail for a small number of steps and iterate it several times. As it will be explained later, the rebound attack often utilizes sparse differential trails for an outbound phase, thus it is desired to start and end the iterative trail with a sparse difference.

Sponge mode: Considering that differences can be injected only through the message input to r bits of the state, the differential trail must start from and end with r-bit rate specified in Sect. 2. Hence, this is another iterative differential trail in a branch-wise level.

In the sponge mode, a half of the rate exists in the left half of the state, e.g. B_6, B_7 for sLiSCP-192, and the other half exists in the right half, e.g. B_{18}, B_{19}. We found that injecting differences in both halves decreases the probability quickly especially to satisfy the constraint that the output difference can only exist in r-bit rate. In the end, for both targets, our goal is to find an iterative difference that starts and ends with single active branch denoted by $(0, 0, 0, \alpha)$. Such trail can be found for 6 steps. Its schematic diagram is shown in Fig. 6 in Appendix A.

Maximizing the Search Space. By considering the attacks on the sponge mode, α in the output difference can be replaced with another one denoted by γ, which relaxes the constraint and may increases the probability of the trail. We then found that by fixing the differential propagation in the first and the last steps to $\alpha \rightarrow \beta$ and $\gamma \rightarrow \delta$, all the internal state differences are fixed, i.e. the search space is maximized. To attack the permutation, α and γ can take any m-bit difference, while to attack the sponge-mode, only $r/2$-bits can have differences. To discuss the differential trail on the sponge mode, it is convenient to denote the state (X_0, X_1, X_2, X_3) by using 8 $m/2$-bit words S_i as $(S_0, S_1, S_2, S_3, S_4, S_5, S_6, S_7)$, then the difference can only be injected to S_2 and S_6 in the sponge-based mode. The 6-step trail in the word-wise level is shown in Fig. 2. A coloured box indicates the propagation of nonzero differences.

The probability of the trail is

$$\Pr\left(\alpha \rightarrow \beta\right)^2 \times \Pr\left(\beta \rightarrow \gamma\right) \times \Pr\left(\gamma \rightarrow \delta\right)^2 \times \Pr\left(\delta \rightarrow \alpha\right).$$

As a consequence, we reduce the search problem for the entire sLiSCP permutation into the problem of 4 parallel searches on 6-round Simeck-48 or 8-round

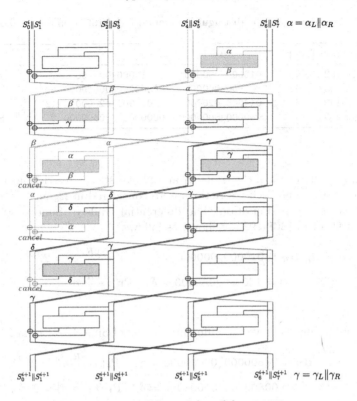

Fig. 2. 6-step differential trail for sLiSCP

Simeck-64, which seems feasible but requires clever coding to find the best combination of the results from the 4 parts. Interestingly, those 4 propagations form a circulation $\alpha \rightarrow \beta \rightarrow \gamma \rightarrow \delta \rightarrow \alpha$. Then, 4 differential propagations can be searched in a sequential way by regarding those 4 propagations as an iterative differential trail against 24-round Simeck-48 and 32-round Simeck-64, which is now feasible and easy to optimize the combined results with existing automatic search tools. We follow the automatic search model of SIMON and SPECK [22,24], due to their similar structures with Simeck. In other word, we have taken some dependencies in the round functions of Simeck into account in the automatic search. In addition, we experimentally verified the probabilities of some characteristics in 6/8-round Simeck found by the tool, and the results match the theoretical predictions of the differential probability. As an example, we show the detail of a 6-round differential trail for Simeck-48 in Table 10 in Appendix.

Search Results. Table 5 shows an overview of the distinguishers we found for 6-step sLiSCP. The differences α and γ are the input and output differences in the trails, which also define the differential for sLiSCP. The reference shows the applications of the distinguishers in this paper.

Table 5. An overview of the distinguishers found for sLiSCP-256 and sLiSCP-192

ID	Version	#steps	α	γ	Pr	ref.
Ω_1	sLiSCP-192	6	010000‖000000	010000‖000000	$2^{-103.96}$	Sect. 4,5
Ω_2	sLiSCP-256	6	08800000‖00000000	08800000‖00000000	$2^{-112.14}$	Sect. 4,5
Ω_3	sLiSCP-192	6	014000‖020000	014000‖020000	$2^{-88.8}$	Sect. 6
Ω_4	sLiSCP-256	6	00000000‖80000000	00000000‖80000000	$2^{-112.14}$	Sect. 6

The input differences of the trails Ω_1 and Ω_2 satisfy the restrictions from the sponge mode, while no such restrictions were considered in Ω_3 and Ω_4 towards the analysis on the permutation. The differential trails of Simeck^6-48 in the differential Ω_1 of sLiSCP-192 are shown as follows.

$$\alpha = \alpha_L\|\alpha_R \triangleq 010000\|000000 = \gamma, \quad \Pr[\alpha \xrightarrow{6R} \beta] = 2^{-20},$$
$$\beta \triangleq \text{1d0000}\|060000 = \delta, \quad \Pr[\beta \xrightarrow{6R} \alpha] = 2^{-18},$$

In the following, we have the differential trails of Simeck^8-64 in sLiSCP-256.

$$\alpha = \alpha_L\|\alpha_R \triangleq 08800000\|00000000 = \gamma, \quad \Pr[\alpha \xrightarrow{8R} \beta] = 2^{-22},$$
$$\beta \triangleq 00800000\|00000000 = \delta, \quad \Pr[\beta \xrightarrow{8R} \alpha] = 2^{-22}.$$

Without the difference restriction, the probability of the trails can be improved for Simeck-48, such as Ω_3 of sLiSCP-192, which is shown below.

$$\alpha = \alpha_L\|\alpha_R \triangleq 014000\|020000 = \gamma, \quad \Pr[\alpha \xrightarrow{6R} \beta] = 2^{-12},$$
$$\beta \triangleq 014000\|008000 = \delta, \quad \Pr[\beta \xrightarrow{6R} \alpha] = 2^{-26}.$$

As for sLiSCP-256, even though there exist 6-step trails with larger probability than the optimal trail in Ω_2, the 6-step distinguisher Ω_2 has an overall advantage by taking the differential effect into account. Yet we still found a new distinguisher shown below Ω_4, which is similar to Ω_2. Considering the lower Hamming weight of α than Ω_2, this is more suitable for rebound attacks.

$$\alpha = \alpha_L\|\alpha_R \triangleq 00000000\|80000000 = \gamma, \quad \Pr[\alpha \xrightarrow{8R} \beta] = 2^{-22},$$
$$\beta \triangleq 00000000\|80000008 = \delta, \quad \Pr[\beta \xrightarrow{8R} \alpha] = 2^{-22}.$$

Interesting, we did not find any case that using different α and γ increases the probability. (In general, it occurs e.g. we confirmed the increase of the probability for different rate positions.) To make the paper simple, hereafter we use α and β instead of γ and δ, respectively.

Furthermore, the probability evaluation in Table 5 takes into account the effect of the differentials within Simeck. Specifically, we enumerate the trails

Table 6. The distribution of trails in the differential $(014000\|020000 \xrightarrow{6R} 014000\|008000)$ of Simeck-48. The differential probability is approximately $2^{-11.3}$.

$-\log(p)$	12	13	14	15	16	17	18	19	20
#char	1	0	2	1	0	0	0	0	0

Table 7. The distribution of trails in the differential $(014000\|008000 \xrightarrow{6R} 014000\|020000)$ of Simeck-48. The differential probability is approximately $2^{-21.8}$.

$-\log(p)$	26	27	28	29	30	31	32	33	34	35	36	37	38	39	40	41	42
#char	3	0	13	14	35	59	102	168	255	452	675	1021	1454	1907	2454	3081	3608

$-\log(p)$	43	44	45	46	47	48	49	50	51	52	53	54
#char	4141	4219	3859	3154	2280	1425	754	333	122	36	8	1

within the differentials of Simeck, which leads to a refined estimation of the probabilities for the distinguishers in sLiSCP. the distribution of trails according to the probability in $(014000\|020000 \xrightarrow{6R} 014000\|008000)$ is shown in Table 6. The differential probability is $2^{-11.3}$. Similarly in Table 7, we show the trails in the differential $(014000\|008000 \xrightarrow{6R} 014000\|020000)$, where the probability is $2^{-21.8}$. With the differential effect taken into account, the probability of Ω_3 is approximately $2^{-88.8}$. For simplicity, we omitted the details of the trails in other distinguishers, and summarise the probabilities of the differentials in Simeck in Table 8.

Table 8. An overview for the probabilities of the obtained differentials

Version	Differential	Probability
Simeck^6-48	$010000\|000000 \to 1d0000\|060000$	$2^{-17.85}$
	$1d0000\|060000 \to 010000\|000000$	$2^{-16.28}$
Simeck^6-48	$014000\|020000 \to 014000\|008000$	$2^{-11.3}$
	$014000\|008000 \to 014000\|020000$	$2^{-21.8}$
Simeck^8-64	$08800000\|00000000 \to 00800000\|00000000$	$2^{-18.69}$
	$00800000\|00000000 \to 08800000\|00000000$	$2^{-18.69}$
Simeck^8-64	$00000000\|80000000 \to 00000000\|80000008$	$2^{-18.69}$
	$00000000\|80000008 \to 00000000\|80000000$	$2^{-18.69}$

4 6-Steps Forgery in AE Mode

In this section, the differentials explained in Sect. 3 are exploited for a forgery attack against 6-steps sLiSCP-192/112 and sLiSCP-256/128 in the AE mode. We apply the approach called "LOCAL attack" that was proposed by Khovratovich and Rechberger [21] and independently found by Wu et al. [29] against ALE [13].

4.1 Forgery

Let '$\|$' denote a concatenation. The attacker first observes a ciphertext having at least two encrypted message blocks $C_0\|C_1$. The ciphertext has a form $(N, A, C_0\|C_1, T)$, where N is a nonce, A is an associated data and T is a tag.

The attacker injects the difference specified in Sect. 3 to C^0 and C^1, namely $\overline{C^0} = C^0 \oplus (0\|\alpha_L)$ and $\overline{C^1} = C^1 \oplus (0\|\alpha_L)$. During the decryption, the difference injected by $\overline{C^0}$ makes the difference of S_6 to be α_L and this propagates through 6 steps so that it can be canceled by the difference from $\overline{C^1}$ with probability $2^{-103.96}$ for sLiSCP-192/112 and $2^{-112.14}$ for sLiSCP-256/128 (See Table 5). Hence the attacker makes decryption queries $(N, A, \overline{C^0}\|\overline{C^1}, T)$, which pass with the above probabilities.

The complexity of the attack against sLiSCP-192/112 is either $2^{103.96}$ data and $2^{103.96}$ verification attempts to achieve high success probability or 1 data and 1 verification attempts to achieve success probability of $2^{-103.96}$. The same applies to sLiSCP-256/128 by replacing $2^{103.96}$ with $2^{112.14}$.

4.2 Extension to State Recovery and Plaintext Recovery

In the duplex AE, the internal state value is always partially leaked as a ciphertext. Along with the information that a pair $C^0\|C^1$ and $\overline{C^0}\|\overline{C^1}$ satisfies the differential propagation, the attacker can recover the internal state as long as the number of candidate values of the internal state is sufficiently reduced. We show that the state recovery attack can be applied to sLiSCP-192.

We enumerate all the solutions of the first 4 active Simeck functions in Fig. 2. The differential for the first step is $\alpha \rightarrow \beta$ that is satisfied with probability $2^{-17.85}$. By examining all 2^{48} input values, $2^{48-17.85} = 2^{30.15}$ solutions will be found. We then further check the match with 24-bits of S_6 leaked by the key stream. $2^{30.15-24} = 2^{6.15}$ values match the observed key stream. In other words, the possible values of the 48-bit word $S_6\|S_7$ is now reduced to $2^{6.15}$ choices.

Similarly, the differential for the active Simeck function in the second step is $\beta \rightarrow \alpha$ that is satisfied with probability $2^{-16.28}$, the differential for the right function in step 3 is $\alpha \rightarrow \beta$ and the left function in step 3 is $\alpha \rightarrow \beta$ both are satisfied with probability $2^{-17.85}$. Hence, once the differential is satisfied, the number of possible state values for those Simeck^6-48 is $2^{31.72}$, $2^{30.15}$ and $2^{30.15}$ respectively. For any combination of paired values of those 4 Simeck functions, the 192-bit state values is uniquely fixed. In other words, the possible choices of the 192-bit state value are limited to the combination of those 4 Simeck functions. Hence the number of possible 192-bit states is $2^{6.15+31.72+30.15+30.15} = 2^{98.17}$.

Suppose that in the forgery attack, the encrypted message blocks is at least 6 blocks, and thus we make $\mathcal{D}(N, A, \overline{C^0}\|\overline{C^1}\|C_2\|C_3\|C_4\|C_5, T)$ in the forgery attack. Then the 128-bit value $C_2\|C_3\|C_4\|C_5$ can be used to filter out wrong candidates of $2^{98.17}$ choices of the 192-bit internal state.

In the end, for the state recovery, the data complexity increases to $6/2 \cdot 2^{103.96} = 2^{105.54}$. The computational complexity is $2^{105.54}$ memory access and $2^{98.17}$ 6-step sLiSCP-192 operations.

5   6-Steps Collision Attacks in Hashing Mode

We again use the 6-steps differential trail in Fig. 2. The forgery attacks in Sect. 4 are rather straightforward applications of the detected differentials. However, in the hash setting, the claimed bit-security is smaller, i.e. 80 bits (resp. 96 bits) for sLiSCP-192 (resp. sLiSCP-256), thus the naive approach with complexity $2^{103.96}$ (resp. $2^{112.14}$) is worse than the brute-force attack.

In the hash setting, attackers have access to the internal state value and can choose message values to control the differential propagation. This allows attackers to find collisions faster than the claimed bit-security for 6 steps.

5.1   Overall: Four-Block Collision Strategy

Our attacks find four-block colliding messages, namely $M^0\|M^1\|M^2\|M^3$ and $M^0\|M^1\|(M^2 \oplus 0\|\alpha_L)\|(M^3 \oplus 0\|\alpha_L)$ that produce the same hash digest.

No message difference is injected in the first and second message block. The purpose of those blocks is to set the state value that is advantageous to satisfy the 6-step differential trail in the third block. In short, the attacker precomputes all paired values that satisfy the differential propagation $\alpha \to \beta$ in the first step in Fig. 2 and $\beta \to \alpha$ in the second step. This allows the attackers to search for $M^0\|M^1$ producing the good values for the internal state after 2 blocks, denoted by $S_0^2\|S_1^2\|\cdots\|S_7^2$. Note that the reason why we need 2 blocks rather than 1 block is that degrees of freedom of a single message block, 2^{32} for sLiSCP-192 and 2^{64} for sLiSCP-256, are too small to find a colliding message pair.

The third block propagates differences as shown in Fig. 2 so that the output difference from the third block can be canceled out by injecting another message difference from the fourth message block.

5.2   Attack Procedure for sLiSCP-256

We first explain the attack for sLiSCP-256 that is instantiated with Simeck^8-64. We denote the left and right functions in step i, where $i \in \{0, 1, \cdots, 5\}$, by Simeck^8-64^{iL} and Simeck^8-64^{iR}, respectively. The illustration of the attack is shown in Fig. 7 in Appendix.

Precomputation

- For all $x_0 \in \{0,1\}^{64}$, compute Simeck^8-64$^{0R}(x_0) \oplus$ Simeck^8-64$^{0R}(x_0 \oplus \alpha)$ to check if the result is β or not. Because $\Pr[\alpha \xrightarrow{8R} \beta] = 2^{-18.69}$, we have $2^{64-18.69} = 2^{45.31}$ choices of x_0. Let y_0 be the corresponding output value for x_0. Those $2^{45.31}$ choices of (x_0, y_0) are stored in a table T^{0R}.

Let x_0^L and x_0^R be the left and right halves of x_0, namely $x_0 = x_0^L \| x_0^R$. T^{0R} is further sorted with respect to the 32-bit value of x_0^R. Because we have $2^{45.31}$ choices in T^{0R}, we expect $2^{45.31-32} = 2^{13.31}$ choices of x_0^L for each x_0^R. In the end, a table T^{0R} of size $2^{45.31}$ is divided into 2^{32} tables $T_i^{0R}, i = 0, 1, \cdots, 2^{32} - 1$, of size $2^{13.31}$ that store $2^{13.31}$ values of x_0^L for $x_0^R = i$.

- Do the same for Simeck^8-64^{1L}. Namely, for all $x_1 \in \{0,1\}^{64}$, compute Simeck^8-64$^{1L}(x_1) \oplus$ Simeck^8-64$^{1L}(x_1 \oplus \beta)$ to check if the result is α or not. Because $\Pr[\beta \xrightarrow{8R} \alpha] = 2^{-18.69}$, we have $2^{64-18.69} = 2^{45.31}$ choices of x_1. Let y_1 be the corresponding output value for x_1. Those $2^{45.31}$ choices of (x_1, y_1) are stored in a table T^{1L}.

The First Two Steps of the Differential. Choose $M_0 \| M_1$ uniformly at random and compute the second block output $S_0^2 \| S_1^2 \| \cdots \| S_7^2$. Thanks to the precomputation of Simeck^8-64^{0R}, for a given S_7^2, there are $2^{13.31}$ choices of x_0^L such that $x_0^L \| S_7^2$ satisfies the differential propagation $\alpha \to \beta$ for the first step. Moreover, the corresponding output y_0 is already stored in the table.

Hence, for a given S_4^2, S_5^2, S_7^2 and $2^{13.31}$ choices of y_0, compute $(S_4^2 \| S_5^2) \oplus y_0$ and check if this matches x_1 in the table T^{1L}. Considering that $2^{45.31}$ choices of x_1 are stored in T^{1L}, the probability of the match after $2^{13.31}$ iterations of y_0 is $2^{-64+45.31+13.31} = 2^{-5.38}$. Therefore, by choosing $2^{5.38}$ choices of $M_0 \| M_1$, we can find $M_0 \| M_1$ and $M_2^R \leftarrow x_0^L \oplus S_6^2$ such that the differential propagation for the first two steps are satisfied.

The Last Four Steps of the Differential. The attacker then uses the 32-bit value of M_2^L as degrees of freedom to satisfy the remaining 4 steps. The probability for the 4 steps is $2^{-18.69 \times 4} = 2^{-74.76}$. After examining 2^{32} choices of M_2^L, all the propagations are satisfied with probability $2^{-74.76+32} = 2^{-42.76}$.

Hence, by iterating the attack procedure so far $2^{42.76}$ times, the attacker can find a desired message pair $M_0 \| M_1 \| M_2$ and $M_0 \| M_1 \| (M_2 \oplus 0 \| \alpha_L)$. Then, the output difference from the third block can be easily canceled by the message difference for the fourth block.

Complexity Analysis. Complexity of the precomputation phase is $2 \cdot 2^{64} = 2^{65}$. It requires a memory to store $2 \cdot 2^{45.31} = 2^{46.31}$ values. The complexity to satisfy the 6-step differential up to the first two steps is $2^{5.38+13.31} = 2^{18.69}$. The complexity to satisfy all the 6-step differential is $2^{65} + 2^{42.76}(2^{18.69} + 2^{32}) \approx 2^{74.76}$. This is faster than the generic attack complexity of 2^{96}.

5.3   Attack Procedure for sLiSCP-192

The attack for sLiSCP-192 is basically the same as one for Simeck-64. The only differences are the state size and the probability of the differentials. We briefly explain the attack for sLiSCP-192.

Precomputation

– Examine $x_0 \in \{0,1\}^{48}$ input values to $\text{Simeck}^6\text{-}48^{0R}$ to pick up all values satisfying the differential propagation $\alpha \xrightarrow{6R} \beta$ that can be satisfied with probability $2^{-17.85}$. As a result, $2^{30.15}$ choices of (x_0, y_0) are stored in a table T^{0R}, and there are about $2^{30.15-24} = 2^{6.15}$ choices of x_0^L for each of x_0^R.
– For $\text{Simeck}^6\text{-}48^{1L}$, the probability of the differential $\beta \xrightarrow{6R} \alpha$ is $2^{-16.28}$. We obtain $2^{31.72}$ choices of x_1 satisfying this differential propagation for T^{1L}.

The First Two Steps of the Differential. Choose $M_0\|M_1$ and the corresponding $S_0^2\|S_1^2\|\cdots\|S_7^2$. For a given S_7^2, there are $2^{6.15}$ choices of x_0^L and the corresponding y_0. Then, for a given $S_4^2\|S_5^2, S_7^2$ and $2^{6.15}$ choices of y_0, compute $S_4^2, S_5^2 \oplus y_0$ and check if this matches x_1. Considering that $2^{31.72}$ choices of x_1 are stored, the probability of the match after $2^{6.15}$ iterations of y_0 is $2^{-48+31.72+6.15} = 2^{-10.13}$. Therefore, by choosing $2^{10.13}$ choices of $M_0\|M_1$, we find $M_0\|M_1$ and M_2^R satisfying the differential trail for the first two steps.

The Last Four Steps of the Differential. The attacker uses 24-bit values of M_2^L as degrees of freedom to satisfy the remaining 4 steps. The probability for the 4 steps is $2^{-17.85*3-16.28} = 2^{-69.83}$. After examining 2^{24} choices of M_2^L, the probability of the remaining 4 steps is $2^{-69.83+24} = 2^{-45.83}$. Hence, by iterating the attack procedure so far $2^{45.83}$ times, a collision is generated.

Complexity Analysis. Complexity of the precomputation phase is $2 \cdot 2^{48} = 2^{49}$. It requires a memory to store $2^{30.15} + 2^{31.72} = 2^{32.14}$ values. The complexity to satisfy the 6-step differential up to the first two steps is $2^{10.13+6.15} = 2^{16.28}$. The complexity to satisfy all the 6-step differential is $2^{49} + 2^{45.83}(2^{16.28} + 2^{24}) \approx 2^{69.83}$. This is faster than the generic attack complexity of 2^{80}.

Table 9. Configuration for rebound attacks

Steps (i)	0	1	2	3	4	5	6	7	8	9	10	11	12	13	14
Propagation in Simeck^{iR}	0	0	I	0	I	0	0	0	I	0	I	0	0	0	(I)
Propagation in Simeck^{iL}	I	0	0	II	I	II	I	0	0	II	I	II	I	0	0
Configuration		F_b			F_{in}						F_f				

'I' and 'II' denote the differential trail $\alpha \to \beta$ and $\beta \to \alpha$, respectively. "(I)" in step 14 denotes that the attacker accepts any output difference from this Simeck function without paying any cost. For sLiSCP-192, 'I' and 'II' are satisfied with probability $2^{-11.3}$ and $2^{-21.8}$, respectively. For sLiSCP-256, both are satisfied with probability $2^{-18.7}$.

6   15-Steps Rebound Attacks Against sLiSCP Permutation

Because sLiSCP is a cryptographic permutation, we also discuss its security as a permutation. We apply the rebound attack [23,25] to show that the differential-based approach can detect non-ideal behaviours for a large number of steps.

Goal of Rebound Attacks. Let x_i and y_i be an input and output of the sLiSCP permutation, respectively, namely $y_i = \text{sLiSCP}(x_i)$. The goal of the rebound attack is to find (x_1, y_1) and (x_2, y_2) where $x_1 \oplus x_2$ and $y_1 \oplus y_2$ belong to a predefined input subspace and output subspace, respectively. If an attacker can find such (x_1, y_1) and (x_2, y_2) against the target permutation faster than a random permutation, the target construction is regarded as non-ideal. This framework is called *limited-birthday distinguisher* (LBD) [16].

The generic attack complexity of LBD was proven by Iwamoto et al. [20]. Let $\mathcal{X}$ and $\mathcal{Y}$ be closed sets of input and output differences. Let also n be a permutation size. Then the generic attack complexity to solve LBD is

$$2^{n+1}/|\mathcal{X}| \cdot |\mathcal{Y}|. \tag{1}$$

An attacker builds a differential trail and divides the target permutation F into three consecutive parts F_b, F_{in}, and F_f, that is, $F = F_f \circ F_{in} \circ F_b$. The attacker first enumerates all the paired values satisfying the differential trail for F_{in}. This is called *an inbound phase* and the collected solutions are called *starting points*. Then the attacker propagates each starting point to F_f and F_b to probabilistically satisfy the differential trails. This is called *an outbound phase*, which is a brute force search by using starting points as degrees of freedom.

Overall Strategy. The most important part of the rebound attack is searching for efficient differentials. We use the 6-step differentials shown in Fig. 2 that was designed to be iterated multiple times. Because the analysis target is a permutation, we do not have to consider the limitation from the message injection positions in the sLiSCP mode. Thus we use the differentials Ω_3 in Table 5 for sLiSCP-192 and Ω_4 for sLiSCP-256.

The distribution of active Simeck functions for 15 steps is shown in Table 9. As in Table 9 and Fig. 8 in Appendix, we locate the inbound phase from steps 4 to 6. This is because if we fix values for 4 active Simeck functions, the entire state value will be fixed. We choose 4 active Simeck functions to cover the lowest probability part, so that the probability of the outbound phase is maximized.

6.1   Attack Procedure for sLiSCP-192

Inbound Phase. We first enumerate all the solutions for the active Simeck-48 functions in the inbound phase. For example, in Step 3, for all $x \in \{0, 1\}^{48}$, compute $\text{Simeck}^6\text{-}48^{3L}(x) \oplus \text{Simeck}^6\text{-}48^{3L}(x \oplus \beta)$ matches the output difference α. If so, we store the solutions in a table T^{3L}. Because the probability of the differential $\beta \xrightarrow{6steps} \alpha$ is $2^{-21.8}$, we expect $2^{48-21.8} = 2^{26.2}$ solutions.

Apply the same procedure for $\text{Simeck}^6\text{-}48^{4L}$, $\text{Simeck}^6\text{-}48^{4R}$, $\text{Simeck}^6\text{-}48^{5L}$ to store the solutions to tables T^{4L}, T^{4R}, and T^{5L}. Considering that the probability of the differential $\alpha \xrightarrow{6steps} \beta$ is $2^{-11.3}$, we expect $2^{48-11.3} = 2^{36.7}$ solutions for T^{4L}, $2^{36.7}$ solutions for T^{4R} and $2^{26.2}$ solutions for T^{5L}.

Outbound Phase. If we fix one solution for each of four active Simeck functions in the inbound phase, the entire 192-bit state value is uniquely fixed. Hence, we propagate the values to F_b and F_f to check if the outbound phase is satisfied.

The number of total starting points is $2^{(2\times 36.7)\times(2\times 26.2)} = 2^{125.8}$, while the probability for F_f is $2^{(6\times -11.3)\times(2\times -21.8)} = 2^{-111.4}$, and the probability for the F_b is $2^{2\times -11.3} = 2^{-22.6}$, in which the total probability is 2^{-134}. Hence, the degrees of freedom is not sufficient to fully satisfy the 15-step differentials

$$(\beta, \alpha, 0, 0) \xrightarrow{15steps} (0, \beta, \alpha, 0).$$

Hence, we relax the differential and accept any 48-bit difference in the second word of the output difference, namely,

$$(\beta, \alpha, 0, 0) \xrightarrow{15steps} (0, *, \alpha, 0).$$

This increases the probability of the outbound phase to $2^{-134+11.3} = 2^{-122.7}$, which can be satisfied with $2^{125.8}$ starting points.

Complexity Evaluation. The inbound phase requires $4 \cdot 2^{48} = 2^{50}$ computations and a memory to store $2^{26.2} + 2^{36.7} + 2^{36.7} + 2^{26.2}$ words for T^{3R}, T^{4L}, T^{4R}, and T^{5L}, which is about $2^{37.7}$ words. The outbound phase requires $2^{122.7}$ computations to satisfy the differential propagations. In the end, the complexity of the attack is $2^{122.7}$ computations and $2^{37.7}$ memory amount.

The complexity to find the same paired values in a random function is much higher. Indeed, the subspace of the input difference is fixed to one choice, thus $|\mathcal{X}| = 1$. The subspace of the output difference is fixed but for the second word, thus $|\mathcal{Y}| = 2^{48}$. From Eq. (1), the generic attack complexity is $2^{192+1}/(1 \cdot 2^{48}) = 2^{145}$, which is higher than our rebound attack complexity.

6.2   Attack Procedure for sLiSCP-256

We again divide the target 15-steps as shown in Table 9. The evaluation for sLiSCP-256 is much simpler than the case of sLiSCP-192 because the probabilities of the both differentials $\alpha \to \beta$ and $\beta \to \alpha$ are $2^{-18.7}$.

In the inbound phase, we enumerate all the solutions of four active Simeck-64 functions. We obtain $2^{64-18.7} = 2^{45.7}$ solutions for each that are stored in four tables of size $2^{45.7}$. Considering all the combination of the solutions, we can generate up to $2^{4 \times 45.7} = 2^{182.8}$ starting points.

In the outbound phase, F_b and F_f contain 2 and 8 active Simeck-64 functions, respectively, thus the entire probability is $2^{10 \times -18.7} = 2^{-187.0}$. Here, we again accept any output difference in the last step, which increases the probability to $2^{9 \times -18.7} = 2^{-168.3}$ and makes the input and output differences of the 15-step sLiSCP-256 as $(\beta, \alpha, 0, 0)$ and $(0, *, \alpha, 0)$.

The complexity of the rebound attack is $2^{168.3}$ computations and memory to store $4 \cdot 2^{45.7} = 2^{47.7}$ values. The complexity to satisfy the same input and output differences against a random permutation is $2^{256+1}/(1 \cdot 2^{64}) = 2^{193}$, which is higher than our rebound attack.

7   Concluding Remarks

In this paper, we investigated the security of sLiSCP permutation, especially the first security analysis in the AE and hash settings defined as the sponge-based construction. We first explained our differential trail search strategy that reduces the search problem of the entire permutation to 24-round Simeck-48 and 32-round Simeck-64. This allowed us to run an existing tool. Based on the detected trail, we performed forgery and state-recovery for 6-steps AE, collision attacks on 6-steps hash and rebound distinguishers on 15-steps permutation. We believe that our several analyses respecting the constraints by the mode will provide a better understanding of the security of sLiSCP.

Acknowledgements. We thank the anonymous reviewers for their valuable comments. This work was initiated during the 7'th Asian Workshop on Symmetric Key Cryptography, we would like to thank the organisers of ASK 2017. Yunwen Liu is supported by the Research Fund KU Leuven C16/18/004, grant agreement No. H2020-MSCA-ITN-2014-643161 ECRYPT-NET, China Scholarship Council (CSC 201403170380) and National Natural Science Foundation (No. 61672530). Ling Song is supported by the Youth Innovation Promotion Association CAS and the National Natural Science Foundation of China (Grants No. 61802399, 61472415, 61732021 and 61772519). Gaoli Wang is supported by the National Natural Science Foundation of China (No. 61572125) and National Cryptography Development Fund (No. MMJJ20180201).

A Appendix

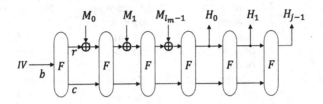

Fig. 3. sLiSCP hashing mode

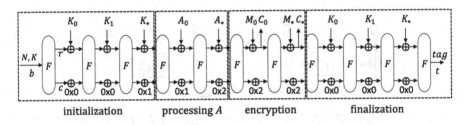

Fig. 4. sLiSCP AE mode

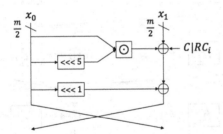

Fig. 5. 1-round of Simeck

Table 10. A 6-round differential trail of Simeck-48 with probability 2^{-12}.

Round	Left difference	Right difference	Probability
0	014000	020000	
1	008000	014000	2^{-4}
2	004000	008000	2^{-2}
3	000000	004000	2^{-2}
4	004000	000000	1
5	008000	004000	2^{-2}
6	014000	008000	2^{-2}
Total			2^{-12}

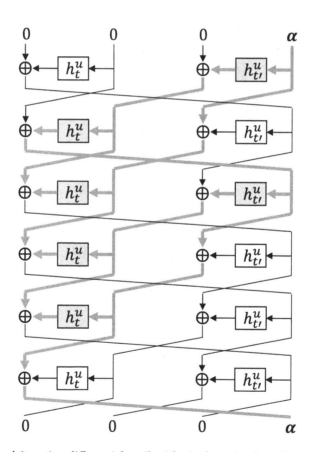

Fig. 6. 6-round iterative differential trail with single active branch. A coloured box indicates the propagation of nonzero differences. (Color figure online)

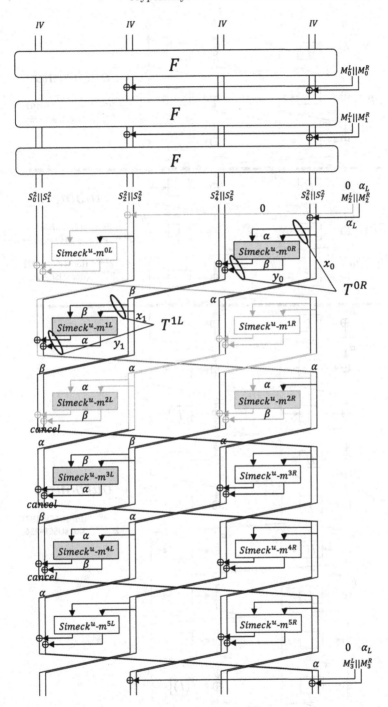

Fig. 7. 6-step collision attack. $\alpha = \alpha_L \| \alpha_R$, and α_R is set to 0. Blue lines show the impact of modifying M_2^L up to step 2, which does not impact to the active Simeck functions in steps 0 and 1, and impacts to all the Simeck functions in steps 2 to 5. (Color figure online)

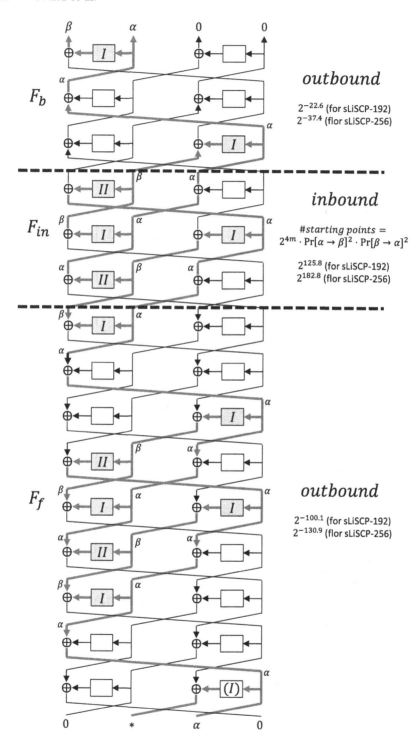

Fig. 8. Differential trail for 15-step rebound attack (Color figure online)

References

1. AlTawy, R., Rohit, R., He, M., Mandal, K., Yang, G., Gong, G.: sLiSCP: simeck-based permutations for lightweight sponge cryptographic primitives. In: Adams, C., Camenisch, J. (eds.) SAC 2017. LNCS, vol. 10719, pp. 129–150. Springer, Cham (2018). https://doi.org/10.1007/978-3-319-72565-9_7

2. AlTawy, R., Rohit, R., He, M., Mandal, K., Yang, G., Gong, G.: sLiSCP-light: towards lighter sponge-specific cryptographic permutations (2018). https://cacr.uwaterloo.ca/techreports/2018/cacr2018-01.pdf

3. Aumasson, J., Henzen, L., Meier, W., Naya-Plasencia, M.: Quark: a lightweight hash. J. Cryptol. **26**(2), 313–339 (2013). https://doi.org/10.1007/s00145-012-9125-6

4. Aumasson, J.-P., Jovanovic, P., Neves, S.: NORX: parallel and scalable AEAD. In: Kutyłowski, M., Vaidya, J. (eds.) ESORICS 2014. LNCS, vol. 8713, pp. 19–36. Springer, Cham (2014). https://doi.org/10.1007/978-3-319-11212-1_2

5. Babbage, S., Dodd, M.: The MICKEY stream ciphers. In: Robshaw, M., Billet, O. (eds.) New Stream Cipher Designs. LNCS, vol. 4986, pp. 191–209. Springer, Heidelberg (2008). https://doi.org/10.1007/978-3-540-68351-3_15

6. Beaulieu, R., Shors, D., Smith, J., Treatman-Clark, S., Weeks, B., Wingers, L.: The SIMON and SPECK families of lightweight block ciphers. IACR Cryptology ePrint Archive 2013, 404 (2013). http://eprint.iacr.org/2013/404

7. Bertoni, G., Daemen, J., Peeters, M., Van Assche, G.: On the indifferentiability of the sponge construction. In: Smart, N. (ed.) EUROCRYPT 2008. LNCS, vol. 4965, pp. 181–197. Springer, Heidelberg (2008). https://doi.org/10.1007/978-3-540-78967-3_11

8. Bertoni, G., Daemen, J., Peeters, M., Van Assche, G.: Duplexing the sponge: single-pass authenticated encryption and other applications. In: Miri, A., Vaudenay, S. (eds.) SAC 2011. LNCS, vol. 7118, pp. 320–337. Springer, Heidelberg (2012). https://doi.org/10.1007/978-3-642-28496-0_19

9. Bertoni, G., Daemen, J., Peeters, M., Van Assche, G.: Cryptographic sponge functions. Submission to NIST (Round 3) (2011). http://sponge.noekeon.org/CSF-0.1.pdf

10. Bertoni, G., Daemen, J., Peeters, M., Van Assche, G., Van Keer, R.: CAESAR submission: Ketje v2. Candidate of CAESAR Competition, September 2016

11. Bogdanov, A., Knezevic, M., Leander, G., Toz, D., Varici, K., Verbauwhede, I.: SPONGENT: a lightweight hash function. In: Preneel and Takagi [27], pp. 312–325. http://dx.doi.org/10.1007/978-3-642-23951-9_21

12. Bogdanov, A., et al.: PRESENT: an ultra-lightweight block cipher. In: Paillier, P., Verbauwhede, I. (eds.) CHES 2007. LNCS, vol. 4727, pp. 450–466. Springer, Heidelberg (2007). https://doi.org/10.1007/978-3-540-74735-2_31

13. Bogdanov, A., Mendel, F., Regazzoni, F., Rijmen, V., Tischhauser, E.: ALE: AES-based lightweight authenticated encryption. In: Moriai, S. (ed.) FSE 2013. LNCS, vol. 8424, pp. 447–466. Springer, Heidelberg (2014). https://doi.org/10.1007/978-3-662-43933-3_23

14. De Cannière, C., Preneel, B.: TRIVIUM. In: Robshaw, M., Billet, O. (eds.) New Stream Cipher Designs. LNCS, vol. 4986, pp. 244–266. Springer, Heidelberg (2008). https://doi.org/10.1007/978-3-540-68351-3_18

15. Dobraunig, C., Eichlseder, M., Mendel, F., Schläffer, M.: Ascon v1.2. Submission to the CAESAR competition. Submission to NIST (Round 3) (2016). http://competitions.cr.yp.to/round3/asconv12.pdf

16. Gilbert, H., Peyrin, T.: Super-sbox cryptanalysis: improved attacks for AES-like permutations. In: Hong, S., Iwata, T. (eds.) FSE 2010. LNCS, vol. 6147, pp. 365–383. Springer, Heidelberg (2010). https://doi.org/10.1007/978-3-642-13858-4_21

17. Guo, J., Peyrin, T., Poschmann, A.: The PHOTON family of lightweight hash functions. In: Rogaway, P. (ed.) CRYPTO 2011. LNCS, vol. 6841, pp. 222–239. Springer, Heidelberg (2011). https://doi.org/10.1007/978-3-642-22792-9_13

18. Guo, J., Peyrin, T., Poschmann, A., Robshaw, M.J.B.: The LED block cipher. In: Preneel and Takagi [27], pp. 326–341. http://dx.doi.org/10.1007/978-3-642-23951-9

19. Hell, M., Johansson, T., Maximov, A., Meier, W.: The grain family of stream ciphers. In: Robshaw, M., Billet, O. (eds.) New Stream Cipher Designs. LNCS, vol. 4986, pp. 179–190. Springer, Heidelberg (2008). https://doi.org/10.1007/978-3-540-68351-3_14

20. Iwamoto, M., Peyrin, T., Sasaki, Y.: Limited-birthday distinguishers for hash functions. In: Sako, K., Sarkar, P. (eds.) ASIACRYPT 2013. LNCS, vol. 8270, pp. 504–523. Springer, Heidelberg (2013). https://doi.org/10.1007/978-3-642-42045-0_26

21. Khovratovich, D., Rechberger, C.: The LOCAL attack: cryptanalysis of the authenticated encryption scheme ALE. In: Lange, T., Lauter, K., Lisoněk, P. (eds.) SAC 2013. LNCS, vol. 8282, pp. 174–184. Springer, Heidelberg (2014). https://doi.org/10.1007/978-3-662-43414-7_9

22. Kölbl, S., Leander, G., Tiessen, T.: Observations on the SIMON block cipher family. In: Gennaro, R., Robshaw, M. (eds.) CRYPTO 2015. LNCS, vol. 9215, pp. 161–185. Springer, Heidelberg (2015). https://doi.org/10.1007/978-3-662-47989-6_8

23. Lamberger, M., Mendel, F., Schläffer, M., Rechberger, C., Rijmen, V.: The rebound attack and subspace distinguishers: application to whirlpool. J. Cryptol. **28**(2), 257–296 (2015)

24. Liu, Y., De Witte, G., Ranea, A., Ashur, T.: Rotational-XOR cryptanalysis of reduced-round SPECK. IACR Trans. Symmetric Cryptol. **2017**(3), 24–36 (2017)

25. Mendel, F., Rechberger, C., Schläffer, M., Thomsen, S.S.: The rebound attack: cryptanalysis of reduced Whirlpool and Grøstl. In: Dunkelman, O. (ed.) FSE 2009. LNCS, vol. 5665, pp. 260–276. Springer, Heidelberg (2009). https://doi.org/10.1007/978-3-642-03317-9_16

26. NIST: Lightweight Cryptography, April 2018. https://csrc.nist.gov/projects/lightweight-cryptography

27. Preneel, B., Takagi, T. (eds.): CHES 2011. LNCS, vol. 6917. Springer, Heidelberg (2011). https://doi.org/10.1007/978-3-642-23951-9

28. Todo, Y.: Structural evaluation by generalized integral property. In: Oswald, E., Fischlin, M. (eds.) EUROCRYPT 2015. LNCS, vol. 9056, pp. 287–314. Springer, Heidelberg (2015). https://doi.org/10.1007/978-3-662-46800-5_12

29. Wu, S., Wu, H., Huang, T., Wang, M., Wu, W.: Leaked-state-forgery attack against the authenticated encryption algorithm ALE. In: Sako, K., Sarkar, P. (eds.) ASIACRYPT 2013. LNCS, vol. 8269, pp. 377–404. Springer, Heidelberg (2013). https://doi.org/10.1007/978-3-642-42033-7_20

30. Yang, G., Zhu, B., Suder, V., Aagaard, M.D., Gong, G.: The Simeck family of lightweight block ciphers. In: Güneysu, T., Handschuh, H. (eds.) CHES 2015. LNCS, vol. 9293, pp. 307–329. Springer, Heidelberg (2015). https://doi.org/10.1007/978-3-662-48324-4_16

Finding Integral Distinguishers with Ease

Zahra Eskandari[1], Andreas Brasen Kidmose[2], Stefan Kölbl[2,3(✉)],
and Tyge Tiessen[2]

[1] Department of Computer Engineering, Ferdowsi University of Mashhad,
Mashhad, Iran
`zahra.eskandari@mail.um.ac.ir`
[2] DTU Compute, Technical University of Denmark, Kongens Lyngby, Denmark
`abki@dtu.dk, stek@mailbox.org`
[3] Cybercrypt, Copenhagen, Denmark

Abstract. The division property method is a technique to determine
integral distinguishers on block ciphers. While the complexity of finding
these distinguishers is higher, it has recently been shown that MILP and
SAT solvers can efficiently find such distinguishers. In this paper, we provide a framework to automatically find those distinguishers which solely
requires a description of the cryptographic primitive. We demonstrate
that by finding integral distinguishers for 30 primitives with different
design strategies.

We provide several new or improved bit-based division property distinguishers for CHACHA, CHASKEY, DES, GIFT, LBLOCK, MANTIS,
QARMA, ROADRUNNER, SALSA and SM4. Furthermore, we present an
algorithm to find distinguishers with lower data complexity more efficiently.

Keywords: Integral attacks · Division property · Tool

1 Introduction

Block ciphers, stream ciphers, and hash functions are the fundamental symmetric cryptographic primitives that are at the base of almost all cryptographic
protocols. One of the most successful set of techniques to evaluate their security
are techniques based on higher-order derivatives.

Higher-order derivatives were first considered in the context of symmetric
cryptography by Lai [Lai94] and shown by Knudsen [Knu95] to attack weaknesses not covered by differential cryptanalysis, and successfully used to break
a cipher design [JK97]. A higher-order derivative in the context of cryptography
is the discrete equivalent of higher-order derivatives of multivariate continuous
functions. The cryptographic primitive can be seen as a vectorial Boolean function where a higher-order derivative evaluates this function at a given point with
respect to some directions/subspace. Such a derivative can for example be used
to find the coefficients of the monomials of the algebraic normal form (ANF) of
a cryptographic primitive.

© Springer Nature Switzerland AG 2019
C. Cid and M. J. Jacobson, Jr. (Eds.): SAC 2018, LNCS 11349, pp. 115–138, 2019.
https://doi.org/10.1007/978-3-030-10970-7_6

An important category of higher-order attacks is integral cryptanalysis. This type of cryptanalysis appeared first in the Square attack [DKR97a], and was later generalised to be apply to other ciphers as well ([KW02, BS10]). In integral cryptanalysis, the goal is to find a set of input bits and a set of output bits, such that when taking the sum over a set of input messages taking all possible values in the selected input bits and arbitrary but constant values in the other input bits, the sum will be balanced in the selected output bits. This can be described as a higher-order derivative that can be taken at any point and evaluates to zero in the specified output bits.

Originally such property was derived using arguments based on the structure of the primitive but Yosuke Todo demonstrated in his EUROCRYPT 2015 paper [Tod15b] a novel method to derive integral distinguishers using the so-called division property formalism whose effectiveness he demonstrated with an attack on full-round Misty [Tod15a]. The technique originally being used on words of at least four bits, has since been applied to bit-based designs as well, albeit at a higher computational cost [TM16].

Another type of higher-order attacks are so-called cube attacks [Vie07, DS09]. In these attacks the cryptographic primitive is viewed as a vectorial Boolean function in both public and secret input bits. By finding coefficients of terms in the public bits that are linear in the secret bits, it is possible to derive a set of linear equations that we can solve to extract the secret input bits. This technique has successfully been applied to stream ciphers and hash functions [DS11, DMP+15].

Contributions. This paper presents a new framework to analyse the security of cryptographic primitives with respect to the bit-based division property by providing a simple way to find distinguishers and testing the number of rounds required for no such distinguisher to exist. We take a look at how finding division property distinguishers can be efficiently automated. To this end, we elaborate how the bit-based division property can be mapped to conditions on the state bits which in turn maps easily to a SAT problem.

Our tool focuses especially on the usability and allows to describe the cryptographic primitives at a high level by providing commonly used operations like S-boxes, linear layers, bit-permutations or modular addition. This completely removes the need of constructing any domain specific models like previous search strategies [XZBL16, SWW17, ZR17].

In order to demonstrate the usability of our tool we implemented 30 primitives following different design strategies. We then use our tool to find several new integral distinguishers, provide a bound for which number no such distinguishers exist in our model and also evaluate for which design strategies our approach becomes computationally infeasible.

In particular we find the following new results:

– We provide the first bit-based integral distinguishers for the permutations used in CHACHA (6 rounds), CHASKEY (4 rounds) and SALSA (6 rounds). We further show that for one more round no distinguisher of this type exists.

- For DES we show that by using the bit-based division property we can improve upon the word-based division property distinguishers by Todo [Tod15b] and add one round. We also show that for 8 rounds no such distinguishers exist.
- We present the first integral distinguisher for both MANTIS (3 forward, 2 backward rounds) and several variants of QARMA (2 forward, 2 backward rounds).
- For the SM4 block cipher we can show a distinguisher for 12 rounds and that no bit-based division property distinguisher exists for 13 rounds. This improves the best previously known integral distinguisher by 4 rounds [LJH+07].
- We find a distinguisher for 17 rounds of LBLOCK, which improves the best previously known results by one round [XZBL16].
- We present 9-round distinguishers for GIFT-64 which improve upon the data complexity of the distinguishers provided by the designers [BPP+17].
- For ROADRUNNER we are able to extend the distinguishers found by the authors [BS15] by one additional round.

For several other primitives we provide a bound at which no bit-based division property distinguishers exists in our model. Furthermore, we present an efficient algorithm to find distinguishers with reduced data complexity by only covering the search space which can actually lead to distinguishers.

Software. We place the tool developed for this paper into the public domain and it is available at https://github.com/kste/solvatore.

Related Work. The division property has been applied to a large variety of cryptographic primitives and has led to significant improvements [Tod15b, Tod15a] over classical integral attacks in some cases. With the extension of the division property to bit-based designs [TM16] the technique can be applied to a larger class of cryptographic primitives. However finding distinguishers with this approach is a difficult task and requires a lot of effort.

The first automated approach for finding bit-based division property distinguishers was presented in [SWW16] and is based on reducing the problem to mixed integer linear programming (MILP). This simplifies the search for distinguishers and allows to apply the bit-based division property to a larger class of cryptographic primitives. Another automated approach based on constraint programming has been proposed in [SGL+17] to find integral distinguishers for PRESENT. In the paper the authors show that this approach can have a better performance than the MILP based technique. The search for ARX and word-based division property has been dealt with in [SWW17] by using SAT resp. SMT solvers.

2   Division Property and Division Trails

The methodology of division properties was devised by Yosuke Todo in his EUROCRYPT 2015 paper [Tod15b]. We elaborate this methodology here in the setting where the words are single bits, i.e., when applied as bit-based division property. While using the original formalism, we will look at it from a slightly different angle to simplify the discussion. For the division property over larger word sizes, we refer to the original paper.

2.1   Background

The formalism of division properties belongs to the family of attack vectors collectively named integral cryptanalysis. The goal of integral cryptanalytic techniques is to find a set of input texts such that the sum of the resulting output texts evaluates to zero in some of the bits. If such a property can be found it directly yields a distinguisher which often can be turned into a key recovery attack.

The most common sets of input texts that are used are those that are equal in some bit positions and take all possible combination of values in the remaining bit positions. The first attack that successfully used this attack vector is the Square attack [DKR97a] on the block cipher Square that is equally applicable to the Advanced Encryption Standard (AES).

There are two main methods that are used to derive an integral distinguisher: structural properties and algebraic degree bounds. In the Square attack and subsequent generalizations [BS01] the integral property could be derived by only looking at structural properties of the cipher such as the SPN or Feistel structure without taking much of the cipher details into consideration (such as concrete S-box, concrete linear layer).

Later it was recognised that these kinds of integral distinguishers correspond to discrete derivatives [Lai94] where the derivative is taken with respect to the active input bits, i.e., those that are varied. As such the structural techniques are a way to determine output bits whose polynomial representations do not contain terms that include all active input bits simultaneously. Taking the derivative with respect to these active input bits will thus necessarily evaluate to zero in these output bits.

The second major technique that is used to derive integral distinguishers uses this view of integral distinguishers as derivatives. By determining upper bounds on the algebraic degree of the polynomials of the output bits, we can determine that derivatives of sufficient degree have to evaluate to zero. Similar to the structural method, the methods used to bound the degree usually ignore large parts of the implementation details, for example by just looking at the degree of rounds and multiplying these.

The division property is an improvement with respect to this situation as it manages to take more implementation details of the cipher into consideration. The downside to this is an increased cost of finding the distinguishers.

2.2   Formalism of Bit-Based Division Properties

In the bit-based division property methodology, the goal is to find, given a set of chosen active input bits, those output bits whose polynomial representations do not contain terms that feature all of these active bits simultaneously. While this could principally be done by simply calculating the exact polynomial representations of the output bits, this is computationally infeasible in all but toy examples. With division properties we use an approximation instead that guarantees to only find valid distinguishers but might fail to find all distinguishers.

In this approximation, we continually track which bits of the state would need to be multiplied to generate a bit whose polynomial representation can contain terms of all active bits. Let us consider an initial state of four bits (x_0, x_1, x_2, x_3) where we activate bits x_1 and x_2, i.e., we are interested in which state bits we would need to multiply to create a term that contains both bits. For this initial state the minimal way of generating such a term is by multiplying those two bits directly. We write this combination as the choice vector $(0, 1, 1, 0)$.[1]

If we now add x_1 to x_3, we get the new state $(x_0, x_1, x_2, x_3 + x_1)$. Now we can generate a term that contains both x_1 and x_2 in two different minimal ways: first again by multiplying the second and third bit or by multiplying the third and the last bit. These correspond to the choice vectors $(0, 1, 1, 0)$ and $(0, 0, 1, 1)$.[2] The only original choice vector $(0, 1, 1, 0)$ has thus been transformed to two choice vectors by the application of the addition.

If we now applied another operation to this state, each of the choice vectors is transformed to other minimal choice vectors, and by iterating this process a tree of minimal choice vectors is spanned whose final nodes are the minimal choice vectors of output bits whose multiplication can create a term that contains all active input bits.

To determine whether a minimal choice vector can be reached from the initial choice vector of active bits, we need to determine whether a path exists in this tree from the initial choice vector to the output choice vector. We will refer to such path as a *division trail*. In particular, to determine whether a specific output bit is zero when evaluating the derivative with respect to the active bits, we need to determine whether the choice vector that only chooses this output bit is reachable. If it is not reachable, we know that this output bit cannot have terms in its polynomial representation that contain all active bits simultaneously and thus the derivative has to evaluate to zero. Should the choice vector be reachable though, nothing definite can be said about the derivative.

2.3 Rules of Choice Vector Propagation

To trace a division trail of minimal choice vectors, we need to know how these minimal choice vectors of state bits are transformed to new choice vectors under the application of operations. In the following we will shortly discuss the application of XOR, AND, bit-copying and S-boxes. As the influence of the operations is local, it is sufficient to restrict the discussion to those bits involved in the operation.

---

[1] In the original paper, this was written slightly more verbosely as $\mathcal{D}^4_{(0,1,1,0)}$.

[2] In the original paper, this would be written as $\mathcal{D}^4_{(0,1,1,0),(0,0,1,1)}$.

Bit-Copying. Let us take a look at the scenario where we have two state bits, and the value of the first bit is copied to the second bit. There are four possible original choice vectors: $(0,0)$, $(1,0)$, $(0,1)$, and $(1,1)$. The first choice vector implies that to generate a term that can contain all active bits, we don't need to multiply any of the two bits. So clearly we still do not need to multiply any of the bits after copying the first bit onto the second, leading to the transition $(0,0) \to (0,0)$.

In the case of $(1,0)$, we need the first bit in the product to generate a term with all active bits but the second one is not required. Thus after copying, we can choose either the first or the second bit (both would also be possible but not minimal). We thus have the two transitions: $(1,0) \to (1,0)$ and $(1,0) \to (0,1)$.

Now in the case of $(1,0)$ and $(1,1)$, the second bit is needed in the product to create a term with all active bits. As it is copied over, it is no longer possible after copying to create this term and thus no valid transitions exist.

XOR. Now for the case where there are two state bits and the first is XORed onto the second. Again we have to look at the four cases $(0,0)$, $(1,0)$, $(0,1)$, and $(1,1)$. As with bit copying, in the case of $(0,0)$, the bits are not necessary in the product, so they are not necessary after the addition as well. This leads to the transition $(0,0) \to (0,0)$.

In the case of $(1,0)$, the first bit value is needed in the product. After the addition, the bit value is also present as part of the sum in the second bit. We can thus either choose the first or the second bit in the product, leading to the transitions $(1,0) \to (1,0)$ and $(1,0) \to (0,1)$.

When we have the case $(0,1)$, the second bit value is needed in the product. As it is still only present in the second bit after the addition, the only valid transition here is $(0,1) \to (0,1)$.

Finally, in the case of $(1,1)$, the product of both bits is needed to create a term with all active bits. Although the second bit contains both original bit values after the addition, it only does so as a sum while we need the product of both. Thus also after the addition, we have to choose both bits, leading to the transition $(1,1) \to (1,1)$.

AND. If we now have again two state bits and we multiply the first onto the second, the situation is analogous to the case of the XOR except if the choice vector before the multiplication is $(1,1)$. In this case the product of both bit values is needed to create a term of all active bits. As the multiplication creates exactly this product in the second bit, the only minimal transition here is $(1,1) \to (0,1)$.

S-Boxes. The easiest way to see how choice vectors are transformed by an S-box is to look at the polynomial representation of the S-box, i.e., the algebraic normal form (ANF). It is tedious but straightforward to deduce the valid output choice vectors for a given input choice vector using the ANF. It can hence be easily automated and we only need to do this once for an S-box.

3   Solvatore - Automated Finding of Integral Properties

Finding integral distinguishers using division properties is a difficult task. Especially for bit-based designs the analysis often requires extensive manual work which is prone to errors. Automatic tools can be very useful and simplify the analysis of cryptographic primitives, allowing us to explore a larger set of attack vectors. On the other hand they can also be very useful in the design process of cryptographic primitives, to optimise parameters and quickly test different design strategies.

In the following, we present our automated tool SOLVATORE, which simplifies the search for bit-based division property distinguishers by providing a framework for implementing a large variety of cryptographic primitives. One of the main focuses of the framework is to not only automate finding the bit-based division property distinguishers, as done in previous work [XZBL16,SWLW16,SWW17], but also to completely abstract away the need for dealing with generating models for the primitives or requiring any domain specific knowledge. This makes it much simpler and less error-prone compared to other approaches to add new primitives to the framework and in general it is far easier to implement a primitive in our tool than writing a standard C implementation as many details can be omitted.

Currently our framework supports the following operations to construct cryptographic primitives:

- Bit operations: bit-copying, **and**, and **xor**.
- Arbitrary S-boxes.
- Linear layers using matrix multiplication over arbitrary fields.
- Modular Addition.
- Bit-permutations.
- Generic cell permutations for ShiftRows or MIDORI-like constructions.

As an example the full description of PRESENT is given in Appendix A which only requires to define the S-box, bit-permutation and on which bits those are applied. In order to analyse the security of PRESENT against the bit-based division property our tool provides functions for checking whether an output bit is balanced for a given choice vector.

In the following we show how we can reduce the problem of finding a division trail to a satisfiability problem. For this we have to construct a Boolean formula which is satisfiable if and only if it forms a valid division trail.

3.1   Modeling Division Property Propagation with SAT

The *Boolean satisfiability problem* (SAT) is a well known problem from computer science. The problem is to decide whether there exists an assignment of variables in a Boolean formula in conjunctive normal form (CNF) such that the formula evaluates to **true**. While the problem is known to be NP-complete, the SAT instances we will construct here are very structured and can often be solved quickly in practice by modern SAT solvers. In the following we show how to reduce the problem of finding division trails to a SAT problem and how this can be useful in the cryptanalysis of cryptographic primitives.

First, we introduce a variable for each bit of the choice vector $S^i = (s_0, \ldots, s_{n-1})$ after the ith operation applied to the state where n is the size of the state. The next step is to define how the choice vector can propagate through different Boolean functions which occur in the round functions of cryptographic primitives. The rules for this have been explained in Subsect. 2.3 and have also been studied in [Tod15a, Tod15b]. We therefore focus here on how we can construct a Boolean formula in CNF which is SAT if and only if the assignment of the variables forms a valid transition of choice vectors.

Bit-Copying. The **copy** operation copies a bit a to an output bit b, and all valid transitions of choice vectors are given by

$$\mathbf{copy}(a_{\text{old}}, b_{\text{old}}) \rightarrow \{(a_{\text{new}}, b_{\text{new}})\}$$
$$\mathbf{copy}(0, 0) \mapsto \{(0, 0)\}$$
$$\mathbf{copy}(1, 0) \mapsto \{(1, 0), (0, 1)\}.$$

The set of clauses C_{copy} which form a Boolean formula which is SAT iff $(a_{\text{old}}, b_{\text{old}}) \xrightarrow{\text{copy}} (a_{\text{new}}, b_{\text{new}})$ is given by

$$C_{\text{copy}} = \{(\neg b_{\text{old}}), (\neg a_{\text{old}} \vee b_{\text{new}} \vee a_{\text{new}}), (a_{\text{old}} \vee \neg b_{\text{new}}), \\ (a_{\text{old}} \vee \neg a_{\text{new}}), (\neg a_{\text{new}} \vee \neg b_{\text{new}})\}. \tag{1}$$

And. The **and** operation corresponds to the result of $a \wedge b \rightarrow b$. The valid transitions are given by

$$\mathbf{and}(a_{\text{old}}, b_{\text{old}}) \rightarrow \{(a_{\text{new}}, b_{\text{new}})\}$$
$$\mathbf{and}(0, 0) \mapsto \{(0, 0)\}$$
$$\mathbf{and}(0, 1) \mapsto \{(0, 1)\}$$
$$\mathbf{and}(1, 0) \mapsto \{(1, 0), (0, 1)\}$$
$$\mathbf{and}(1, 1) \mapsto \{(0, 1)\}.$$

Just as for the **copy** operation, translating this to a SAT sentence is straightforward and gives the following set of clauses

$$C_{\text{and}} = \{(a_{\text{old}} \vee \neg a_{\text{new}}), (\neg b_{\text{old}} \vee b_{\text{new}}), (\neg b_{\text{new}} \vee \neg a_{\text{new}}), \\ (\neg a_{\text{old}} \vee b_{\text{new}} \vee a_{\text{new}}), (a_{\text{old}} \vee b_{\text{old}} \vee \neg b_{\text{new}}). \tag{2}$$

Xor. The **xor** operation corresponds to the result of $a \oplus b \rightarrow b$. The valid transitions are given by

$$\mathbf{xor}(a_{\text{old}}, b_{\text{old}}) \rightarrow \{(a_{\text{new}}, b_{\text{new}})\}$$
$$\mathbf{xor}(0,0) \mapsto \{(0,0)\}$$
$$\mathbf{xor}(0,1) \mapsto \{(0,1)\}$$
$$\mathbf{xor}(1,0) \mapsto \{(1,0),(0,1)\}$$
$$\mathbf{xor}(1,1) \mapsto \{(1,1)\}$$

which corresponds to the following clauses

$$C_{\text{xor}} = \{(a_{\text{old}} \vee \neg a_{\text{new}}), (\neg b_{\text{old}} \vee b_{\text{new}}), (b_{\text{old}} \vee \neg b_{\text{new}} \vee \neg a_{\text{new}}),$$
$$(\neg a_{\text{old}} \vee a_{\text{new}} \vee b_{\text{new}}), (b_{\text{old}} \vee a_{\text{old}} \vee \neg b_{\text{new}}), \tag{3}$$
$$(\neg b_{\text{old}} \vee \neg a_{\text{old}} \vee a_{\text{new}}).$$

S-boxes. As described in Subsect. 2.3, the transition rules for S-boxes can easily be deduced automatically. The rules create a truth table for involved variables which can be transformed to a CNF using standard methods.

Linear Layers. Many popular designs, like the AES, use a complex linear layer in order to get good diffusion. These linear layers are often represented as $d \times d$ matrices over some field $\mathbb{F}_2^k$. In order to model the trail propagation we can represent these transformations as $kd \times kd$ matrices over $\mathbb{F}_2$, which then can be decomposed into the basic **copy** and **xor** operations.

In order to simplify the description of such linear layers in our tool, we implemented this decomposition and it is only required to provide the irreducible polynomial for the field $\mathbb{F}_2^k$ and the matrix. From the irreducible polynomial it is possible to deduce the $k \times k$ matrices that represent the elements of $\mathbb{F}_k$ as matrices over $\mathbb{F}_2$. Substituting these matrices in the original matrix over $\mathbb{F}_k$ now creates the $nk \times nk$ binary matrix.

Modular Addition. Modular addition is used as a non-linear component in ARX-ciphers like HIGHT, LEA, and SPECK. We can use the same approach as [SWLW16] to decompose the modular addition into **xor** and **and**. Let z, x, y be n bit-variables with z_i, y_i, x_i as the ith bits, counting from the least significant bit, and $z = x \boxplus y$. The modular addition modulo 2^n is given by:

$$z_i = x_i \oplus y_i \oplus c_i$$
where
$$c_i = x_{i-1}y_{i-1} \oplus (x_{i-1} \oplus y_{i-1})c_{i-1} \text{ for } i > 0$$
$$c_0 = 0$$

So far we have assumed that both x, y are variables, however in some ciphers one of them is a constant, e.g. a round key. Since we can ignore **xor** and **and** with a constant we get the following expressions.

$$z_i = x_i \oplus c_i$$

where

$$c_i = x_{i-1} \oplus x_{i-1}c_{i-1} \text{ for } i > 0$$
$$c_0 = 0$$

Similar, if we want to find a distinguisher on a cipher like Bel-T or the inverse of an ARX-cipher we also need modular subtraction. To do modular subtraction we can use the fact that

$$x \boxminus y = x \boxplus (-y) = x \boxplus (2^n - y) = x \boxplus ((2^n - 1) - y) \boxplus 1 = x \boxplus \overline{y} \boxplus 1 \quad (4)$$

Since the NOT operation has no effect on whether a bit is balanced or not we can omit it to get $x \boxminus y = x \boxplus y \boxplus 1$. This means that we can do modular subtraction with one modular addition and one constant addition.

3.2 Finding Integral Distinguishers

In order to find useful integral properties of a cipher, we have to propagate an initial choice vector S^0 and check whether it is impossible to reach certain choice vectors S^r after r rounds. If we can show that an output choice vector that is everywhere zero except for a single **1** in one bit is unreachable, we know that this bit has to be balanced.

In particular we are often interested in whether any bit in the output will be balanced. This corresponds to showing that at least one of the vectors in the set

$$S^r \in \{w \in \mathbb{F}_2^n \mid \mathbf{hw}(w) = 1\}. \quad (5)$$

is unreachable, where $\mathbf{hw}(x)$ is the Hamming weight of the vector.

Contrarily, we can also use this approach to show the absence of a bit-based division property distinguisher in our model. Checking all possible options for the starting choice vector would be (for most primitives) computationally infeasible. Fortunately it is sufficient to show for all starting choice vectors in the set

$$S^0 \in \{w \in \mathbb{F}_2^n \mid \mathbf{hw}(w) = n - 1\}. \quad (6)$$

that all choice vectors in the set in Eq. 5 are reachable. This works because the balancedness of the output bits is preserved when we exchange the input choice vector with any vector greater than it (with respect to the above ordering).

We will use the following notation to simplify the description of the distinguishers found later in the paper. The set of *active* bits will be denoted as

$$A = \{i \mid S_i^0 = 1, \ i = 0, \dots, n - 1\} \quad (7)$$

and correspondingly the set of *constant* bits as

$$\overline{A} = \overline{\{i \mid S_i^0 = 1, \ i = 0, \dots, n - 1\}} = \{i \mid S_i^0 = 0, \ i = 0, \dots, n - 1\}. \quad (8)$$

The set of bits which are balanced at the output is denoted as B. We can now describe a distinguisher, for a function f, as

$$A \xrightarrow{f} B. \tag{9}$$

If a valid division trail from A to B exists we will also use the more compact notation $\mathbf{DP}(A) = B$ if the function is clear from context.

Note that while the notation for the set of active bits at the input and the balanced bits at the output looks very similar it conveys a very different meaning in the context of the division property. For a range of bits $s_i, s_{i+1}, \ldots, s_j$ we will use the notation s_{i-j}.

4   Distinguishers and Bounds

We implemented a variety of cryptographic primitives in SOLVATORE to demonstrate the versatility of our tool and the ease of adding primitives with different design principles.

- **SPN**: GIFT, LED, MIDORI, PHOTON, PRESENT, SKINNY, SPONGENT
- **ARX**: BELT, CHACHA, CHASKEY, LEA, HIGHT, SALSA, SPARX, SPECK
- **Feistel**: DES, LBLOCK, MISTY, ROADRUNNER, SKIPJACK, SM4, TWINE
- **Reflection**: MANTIS, PRINCE, QARMA
- **Bit-sliced**: ASCON, RECTANGLE
- **LFSR-based**: BIVIUM, TRIVIUM, KREYVIUM

We will first go over the general methodology and after that over the results on the different primitive classes obtained using SOLVATORE. This includes both bit-based division property distinguishers and finding the number of rounds at which no such distinguisher exists anymore. All results have been obtained on an Intel Core i7-4770S running Ubuntu 17.10 using the Python interface to CryptoMiniSat 5.0.1. Several examples for distinguishers we found are given in Appendix B.

4.1   Methodology

Finding a Bound. As a first step we try to find the number of rounds r^* at which no bit-based division property distinguisher in our model exists. This is done by testing all set of active bits of type

$$A_j = \{i \mid i \in \mathbb{Z}_n \setminus j\} \quad \forall j \in \mathbb{Z}_n. \tag{10}$$

This corresponds to all vectors where a single bit is constant. If for all possible choices the set of balanced bits $B_j = \mathbf{DP}(A_j)$ is empty we know that no such distinguisher exists for r^* rounds.

Reducing Data Complexity. In order to reduce the data complexity for the distinguishers covering the most rounds we use different strategies. The naive approach would be to increase the number of constant bits c, try out all possible combinations and check whether the resulting set of balanced bits B is not empty. This might work in some cases however the complexity increases very quickly as we have to test all $\binom{n}{c}$ possible choices.

This can be improved by only testing those combinations of constant bits which can actually lead to non-empty sets B. First, we compute the set of constant bits

$$G_1 = \{j \mid \mathbf{DP}(A_j) = B_j \wedge (|B_j| > 0) \quad \forall j \in \mathbb{Z}_n\} \tag{11}$$

for which at least one of the bits after r rounds is balanced, similar to the case where we try to find the bound. Next, we look at all combinations of two elements of G_1 which share at least one balanced bit

$$G_2 = \{\{i,j\} \mid (i \neq j) \wedge (|\mathbf{DP}(A_i) \cap \mathbf{DP}(A_j)|) > 0, \forall i, j \in G_1\}. \tag{12}$$

We can continue the last step in a similar way until G_i is empty by testing all combinations of the sets of bits in G_i repeatedly. Note that in the next step we would not have single indices but sets of indices and we therefore look whether the union of these sets of constant bits lead to a non-empty set B. Another advantage of this approach is that we only need to test those bits for the balancedness property which were already balanced in the last iteration.

In each step the elements in G_i are a set of constant bits which will have at least one balanced bit in the output after r rounds. This approach improves the complexity of finding distinguishers with lower data complexity significantly, but often it is still computationally infeasible to find an optimal distinguisher. For more structured designs it often helps to look at the word level and only look at maximizing the number of constant words as there are fewer combinations which we have to check.

4.2   SPN

We will use 9 rounds of SPONGENT-88 as an example to show the benefits of the optimised search for a distinguisher with lower data complexity. In order to estimate the complexity we will count for how many choice vectors we would have to compute the set of balanced bits B. Using the optimised search we only have to test 1819 choice vectors (see Table 1) to find distinguishers with up to 4

Table 1. Results from the optimised search for SPONGENT-88. Combinations are the number of pairs (i,j) in the sets G_i which share bits in their corresponding sets B_i and B_j.

	G_1	G_2	G_3	G_4	G_5
Size ($|G|$)	43	40	25	1	0
Combinations	878	643	234	0	-

Table 2. Overview of our distinguishers and bounds for SPN-based designs.

Cipher	Rounds	Active bits	Balanced bits
GIFT-64	9	61	5
	9	62	11
	9	63	30
	10	No distinguisher	
GIFT-128	11	127	32
	12	No distinguisher	
LED	5	60	64
	8	No distinguisher	
MIDORI-64	6	48	16
	8	No distinguisher	
MIDORI-128	5	104	128
PHOTON-100	4	12	100
	5	99	100
PHOTON-144	4	24	144
PHOTON-196	4	28	196
PHOTON-256	4	32	256
PRESENT	9	60	1
	10	No distinguisher	
SKINNY-64	10	48	9
	11	No distinguisher	
SPONGENT-88	9	84	3
	9	87	54
	10	No distinguisher	
SPONGENT-136	10	132	8
	10	135	93
	11	No distinguisher	
SPONGENT-176	12	No distinguisher	

constant bits and exclude any distinguisher with 5 constant bits. Using the naive approach we would have to test 679120 choice vectors to find all distinguishers up to 4 bits and check $\binom{128}{5}$ combinations to exclude the existence of any further distinguishers.

For SKINNY-64 we can find a distinguisher with the same data complexity as the one given by the authors [BJK+16] with one additional balanced bit and show that no distinguishers exist for 11 rounds.

For GIFT-64 we use our optimal approach and no better distinguisher exists. We can find a 9-round distinguisher similar to the one by the authors [BPP+17], but also distinguishers with a lower data complexity. For GIFT-128 finding distinguishers takes significantly longer and we were only able to find a distinguisher with high data complexity similar to the original one.

For several variants of PHOTON we can find distinguishers with low data complexity by searching for combinations of constant words. However for more rounds the search time increases quickly and we are not able to improve any results. The complex linear layer generates a large number of clauses which seems to be the main limiting reason (Table 2).

4.3   ARX

First we look at the permutation used in the CHASKEY MAC [MMH+14]. We can find a distinguisher for 3 rounds with only two constant words, one with high complexity for 4 rounds and show that no bit-based division property distinguishers for 5 rounds exist. This confirms the claim by the authors that CHASKEY is likely to resist this type of attacks. Considering the construction used for the MAC it seems infeasible to mount an attack based on the 4-round distinguisher.

The large state of SALSA and CHACHA make it difficult to adopt our approach for reducing the data complexity. We therefore keep whole words constants and try to find the maximum number. For 6 rounds of SALSA the only distinguisher which exists keeps the first word constant and the one for CHACHA has only a single constant bit. In both cases no distinguisher exists for 7 rounds. On the actual mode in which SALSA and CHACHA are used as a stream cipher we can only control the 64-bit nonce in a single block. In this setting there are no bit-based division property distinguisher for 4 rounds of SALSA and 2 rounds of CHACHA.

We can also confirm the results from [SWW17] using our optimal search algorithm for HIGHT, LEA and SPECK. We noticed that SOLVATORE performs significantly better for finding these distinguishers even though we use the same SAT solver. It only took us 28/195/51 seconds compared to 15/30/6 minutes for finding the optimal distinguishers for HIGHT/LEA/SPECK. This gap could be explained by the slightly different model resp. using a better search strategy.

BEL-T is a block cipher which has been adopted as a national standard in the Republic of Belarus and combines S-boxes with modular addition. There is only a very limited amount of cryptanalysis available [JP15] (also provides an English description of the algorithm). We provide the first analysis with respect to integral attacks for BEL-T and can find a fairly efficient distinguisher for 2 rounds while showing that none exist for 3 rounds.

In the case of SPARX we can confirm the results by the authors [DPU+16]. The full summary of the results for ARX-based primitives can also be found in Table 3.

4.4   Feistel

For DES we improve the best bit-based division property distinguisher [Tod15b] by one round. The original distinguisher for DES also uses the division property but only word-based which makes this improvement possible.

One of the most successful applications of the division property is the full break of MISTY [Tod17]. It is also based on the analysis on the word level so one might suspect that it can be improved by looking at the bit-based division property. We tried to find the same distinguishers as in the original attack automatically however the complexity seems too high without further optimizations. We could only find a distinguisher for 3 rounds.

Table 3. Overview of our distinguishers and bounds for ARX-based designs.

Cipher	Rounds	Active bits	Balanced bits
CHACHA	6	511	138
	7	No distinguisher	
CHASKEY	3	64	6
	4	127	5
	5	No distinguisher	
LEA	8	126	16
	8	118	1
	9	No distinguisher	
HIGHT	18	63	2
	19	No distinguisher	
SALSA	6	480	129
	7	No distinguisher	
SPECK-32	6	31	1
	7	No distinguisher	
SPECK-48	6	45	1
	7	No distinguisher	
SPECK-64	6	61	1
	7	No distinguisher	
SPECK-96	6	93	1
	7	No distinguisher	
SPECK-128	6	125	1
	7	No distinguisher	
BELT	2	45	5
	3	No distinguisher	
SPARX-64	3	32	32
	4	No distinguisher	
SPARX-128	4	96	64
	5	No distinguisher	

The best integral distinguisher on SM4 covers 8 rounds [LJH+07]. By using the bit-based division property we can improve those distinguishers to 12 rounds, although at a high complexity. We further can show that no such distinguishers exist for 13 rounds.

In the case of LBLOCK we are able to extend the distinguisher found with MILP [XZBL16] by one additional round and for ROADRUNNER we can find a 5-round distinguisher which also covers one more round than the best known distinguisher [BS15].

Table 4. Overview of our distinguishers and bounds for Feistel networks.

Cipher	Rounds	Active bits	Balanced bits
DES	7	60	8
	8	No distinguisher	
LBLOCK	17	63	4
	18	No distinguisher	
MISTY	3	32	64
ROADRUNNER	5	58	8
	6	No distinguisher	
SKIPJACK	$19(A^8B^8A^3)$	47	16
	$20(A^8B^8A^4)$	56	8
	$21(A^8B^8A^5)$	No distinguisher	
SIMON32	14	31	16
	15	No distinguisher	
SIMON48	16	47	24
	17	No distinguisher	
SIMON64	18	63	22
	19	No distinguisher	
SIMON96	22	95	5
	23	No distinguisher	
SIMON128	26	127	3
	27	No distinguisher	
SIMECK32	15	31	7
	16	No distinguisher	
SIMECK48	18	47	5
	19	No distinguisher	
SIMECK64	21	63	5
	22	No distinguisher	
SM4	12	126	32
	13	No distinguisher	
TWINE	16	63	32
	17	No distinguisher	

For all variants of SIMON and SIMECK we can reproduce the results from [XZBL16], show that these have the lowest data complexity and that there are no distinguisher in our model for more rounds (Table 4).

4.5   Reflection

Block ciphers based on the reflection design strategy, introduced by PRINCE, are a popular choice for low-latency designs. We will denote the number of rounds as $f + b$, where f are the rounds before the middle layer and b the rounds after the middle layer (see Table 5).

Table 5. Results on reflection ciphers.

Cipher	Rounds	Active bits	Balanced bits
MANTIS	2 + 2	12	16
	3 + 2	32	16
	3 + 3	No distinguisher	
PRINCE	1 + 1	12	64
	2 + 1	32	64
	1 + 2	32	64
	2 + 2	No distinguisher	
QARMA-64/σ_0	2 + 2	48	16
	3 + 3	No distinguisher	
QARMA-64/σ_1	2 + 2	52	64
	3 + 3	No distinguisher	
QARMA-64/σ_2	2 + 2	52	64
	3 + 3	No distinguisher	
QARMA-128/σ_0	2 + 2	96	128
	3 + 3	No distinguisher	
QARMA-128/σ_1	2 + 2	96	128
	3 + 3	No distinguisher	
QARMA-128/σ_2	2 + 2	120	128
	3 + 3	No distinguisher	

Table 6. Results on bit-sliced ciphers.

Cipher	Rounds	Active bits	Balanced bits
ASCON	5	16	320
RECTANGLE	9	60	
	10	No Distinguisher	

For PRINCE we can find a bit-based division property distinguisher with the same complexity as the best higher-order differential given in [RR16] and show that for one additional round none exist. Very similar distinguisher also exist for MANTIS with the only difference being that one can extend those by one round in forward and backwards direction. The distinguishers for QARMA can cover a similar number of rounds although at a much higher data complexity.

4.6 Bit-Sliced

In this category we look at two LS-designs (see Table 6). The permutation used in the authenticated encryption scheme ASCON and the block cipher RECTANGLE. For ASCON we can improve the data complexity of the 5 round distin-

Table 7. Results on LFSR-based stream ciphers.

Cipher	Rounds	Active bits	Balanced bits
BIVIUM	681	79	1
TRIVIUM	707	79	1
KREYVIUM	713	127	1

guisher [Tod15b] by a factor of 4, however for more rounds we could not improve any results as the computations takes too long. For RECTANGLE we are able to show that no distinguisher exists for 10 rounds and find the already known 9-round distinguisher from [XZBL16].

4.7 LFSR-Based

We looked at three LFSR-based stream ciphers which share a similar structure. The active bits are taken over the choice of IV and our distinguishers here checks whether the output bit of the key stream is balanced after r rounds. It is very likely that there are more bits balanced in the state, but we can only distinguish the key stream if the resulting key stream bit is also balanced.

While we could find some distinguishers the time it takes to find a balanced output bit of the keystream quickly increases and other approaches seem to be more promising for constructing distinguisher based on the division property for this type of ciphers [TIHM17] (Table 7).

4.8 Overview

Using SOLVATORE we were able to demonstrate several new distinguishers, reduce the data complexity and show at which number of rounds a primitive becomes resistant against bit-based division property. In Fig. 1 we give an

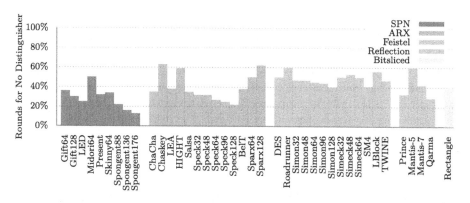

Fig. 1. Overview of the fraction of rounds required before we can show that no bit-based division property distinguishers exist in our model.

overview of the number of rounds required before no bit-based division property distinguisher exists in relation to the full number of rounds of the primitive. It can be seen that most ciphers provide a fairly large security margin against these type of attacks and also for many of these designs there are indeed better distinguishers based on other techniques like differential and linear cryptanalysis.

The performance of SOLVATORE varies a lot from the designs and for some it is not feasible to find good distinguishers. For instance we also implemented both AES and KECCAK in our tool, but we could only obtain very limited results which could not improve upon the state-of-the-art.

5   Conclusion and Future Work

In this work we presented a new framework to automatically find division property distinguishers for a large class of cryptographic primitives by reducing the problem to SAT. We also provide a cryptanalysis tool implementing this approach, providing a simple way to describe primitives, allowing both designers and cryptanalysts to evaluate cryptographic primitives against this attack vector.

Using this tool we present several new or improved bit-based division property distinguishers for CHACHA, CHASKEY, DES, GIFT, LBLOCK, MANTIS, QARMA, ROADRUNNER, SALSA and SM4.

Furthermore, we provide an improved algorithm for finding distinguisher with an optimal data complexity and show for several primitives that no bit-based division property distinguisher can exist for more rounds.

A   Implementation of Present

The following example shows how one can implement the PRESENT cipher in our framework to analyse its properties against bit-based division property attacks.

```
from cipher_description import CipherDescription

present_sbox = [0xC, 0x5, 0x6, 0xB, 0x9, 0x0, 0xA, 0xD,
                0x3, 0xE, 0xF, 0x8, 0x4, 0x7, 0x1, 0x2]
present_permutations = [\
    ['s1', 's16', 's4'], ['s2', 's32', 's8'], ['s3', 's48', 's12'],
    ['s5', 's17', 's20'],['s6', 's33', 's24'], ['s7', 's49', 's28'],
    ['s9', 's18', 's36'], ['s10', 's34', 's40'], ['s11', 's50', 's44'],
    ['s13', 's19', 's52'], ['s14', 's35', 's56'], ['s15', 's51', 's60'],
    ['s22', 's37', 's25'], ['s23', 's53', 's29'], ['s26', 's38', 's41'],
    ['s27', 's54', 's45'], ['s30', 's39', 's57'], ['s31', 's55', 's61'],
    ['s43', 's58', 's46'], ['s47', 's59', 's62']]

present = CipherDescription(64)
present.add_sbox('S-box', present_sbox)
for i in range(16):
    bits = ["s{}".format(4*i + 0),
            "s{}".format(4*i + 1),
            "s{}".format(4*i + 2),
            "s{}".format(4*i + 3)]
    present.apply_sbox('S-box', bits, bits)
for p in present_permutations:
    present.apply_permutation(p)
```

Using this description of the PRESENT block cipher we can mount our analysis. The following code checks whether no bit-based division property distinguisher exists for 10 rounds of PRESENT.

```
from itertools import combinations
from solvatore import Solvatore
from cipher_description import CipherDescription
from ciphers import present

cipher = present.present
rounds = 10

solver = Solvatore()
solver.load_cipher(cipher)
solver.set_rounds(rounds)

# Look over all combination for one non active bit
for bits in combinations(range(64), 1):
    nonactive_bits = bits
    active_bits = {i for i in range(64) if i not in nonactive_bits}

    # Find all balanced bits
    balanced_bits = []
    for i in range(cipher.state_size):
        if solver.is_bit_balanced(i, rounds, active_bits):
            balanced_bits.append(i)

    if len(balanced_bits) > 0:
        print("Found distinguisher!")
        print(active_bits, balanced_bits)
```

B   Overview of Distinguishers

In the following we list some of the new distinguishers we found.

B.1   ChaCha

$$\overline{\{0\}} \xrightarrow{\text{9-round}} \{32 - 68, 192 - 223, 352 - 415, 424 - 428\} \tag{13}$$

B.2   Chaskey

$$\overline{\{64 - 127\}} \xrightarrow{\text{3-round}} \{80 - 85\} \tag{14}$$

$$\overline{\{96\}} \xrightarrow{\text{4-round}} \{80 - 81\} \tag{15}$$

B.3   DES

$$\overline{\{50 - 52, 63\}} \xrightarrow{\text{7-round}} \{0, 3, 9, 10, 18, 19, 25, 28\} \tag{16}$$

B.4   GIFT-64

$$\overline{\{0-2\}} \xrightarrow{\text{9-round}} \{3,7,27,43,59\} \tag{17}$$

B.5   LBlock

$$\overline{\{34\}} \xrightarrow{\text{17-round}} \{2,3,30,31\} \tag{18}$$

B.6   Mantis

$$\overline{\{0-7,16-23,40-47,56-63\}} \xrightarrow{3+2 \text{ rounds}} \{2,6,10,14,18,22,26,30 \\ 34,38,42,46,50,54,58,62\} \tag{19}$$

B.7   QARMA

QARMA-64/σ_0

$$\overline{\{0-3,20-23,40-43,60-63\}} \xrightarrow{2+2 \text{ rounds}} \{1,5,9,13,17,21,25,29, \\ 33,37,41,45,49,53,57,61\} \tag{20}$$

QARMA-64/σ_1

$$\overline{\{0-3,20-23,40-43\}} \xrightarrow{2+2 \text{ rounds}} \{0-63\} \tag{21}$$

QARMA-64/σ_2

$$\overline{\{0-3,20-23,40-43\}} \xrightarrow{2+2 \text{ rounds}} \{0-63\} \tag{22}$$

QARMA-128/σ_0

$$\overline{\{0-15,32-47\}} \xrightarrow{2+2 \text{ rounds}} \{0-127\} \tag{23}$$

QARMA-128/σ_1

$$\overline{\{0-15,32-47\}} \xrightarrow{2+2 \text{ rounds}} \{0-127\} \tag{24}$$

QARMA-128/σ_2

$$\overline{\{0-7\}} \xrightarrow{2+2 \text{ rounds}} \{0-127\} \tag{25}$$

B.8   RoadRunner

$$\overline{\{0,1,8,9,16,17\}} \xrightarrow{\text{5-round}} \{32,33,40,41,48,49,56,57\} \tag{26}$$

B.9   Salsa

$$\overline{\{0-31\}} \xrightarrow{\text{6-round}} \{128-255, 295\} \tag{27}$$

B.10   SM4

$$\overline{\{96, 97\}} \xrightarrow{\text{12-round}} \{0-31\} \tag{28}$$

References

[BJK+16] Beierle, C., et al.: The SKINNY family of block ciphers and its low-latency variant MANTIS. In: Robshaw, M., Katz, J. (eds.) CRYPTO 2016. LNCS, vol. 9815, pp. 123–153. Springer, Heidelberg (2016). https://doi.org/10.1007/978-3-662-53008-5_5

[BPP+17] Banik, S., Pandey, S.K., Peyrin, T., Sasaki, Y., Sim, S.M., Todo, Y.: GIFT: a small present. In: Fischer, W., Homma, N. (eds.) CHES 2017. LNCS, vol. 10529, pp. 321–345. Springer, Cham (2017). https://doi.org/10.1007/978-3-319-66787-4_16

[BS01] Biryukov, A., Shamir, A.: Structural cryptanalysis of SASAS. In: Pfitzmann, B. (ed.) EUROCRYPT 2001. LNCS, vol. 2045, pp. 395–405. Springer, Heidelberg (2001). https://doi.org/10.1007/3-540-44987-6_24

[BS10] Biryukov, A., Shamir, A.: Structural cryptanalysis of SASAS. J. Crypt. **23**(4), 505–518 (2010)

[BS15] Baysal, A., Şahin, S.: RoadRunneR: a small and fast bitslice block cipher for low cost 8-bit processors. In: Güneysu, T., Leander, G., Moradi, A. (eds.) LightSec 2015. LNCS, vol. 9542, pp. 58–76. Springer, Cham (2016). https://doi.org/10.1007/978-3-319-29078-2_4

[DKR97a] Daemen, J., Knudsen, L., Rijmen, V.: The block cipher Square. In: Biham, E. (ed.) FSE 1997. LNCS, vol. 1267, pp. 149–165. Springer, Heidelberg (1997). https://doi.org/10.1007/BFb0052343

[DMP+15] Dinur, I., Morawiecki, P., Pieprzyk, J., Srebrny, M., Straus, M.: Cube attacks and cube-attack-like cryptanalysis on the round-reduced keccak sponge function. In: Oswald, E., Fischlin, M. (eds.) EUROCRYPT 2015. LNCS, vol. 9056, pp. 733–761. Springer, Heidelberg (2015). https://doi.org/10.1007/978-3-662-46800-5_28

[DPU+16] Dinu, D., Perrin, L., Udovenko, A., Velichkov, V., Großschädl, J., Biryukov, A.: Design strategies for ARX with provable bounds: SPARX and LAX. In: Cheon, J.H., Takagi, T. (eds.) ASIACRYPT 2016. LNCS, vol. 10031, pp. 484–513. Springer, Heidelberg (2016). https://doi.org/10.1007/978-3-662-53887-6_18

[DS09] Dinur, I., Shamir, A.: Cube attacks on tweakable black box polynomials. In: Joux, A. (ed.) EUROCRYPT 2009. LNCS, vol. 5479, pp. 278–299. Springer, Heidelberg (2009). https://doi.org/10.1007/978-3-642-01001-9_16

[DS11] Dinur, I., Shamir, A.: Breaking grain-128 with dynamic cube attacks. In: Joux, A. (ed.) FSE 2011. LNCS, vol. 6733, pp. 167–187. Springer, Heidelberg (2011). https://doi.org/10.1007/978-3-642-21702-9_10

[JK97]   Jakobsen, T., Knudsen, L.R.: The interpolation attack on block ciphers. In: Biham, E. (ed.) FSE 1997. LNCS, vol. 1267, pp. 28–40. Springer, Heidelberg (1997). https://doi.org/10.1007/BFb0052332

[JP15]   Jovanovic, P., Polian, I.: Fault-based attacks on the Bel-t block cipher family. In: DATE, pp. 601–604. ACM (2015)

[Knu95]   Knudsen, L.R.: Truncated and higher order differentials. In: Preneel, B. (ed.) FSE 1994. LNCS, vol. 1008, pp. 196–211. Springer, Heidelberg (1995). https://doi.org/10.1007/3-540-60590-8_16

[KW02]   Knudsen, L., Wagner, D.: Integral cryptanalysis. In: Daemen, J., Rijmen, V. (eds.) FSE 2002. LNCS, vol. 2365, pp. 112–127. Springer, Heidelberg (2002). https://doi.org/10.1007/3-540-45661-9_9

[Lai94]   Lai, X.: Higher order derivatives and differential cryptanalysis. In: Blahut, R.E., Costello, D.J., Maurer, U., Mittelholzer, T. (eds.) The Springer International Series in Engineering and Computer Science Communications and Information Theory, vol. 276, pp. 227–233. Springer, Boston (1994)

[LJH+07]   Liu, F., et al.: Analysis of the SMS4 block cipher. In: Pieprzyk, J., Ghodosi, H., Dawson, E. (eds.) ACISP 2007. LNCS, vol. 4586, pp. 158–170. Springer, Heidelberg (2007). https://doi.org/10.1007/978-3-540-73458-1_13

[MMH+14]   Mouha, N., Mennink, B., Van Herrewege, A., Watanabe, D., Preneel, B., Verbauwhede, I.: Chaskey: an efficient MAC algorithm for 32-bit microcontrollers. In: Joux, A., Youssef, A. (eds.) SAC 2014. LNCS, vol. 8781, pp. 306–323. Springer, Cham (2014). https://doi.org/10.1007/978-3-319-13051-4_19

[RR16]   Rasoolzadeh, S., Raddum, H.: Faster key recovery attack on round-reduced PRINCE. In: Bogdanov, A. (ed.) LightSec 2016. LNCS, vol. 10098, pp. 3–17. Springer, Cham (2017). https://doi.org/10.1007/978-3-319-55714-4_1

[SGL+17]   Sun, S., et al.: Analysis of aes, skinny, and others with constraint programming. IACR Trans. Symmetric Cryptol. 1, 2017 (2017)

[SWLW16]   Sun, L., Wang, W., Liu, R., Wang, M.: MILP-aided bit-based division property for ARX-based block cipher. Cryptology ePrint Archive, Report 2016/1101 (2016). http://eprint.iacr.org/2016/1101

[SWW16]   Sun, L., Wang, W., Wang, M.: MILP-aided bit-based division property for primitives with non-bit-permutation linear layers. IACR Cryptology ePrint Archive 2016:811 (2016)

[SWW17]   Sun, L., Wang, W., Wang, M.: Automatic search of bit-based division property for ARX ciphers and word-based division property. Cryptology ePrint Archive, Report 2017/860 (2017). https://eprint.iacr.org/2017/860

[TIHM17]   Todo, Y., Isobe, T., Hao, Y., Meier, W.: Cube attacks on non-blackbox polynomials based on division property. In: Katz, J., Shacham, H. (eds.) CRYPTO 2017. LNCS, vol. 10403, pp. 250–279. Springer, Cham (2017). https://doi.org/10.1007/978-3-319-63697-9_9

[TM16]   Todo, Y., Morii, M.: Bit-based division property and application to SIMON family. In: Peyrin, T. (ed.) FSE 2016. LNCS, vol. 9783, pp. 357–377. Springer, Heidelberg (2016). https://doi.org/10.1007/978-3-662-52993-5_18

[Tod15a]   Todo, Y.: Integral cryptanalysis on full MISTY1. In: Gennaro, R., Robshaw, M. (eds.) CRYPTO 2015. LNCS, vol. 9215, pp. 413–432. Springer, Heidelberg (2015). https://doi.org/10.1007/978-3-662-47989-6_20

[Tod15b]   Todo, Y.: Structural evaluation by generalized integral property. In: Oswald, E., Fischlin, M. (eds.) EUROCRYPT 2015. LNCS, vol. 9056, pp.

287–314. Springer, Heidelberg (2015). https://doi.org/10.1007/978-3-662-46800-5_12

[Tod17] Todo, Y.: Integral cryptanalysis on full MISTY1. J. Cryptology **30**(3), 920–959 (2017)

[Vie07] Michael Vielhaber. Breaking ONE.FIVIUM by AIDA an algebraic IV differential attack. IACR Cryptology ePrint Archive, 2007:413 (2007)

[XZBL16] Xiang, Z., Zhang, W., Bao, Z., Lin, D.: Applying MILP method to searching integral distinguishers based on division property for 6 lightweight block ciphers. In: Cheon, J.H., Takagi, T. (eds.) ASIACRYPT 2016. LNCS, vol. 10031, pp. 648–678. Springer, Heidelberg (2016). https://doi.org/10.1007/978-3-662-53887-6_24

[ZR17] Wenying, Z., Rijmen, V.: Division cryptanalysis of block ciphers with a binary diffusion layer. Cryptology ePrint Archive, Report 2017/188 (2017). https://eprint.iacr.org/2017/188

Towards Key-Dependent Integral and Impossible Differential Distinguishers on 5-Round AES

Kai Hu[1,3], Tingting Cui[2], Chao Gao[4], and Meiqin Wang[1(✉)]

[1] Key Laboratory of Cryptologic Technology and Information Security,
Ministry of Education, Shandong University, Jinan 250100, China
hukai@mail.sdu.edu.cn, mqwang@sdu.edu.cn
[2] School of Cyberspace, Hangzhou Dianzi University, Hangzhou 310000, China
cuitingting@hdu.edu.cn
[3] Shandong Computer Science Center (National Supercomputer Center in Jinan),
Jinan 250100, China
[4] Affiliated Hospital of Shandong University of Traditional Chinese Medicine,
Jinan 250100, China
szygaochao@163.com

Abstract. Reduced-round AES has been a popular underlying primitive to design new cryptographic schemes and thus its security including distinguishing properties deserves more attention. At Crypto'16, a key-dependent integral distinguisher on 5-round AES was put forward, which opened up a new direction to take more insights into the distinguishing properties of AES. After that, two key-dependent impossible differential (ID) distinguishers on 5-round AES were proposed at FSE'16 and CT-RSA'18, respectively. It is strange that the current key-dependent integral distinguisher requires significantly higher complexities than the key-dependent ID distinguishers, even though they are constructed with the same property of MixColumns ($2^{128} \gg 2^{98.2}$). Proposers of the 5-round key-dependent distinguishers claimed that the corresponding integral and ID distinguishers can only work under chosen-ciphertext and chosen-plaintext settings, respectively, which is very different from the situations of traditional key-independent distinguishers.

In this paper, we first construct a novel key-dependent integral distinguisher on 5-round AES with 2^{96} chosen plaintexts, which is much better than the previous key-dependent integral distinguisher that requires the full codebook proposed at Crypto'16. Secondly, We show that both distinguishers are valid under either chosen-plaintext setting or chosen-ciphertext setting, which is different from the claims of previous cryptanalysis. However, under different settings, complexities of key-dependent integral distinguishers are very different while those of the key-dependent ID distinguishers are almost the same. We analyze the reasons for it.

Keywords: AES · Key-dependent · Integral · Impossible differential

© Springer Nature Switzerland AG 2019
C. Cid and M. J. Jacobson, Jr. (Eds.): SAC 2018, LNCS 11349, pp. 139–162, 2019.
https://doi.org/10.1007/978-3-030-10970-7_7

1 Introduction

1.1 Background

In symmetric-key cryptanalysis, one usually starts by identifying a distinguisher on the reduced-round target cipher and then proceeds with the key-recovery attack for more rounds. Besides the key recovery, the distinguishing property of some cryptographic schemes itself has been more and more important because many of new ciphers are designed based on well-studied schemes. Among these underlying primitives, reduced-round Advanced Encryption Standard (AES) [4] is a very popular choice. In one hand, the security of reduced-round AES has been analyzed a lot and in the other hand, processor manufactures provided single round instruction for AES, which much encourages researchers to rely on them for new designs. For example, the authentication encryption algorithm AEGIS [14] uses four rounds of AES in the state update functions and ELmd [5] suggests using some reduced-round including 5-round AES. Although the security of these schemes does not completely depend on the basic primitives, it is useful to understand them more deeply by studying the reduced-round AES.

Many distinguishers on reduced-round AES have been proposed and used to evaluate its security for different number of rounds. Traditional distinguishers can only cover four or less rounds [1,2,4,6,8,10]. At Crypto'16, Sun et al. proposed the first 5-round zero-correlation (ZC) linear hull and transformed it into a 5-round integral distinguisher. Then, with the statistical integral technique presented at FSE'16 [13], Cui et al. gave an attack on 5-round AES [3]. In [7,8], 5-round ID distinguishers were put forward by Grassi et al. In all, the 5-round ZC linear hull, integral, statistical integral and ID distinguishers are all key-dependent, which are valid only if the conditions of keys are satisfied. Later, the first key-independent 5-round distinguisher, named multiple-of-n distinguisher, was given in [9]. This distinguisher has a key-dependent variant based on the multiple-of-n property [7]. More recently, an interesting adaptive chosen-plaintext-ciphertext distinguisher Yoyo was proposed to mount a distinguishing attack [11] on reduced-round AES.

This paper focuses on the key-dependent distinguishers on 5-round AES. Key-dependent distinguishers can be regarded as "something in the middle" between secret-key distinguishers and key recovery attacks. Although the complexities of the key-dependent integral and ID distinguishers are higher than that of the multiple-of-n or Yoyo distinguisher, more insights for structural properties of AES such as the details of MixColumns (MC) matrix can be identified, which is based on the fact that all public key-dependent distinguishers on 5-round AES are based on the details of coefficients of this matrix.

Among key-dependent distinguishers on 5-round AES, there is a big gap between the complexities of the integral and ID distinguishers. Even with the same property (Property 1 which we will introduce in Sect. 2.3) of MC matrix, the integral distinguisher requires the whole codebook, while the ID distinguisher just needs $2^{98.2}$ chosen plaintexts. Moreover, it is claimed that the integral distinguisher proceeds only under chosen-ciphertext setting in [12] and the ID distinguishers work only under chosen-plaintext model in [7,8], because these two kinds of distinguishers are based or Property 1 or Property 2 of MC matrix (introduced in Sect. 2) but MC^{-1} matrix does not have such properties.

It is strange that the key-dependent integral and ID distinguishers can work only under specific scenarios, which is a limitation for key-dependent distinguishers. This paper investigates the principles behind the phenomenon and try to remove the limitations. The key-dependent integral distinguisher proposed at Crypto'16 requires the whole codebook and 2^{128} memory accesses. However, a distinguisher that requires the full codebook is usually thought as a trivial attack. Thus, we hope to reduce the complexities of the key-dependent integral distinguisher.

1.2 Contributions

The contributions of this paper are two-fold as follows:

Improved Key-Dependent Integral Distinguisher on 5-Round AES.
Key-dependent integral distinguisher on 5-round AES [12] is derived by setting the constraints on the ciphertexts and requires the whole codebook. We construct a new integral distinguisher with only 2^{96} chosen plaintexts. Both our distinguisher and the one in [12] take advantage of the same property of MC matrix. In addition, our distinguisher works under the chosen-plaintext setting instead of the chosen-ciphertext setting. The complexities of chosen-plaintext and chosen-ciphertext key-dependent integral distinguishers are very different. We find that the reason lies on the addition of the last round key. Under chosen-ciphertext setting, we have to guess one byte of key information to achieve the attack while we avoid it under the chosen-plaintext setting.

Key-Dependent ID Distinguishers on 5-Round AES Under Chosen-Ciphertext Setting. We transform the chosen-plaintext key-dependent ID distinguishers into chosen-ciphertext ones, which extends the attacks presented in [7,8]. Both the distinguisher with $2^{98.2}$ chosen plaintexts in [8] and the one with $2^{76.4}$ chosen plaintexts in [7] can be transformed into new ID distinguishers with $2^{99.6}$ and $2^{76.5}$ chosen ciphertexts, respectively. The key-dependent ID distinguishers have slightly different complexities under different attacking scenarios. As the case for integral distinguishers, we analyze the influences of the key addition operation which the key-dependent ID distinguishers depend on.

The complexities of key-dependent integral and ID distinguishers under different models are listed in Table 1.

1.3 Outline of This Paper

In Sect. 2, some preliminaries are given. Then, we present new key-dependent integral distinguishers on 5-round AES in Sect. 3. In Sect. 4, we give the ID distinguishers on 5-round AES under chosen-ciphertext setting. At last, we conclude this paper in Sect. 5.

Table 1. Key-dependent integral and ID distinguishers on 5-round AES.

Distinguisher	Property of MC	Scenario	Data	Time (MA)	Reference
Integral	Property 1	CC	2^{128}	2^{128}	[12]
		CP	$\mathbf{2^{96}}$	$\mathbf{2^{96}}$	Sect. 3
ID	Property 1	CP	$2^{98.2}$	2^{107}	[8]
		CC	$\mathbf{2^{99.6}}$	$\mathbf{2^{103.6}}$	Sect. 4
ID	Property 2	CP	$2^{76.4}$	$2^{81.5}$	[7]
		CC	$\mathbf{2^{76.5}}$	$\mathbf{2^{80.5}}$	Sect. 4

– CP: Chosen-Plaintext CC: Chosen-Ciphertext MA: Memory Access

2   Preliminaries

2.1   Notations

To make the description clear and concise, we list some notations used in this paper as follows.

- P: plaintext;
- C: ciphertext;
- K^r: round key of the r-th round and the whitening key is K^0;
- $X^{r,OP}$: the state after OP operation of the r-th round. e.g. $X^{4,MC}$ is the state after the MixColumns operation of the fourth round function, the state after the whitening key addition is denoted as $X^{0,AK}$;
- $X_{i,j}$, $i,j = 0,1,2,3$: the byte in the i-th row and j-th column of the state X.
- OP_r: the OP operation of the r-th round, AK_0 means the AddRoundKey operation with the whitening key.

2.2   Description of AES

AES [4] is a 128-bit iterative block cipher that adopts substitution-permutation network (SPN). It has three versions according to the size of key, namely AES-128, -192 and -256, respectively, whose total rounds N_r are 10, 12 and 14 individually. The 128-bit internal state of AES can be regarded as a 4×4 matrix, each cell of which is an 8-bit value. All operations in AES are defined in the finite field $GF(2^8)$ whose irreducible polynomial is $m(x) = x^8 + x^4 + x^3 + x + 1$. Each round function $R(x) = AK \circ MC \circ SR \circ SB(x)$ has four components as follows.

- SubBytes (SB): A nonlinear bijective mapping $S : \mathbb{F}_2^8 \to \mathbb{F}_2^8$ on each byte of the state;
- ShiftRows (SR): Left rotate the i-th row by i bytes, where $i = 0,1,2,3$;
- MixColumns (MC): Left multiply with an MDS matrix over the field $GF(2^8)$ on each column. The matrices used in the MC operation and its reverse operation MC^{-1} are

$$MC = \begin{bmatrix} 0x2\ 0x3\ 0x1\ 0x1 \\ 0x1\ 0x2\ 0x3\ 0x1 \\ 0x1\ 0x1\ 0x2\ 0x3 \\ 0x3\ 0x1\ 0x1\ 0x2 \end{bmatrix} \quad and \quad MC^{-1} = \begin{bmatrix} 0xe\ 0xb\ 0xd\ 0x9 \\ 0x9\ 0xe\ 0xb\ 0xd \\ 0xd\ 0x9\ 0xe\ 0xb \\ 0xb\ 0xd\ 0x9\ 0xe \end{bmatrix} ;$$

– AddRoundKey (AK): XOR with a round key.

We can change the orders of MC and AK operations in some situations, i.e. $R(x) = MC \circ EAK \circ SR \circ SB(x)$, where $MC \circ EAK = AK \circ MC$. Note that there is a whitening key XORed with plaintext before the first round function and the MC operation in the last round is omitted.

For decryption process, N_r reverse rounds are applied to the ciphertext matrix. Each reverse round function applies four reverse operations: InvSubBytes(SB^{-1}), InvShiftRows(SR^{-1}), InvMixColumns(MC^{-1}) and InvAddRoundKey(AK^{-1}).

2.3 Previous Integral and ID Distinguishers on 5-Round AES

In this subsection, we recall the previous key-dependent integral and ID distinguishers on 5 rounds of AES [7,8,12]. The key techniques for these distinguishers are that they take advantage of the properties of MC matrix and manage to extend the known 4-round distinguishers one more round. We conclude the properties as follows.

Property 1. *The matrix of MC operation has two equal coefficients in each row or each column, i.e., the MC matrix of AES has two elements equal to 1 in each row or each column.*

Property 2. *The matrix of MC operation has two rows satisfying Eq. (1) or two columns satisfying Eq. (2).*

$$\begin{cases} MC[i_1,j] \oplus MC[i_1,k] \oplus MC[i_1,l] = 0, \\ MC[i_2,j] \oplus MC[i_2,k] \oplus MC[i_2,l] = 0. \end{cases} \tag{1}$$

$$\begin{cases} MC[j,i_1] \oplus MC[k,i_1] \oplus MC[l,i_1] = 0, \\ MC[j,i_2] \oplus MC[k,i_2] \oplus MC[l,i_2] = 0. \end{cases} \tag{2}$$

where $i_1 \neq i_2$, $j \neq k \neq l$, $0 \leq i_1, i_2, j, k, l \leq 3$.

Integral Distinguisher on 5-Round AES [12]. The 5-round integral distinguisher is transformed from a 5-round ZC linear hull based on Property 1 by setting a specific condition on ciphertexts. The ZC linear hull is illustrated in Proposition 1 and Fig. 4 in Appendix D.

Proposition 1. *Divide the whole ciphertext-plaintext space into 2^8 sets according to the value of $C_{0,0} \oplus C_{1,3}$ as*

$$V_\Delta = \{(C, P)|C_{0,0} \oplus C_{1,3} = \Delta, \Delta \in \mathbb{F}_2^8\}.$$

If the input mask Γ_{in} on ciphertext and output mask Γ_{out} on plaintext are as follows,

$$\Gamma_{in} = (\alpha_{i,j}), 0 \leqslant i, j \leqslant 3, \quad \alpha_{i,j} = \begin{cases} a, & \text{if } (i,j) \in \{(0,0),(1,3)\}; \\ 0, & \text{otherwise.} \end{cases}$$

$$\Gamma_{out} = (\beta_{i,j}), 0 \leqslant i, j \leqslant 3, \quad \beta_{i,j} = \begin{cases} nonzero, & \text{if } (i,j) = (0,0); \\ 0, & \text{otherwise.} \end{cases}$$

where $a \in \mathbb{F}_2^8 \backslash \{0\}$.

Then $(\Gamma_{in} \rightarrow \Gamma_{out})$ is a 5-round ZC linear hull when the ciphertexts are chosen from one specific set of $V_\Delta, \Delta = K_{0,0}^5 \oplus K_{1,3}^5$.

Bogdanov *et al.* proposed a link between ZC linear hull and integral distinguisher in [2], which is summarized in Theorem 1.

Theorem 1 (From [2]). *Assume $H : \mathbb{F}_2^s \times \mathbb{F}_2^t \rightarrow \mathbb{F}_2^u \times \mathbb{F}_2^v$ is (part of) a cipher, without loss of generality, we can decompose the cipher and define the part cipher as*

$$H(x,y) = \begin{pmatrix} H_1(x,y) \\ H_2(x,y) \end{pmatrix}, H_1 : \mathbb{F}_2^s \times \mathbb{F}_2^t \rightarrow \mathbb{F}_2^u, H_2 : \mathbb{F}_2^s \times \mathbb{F}_2^t \rightarrow \mathbb{F}_2^v.$$

If we fix the t bits of input value as λ and consider only u bits of the output value, we can construct another function $T_\lambda(x) : \mathbb{F}_2^s \rightarrow \mathbb{F}_2^u$ as follows

$$T_\lambda(x) = H_1(x, \lambda).$$

When the input and output linear masks a and b are independent, the approximation $b \cdot H(x) \oplus a \cdot x$ has correlation zero for any $a = (a_1, 0)$ and any $b = (b_1, 0) \neq 0$ (zero-correlation) if and only if the function T_λ is balanced for any λ (integral).

With Theorem 1, one ZC linear hull on 5-round AES can be transformed into an integral distinguisher, which is shown in Proposition 2.

Proposition 2. *Divide the whole ciphertext-plaintext space into 2^8 sets*

$$V_\Delta = \{(C, P)|C_{0,0} \oplus C_{1,3} = \Delta, \Delta \in \mathbb{F}_2^8\}.$$

There is always one Δ such that

$$T_\Delta = \sum_{(C,P) \in V_\Delta} P = 0.$$

Note that this 5-round integral distinguisher requires the full codebook.

ID Distinguishers on 5-Round AES [7,8]. The first ID distinguisher on 5-round AES [8] is similar to the 5-round integral one [12]. It manages to extend the traditional 4-round impossible distinguisher one more round. This 5-round ID (see Fig. 5 in Appendix D) is summarized in Proposition 3.

Proposition 3. *For plaintexts in the sets*

$$V_\Delta = \{(P^l, C^l), l = 0, 1, 2, \cdots, 255 | P^l_{0,0} \oplus P^l_{1,1} = \Delta, \quad \forall l \quad and$$
$$P^l_{i,j} = P^m_{i,j} \quad \forall(i,j) \notin \{(0,0), (1,1)\} \quad and \quad l \neq m\},$$

there is always one Δ such that the difference of any two corresponding ciphertexts after 5-round AES encryption cannot be inactive in three reverse-diagonals at the same time.

This ID distinguisher requires $2^{98.2}$ chosen plaintexts with success rate 95%.

The second ID distinguisher based on Property 2 was proposed in [7], which requires $2^{76.4}$ chosen plaintexts. It is illustrated in Proposition 4 and shown in Fig. 6 in Appendix D.

Proposition 4. *For plaintexts in the sets*

$$A_{(\Delta_1, \Delta_2)} = \{(P^l, C^l) \, l = 0, 1, \cdots, 255| \; P^l_{0,0} \oplus P^l_{1,1} = \Delta_1 \quad \forall i, P^l_{0,0} \oplus P^l_{2,2} = \Delta_2 \quad \forall i$$
$$and \quad P^l_{i,j} = P^m_{i,j} \quad \forall(i,j) \notin \{(0,0), (1,1), (2,2)\} \quad and \quad l \neq m\} \quad .$$

there is always one tuple of (Δ_1, Δ_2) that the difference of ciphertexts after 5-round AES encryption cannot be inactive in two reverse-diagonals in the same time.

This distinguisher requires $2^{76.4}$ chosen plaintexts with success rate 95%.

3 Improved Integral Distinguishers on AES

The 5-round integral distinguisher based on Property 1 proposed in [12] requires the whole codebook, which will limit its contribution. However, we can improve this distinguisher by significantly reducing data and time complexities. In Sect. 3.1, we put forward an improved 5-round integral distinguisher based on Property 1 with 2^{96} chosen plaintexts, which is the longest integral distinguisher on AES as far as we know. In fact, our attack can be regarded as a chosen-plaintext counterpart of the distinguisher in [12]. Interestingly, the data complexities are very different between the two distinguishers. In Sect. 3.2, we discuss the reason why there is such a big gap between the data complexities. Originally, we plan to construct the key-dependent integral distinguisher based on Property 2 which was already used in building the key-dependent ID distinguisher, but we fail to do it. We discuss the reasons for it in Appendix A.

3.1 Improved Key-Dependent Integral Distinguisher on 5-Round AES

The 5-round integral distinguisher in [12] requires the whole codebook while the ID distinguisher in [8] needs only $2^{98.2}$ chosen plaintexts. Both distinguishers use Property 1 of MC matrix. There is a big gap for complexities between them. In this section, we will propose an improved integral distinguisher to eliminate or narrow this gap.

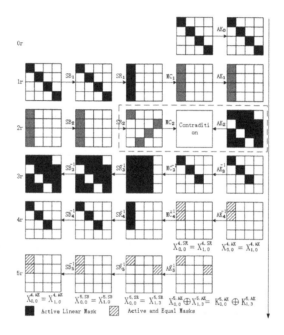

Fig. 1. 5-round ZC linear hull of AES.

In order to improve the 5-round integral distinguisher, we first construct a novel 4-round integral distinguisher on AES summarized in Lemma 1, which is transformed from a 4-round ZC linear hull shown in Fig. 1 (from Round 1 to Round 4), whose input mask Γ_{in} and output mask Γ_{out} are as follows.

$$\Gamma_{in} = (\alpha_{i,j}), 0 \leqslant i,j \leqslant 3, \quad \alpha_{i,j} = \begin{cases} nonzero & \text{if } (i,j) \in \{(0,0),(1,1),(2,2),(3,3)\}, \\ 0 & \text{otherwise} \end{cases}.$$
$$\tag{3}$$

$$\Gamma_{out} = (\beta_{i,j}), 0 \leqslant i,j \leqslant 3, \quad \beta_{i,j} = \begin{cases} b & \text{if } (i,j) \in \{(0,0),(1,0)\}, \\ 0 & \text{otherwise} \end{cases}, b \in \mathbb{F}_2^8. \tag{4}$$

Lemma 1. *For 4-round AES with MC operation in the last round, if we take all 2^{96} plaintexts P by fixing $(P_{0,0}, P_{1,1}, P_{2,2}, P_{3,3})$ as constant, each value of $C_{0,0} \oplus C_{1,0} \in \mathbb{F}_2^8$ of ciphertexts appears 2^{88} times.*

Proof. As shown in Fig. 1, Γ_{in} and Γ_{out} (Eqs. (3) and (4)) are independent and lead to a ZC linear hull on 4 rounds of AES. According to Theorem 1,

- Γ_{in} can be denoted as $(a, 0)$, where a can be any value in $\mathbb{F}_2^{32}$;
- Γ_{out} can be denoted as $(b, b, 0)$, where b can be any value in $\mathbb{F}_2^8 \backslash \{0\}$.

Since it is required that Γ_{out} should be any value except 0, we proceed with some transformations on the output of 4-round AES in order to satisfy the conditions of Theorem 1.

Firstly, we can rewrite 4-round AES as a function H with two inputs and three outputs:

$$H(x, y) = (H_1(x, y), H_2(x, y), H_3(x, y)).$$

where $x = (P_{0,0}, P_{1,1}, P_{2,2}, P_{3,3})$, y is the concatenated value of other 12 bytes of plaintext, $(H_1(x, y), H_2(x, y)) = (C_{0,0}, C_{1,0})$ and $H_3(x, y)$ is the concatenated value of other 14 bytes.

We can produce a new function H' based on the function H with the same inputs:

$$H'(x, y) = (H_1(x, y) \oplus H_2(x, y), H_3(x, y)).$$

Then for the new function H', we derive that the linear approximation with $\Gamma_{in} = (a, 0)$ and $\Gamma_{out} = (b, 0)$ has correlation zero, where a can be any value in $\mathbb{F}_2^{32}$ and b can be any value in $\mathbb{F}_2^8 \backslash \{0\}$.

With Theorem 1, we can transform the ZC linear approximation on H' into an integral distinguisher, i.e. if we take all 2^{96} plaintexts P by fixing $(P_{0,0}, P_{1,1}, P_{2,2}, P_{3,3})$ as constant, the values of $H_1(x, y) \oplus H_2(x, y)$ are balanced, which means that each value of $C_{0,0} \oplus C_{1,0} \in \mathbb{F}_2^8$ of ciphertexts appears 2^{88} times. $\qquad \square$

Based on Lemma 1, we can add one more round behind the 4-round integral distinguisher to deduce a 5-round integral distinguisher by the idea of Lemma 2 as follows.

Lemma 2. *For one-round AES without MC operation (i.e. $AK \circ SR \circ SB$), if we take N plaintexts P where N_1 plaintexts satisfy $P_{0,0} \oplus P_{1,0} = 0$, then there must be at least one $\delta \in \mathbb{F}_2^8$ such that the number of ciphertexts C satisfying $C_{0,0} \oplus C_{1,3} = \delta$ is exactly N_1 with probability 1.*

Proof. Due to the bijective mapping S-box S, we have

$$S(P_{0,0}) \oplus S(P_{1,0}) = \begin{cases} 0, & \text{if } P_{0,0} \oplus P_{1,0} = 0, \\ nonzero, & \text{if } P_{0,0} \oplus P_{1,0} \neq 0. \end{cases}$$

After SB operation, there are exactly N_1 values of $X^{1,SB}$ satisfying $X_{0,0}^{1,SB} \oplus X_{1,0}^{1,SB} = 0$, which leads $C_{0,0} \oplus C_{1,3} = K_{0,0}^1 \oplus K_{1,3}^1$ as well. Let $\delta = K_{0,0}^1 \oplus K_{1,3}^1$, thus $C_{0,0} \oplus C_{1,3} = \delta$ happens exactly N_1 times. $\qquad \square$

With Lemmas 1 and 2, our new 5-round integral distinguisher on AES is summarized in Proposition 5.

Proposition 5. *Taking all 2^{96} plaintexts P by fixing $(P_{0,0}, P_{1,1}, P_{2,2}, P_{3,3})$ as constant, after 5-round AES encryption, there is at least one $\delta \in \mathbb{F}_2^8$ such that the number of ciphertexts satisfying $C_{0,0} \oplus C_{1,3} = \delta$ is exactly 2^{88}. Meanwhile, for any random permutation, the same event happens with probability only about $2^{-40.7}$.*

Proof. For 5-round AES, $X_{0,0}^{4,AK} \oplus X_{1,0}^{4,AK} = 0$ happens 2^{88} (out of 2^{96}) times according to Lemma 1. Then due to Lemma 2, $N = 2^{96}$ and $N_1 = 2^{88}$, so there is one δ such that $C_{0,0} \oplus C_{1,3} = \delta$ happens exactly 2^{88} times.

For a random permutation, the number N_δ of ciphertexts satisfying $C_{0,0} \oplus C_{1,3} = \delta$ for a fixed δ follows the binomial distribution

$$N_\delta \sim \mathcal{B}(2^{96}, 2^{-8}).$$

According to the Central Limit Theorem, the normal distribution can approximate the binomial distribution in this situation. Now

$$N_\delta \sim \mathcal{N}(2^{88}, 2^{96} \times 2^{-8} \times (1 - 2^{-8})).$$

Therefore, $p(N_\delta = 2^{88}) \approx 2^{-48.64}$. Because of 2^8 possible values for δ, the probability that there is at least one value for δ satisfying $N_\delta = 2^{88}$ is $1 - (1 - p(N_\delta = 2^{88}))^{2^8} \approx 2^{-40.7}$. $\qquad\square$

The whole process of the integral distinguishing attack on 5-round AES is illustrated in Algorithm 1.

Algorithm 1. Improved 5-Round Integral Distinguisher on AES

Input: 2^{96} plaintexts P^i, $i = 0, 1, 2, \ldots, 2^{96} - 1$
Output: 5-Round AES or Random Permutation
1 Set one 8-bit vector counter $V[256]$ and initialize it as zero;
2 **for** *Each P^i of 2^{96} plaintexts* **do**
3 | Query its ciphertext C^i and calculate $\delta = C_{0,0}^i \oplus C_{1,3}^i$;
4 | Let $V[\delta] = V[\delta] + 1$;
5 **for** *Each $\delta \in \mathbb{F}_2^8$* **do**
6 | **if** $V[\delta]{=}2^{88}$ **then**
7 | | **return** 5-Round AES; .
8 **return** Random Permutation;

In Algorithm 1, the data complexity is 2^{96} chosen plaintexts and the time complexity is about 2^{96} memory accesses. Since we set a 2^8 vector counter, the memory requirements are 2^8 which can be ignored. The type-II error probability (the probability to wrongfully accept a random permutation as AES) is $2^{-40.7}$.

3.2 Gap for Complexities Between Chosen-Plaintext and Chosen-Ciphertext Integral Distinguishers

Interestingly, there exists a gap between the complexities of chosen-plaintext and chosen-ciphertext integral distinguishers although they are constructed from a same (or similar) ZC linear hull.

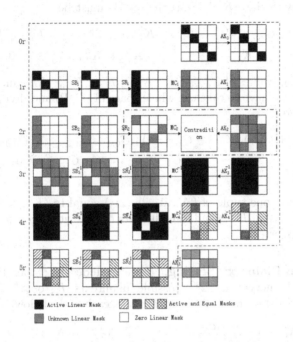

Fig. 2. 5-round integral distinguisher with(out) AK_5.

In the chosen-ciphertext integral distinguisher, we need to guess one byte of $K_{0,0}^5 \oplus K_{1,3}^5$, which increases the complexities by a factor of 2^8. This inspires us that the AK operation which the integral distinguisher depends on, i.e. AK_5, influences the complexities. In this subsection, we investigate the influences of AK_5 on complexities by considering chosen-ciphertext and chosen-plaintext integral distinguishers on 5-round AES with and without AK_5, respectively. Notice that we use a general variant of the key-dependent integral distinguisher with four active masks on plaintext bytes (see Fig. 2).

Under Chosen-Ciphertext Setting. If we omit the operation AK_5 (the enclosure area by dotted line in Fig. 2) and decrypt from $X^{5,SR}$ in subspace $V_{X^{5,SR}}$ as follows to the plaintext P

$$V_{X^{5,SR}} = \{(X^{5,SR}, P) \mid X_{0,0}^{5,SR} = X_{1,3}^{5,SR}, X_{0,1}^{5,SR} = X_{3,2}^{5,SR}, X_{2,0}^{5,SR} = X_{3,3}^{5,SR},$$
$$X_{1,2}^{5,SR} = X_{2,1}^{5,SR}, X_{i,j}^{5,SR} \in \mathbb{F}_2^8, 0 \leqslant i,j \leqslant 3\},$$

we can construct a chosen-ciphertext integral distinguisher whose corresponding plaintexts satisfy the balance property, i.e. each possible value of plaintext byte has the same number of occurrences. Since the size of $V_{X^{5,SR}}$ is 2^{96}, this integral distinguisher requires data complexity 2^{96} chosen ciphertexts.

If the operation AK_5 is included into the distinguisher (whole area in Fig. 2), we have to take a subspace of ciphertexts V_C which can produce $V_{X^{5,SR}}$ after the proceeding with the AK_5^{-1}. Thus the set V_C must be

$$V_C = \{(C, P) \mid C_{0,0} \oplus C_{1,3} = K_{0,0}^5 \oplus K_{1,3}^5, C_{0,1} \oplus C_{3,2} = K_{0,1}^5 \oplus K_{3,2}^5,$$
$$C_{2,0} \oplus C_{3,3} = K_{2,0}^5 \oplus K_{3,3}^5, C_{1,2} \oplus C_{2,1} = K_{1,2}^5 \oplus K_{2,1}^5, C_{i,j} \in \mathbb{F}_2^8, 0 \leqslant i, j \leqslant 3\}.$$

However, the exact values of $K_{0,0}^5 \oplus K_{1,3}^5$, $K_{0,1}^5 \oplus K_{3,2}^5$, $K_{2,0}^5 \oplus K_{3,3}^5$ and $K_{1,2}^5 \oplus K_{2,1}^5$ are unknown, so we have to take the whole space of (C, P) and divide it into 2^{32} subspaces as follows:

$$V_{\Delta_0, \Delta_1, \Delta_2, \Delta_3} = \{(C, P) \mid C_{0,0} \oplus C_{1,3} = \Delta_0, C_{0,1} \oplus C_{3,2} = \Delta_1, C_{2,0} \oplus C_{3,3} = \Delta_2,$$
$$C_{1,2} \oplus C_{2,1} = \Delta_3, C_{i,j} \in \mathbb{F}_2^8, 0 \leqslant i, j \leqslant 3\},$$

with $\Delta_i \in \mathbb{F}_2^8, 0 \leqslant i \leqslant 3$.

There is always one tuple of $(\Delta_0, \Delta_1, \Delta_2, \Delta_3)$ equal to $(K_{0,0}^5 \oplus K_{1,3}^5, K_{0,1}^5 \oplus K_{3,2}^5, K_{2,0}^5 \oplus K_{3,3}^5, K_{1,2}^5 \oplus K_{2,1}^5)$ and thus the data complexity becomes 2^{128} instead of 2^{96} chosen ciphertexts.

Under Chosen-Plaintext Setting. If we exclude AK_5 operation from 5-round AES and encrypt all 2^{96} possible plaintexts P to $X^{5,SR}$ by fixing $(P_{0,0}, P_{1,1}, P_{2,2}, P_{3,3})$ as constant. From Sect. 3.1, each of the following four events

1. $X_{0,0}^{5,SR} \oplus X_{1,3}^{5,SR} = 0,$ 2. $X_{0,1}^{5,SR} \oplus X_{3,2}^{5,SR} = 0,$

3. $X_{2,0}^{5,SR} \oplus X_{3,3}^{5,SR} = 0,$ 4. $X_{1,2}^{5,SR} \oplus X_{2,1}^{5,SR} = 0,$

occurs 2^{88} times with probability 1. We can distinguish AES from a random permutation with 2^{96} chosen plaintexts.

Again we take AK_5 operation into consideration, each of four events

1. $C_{0,0} \oplus C_{1,3} = K_{0,0}^5 \oplus K_{1,3}^5,$ 2. $C_{0,1} \oplus C_{3,2} = K_{0,1}^5 \oplus K_{3,2}^5,$

3. $C_{2,0} \oplus C_{3,3} = K_{2,0}^5 \oplus K_{3,3}^5,$ 4. $C_{1,2} \oplus C_{2,1} = K_{1,2}^5 \oplus K_{2,1}^5,$

occurs 2^{88} times with probability 1, respectively.

Though we do not know any information about the secret key, we can predict there is always one tuple of $(\Delta_0', \Delta_1', \Delta_2', \Delta_3')$ ensuring each of the four experiences

1. $C_{0,0} \oplus C_{1,3} = \Delta_0',$ 2. $C_{0,1} \oplus C_{3,2} = \Delta_1',$

3. $C_{2,0} \oplus C_{3,3} = \Delta_2',$ 4. $C_{1,2} \oplus C_{2,1} = \Delta_3',$

to occur 2^{88} times (when $(\Delta_0', \Delta_1', \Delta_2', \Delta_3')$ are just the four XOR values of K_5). Yet any one event occurs with probability about $2^{-40.7}$ for a random permutation. So 2^{96} chosen plaintexts are enough to proceed this distinguishing attack.

At last, we summarize the reasons resulting in the gap from two cases between chosen-plaintext and chosen-ciphertext integral distinguishers. If AK_5 is omitted, the data complexities of the two distinguishers under both settings are the same. If AK_5 is included, the chosen-ciphertext integral distinguisher has to take the whole codebook while the chosen-plaintext integral distinguisher does not increase the data complexity. To make it more clear, we compare the data complexities of them in Table 2.

Table 2. Data complexities of integral distinguishers with(out) AK_5.

Setting	Target	Data complexity	Time (MA)
CC	5-round AES without AK_5	2^{96}	2^{96}
	5-round AES with AK_5	2^{128}	2^{128}
CP	5-round AES without AK_5	2^{96}	2^{96}
	5-round AES with AK_5	2^{96}	2^{96}

– CC: Chosen-Ciphertext CP: Chosen-Plaintext MA: Memory Access

4 ID Distinguishers on 5-Round AES Under Chosen-Ciphertext Setting

Until now there have been two key-dependent ID distinguishers on 5-round AES in [7,8] by utilizing the Property 1 and 2 of MC matrix respectively. In this section we put forward two ID distinguishers on 5-round AES under chosen-ciphertext model in Sects. 4.1 and 4.2 respectively, which are transformed from the ones under chosen-plaintext setting. Their data complexities are $2^{99.6}$ and $2^{76.5}$ chosen ciphertexts, which are slightly different from those of the original ones with $2^{98.2}$ and $2^{76.4}$ chosen plaintexts, respectively. We analyze the reasons in Appendix C.

4.1 ID Distinguisher on 5-Round AES Based on Property 1 of MC

In this subsection, we first propose 16 key-dependent IDs for 5-round AES shown in Proposition 6 and we list one of them in Fig. 5. With these IDs, a distinguisher under chosen-ciphertext model is put forward with data complexity $2^{99.6}$ chosen ciphertexts.

Proposition 6. *If the difference of ciphertext pair (C^1, C^2) is nonzero at the four bytes $(C_{0,3}, C_{1,2}, C_{2,1}, C_{3,0})$ and zero at other 12 bytes, after 5-round AES decryption, the corresponding plaintext pair (P^1, P^2) never satisfies each of the following 16 cases:*

$$P_{s,t}^1 \oplus P_{s+1,t+1}^1 = P_{s,t}^2 \oplus P_{s+1,t+1}^2 = K_{s,t}^0 \oplus K_{s+1,t+1}^0,$$
$$P_{l,m}^1 \oplus P_{l,m}^2 = 0, (l,m) \neq (s,t), (s+1,t+1),$$

where $0 \leqslant s, t \leqslant 3$.[1]

Proof. Proof by contradiction. Assume that there is one ciphertext pair (C^1, C^2) leading to such plaintext pair (P^1, P^2). From the forward direction, since there exists one (s, t) such that (P^1, P^2) satisfies $P_{s,t}^1 \oplus P_{s+1,t+1}^1 = P_{s,t}^2 \oplus P_{s+1,t+1}^2 = K_{s,t}^0 \oplus K_{s+1,t+1}^0$, we have $\Delta X_{s,t}^{1,SB} = \Delta X_{s+1,t+1}^{1,SB}$. Due to the Property 1 of MC matrix, there are only three nonzero bytes of difference $\Delta X^{1,MC}$ in one column, which leads to at least one zero byte on each column of $\Delta X^{3,SR}$. From the backward direction, (C^1, C^2) results in at most one nonzero byte for each column of $\Delta X^{3,MC}$. Since the branch number of MC matrix is 5, each column of $\Delta X^{3,MC}$ has at least two zero bytes. This yields a contradiction and shows that they are IDs. □

Taking $(s, t) = (0, 0)$ as an example, we illustrate Proposition 6 in Fig. 5. Actually, the value of $K_{s,t}^0 \oplus K_{s+1,t+1}^0$ is secret, so we cannot directly check whether $P_{s,t}^1 \oplus P_{s+1,t+1}^1 = P_{s,t}^2 \oplus P_{s+1,t+1}^2 = K_{s,t}^0 \oplus K_{s+1,t+1}^0$ or not. In the following, we will define good pair to further identify if there exist solutions for $K_{s,t}^0 \oplus K_{s+1,t+1}^0$ by using the ID characteristic.

Definition 1 (Good Pair). *One pair* (P^1, P^2) *is a good pair related to* (s, t) *if it satisfies the following conditions:*

$$P_{s,t}^1 \oplus P_{s+1,t+1}^1 = P_{s,t}^2 \oplus P_{s+1,t+1}^2,$$
$$P_{l,m}^1 \oplus P_{l,m}^2 = 0, (l, m) \neq (s, t), (s+1, t+1),$$

where (s, t), $0 \leqslant s, t \leqslant 3$.

No matter how many ciphertext pairs as the form in Proposition 6 we take, for each (s, t) there always exists one value $\delta_{s,t} \in \mathbb{F}_2^8$ that $P_{s,t}^1 \oplus P_{s+1,t+1}^1 = P_{s,t}^2 \oplus P_{s+1,t+1}^2 = \delta_{s,t} = K_{s,t}^0 \oplus K_{s+1,t+1}^0$ never happens for each good pair.

According to the fact above, we put forward an ID distinguishing attack on 5-round AES under chosen-ciphertext model, see Algorithm 2. For each of 16 (s, t), $0 \leq s, t, \leq 3$, we take N_s structures of ciphertexts that each one includes 2^{32} ciphertexts by traversing all values of bytes $(C_{0,3}, C_{1,2}, C_{2,1}, C_{3,0})$ and fixing other bytes as constant, to find all good pairs and record their $P_{s,t}^1 \oplus P_{s+1,t+1}^1$ in a vector counter $V_{s,t}$. For 5-round AES, there is always a value δ_{st} never happening in $V_{s,t}$ for each (s, t). The probability that there is always a value $\delta_{s,t}$ never happening in $V_{s,t}$ for each (s, t) for a random permutation is calculated in Proposition 7.

Proposition 7. *For a random permutation, for each of 16* (s, t), $0 \leq s, t \leq 3$, *the probability that there always exists at least one value* $\delta_{s,t} = P_{s,t}^1 \oplus P_{s+1,t+1}^1$ *never appearing for any one of N random good pairs is* $2^{128} \times (1 - 2^{-8})^{16N}$.

---

[1] The addition used in subscripts of the equations are actually addition modulo 4. For example, when $t = 3$, $t + 1 = 0$.

Proof. For a random permutation and any given value of (s, t), the event that there is at least one value for $\delta_{s,t} = P^1_{s,t} \oplus P^1_{s+1,t+1}$ never occurring for any one of N random good pairs happens with the following probability

$$p_{s,t} = 2^8 \times (1 - 2^{-8})^N,$$

then the probability that this event happens for all 16 values of (s, t) is $p^{16}_{s,t} = 2^{128} \times (1 - 2^{-8})^{16N}$. □

Algorithm 2. 5-Round ID Distinguisher under Chosen-Ciphertext Model Based on Property 1

    Input: N_s structures of ciphertexts and corresponding plaintexts
    Output: 5-Round AES or Random Permutation
1 **for** *Each $s \in \{0, 1, 2, 3\}$* **do**
2   **for** *Each $t \in \{0, 1, 2, 3\}$* **do**
3   Initialize 256 indicators $V[256]$ as false;
4   **for** *Each one of N_s structures* **do**
            `// Each structure includes` 2^{32} `ciphertexts.`
5   Initialize a table $T[2^{32}]$;
6   Query the corresponding 2^{32} plaintexts and put them into T;
7   Sort T according to the value of 14 bytes except the (s, t)-th and $(s+1, t+1)$-th bytes;
8   Traverse all items of T and find adjacent plaintexts to combine good pairs;
            `// About` $N = N_s \times 2^{63} \times 2^{-120}$ `good pairs are found.`
9   **for** *Each (P^1, P^2) of N good pairs* **do**
10   Let $V[P^1_{s,t} \oplus P^1_{s+1,t+1}] = $ true;
11   **if** *all 256 indicators are true* **then**
12   **return** Random Permutation;

13 **return** 5-Round AES;

By setting the type-II error probability as 5%, it means that the success rate is 95%, then, $N \approx 2^{10.6}$ good pairs are required for each $(s, t), 0 \le s, t \le 3$. Since the probability to find a good pair from random ones is 2^{-120}, we have $N_s = 2^{67.6}$ by using $N_s \times 2^{63} \times 2^{-120} = N$. As a result, the data complexity is $2^{99.6}$ chosen ciphertexts. From Algorithm 2, Step 6 needs $16 \times N_s \times 2^{32} = 2^{103.6}$ memory accesses. Since the time to sort a table of size 2^n is $O(2^n log(2^n))$, Step 7 needs about $16 \times N_s 2^{32} log(2^{32})$. Then the time complexities of Step 8 and Steps 9–10 are $16 \times N_s \times 2^{32} = 2^{103.6}$ and $16 \times N_s \times N = 2^{82.2}$ memory accesses, respectively. Totally, the time complexity is about $2^{103.6}$ memory accesses. The memory requirements are 2^{32} to construct table T.

4.2 ID Distinguisher on 5-Round AES Based on Property 2 of MC

Similar to the method of constructing ID distinguisher on 5-round AES under chosen-ciphertext model in Sect. 4.1, we also can get an ID distinguisher under chosen-ciphertext model by using Property 2 of MC matrix transformed from the distinguisher in [7], see Proposition 8.

Proposition 8. *If the difference of ciphertext pair* (C^1, C^2) *is nonzero at the eight bytes* $(C_{0,3}, C_{1,2}, C_{2,1}, C_{3,0}, C_{0,2}, C_{1,2}, C_{2,0}, C_{3,3})$ *and zero at other 8 bytes, after 5-round AES decryption, the corresponding plaintext pair* (P^1, P^2) *never satisfies any one of the following 16 cases:*

$$P^1_{s,t} \oplus P^1_{s+1,t+1} = P^2_{s,t} \oplus P^2_{s+1,t+1} = K^0_{s,t} \oplus K^0_{s+1,t+1},$$
$$P^1_{s,t} \oplus P^1_{s+2,t+2} = P^2_{s,t} \oplus P^2_{s+2,t+2} = K^0_{s,t} \oplus K^0_{s+2,t+2},$$
$$P^1_{l,m} \oplus P^2_{l,m} = 0, (l,m) \neq (s,t), (s+1,t+1), (s+2,t+2),$$

where $0 \leqslant s, t \leqslant 3$.

However, for a random permutation, under each (s,t), *the probability that there always exists a tuple* $(\delta^1_{s,t}, \delta^2_{s,t})$ *that* $\delta^1_{s,t} = P^1_{s,t} \oplus P^1_{s+1,t+1}$ *and* $\delta^2_{s,t} = P^1_{s,t} \oplus P^1_{s+2,t+2}$ *never appearing for any one of* N *random good pairs is* $2^{256} \times (1 - 2^{-16})^{16N}$.

We omit the proof here due to its similarity to the distinguisher in Sect. 4.1. The distinguisher is illustrated in Algorithm 3 which is in Appendix B. The data and time complexities are $2^{76.5}$ chosen-ciphertexts and $2^{80.5}$ memory accesses, respectively. The type-II error probability is 5%.

5   Conclusions

In this paper, we study key-dependent integral and ID distinguishers on 5-round AES. A new key-dependent integral distinguisher is constructed with 2^{96} chosen plaintexts, which is more efficient than the previous one that requires the full codebook. Under different settings, the complexities of key-dependent integral distinguishers have a significant gap while those of the key-dependent ID distinguishers are almost the same. We analyze the principles behind the phenomena. If the AK operation which the key-dependent distinguishers depend on is positioned in the end of the distinguishers, the data complexities of integral and ID distinguishers will be almost unchanged no matter whether we consider or not the AK operations. Otherwise, the data complexities will increase significantly when we contain the AK operations in 5-round AES.

Acknowledgement. The authors thank the anonymous SAC 2018 reviewers for careful reading and many helpful comments. This work is supported by National Natural Science Foundation of China (Grant No. 61572293), Key Science Technology Project of Shandong Province (Grant No. 2015GGX101046), and Chinese Major Program of National Cryptography Development Foundation (Grant No. MMJJ2017012).

A   Property 2 and Key-Dependent Integral Distinguisher

In [7], Grassi *et al.* took advantage of Property 2 to build a more efficient ID distinguisher requiring $2^{76.4}$ chosen plaintexts. A question arises: Can we build an integral distinguisher based on Property 2?

Recall the key-dependent ID distinguisher based on Property 2, once the differences of $X_{0,0}^{1,SR}$, $X_{1,0}^{1,SR}$ and $X_{2,0}^{1,SR}$ are identical, differences on $X_{0,0}^{1,MC}$ and $X_{1,0}^{1,MC}$ will be zero with probability 1 (As described in Sect. 2.3). Therefore, in order to construct a key-dependent integral distinguisher with the similar technique we have to enforce the mask on $X^{4,MC}$ to statisfy following condition:

$$\Gamma_{X^{4,MC}} = \Gamma_{X^{4,AK}} = \beta_{i,j}, 0 \leqslant i,j \leqslant 3,$$

$$\beta_{i,j} \begin{cases} b \in F_2^8 \backslash \{0\} & \text{if}(i,j) = \in \{(0,0),(1,0),(2,0)\}, \\ 0 & \text{otherwise.} \end{cases}$$

For the purpose of extending the ZC linear hull one more round, we should carefully select the masks of $\Gamma_{X^{5,SB}}$ and make sure the correlation of $\Gamma_{X^{5,SB}} \rightarrow \Gamma_{X^{4,AK}}$ is 1, i.e. the equation

$$b \cdot (X_{0,0}^{4,AK} \oplus X_{0,0}^{4,AK} \oplus X_{0,0}^{4,AK}) = \Gamma_{X^{5,SB}} \cdot X^{5,SB}$$

always holds for any $X^{5,SB}$. Unfortunately, we cannot find any set of $X^{5,SB}$ or value of $\Gamma_{5,SB}$ to ensure it because of the non-linear property of SB.

B   Algorithm of 5-Round ID Distinguisher Under Chosen-Ciphertext Model Based on Property 2

The Algorithm 3 shows the process that we transfer the chosen-plaintext ID distinguisher based on Property 2 into a chosen-ciphertext one.

C   Gap Between Complexities of Chosen-Plaintext and Chosen-Ciphertext ID Distinguishers

Although the key-dependent integral distinguishers on 5-round AES have different data complexities under chosen-plaintext and chosen-ciphertext models, the complexity of key-dependent chosen-ciphertext ID distinguisher is slightly different from that of the chosen-plaintext one.

Similar to the key-dependent integral distinguishers, we will consider the influences of AK_0 operation, which the key-dependent ID distinguishers depend on. In this subsection, we only take the key-dependent ID distinguisher based on Property 1 for example. Situations are similar for the distinguihser based on Property 2. Notice that here we use a general ID characteristic with more active plaintext bytes (see Fig. 3) to make our analysis more convincing.

Algorithm 3. 5-Round ID Distinguisher under Chosen-Ciphertext Model Based on Property 2

Input: N_s structures of ciphertexts and corresponding plaintexts
Output: 5-Round AES or Random Permutation

1 **for** *Each $s \in \{0, 1, 2, 3\}$* **do**
2   **for** *Each $t \in \{0, 1, 2, 3\}$* **do**
3   Initialize 2^{16} indicators $V[2^{16}]$ as false;
4   **for** *Each one of N_s structures* **do**
             `// Each structure includes` 2^{64} `ciphertexts.`
5   Initialize a table $T[2^{64}]$;
6   Query the corresponding 2^{64} plaintexts and put them into T ;
7   Sort T according to the value of 13 bytes except bytes (s, t) and $(s+1, t+1)$ and $(s+2, s+2)$;
8   Traverse all items of T and find adjacent plaintexts to combine find good pairs;
             `// About` $N = N_s \times 2^{127} \times 2^{-120}$ `good pairs are found.`
9   **for** *Each (P^1, P^2) of N good pairs* **do**
10   Let $V[P^1_{s,t} \oplus P^1_{s+1,t+1}]$ = true;
11   **if** *all 2^{16} indicators are true* **then**
12   **return** Random Permutation;

13 **return** 5-Round AES;

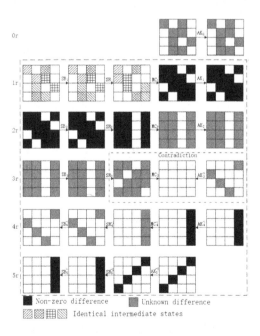

Fig. 3. 5-round impossible distinguisher with(out) AK_5.

Under Chosen-Plaintext Setting. If AK_0 operation is excluded from the 5-round AES (the enclosure area by dotted line in Fig. 3), we encrypt a pair of $(X^{0,AK}, \bar{X}^{0,AK})$ satisfying

- *Condition 1:*

$$\hat{X}_{0,0}^{0,AK} = \hat{X}_{1,1}^{0,AK}, \hat{X}_{1,2}^{0,AK} = \hat{X}_{2,3}^{0,AK},$$
$$\hat{X}_{0,2}^{0,AK} = \hat{X}_{3,1}^{0,AK}, \hat{X}_{0,3}^{0,AK} = \hat{X}_{3,2}^{0,AK},$$

where $\hat{X}$ represents X or $\bar{X}$;
- *Condition 2:*

$$X_{j,k}^{0,AK} = \bar{X}_{j,k}^{0,AK},$$

where $(j,k) \neq (0,0), (1,1), (1,2), (2,3), (0,2), (3,1), (0,3), (3,2),$

It is impossible that the corresponding ciphertext pair of $(C, \bar{C})$ has the active differences in only one reverse-diagonal. Yet for a random permutation, such pair appears with probability $4 \times 2^{-96} = 2^{-94}$. Given 2^{N_1} pairs of $(X^{0,AK}, \bar{X}^{0,AK})$, the probability p_1 that we identify a random permutation as 5-round AES without AK_0 is

$$p_1 = 1 - (1 - 2^{-94})^{2^{N_1}} = 1 - e^{-2^{N_1-94}}.$$

If we set $p_1 \geqslant 95\%$, then $N_1 \geqslant 95.6$.

All the $X^{0,AK}$ satisfying *Condition 1* and *2* compose a structure whose size is 2^{32}. Each structure can produce 2^{63} pairs. To construct $2^{95.6}$ pairs, we need to take $2^{95.6-63}$ different structures. Therefore, the total data complexity is $2^{95.6-63+32} = 2^{64.6}$ chosen plaintexts. To check the specific ciphertext pairs, we insert each ciphertext into a hash table indexed by four bytes in one diagonal and test whether there are two different ciphertexts in the same row of the hash table. Therefore, the time complexity of this attack is $2^{64.6}$ memory accesses.

If the AK_0 operation is taken into consideration, we will encrypt a pair of plaintexts $(P, \bar{P})$ and expect that the difference of corresponding $(C, \bar{C})$ would never be active in only one reverse-diagonal. To ensure it, $(P, \bar{P})$ should satisfy Eqs. (5) and (6):

$$\begin{aligned} \hat{P}_{0,0} \oplus \hat{P}_{1,1} = K_{0,0}^0 \oplus K_{1,1}^0, \hat{P}_{1,2} \oplus \hat{P}_{2,3} = K_{1,2}^0 \oplus K_{2,3}^0, \\ \hat{P}_{0,2} \oplus \hat{P}_{3,1} = K_{0,2}^0 \oplus K_{3,1}^0, \hat{P}_{0,3} \oplus \hat{P}_{3,2} = K_{0,3}^0 \oplus K_{3,2}^0, \end{aligned} \quad (5)$$

where $\hat{P}$ represents P or $\bar{P}$.

$$P_{j,k}^1 = P_{j,k}^2, (j,k) \neq (0,0), (1,1), (1,2), (2,3), (0,2), (3,1), (0,3), (3,2). \quad (6)$$

However, the XOR values of K^0 involved in Eq. (5) are unknown. We traverse 2^{32} possible values to ensure that the right XOR values of key are contained. For each XOR value in our traversing process, we fix other eight bytes of plaintexts involved in Eq. (6) as constant. Then we get 2^{32} structures of plaintexts.

For 5-round AES, structures with the right XOR values, i.e. the four XOR values are equal to the XOR values of a key described in Eq. (5), will never produce ciphertext pairs which have active differences in only one reverse-diagonal, but the structures with the wrong XOR values will do. However, for a random permutation, there will be at least one pair of ciphertexts with active bytes in only one diagonal if we take enough structures for each of 2^{32} XOR values.

The key point of the distinguisher is that we take enough pairs and make sure that we can get ciphertext pairs with active bytes in only one diagonal for each XOR value, if the target is a random permutation. If the probability that we get such a pair for one XOR value is p_1', the probability that we get such pairs for all the 2^{32} XOR values is $(p_1')^{2^{-32}}$.

If we set the probability that we can identify a random permutation as a random permutation at least 95%, we get $p_1' \geqslant (0.95)^{2^{-32}}$.

Given $2^{N_1'}$ pairs from structures one certain XOR value, p_1' can be calculated as follows

$$p_1' = 1 - (1 - 2^{-94})^{2^{N_1'}} = 1 - e^{-2^{N_1'-94}}.$$

Since $p_1' \geqslant (95\%)^{2^{-32}}$, we get $N_1' \geqslant 98.7$.

One structure produces 2^{63} pairs, so we need $2^{98.7-63} = 2^{35.7}$ structures, i.e. $2^{35.7+32} = 2^{67.7}$ chosen plaintexts for each XOR values. We have 2^{32} possible XOR values, so the total complexity is $2^{67.7+32} = 2^{99.7}$ chosen plaintexts. For each XOR value, we encrypt plaintexts and insert the corresponding ciphertexts into a hash table indexed by the four bytes in one diagonal and then check whether there are two ciphertexts in the same row of the hash table. Thus the time complexity is $2^{99.7}$ memory accesses.

Under Chosen-Ciphertext Setting. If AK_0 operation is excluded and we decrypt a pair of ciphertexts $(C, \bar{C})$ with active bytes in only one diagonal to $(X^{0,AK}, \bar{X}^{0,AK})$. For 5-round AES without AK_0, the pair $(X^{0,AK}, \bar{X}^{0,AK})$ will never satisfy *Condition 1* and *2* at the same time while for a random permutation, such pair appears with probability 2^{-128} (2^{-64} for the probability to satisfy *Condition 1* and 2^{-64} for *Condition 2*).

In order to distinguish 5-round AES without AK_0 from a random permutation, we use 2^{N_2} ciphertext pairs, thus the probability p_2 that there will be at least a pair of $(X^{0,AK}, \bar{X}^{0,AK})$ satisfying *Condition 1* and *2* for a random permutation is:

$$p_2 = 1 - (1 - 2^{-94})^{2^{N_2}} = 1 - e^{-2^{N_2-94}}.$$

Setting $p_2 \geqslant 95\%$ we will get $N_2 \geqslant 129.6$.

We fix 12 bytes of three diagonals as constants and take all possible values for other four bytes to compose a structure. Each structure provides 2^{63} pairs with 2^{32} ciphertexts. Thus we need $2^{129.6-63} = 2^{66.6}$ structures and the total data complexity is $2^{66.6+32} = 2^{98.6}$. We decrypt ciphertexts and insert the $X^{0,AK}$

satisfying *Condition 1* into a hash table indexed by eight bytes involved in *Condition 2* and then check whether there are two texts in the same row of the hash table. Thus the time complexity is $2^{98.6}$ memory access.

If AK_0 operation is contained and we decrypt a pair of ciphertexts $(C, \bar{C})$ with active differences in only one diagonal to get corresponding plaintext pair $(P, \bar{P})$, the intermediate state $(X^{0,AK}, \bar{X}^{0,AK})$ will never satisfy *Condition 1* and *2*, thus $(P, \bar{P})$ cannot satisfy Eqs. (5) and (6), neither.

Since we do not know the key information involved in the Eq. (5), we have to collect good pairs and test whether each possible XOR value will occur as described in Sect. 4.1. Given $2^{N'_2}$ ciphertext pairs, we expect to collect $2^{N'_2 - 96}$ good pairs. The probability p'_2 that all the possible XOR values will occur is

$$p'_2 = 1 - 2^{32} \times (1 - 2^{-32})^{2^{N'_2 - 96}}$$

Setting $p'_2 \geqslant 95\%$ we can get $N'_2 \geqslant 132.7$.

Since one structure provides 2^{32} ciphertexts and 2^{63} pairs, we need $2^{132.7-63} = 2^{69.7}$ structures and totally $2^{69.7+32} = 2^{101.7}$ chosen ciphertexts. When proceeding the attack, we decrypt ciphertexts and insert the corresponding plaintexts satisfying *Condition 1* into a hash table indexed by other eight bytes, and check whether there are two plaintexts in the same row. Therefore the time complexity is $2^{101.7}$ memory accesses. The complexity is very similar with the distinguisher without AK_0.

We analyze the reason why the chosen-plaintext and chosen-ciphertext ID distinguishers have a similar data complexity. Without AK_0 operation, the chosen-plaintext distinguisher requires $2^{64.6}$ chosen plaintexts while the chosen-ciphertext distinguisher needs $2^{98.6}$ chosen ciphertexts. However, when we take the AK_0 operation into consideration, the data complexity increases significantly under chosen-plaintext setting while it remains almost unchanged under chosen-ciphertext setting. To make it clear, we list the complexities of these distinguishers in Table 3.

Table 3. Data complexities of 5-round ID distinguishers with(out) AK_0.

Setting	Target	Data complexity	Time (MA)
CP	5-round AES without AK_0	$2^{64.6}$	$2^{64.6}$
	5-round AES with AK_0	$2^{96.7}$	$2^{101.7}$
CC	5-round AES without AK_0	$2^{98.6}$	$2^{98.6}$
	5-round AES with AK_0	$2^{101.7}$	$2^{101.7}$

– CC: Chosen-Ciphertext CP: Chosen-Plaintext MA: Memory Access

D   Figures of the Distinguisher Introduced in Sect. 2

See Figs. 4, 5, and 6.

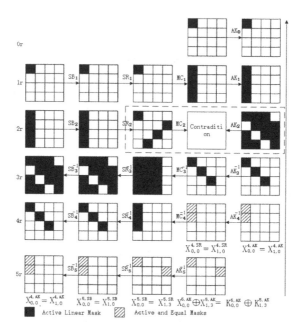

Fig. 4. ZC linear hull of 5-round AES [12].

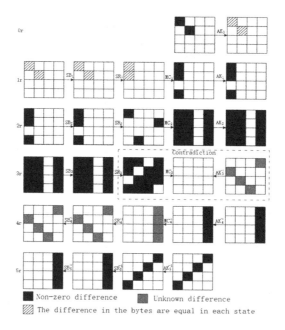

Fig. 5. ID of 5-round AES based on Property 1 [8].

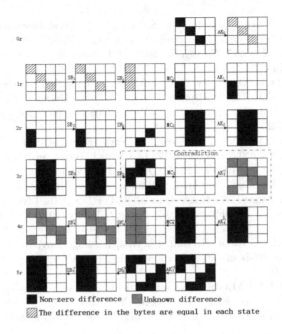

Fig. 6. ID of 5-round AES based on Property 2 [7].

References

1. Biham, E., Keller, N.: Cryptanalysis of reduced variants of Rijndael. In: 3rd AES Conference, vol. 230 (2000)
2. Bogdanov, A., Leander, G., Nyberg, K., Wang, M.: Integral and multidimensional linear distinguishers with correlation zero. In: Wang, X., Sako, K. (eds.) ASIACRYPT 2012. LNCS, vol. 7658, pp. 244–261. Springer, Heidelberg (2012). https://doi.org/10.1007/978-3-642-34961-4_16
3. Cui, T., Sun, L., Chen, H., Wang, M.: Statistical integral distinguisher with multi-structure and its application on AES. In: Pieprzyk, J., Suriadi, S. (eds.) ACISP 2017. LNCS, vol. 10342, pp. 402–420. Springer, Cham (2017). https://doi.org/10.1007/978-3-319-60055-0_21
4. Daemen, J., Rijmen, V.: The Design of Rijndael: AES-The Advanced Encryption Standard. ISC. Springer Science & Business Media, Heidelberg (2013). https://doi.org/10.1007/978-3-662-04722-4
5. Datta, N., Nandi, M.: ELmD v2.0 (2015). Submission to the caesar competition
6. Gilbert, H., Minier, M.: A collision attack on 7 rounds of Rijndael. In: AES Candidate Conference, pp. 230–241 (2000)
7. Grassi, L.: MixColumns properties and attacks on (round-reduced) AES with a single secret S-Box. In: Smart, N.P. (ed.) CT-RSA 2018. LNCS, vol. 10808, pp. 243–263. Springer, Cham (2018). https://doi.org/10.1007/978-3-319-76953-0_13
8. Grassi, L., Rechberger, C., Rønjom, S.: Subspace trail cryptanalysis and its applications to AES. IACR Trans. Symmetric Cryptol. **2016**(2), 192–225 (2016)

9. Grassi, L., Rechberger, C., Rønjom, S.: A new structural-differential property of 5-round AES. In: Coron, J.-S., Nielsen, J.B. (eds.) EUROCRYPT 2017. LNCS, vol. 10211, pp. 289–317. Springer, Cham (2017). https://doi.org/10.1007/978-3-319-56614-6_10

10. Lu, J., Dunkelman, O., Keller, N., Kim, J.: New impossible differential attacks on AES. In: Chowdhury, D.R., Rijmen, V., Das, A. (eds.) INDOCRYPT 2008. LNCS, vol. 5365, pp. 279–293. Springer, Heidelberg (2008). https://doi.org/10.1007/978-3-540-89754-5_22

11. Rønjom, S., Bardeh, N.G., Helleseth, T.: Yoyo tricks with AES. In: Takagi, T., Peyrin, T. (eds.) ASIACRYPT 2017. LNCS, vol. 10624, pp. 217–243. Springer, Cham (2017). https://doi.org/10.1007/978-3-319-70694-8_8

12. Sun, B., Liu, M., Guo, J., Qu, L., Rijmen, V.: New insights on AES-like SPN ciphers. In: Robshaw, M., Katz, J. (eds.) CRYPTO 2016. LNCS, vol. 9814, pp. 605–624. Springer, Heidelberg (2016). https://doi.org/10.1007/978-3-662-53018-4_22

13. Wang, M., Cui, T., Chen, H., Sun, L., Wen, L., Bogdanov, A.: Integrals go statistical: cryptanalysis of full skipjack variants. IACR Cryptology ePrint Archive 2016:178 (2016)

14. Wu, H., Preneel, B.: AEGIS: a fast authenticated encryption algorithm. In: Lange, T., Lauter, K., Lisoněk, P. (eds.) SAC 2013. LNCS, vol. 8282, pp. 185–201. Springer, Heidelberg (2014). https://doi.org/10.1007/978-3-662-43414-7_10

Mind the Gap - A Closer Look
at the Security of Block Ciphers
against Differential Cryptanalysis

Ralph Ankele[1]([envelope])(iD) and Stefan Kölbl[2,3](iD)

[1] Royal Holloway University of London, Egham, UK
ralph.ankele.2015@rhul.ac.uk
[2] DTU Compute, Technical University of Denmark, Kongens Lyngby, Denmark
[3] Cybercrypt, Hellerup, Denmark
stek@mailbox.org

Abstract. Resistance against differential cryptanalysis is an important design criteria for any modern block cipher and most designs rely on finding some upper bound on probability of single differential characteristics. However, already at EUROCRYPT'91, Lai et al. comprehended that differential cryptanalysis rather uses *differentials* instead of single *characteristics*.

In this paper, we consider exactly the gap between these two approaches and investigate this gap in the context of recent lightweight cryptographic primitives. This shows that for many recent designs like Midori, Skinny or Sparx one has to be careful as bounds from counting the number of active S-boxes only give an inaccurate evaluation of the best differential distinguishers. For several designs we found new differential distinguishers and show how this gap evolves. We found an 8-round differential distinguisher for Skinny-64 with a probability of $2^{-56.93}$, while the best single characteristic only suggests a probability of 2^{-72}. Our approach is integrated into publicly available tools and can easily be used when developing new cryptographic primitives.

Moreover, as differential cryptanalysis is critically dependent on the distribution over the keys for the probability of differentials, we provide experiments for some of these new differentials found, in order to confirm that our estimates for the probability are correct. While for Skinny-64 the distribution over the keys follows a Poisson distribution, as one would expect, we noticed that Speck-64 follows a bimodal distribution, and the distribution of Midori-64 suggests a large class of weak keys.

Keywords: Symmetric-key cryptography · Differential cryptanalysis Lightweight cryptography · SAT/SMT solver · IoT · LBlock · Midori Present · Prince · Rectangle · Simon · Skinny · Sparx · Speck · Twine

R. Ankele—This research was partially supported by the European Union's Horizon 2020 research and innovation programme under grant agreement No. H2020-MSCA-ITN-2014-643161 ECRYPT-NET.

C. Cid and M. J. Jacobson, Jr. (Eds.): SAC 2018, LNCS 11349, pp. 163–190, 2019.
https://doi.org/10.1007/978-3-030-10970-7_8

1   Introduction

Differential cryptanalysis, first published by Biham and Shamir [9] to analyse the DES, has become one of the prime attack vectors which any modern symmetric-key primitive has to be resistant against. The idea behind differential cryptanalysis is to find a correlation between the difference of a pair of plaintexts and ciphertexts which holds with high probability. The challenge for an cryptanalyst consists of finding such a correlation or to show that no such correlation exists. A popular approach is to design a cipher in such a way that one can find a bound on the best differential characteristics, either directly e.g., the wide-trail strategy deployed in AES or using methods based on Matsui's algorithm, MILP or SAT.

A differential characteristic specifies all the intermediate differences after each round of the primitive. However, when constructing a differential distinguisher one only cares about the input and output difference. It is often assumed that a single characteristic dominates the probability of such a differential, however this is not true in general and leads to imprecise estimates of the probability in many cases [10,24].

In the work by Lai, Massey and Murphy [33] they showed that if an iterated cryptographic primitive has independent round-keys, it can be considered as a Markov cipher. As differential cryptanalysis considers just the first and last difference and ignores the intermediate values, the probability of such a *differential* can then be computed as the sum of all characteristics, that are formed by the differentials. While this assumes that the rounds are independent, it provides a more precise estimate and the probability of the most probable *differential* will always be greater than the probability of the most probable *characteristic*.

Contributions. We provide a broad study covering different design strategies and investigate the differential gap between single *characteristics* and *differentials* for the block ciphers LBlock, Midori, Present, Prince, Rectangle, Simon, Skinny, Sparx, Speck and Twine. In order to do this, we use an automated approach for enumerating the characteristics with the highest probability contributing to a differential based on SMT solvers [41], which we adopt to different design strategies. This allows us to efficiently enumerate a large set of characteristics contributing to the probability of a differential resulting in a precise estimate for the probability of differentials.

For Skinny-64 we present an 8-round differential distinguisher with a probability of $2^{-56.93}$, while the best single characteristic only suggests a probability of 2^{-72}. For Midori-64 we show that the best characteristic for 8 rounds, with a probability of 2^{-76} can be used to find a differential with a probability of $2^{-60.86}$. Our results show that in the case of many new lightweight ciphers like Midori-64, Skinny-64, and Sparx-64 the probabilities improve significantly and that we can find differential distinguishers which are able to cover more rounds. This suggests that one should be particularly careful with lightweight block ciphers when using simpler approximations like counting the number of active S-boxes.

Our method is generic and can easily be applied to other designs as one only needs to describe the differential behaviour of the round function and can re-use all the components we implemented for doing so. This allows both to find optimal differential characteristics and to enumerate all characteristics contributing to a differential.

Furthermore, we provide experiments to verify that our estimates of the differential probability provide a good approximation. However, we also noticed that the distribution over the choice of keys varies significantly for some design strategies and that commonly made assumptions do not hold for reduced-round versions. While for Skinny-64 the distribution over the keys follows relatively closely what one would expect we noticed that for Midori-64 for a large class of keys there are no pairs following the differential at all, while for very few keys the probability is significantly higher.

Related Work. Daemen and Rijmen firstly studied the probability of differentials for AES in their work on Plateau Characteristics [20]. In their work, they analysed AES on the distribution of differential probability over the choice of keys and showed that all 2-round characteristics have either a zero probability or for a small subset of keys the probability is non-zero. However, they only considered AES, but conjectured that other ciphers with 4-uniform S-boxes will show a similar result. In the case of AES and AES-like ciphers, there has also been a lot of research in studying the expected differential/linear probability (MEDP/MELP) [18,30], that is used to provable bound the security of a block cipher against differential/linear cryptanalysis.

In recent years, many automated tools were proposed that could help designers to prove bounds against differential/linear attacks. Mouha et al. [42] used Mixed Integer Linear Programming (MILP) to count active S-boxes and compute provable bounds. Furthermore, there have been a few approaches of using automated tools to find optimal characteristics, and to collect many characteristics with the same input/output differences. This idea was first introduced by Sun et al. [46] who used MILP. Likewise, tools using SAT/SMT solvers are used where the results were applied to Salsa-20 [41], Norx [5], and Simon [31].

Moreover, there exist several design and attack papers that study the effect of numerous characteristics contributing to the probability of a differential: Mantis [24], Noekeon [29], Salsa [41], Simon/Speck [11,31], Rectangle [54] and Twine [10]. Yet, these are often based on truncated differentials or dedicated algorithms for finding large numbers of characteristics. For example in [25], Eichlseder and Kales attack Mantis-6 by finding a large cluster of differential characteristics. Contrary to the attack on Mantis-5 by Dobraunig et al. [24] where the cluster was found manually, in the attack on Mantis-6, Eichlseder and Kales used a tool based on truncated differentials.

Similar effects have also been observed in the case of linear cryptanalysis, where Abdelraheem et al. [1] showed that the security margins based on the distribution of linear biases are not always accurate. Their work has further been studied and improved by Blondeau and Nyberg [13].

Software. All the models for enumerating the differential characteristics are publicly available at https://github.com/TheBananaMan/cryptosmt.

Outline. The remainder of this paper is structured as follows. After briefly revisiting some of the necessary definitions about differential cryptanalysis in Sect. 2, we provide details about the automated tools that we use in Sect. 3 and describe how to efficiently find differential characteristics for various ciphers. In Sect. 4 we present the results of our analysis on the gap between single differential characteristics and differentials for various cryptographic primitives. We also analyze the best differential attacks, that are published on those ciphers so far, and show if the attacks can be improved by considering the aforementioned differential gaps. Moreover, in Sect. 5 we give details about our experiments of the distribution over keys for the probability of differentials.

2   Differentials and Differential Characteristics

Differential cryptanalysis is one of the most powerful techniques in the analysis of symmetric-key primitives. Many extensions to it have been developed and it has found wide applications on block ciphers, stream ciphers and cryptographic hash functions. In the following, we state some definitions and notations that we will use throughout the paper.

A *block cipher* is a family of permutations parameterised by a *key* $K \in \mathbb{F}_2^k$, that maps a set of *plaintexts* $P \in \mathbb{F}_2^n$ to a set of *ciphertexts* $C \in \mathbb{F}_2^n$

$$E_K : \mathbb{F}_2^k \times \mathbb{F}_2^n \to \mathbb{F}_2^n. \tag{1}$$

Virtually all currently used block ciphers are *iterative* block ciphers, i.e., they are composed of applying a simple round function r times

$$E_K(\cdot) = f_r(\cdot) \circ \ldots \circ f_1(\cdot). \tag{2}$$

The idea of differential cryptanalysis is to look at pairs of plaintexts (p_1, p_2) and the corresponding ciphertexts (c_1, c_2) and try to find a correlation between the differences α and β, where $\alpha = p_1 \oplus p_2$ and $\beta = c_1 \oplus c_2$.

Definition 1. *A differential is a pair of differences* $(\alpha, \beta) \in \mathbb{F}_2^n \times \mathbb{F}_2^n$.

If such a correlation holds with high probability, we can use this to distinguish the block cipher from a random permutation and further use this to mount key-recovery attacks.

Definition 2. *The differential probability of a differential over a block cipher is*

$$DP(\alpha \xrightarrow{E_K} \beta) = \Pr_X(E_K(X) \oplus E_K(X \oplus \alpha) = \beta). \tag{3}$$

where X is a random variable that is uniformly distributed over $\mathbb{F}_2^n$.

For ease of notation we define the *weight* of a differential as $-\log_2(\mathrm{DP}(\cdot))$. Any non-zero differential for a random permutation $F_\$: \mathbb{F}_2^n \to \mathbb{F}_2^n$ will have a differential probability close to 2^{-n}. Therefore one is interested in finding any differential with $\mathrm{DP}(\alpha \xrightarrow{E_K} \beta) \gg 2^{-n}$. In general, it is computationally infeasible to compute the exact value of the DP as this would require to exhaustively search through the whole space of all possible plaintexts. One can use the structure of a block cipher, to obtain a good approximation of the actual DP with less computational effort by tracking the differences through the round functions.

Definition 3. *A differential characteristic is a sequence of differences*

$$Q = (\alpha_1 \xrightarrow{f_1} \alpha_2 \xrightarrow{f_2} \ldots \xrightarrow{f_{r-1}} \alpha_r). \tag{4}$$

Yet, it is still computationally infeasible to compute the exact value of $\mathrm{DP}(Q)$ and the general approach is to assume independence of the rounds. For most designs it is feasible to compute the exact probability of a differential for a single round. One can therefore compute

$$\mathrm{DP}(Q) \approx \prod_{i=1}^{r-1} \Pr_X(\alpha_i \xrightarrow{f_i} \alpha_{i+1}). \tag{5}$$

While this assumption of independent rounds is not true in general, it has been shown to serve as a good approximation in practice. However, if an adversary wants to construct a distinguisher, she actually does not care about any intermediate differences and is only interested in the probability of the differential. The adversary can therefore collect all differential characteristics sharing the same input and output difference to get a better estimate

$$\Pr(\alpha_1 \xrightarrow{E} \alpha_r) = \sum_{\alpha_2,\ldots,\alpha_{r-1}} \Pr_X(\alpha_1 \xrightarrow{f_1} \alpha_2 \xrightarrow{f_2}_{f_1(X)} \cdots \alpha_{r-1} \xrightarrow{f_{r-1}}_{f_{r-1}\circ\ldots\circ f_1(X)} \alpha_r). \tag{6}$$

It is often assumed that the probability of the differential is close to the probability of the best single characteristic. While this might hold for some ciphers this assumption has been shown to be inaccurate in several cases and does not hold for many modern block ciphers [10,24]. We will show later in Sect. 4 that this assumption fails particularly often for some recently designed lightweight block ciphers.

We consider two different criteria for a design: *differential characteristic resistant* (DCR), which means that no single differential characteristic exists with a probability larger than 2^{-n} and *differential resistant* (DR) which means that it should be difficult to find a differential with a probability larger than 2^{-n}. Note that we typically can not avoid that there are differentials with $\mathrm{DP} \geq 2^{-n}$, as if we fix the input difference to α_1 then $\sum_{\alpha_r \neq 0} \Pr(\alpha_1 \xrightarrow{E} \alpha_r) = 1$. This implies that there exists at least one differential with a probability $\mathrm{DP} \geq 2^{-n}$. In the *Wide-Trail Strategy* which was used to design the AES and subsequently many other ciphers, Daemen and Rijmen suggest that it is a sound design strategy to restrict the probability of difference propagation [19]. Nevertheless, this does not result in a proof for security.

Note that in the definitions so far the influence of the keys was ignored. However, the DP for a specific differential strongly depends on the choice of the secret key and it is therefore of interest how this distribution looks like. To solve this problem we could compute the probabilities of a differential over the whole key space, however this is again practically infeasible which leads one to use the *expected differential probability*.

Definition 4. *The* expected differential probability *of a block cipher E_k of an r-round differential (α, β), with a key-size of κ-bits is defined as*

$$\text{EDP}(\alpha \xrightarrow{E} \beta) = 2^{\kappa} \sum_{k \in \mathbb{F}_2^{\kappa}} \Pr_X(\alpha \xrightarrow[X]{E_k} \beta). \tag{7}$$

In order to derive some sort of security proof against differential cryptanalysis often the *Hypothesis of Stochastic Equivalence* [33] is used which states that for all differentials Q it holds that for most keys K the differential probability of a characteristic is close to the expected differential probability, $\text{DP}_K(Q) \approx \text{EDP}(Q)$. In practice this hypothesis does not always hold [16], which we will also see later in Sect. 5.

3 Finding Differential Characteristics Efficiently

While there are many methods based on SAT, MILP or Matsui's algorithm to find differential characteristics and even prove an upper bound on the probability of the best single characteristic, it remains a hard problem to find a good estimate on the probability of the best differential. Even finding those differential characteristics remains a difficult problem for some design strategies and cryptanalysts had to search manually for differentials in some attacks [53]. Nowadays a variety of automated tools [12,35,45] is available which are constantly improved and help cryptanalysts in finding good differential characteristics.

3.1 SAT/SMT Solvers

SAT solvers are used to solve the Boolean satisfiability problem (SAT) and are based on heuristic algorithms. A solver starts from an initial assignment for the literals and then builds a search tree by using systematic backtracking until all conflicting clauses are resolved and either an assignment of variables for a satisfiable set of clauses is returned or the solver decides that this instance is unsatisfiable. The most commonly algorithms used in SAT solvers are based on the original idea of DPLL [21].

SMT solvers are more powerful than SAT solvers in the sense that they can express constraints on a higher abstraction layer and allow simple first-order logic. In general, SMT solvers often translate the problem to SAT and then use an improved version of the DPLL algorithm and backtracking to infer when theory conflicts arise. Moreover, the solver checks the feasibility of conjunctions from the first-order logic predicates as it interacts with the Boolean formulas that are returned by the SAT solver.

There exists a few SAT/SMT solvers that are suitable for our use cases. STP [50] is an SMT solver that uses the CVC and SMTLIB2 language to encode the constraints and then invokes a SAT solver to check for satisfiability of the model. CryptoMiniSat [40] is an advanced SAT solver that supports features like XOR recovery[1] to simplify clauses. As XOR operations are commonly used in cryptography this can be an advantage and potentially reduces the solving time. We also considered other solvers like Boolector [43], which for some instances provide a better performance, however in general this only provides an improvement by a small constant factor and it is hard to identify for which instances one obtains any advantage.

3.2 From Differential Cryptanalysis to Satisfiability Modulo Theories

When using automated tool like SAT/SMT solvers, one can simplify the search for differential characteristics and differentials by modeling the differential behavior of the block cipher. For this we represent all intermediate states of our block cipher as variables which corresponds to the differences and encode the transitions of differences through the round functions as constraints that can be processed by the SMT/SAT solver. An advantage of using SMT over SAT for the modeling is that most SMT solvers support reasoning over bit-vectors which are commonly used in block cipher designs, especially when considering word-oriented ciphers. This both simplifies the modeling of the constraints and can lead to an improved time for solving the given problem instances compared to an encoding in SAT.

Constructing an SMT Model. In this paper, we focus on a tool that uses the CVC language[2] for encoding the differential behavior of block ciphers. Therefore, we encode the constraints imposed by the round function for each round of the block cipher and the probability of the resulting differential transitions. Our main goal here is to construct an SMT model which decides whether

$$\exists Q : \mathrm{DP}(Q) = 2^{-t}, \tag{8}$$

which allows us to find the best differential characteristic Q for a cipher by finding the minimum value t for which the model is satisfiable.

In order to represent the differential behaviour of a cipher we consider any operation in the cipher, e.g., the application of an S-box, matrix multiplication, word-wise operation or bit operation, and add constraints for a valid transition

---

[1] See https://www.msoos.org/2011/03/recovering-xors-from-a-cnf/.

[2] A list of all bitwise and word level functions in CVC is available at: http://stp. github.io/cvc-input-language/.

from an input to an output difference such that any valid assignment to the variables corresponds to a valid differential characteristic in the actual operation. For any non-linear component we introduce additional variables w^j which represent the $\log_2$ probability of the differential transition. The probability of Q is then given by $\sum w^j$. This means that a valid assignment for all these variables directly gives us the differential characteristic Q with all intermediate differences and $\mathrm{DP}(Q) = p$.

In the following we give an overview on how the different components of the ciphers can be modeled in the SMT model. The algorithms to find the optimal differential characteristics and consequently good estimates for the differentials are described in Sect. 3.3.

S-Boxes. Substitution Permutation Network (SPN) ciphers typically use S-boxes, which are non-linear functions operating on a small number of bits. These are often 4- or 8-bit functions and therefore we can compute the differential probability by simply constructing the Difference Distribution Table (DDT), which is a full lookup table of all possible pairs of input/output differences, for each S-box. In our SMT model we represent the input difference to an n-bit S-box as $\alpha = \alpha_1, \ldots, \alpha_n$ respectively the output as $\beta = \beta_1, \ldots, \beta_n$. These variables correspond to the input/output difference to this S-box and we want to constraint them to only allow non-zero probability combinations of input and output differences. We further introduce additional variables $w = w_1, \ldots, w_n$ which are used to represent the probability of the transition. The probability of the transition is encoded as $2^{-wt(w)}$, where $wt(\cdot)$ denotes the Hamming weight of w.

In order to construct the constraints on the variables, we first find all valid transitions and their corresponding probability. We want to construct a CNF which is satisfiable if and only if the assignment corresponds to such a valid characteristic. One simple way to this is by just considering all assignments which are impossible. If a transition is defined as $(a \xrightarrow{S} b)$ and has a probability c then we add the following clause

$$
\begin{aligned}
T = \ & N(a_1, \alpha_1) \vee \ldots \vee N(a_n, \alpha_n) \vee \\
& N(b_1, \beta_1) \vee \ldots \vee N(b_1, \beta_n) \vee \\
& N(c_1, w_1) \vee \ldots \vee N(c_n, w_n)
\end{aligned}
\tag{9}
$$

where

$$
N(x_i, y_i) = \begin{cases} \neg y_i, \text{if } x_i = 0 \\ y_i, \text{if } x_i = 1 \end{cases}.
\tag{10}
$$

This clause is only satisfiable if the variables of the corresponding S-box are not set to the invalid assignment. For example let $a = (1, 0, 1, 1)$, $b = (0, 0, 0, 0)$ and $c = (0, 0, 0, 0)$ then we add the clause

$$
(\neg\alpha_0 \vee \alpha_1 \vee \neg\alpha_2 \vee \neg\alpha_3 \vee \beta_0 \vee \beta_1 \vee \beta_2 \vee \beta_3 \vee w_0 \vee w_1 \vee w_2 \vee w_3).
\tag{11}
$$

We implemented this approach to generate the SMT models for 4- and 8-bit S-boxes, where most of the lightweight ciphers actually use 4-bit S-boxes which allows a very compact description (i.e., to represent the 4-bit S-box of Skinny we need 12 variables and about 3999 clauses in CNF). Note that our method is limited to S-boxes which have a DDT with entries that are a power of 2. For other S-boxes a similar method could be used by using l additional variables for encoding probabilities of the form $2^{-0.5}, 2^{-0.25}, \ldots$ to get an approximation of the actual probability.

Linear Layers. The diffusion layers of Substitution Permutation Networks in lightweight ciphers are often constructed with simple bit-permutations (e.g., Present) or by multiplication with matrices having only binary coefficients (e.g., Midori, Skinny). ARX-based ciphers (e.g., Speck) use the diffusion properties of XOR combined with rotations. Feistel networks (e.g., Simon, LBlock, Twine) also mix the state by switching parts of the states on every Feistel switch.

For modeling rotations and bit-permutations in an SAT/SMT solver, we simply have to re-index the variables accordingly before they are input to another function. This can be achieved using SMT predicates (ASSERT and equality) in the CVC language. Rotations can be realized using predicates for *shifting* words and the word-wise *or* function that are available in the CVC language. The multiplication by a binary matrix can be modeled using the *xor* predicate at the word-level.

ARX Designs. ARX designs use modular additions (modulo 2^n), XOR and rotations. As modular addition is the only non-linear component, that is not already available in the SMT solver, we use an algorithm proposed by Lipmaa and Moriai [36] to efficiently compute the differential probability of modular addition. Let $xdp^+(\alpha, \beta \to \gamma)$ be the XOR differential probability of modular addition, where α, β are input differences and γ is the output difference, then it holds that a differential is valid if and only if:

$$\text{eq}(\alpha \ll 1, \beta \ll 1, \gamma \ll 1) \wedge (\alpha \oplus \beta \oplus \gamma \oplus (\beta \ll 1)) = 0 \qquad (12)$$

where

$$\text{eq}(x, y, z) := (\neg x \oplus y) \wedge (\neg x \oplus z). \qquad (13)$$

The weight of a valid differential is defined as:

$$w(\alpha, \beta, \gamma) := -\log_2 (xdp^+(\alpha, \beta \to \gamma)) = wt'(\neg eq(\alpha, \beta, \gamma)). \qquad (14)$$

where $wt'(\cdot)$ denotes the Hamming weight omitting the most significant bit. We implemented this algorithm to calculate the differential probability of modular additions.

3.3   Finding the Best Characteristics and Differentials

We use the open-source tool CryptoSMT [45] for the automated search of differential characteristics and implemented several missing functionalities for block ciphers (i.e., support for S-boxes as described in Sect. 3.2, and binary diffusion matrices). CryptoSMT is based on the state-of-the-art SAT/SMT solvers, CryptoMiniSat [40] and STP [50].

The tool offers a simple API that allows cryptanalysts and designers to formulate various cryptanalytic problems and solve them with the underlying SAT/SMT solver. We added the models for the block ciphers Skinny, Midori, Rectangle, Present, Prince, Sparx, Twine and LBlock (Note that some of these are block cipher families and we focused on a subset of parameters) to CryptoSMT and use the following two functionalities provided by the tool:

– Decide if a differential characteristic with probability p exists.
– Enumerate all differential characteristics with a probability of p.

Based on this we can achieve our two goals, namely finding the best differential characteristic and estimating the probability of the differential.

Best Differential Characteristic. In order to find the characteristic Q with maximum probability p_{max} for r rounds of a block cipher we start by checking whether our model is satisfiable for a probability of p, starting at $p = 1$. If our model is not satisfiable we continue by checking whether there is a valid assignment for $p = 2^{-1}$. Note that for all our block ciphers the probability of the differential transitions are powers of two and therefore there does not exist any differential characteristic which has a probability p' such that $2^{-(t+1)} < p' < 2^{-t}$ for any integer t. We continue this process until we reach a model which is satisfiable, which gives us an assignment of all variables of the state forming a valid differential characteristic with probability $p_{max} = 2^{-t}$. Considering that we start with probability $p = 1$ and then we constantly increase the weight, and finish as soon as we found an valid assignment, we can ensure that we found the best differential characteristic.

Estimate for the Probability of a Differential. In order to find a good differential we can use a tool assisted approach to compute an approximation for Eq. 6, as shown in [41]. We first obtain the best single characteristic Q with probability $p = 2^{-t}$ which gives us the input difference α_1 and output difference α_r. Subsequently we modify our model and fix the input and output difference to α_1 respectively α_r. Note that this restricts the search space significantly and results in a much faster time for solving any subsequent SMT instances.

The next step is to find all differential characteristics Q, such that $DP(Q) = 2^{-u}$, for $u = t, t + 1, \ldots$, under this new constraints. This allows us to collect more and more terms of the sum in Eq. 6, improving our approximation for the differential. By doing this process we always search for those differential characteristics which contribute the most to the probability of the differential first.

Here we assume that the input and output difference imposed by the best differential characteristic correspond to a good differential. While this assumption might not always hold and some of the differentials we found significantly improve the best differential distinguishers there could still exist better starting points for our search, for example as shown in [32] against the block cipher Simeck.

4   Analysis of the Gap in Lightweight Ciphers

The construction of cryptographic primitives optimized for resource constrained devices has received a lot of attention over the last decade and various design strategies and optimisation targets have been explored. All these primitives exhibit the idea of using simpler operations in order to save costs and therefore often exhibit a simpler algebraic structure compared to other symmetric-key algorithms.

For some design strategies this leads to a significant larger gap between single characteristics and differentials. This gap becomes especially relevant for aggressively optimised designs with minor security margins. Table 1 gives an overview of all the block ciphers we analysed with the methodology outlined in Sect. 3 and their security margins as well as the best known differential attacks.

Table 1. Best attacks and security margins (active S-boxes) for various design strategies for symmetric cryptographic primitives. D/MD/RK/ID/R/TD = differential, multiple differential, related-key, impossible differential, rectangle, truncated differential

Group	Design strategy	Cipher	Block size	Key size	Rounds	Margin (active S-boxes)	Best differential attack	Exploit differentials
SPN	AES-like	Midori	64	128	16	9 rounds	Full rounds (RK) [26]	✗
		Skinny	64	64	32	24 rounds	19 rounds (ID) [38]	✓
		Skinny	64	128	36	28 rounds	23 rounds (ID) [3,38]	✓
		Skinny	64	192	40	32 rounds	27 rounds (R) [38]	✓
	Bit-sliced	Rectangle	64	80/128	25	-	18 rounds (D) [48,54]	Section 4.6
	Present-like	Present	64	80/128	31	12 rounds	26 rounds (D) [37,51]	✓
	Reflection	Prince	64	128	12	-	10 rounds (MD) [17]	✓
	ARX	Sparx	64	128	24	9 rounds	16 rounds (TD) [4]	✓
Feistel	AND-RX	Simon	64	96	42	-	26 rounds (D) [2]	✓
		Simon	64	128	44	-	26 rounds (D) [2]	✓
	ARX	Speck	64	96	26	-	19 rounds (D) [44]	✓
		Speck	64	128	27	-	20 rounds (D) [44]	✓
	GFN	Twine	64	80	36	21 rounds	23 rounds (ID) [10]	✗
		Twine	64	128	36	21 rounds	25 rounds (ID) [10]	✗
	Two-branched	LBlock	64	80	32	17 rounds	24 rounds (ID) [52]	✗

4.1   Designs Strategies

We categorise these lightweight ciphers according to their design strategies as this has the largest influence on the gap. In general one can distinguish between two main design families: Substitution-Permutation Networks (SPN) and Feistel Networks. Within these families we can gather ciphers according to other structural properties. These are for SPN: AES-like, Bit-sliced S-boxes, Bit-based Permutation Layers, Reflection Ciphers, ARX-based and for Feistel: ARX-based, Generalized Feistel Networks and Two-branched.

In our study, we then analyzed the differential gaps for Midori [6], Skinny [8], Rectangle [54], Present [14], Prince [15], Sparx [23], Simon [7], Speck [7], Twine [47], and LBlock [47] where Table 1 categorises the ciphers according their aforementioned structural properties.

4.2 Skinny

Skinny [8] is an AES-like tweakable block cipher, based on the Tweakey framework [28]. The aim of Skinny is to achieve the hardware performance of the AND-RX-cipher Simon and have strong security bounds against differential/linear attacks (this includes the related-key scenario), while also having competitive software performance. The resistance against differential/linear attacks in Skinny is based on counting the minimal number of active S-boxes, in the single-key and related-tweakey models. As the design of Skinny is based on a few very simple but highly efficient cryptographic building blocks it seems intuitive that one can expect that a large number of differential characteristics will contribute to a differential. Recent attacks [3,38] exploited the low branch number of the binary diffusion matrix, as well as properties of the tweakey schedule.

Using our tool-assisted approach we analysed this gap in Skinny-64 (see Fig. 1) and can provide some new insights to the security of Skinny-64. For example the best 8-round single differential characteristic Q^8_{max} suggests a probability of 2^{-72} while the differential D^8 defined by the input/output difference of Q^8_{max} consists of a large cluster of characteristics leading to the differential

$$0\text{x}0104401000\text{C}01\text{C}00 \xrightarrow{8-\text{round Skinny}-64} 0\text{x}0606060000060666 \qquad (15)$$

with a probability larger than $2^{-56.55}$ by taking all 821896 characteristics[3] into account which have DP $> 2^{-99}$. Note that the probabilities and the number of characteristics are obtained with a fixed input/output difference as noted in Eq. 15. This suggests that estimates from active S-boxes should be taken with care as the gap is fairly large. However, the number of rounds in Skinny-64 is chosen very conservatively and it provides a large security margin.

In particular the probability of the differential improves very quickly when adding more characteristics, as the distribution of the number of characteristics with a probability 2^{-t} is very flat over the choice of t (see Fig. 1). For example there are 39699 characteristics with DP $= 2^{-75}$ and 25413 characteristics with DP $= 2^{-76}$ and the probability of the differential only improves marginally by considering more characteristics with a lower probability. On the contrary, for designs like Simon (see Fig. 5) this distribution grows exponentially as the probability of the single characteristics decreases as has also been noted in [31], and one

---

[3] This process took in total 23.5 h on a single core, however after 1 h the estimate for the differential probability improves by less than $2^{-0.9}$.

has to take a much larger number of characteristics into account before getting a good approximation. For a detailed overview over how many characteristics contribute to each differential see Appendix A.

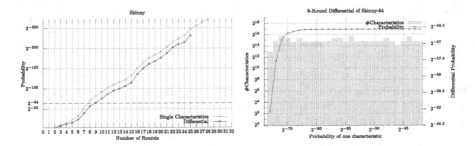

Fig. 1. Probability for the best single characteristics and differentials for Skinny-64 (left), and the distribution of the number of characteristics with a fixed probability contributing to the best 8-round differential for Skinny-64 (right). The green line indicates the probability of the differential when summing up the probability of all characteristics up to this probability, which highlights the small improvement when adding all lower probability characteristics. (Color figure online)

4.3 Midori

Midori is an AES-like lightweight block cipher optimized for low-energy usage using a binary near-MDS matrix combined with a generic cell permutation for diffusion. Despite that Midori-64 has a large number of 2^{32} weak keys, for which Midori-64 can be practically broken with invariant subspace attacks [27], there has been no differential attacks on even reduced versions of Midori, apart from a related-key attack by Gérault and Lafourcade [26].

The gap between the differential probability of a single characteristic and a differential behaves similar to Skinny-64, i.e., counting the active S-boxes gives an inaccurate bound against differential distinguishers. For example we found new differentials for Midori-64 where the 8-round single differential characteristic suggests a probability of 2^{-76} and the corresponding 8-round differential

$$\texttt{0x0A000000A0000005} \xrightarrow{\text{8-round Midori-64}} \texttt{0x000000000000A0AA} \qquad (16)$$

has a probability larger than $2^{-60.86}$ by summing all 693730 characteristics up to a probability of 2^{-114}. Similar to Skinny the distribution of the contributing characteristics is very flat, which means that we quickly approach a good estimate for the probability of the differential (see Fig. 2).

4.4 Sparx

Sparx [23] is based on the *long-trail strategy*, introduced alongside with Sparx, which can be seen as combining the ARX approach with an SPN, allowing to

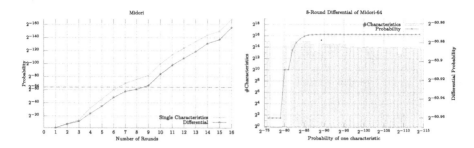

Fig. 2. Probability for the best single characteristics and differentials for various rounds of `Midori`-64 (left), and distribution of the characteristics contributing to the best 8-round differential for `Midori`-64 (right).

provide bounds on the differential resistance of an ARX cipher by counting the active S-boxes. While it is also feasible to prove such a bound using the methodology from Sect. 3, it is often computationally infeasible or the bounds are not very tight [41]. The designers of `Sparx` used the YAARX toolkit [12] to show truncated characteristics, that they used to compute the differential bounds. One of the main design motivations of `Sparx` was that it should be very difficult to find differential characteristics for a large number of rounds for ARX-based ciphers with a state of more than 32 bits [22].

In general ARX ciphers do not have a very strong differential effect compared to the previous lightweight SPN constructions, however as `Sparx` is in-between those it is an interesting target. Our results suggest that `Sparx`-64 has a differential effect comparable to other ARX designs like `Speck`-64 (see Fig. 3). The major limitation for applying our approach to `Sparx` is that the search for optimal differential characteristics on `Sparx` is computationally very costly. While single-characteristics up to 6 rounds can be found in less then 5 min, the 10-round single-characteristic took already 32 days, on a single core[4].

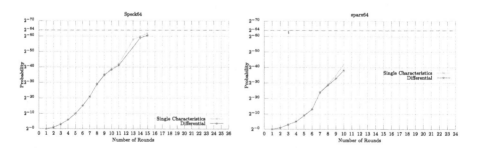

Fig. 3. Comparison of the best single characteristics and differentials for various rounds of `Speck`-64 (left), and `Sparx`-64 (right).

---

[4] Note that this process can not easily be parallelized as most SAT solvers are inherently serial.

4.5 Results for Other Lightweight Ciphers

Table 2 summarizes the gaps between single-characteristics and differentials for all lightweight block ciphers we analyzed. We observed that for most ciphers a large gap between the probability for single-characteristics and differentials exists and that a higher number of rounds is required for the block ciphers to be *differential resistant*. The gaps also increase significantly with the number of rounds, which is not surprising as with more rounds there are more valid differential characteristics for a given input/output difference.

The biggest gap, in term of number of rounds, occurs for Simon-64 with a gap of five rounds. There is also a 2-round gap for ciphers like Present, Midori and Twine. However, it seems that the gap for Simon-64 grows faster, considering that the differentials and characteristics seem to follow an exponential growth as also observed in [31]. In comparison Present, Midori and Twine seem to grow in a linear way. In relation to the number of rounds, the gap for Midori also has quite a significant impact and allows to extend the distinguisher by two rounds. Further we observed that there seem to be nearly no gaps for ciphers like Rectangle and Speck. We illustrate the gaps for the analyzed ciphers in Fig. 4 and we provide Fig. 5 showing the distribution of valid differential characteristics that contribute to the probability of the best differential for each cipher.

Table 2. Gap between the number of rounds required for a cipher to be *differential characteristic resistant* (DCR) and *differential resistant* (DR). Note that DR is only a lower bound and there might still exist better differentials.

Group	Design strategy	Cipher	Block size	Key size	Rounds	DCR	DR
SPN	AES-like	Midori	64	128	16	7	9
		Skinny	64	64/128/192	32	8	9
	Bit-sliced	Rectangle	64	80/128	25	15	15
	Present-like	Present	64	80/128	31	15	17
	Reflection	Prince	64	128	12	6	8
	ARX-based	Sparx	64	128	24	15	15[a]
Feistel	AND-RX	Simon	64	96/128	42	19	24[b]
	ARX	Speck	64	96/128	26	>15	>15[c]
	GFN	Twine	64	80/128	36	14	16
	Two-branched	LBlock	64	80	32	15	16

[a] Single-characteristic differentials of Sparx [23] are proven to reach 15 rounds, while the authors mention that they don't expect the bound to be tight.
[b] The best differentials for Simon-64 reach 23 rounds with $2^{-63.91}$ [39].
[c] The best differentials for Speck-64 reach 15 rounds with $2^{-60.56}$ [44].

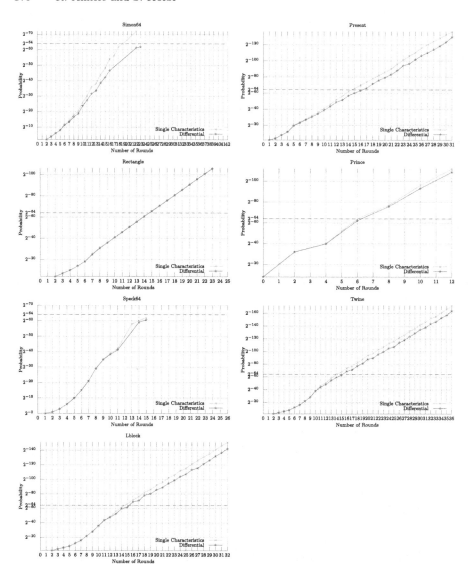

Fig. 4. Probability for the best single characteristics and differentials for various rounds of different block ciphers. 1st row: `Simon-64` (left) and `Present` (right), 2nd row: `Rectangle` (left) and `Prince` (right), 3rd row: `Speck-64` (left) and `Twine` (right), 4th row: `LBlock` (left)

4.6   Application of the Differential Gaps to the Best Published Differential Attacks

In the following, we analyze the best published attacks and discuss improvements of the attacks when possible:

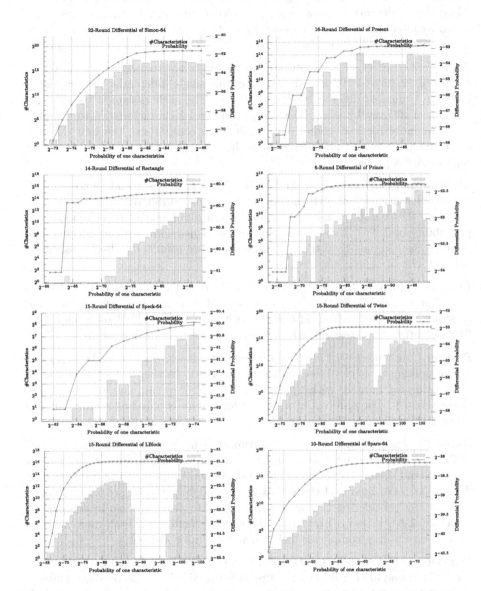

Fig. 5. Distribution of the characteristics contributing to the best differential for various block ciphers. 1st row: Simon-64 (left) and Present (right), 2nd row: Rectangle (left) and Prince (right), 3rd row: Speck-64 (left) and Twine (right), 4th row: LBlock (left) and Sparx-64 (right)

Midori-64. Gérault and Lafourcade [26] proposed related-key differential attacks on full-round Midori-64, where they use 16 15-round and $4 \cdot 14$-round related-key differential characteristics to recover the key. In their attacks they do not exploit differentials. In comparison, the best differential that we found reaches 8 rounds with a probability of $2^{-60.86}$.

Skinny-64. Liu et al. [38] propose related-tweakey rectangle attacks on 26 rounds of Skinny-64-192 and they use optimal single differential characteristics based on truncated differential characteristics. The authors exploit the differential gap of Skinny by using 5000 single differential characteristics to compute the differential for a 22-round distinguisher. In comparison, the best differential characteristic with no differences in the tweak/key that we found reaches 8 rounds with a probability of $2^{-56.55}$.

Rectangle. Zhang et al. [54] studied the differential effect and showed an 18-round differential attack, where they used a 14-round differential with a probability of $2^{-62.83}$. In our analysis we found a better differential for 14 rounds with probability of $2^{-60.63}$ by summing up 40627 single-characteristics which would improve the complexity of these attacks. For more rounds the distinguisher are below 2^{-64}.

Present. Liu and Jin [37] presented an 18-round attack based on slender-sets. Wang et al. [51] further presented normal differential attacks on 16-round Present where they used a differential with probability $2^{-62.13}$ by summing up 91 differential characteristics which is comparable to our differentials.

Prince. Canteaut et al. [17] showed differential attacks on 10 rounds of Prince, by considering multiple differential characteristics. In their attack they use 12 differentials for 6 rounds with a probability of $2^{-56.42}$ by summing up 1536 single-characteristics. The differential we found for 6 rounds only has a probability of about 2^{-62}, but does not lead to further improvements of the attack.

Sparx-64. Ankele and List [4] studied truncated differential attacks on 16 rounds of Sparx-64/128 and used single differential characteristics, for the first part of the 14-round distinguisher, and truncated the second part of the distinguisher. The designers of Sparx-64 claim that Sparx is differential secure for 15 rounds, however, by considering the differential effect of Sparx-64, also in comparison with Speck-64, it seems likely that there exist differentials with more than 15 rounds with a data complexity below using the full codebook.

Simon-64. Abed et al. [2] presented differential attacks on Simon-64, where they used a 21-round distinguisher with a probability of $2^{-61.01}$. Better distinguishers are reported by [39] for 23 rounds with a probability of $2^{-63.91}$. The differentials we found are in line with previous results.

Speck-64. Song et al. [44] presented 20-round attacks on Speck-64 by constructing a distinguisher from two short characteristics where they concatenated the two characteristics to a 15-round characteristic with probability $2^{-60.56}$. The distinguishers used in the attack are already based on differentials and the differentials we found do not lead to any improvement.

Twine. Biryukov et al. [10] showed a 25-round impossible differential attack and a truncated differential attack on 23 rounds by chaining several iterated 4-round characteristics together. In the paper the authors also considered differentials for 12 rounds with a probability of $2^{-52.08}$ and 16 rounds with probability $2^{-67.59}$. The best differential that we found reaches 15 rounds with a probability of $2^{-62.89}$.

LBlock. Wang et al. [52] published a 24-round impossible differential attack on LBlock. Due to the nature of impossible differential attacks, characteristics with probability 1 are used for constructing these. The best differential that we found reaches 15 rounds with a probability of $2^{-61.43}$.

5 Experimental Verification and the Influence of Keys

In Sect. 2 we made several assumptions in order to compute $DP(Q)$ and in this section we compare the theoretical estimates with experiments for reduced-round versions. This serves two purpose: First we want to see how close our estimate for $DP(\alpha, \beta)$ is and secondly we want to see the distribution over the choice of keys. Specifically, we are interested in the number of pairs

$$\delta_K(\alpha, \beta) = \#\{x \in \mathbb{F}_2^n \mid E_K(x) \oplus E_K(x \oplus \alpha) = \beta\}. \tag{17}$$

This number of *good* pairs will vary over the choice of the key. For a random process we would expect that the number of valid pairs is about $DP \cdot 2^n$ and follows a Poisson distribution.

Definition 5. *Let X be a Poisson distributed random variable representing the number of pairs (a, b) with values in $\mathbb{F}_2^n$ following a differential $D = (\alpha \xrightarrow{f} \beta)$, that means $f(a) \oplus f(a \oplus \alpha) = \beta$, then*

$$\Pr(X = l) = \frac{1}{2}(2^n p)^l \frac{e^{-(2^n p)}}{l!} \tag{18}$$

where p is the probability of the differential.

In the following, we experimentally verify differentials for Skinny, Speck and Midori for a large number of random pairs of plaintexts and a random choice of keys to see how good this approximation is.

5.1 Skinny

As a first example we look at Skinny-64. We use the 6-round differential

$$D = (\texttt{0x0000010010000041}, \texttt{0x4444004040044044})$$

for Skinny-64. The best characteristic which is part of D has a probability of 2^{-32} and by collecting all characteristics (100319) contributing to this differential

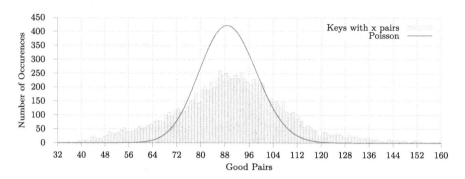

Fig. 6. Distribution of $\delta_K(D)$ over a random choice of K for 6-round `Skinny-64`.

we estimate $\mathrm{DP}(D) \approx 2^{-23.52}$. We try out 2^{30} randomly selected pairs for 10000 keys and count the number of pairs following D. From our estimate we would expect that on average we get about 89 pairs for a key.

As one can see from Fig. 6 our estimate of $\mathrm{DP}(D)$ provides a good approximation for the distribution over the keys, although the distribution has a larger variance than we expected.

5.2 Speck

For `Speck-64` we look at the differential

$$D = ((\texttt{0x40004092}, \texttt{0x10420040}), (\texttt{0x8080A080}, \texttt{0x8481A4A0}))$$

over 7 rounds. The best characteristic in D has a probability of 2^{-21} and this only slightly improves to about $2^{-20.95}$ using 6 additional characteristics. We again run our experiments for 2^{30} randomly selected pairs for 10000 keys and count the number of pairs following D. On average we would expect 530 pairs.

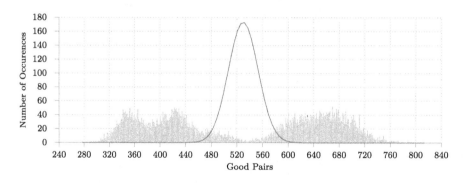

Fig. 7. Distribution of $\delta_K(D)$ over a random choice of K for 7-round `Speck-64`.

In Fig. 7 it can be seen that for 7-round `Speck-64` the distribution is bimodal and we over- respectively underestimate the number of valid pairs for most keys.

5.3   Midori

For Midori-64 we look at the differential

$$D = (\texttt{0x0200200000020000}, \texttt{0x0202220020020020})$$

over 4 rounds. The best characteristic in D has a probability of 2^{-32} and this improves to about $2^{-23.79}$ using 896 additional characteristics. We again run our experiments for 2^{30} randomly selected pairs for 3200 keys and count the number of pairs following D. On average we would expect about 74 pairs.

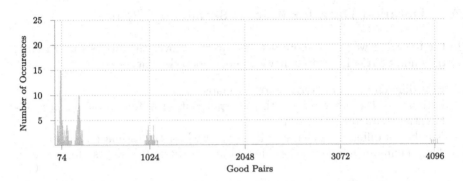

Fig. 8. Distribution of $\delta_K(D)$ over a random choice of K for 4-round Midori-64. We omitted the 2545 keys with 0 good pairs in this plot.

In Fig. 8 it can be seen that for 4-round Midori-64 the distribution is very different from the previous cases. For some keys the probability is significantly higher and for about 80% of the keys we get 0 good pairs. This means that for a large fraction of keys we actually found an impossible differential and one should be careful when constructing differential distinguishers for Midori. In particular it would be interesting to classify this set of impossible keys and we leave this as an open problem. Moreover, this also implies the existance of a large class of weak keys, that has also been observed in the invariant subspace attacks on Midori-64 [27,34,49].

6   Conclusions

In this work we showed for several lightweight block ciphers that the gap between single characteristics and differentials can be surprisingly large. This leads to significantly higher probability of differentials in several designs and allows us to have differential distinguishers covering more rounds.
    We provided a simple framework to automate the process of collecting many differential characteristics that are contributing to the probability of a differential. We hope this will encourage future designs of cryptographic primitives to

apply our methodology in order to provide better bounds on the security against differential cryptanalysis.

Further we verified differentials for a reduced number of rounds experimentally and showed that our improved estimates of the probability of differentials of Skinny closely resembles what happens in experiments. However, we can also observe that some commonly made assumptions on the distribution of good pairs following a differential over the choice of keys has to be made very carefully. For instance, the results for Speck and Midori indicate that one needs to be very careful in presuming that the estimates apply to all key values.

A   Detailed Data for Midori, Skinny and Sparx

In the following we give a more detailed overview over the analysis on Midori, Skinny and Sparx. In particular we give the following metrics

- Best differential characteristic for r rounds.
- Estimate of the differential with the input/output difference of the best differential characteristic found.
- Number of differential characteristics we used for the estimate.
- The maximum *weight* of the differential characteristics we use for the estimate.
- Search time to find the best single differential characteristic and all the differential characteristics for the best differential (Tables 3 and 4).

Table 3. Detailed results on the differentials found for Midori-64.

r	Pr_{char}	Pr_{diff}	#Characteristics	Max weight	Time$_{char}$	Time$_{diff}$
4	2^{-32}	$2^{-23.79}$	896	36	31 m 36 s	2 m 4 s
5	2^{-46}	$2^{-35.13}$	55168	54	56 m 42 s	1 h 10 m
6	2^{-60}	$2^{-48.36}$	11072	71	1 h 54 m	29 m
7	2^{-70}	$2^{-57.43}$	28588	99	3 h 12 m	1 h 32 m
8	2^{-76}	$2^{-60.87}$	693730	114	1 h 6 m	23 h 36 m
9	2^{-82}	$2^{-66.52}$	104694	90	56 m	3 h 12 m
10	2^{-100}	$2^{-83.86}$	120181	106	5 h 12 m	4 h 36 m
11	2^{-114}	$2^{-98.04}$	87055	119	10 h 56 m	3 h 18 m
12	2^{-124}	$2^{-108.59}$	88373	131	1 d 02 h	4 h 54 m
13	2^{-134}	$2^{-118.70}$	56596	139	22 h 02 m	3 h 06 m
14	2^{-144}	$2^{-131.18}$	13932	149	1 d 16 h	9 h 36 m
15	2^{-150}	$2^{-137.07}$	25680	155	20 h 30 m	1 h 48 m
16	2^{-168}	$2^{-155.58}$	11815	172	3 d 21 h	1 h 12 m

Table 4. Detailed results on the differentials found for **Skinny**-64.

r	Pr_{char}	Pr_{diff}	#Characteristics	Max weight	$Time_{char}$	$Time_{diff}$
6	2^{-32}	$2^{-23.51}$	100319	45	22 m 54 s	1 h 38 m
7	2^{-52}	$2^{-39.49}$	141800	58	1 h 03 m	5 h 13 m
8	2^{-72}	$2^{-56.55}$	821896	98	1 h 24 m	23 h 20 m
9	2^{-82}	$2^{-65.36}$	277464	89	1 h 06 m	29 h 25 m
10	2^{-92}	$2^{-75.98}$	66438	92	1 h 42 m	2 h 59 m
11	2^{-102}	$2^{-86.63}$	64339	103	2 h 36 m	3 h 14 m
12	2^{-110}	$2^{-95.00}$	62382	113	3 h 12 m	3 h 37 m
13	2^{-116}	$2^{-100.06}$	165079	124	2 h 42 m	24 h 42 m
14	2^{-122}	$2^{-106.71}$	100457	127	3 h 30 m	10 h 25 m
15	2^{-132}	$2^{-114.65}$	326404	142	7 h 23 m	37 h 21 m
16	2^{-150}	$2^{-135.41}$	24598	150	30 h 35 m	1 h 44 m
17	2^{-164}	$2^{-150.07}$	21524	165	60 h 09 m	1 h 53 m
18	2^{-176}	$2^{-161.64}$	20903	177	92 h 04 m	1 h 54 m
19	2^{-184}	$2^{-168.27}$	54245	185	60 h 22 m	3 h 38 m
20	2^{-192}	$2^{-176.74}$	39169	193	60 h 10 m	2 h 59 m
...						

B Differentials for Midori, Skinny and Sparx

In the following we give the best differentials that we found for Midori, Skinny and Sparx. The differentials for many other lightweight ciphers together with the source code to generate the differential models is publicly available at: https://github.com/TheBananaMan/cryptosmt (Tables 5, 6, 7 and 8).

Table 5. Detailed results on the differentials found for **Sparx**-64.

r	Pr_{char}	Pr_{diff}	#Characteristics	Max weight	$Time_{char}$	$Time_{diff}$
1	1	1	1	1	0.02 s	0.03 s
2	2^{-1}	2^{-1}	1	2	0.1 s	0.07 s
3	2^{-3}	2^{-3}	1	4	0.5 s	0.09 s
4	2^{-5}	$2^{-4.99}$	8	49	2.4 s	3.36 s
5	2^{-9}	$2^{-8.99}$	12944	58	25 s	2 m 12 s
6	2^{-13}	$2^{-12.99}$	70133	51	3 m 48 s	3 h 06 m
7	2^{-24}	$2^{-23.95}$	56301	60	47 h 48 m	28 m
8	2^{-29}	$2^{-28.53}$	37124	60	15 d 5 h	17 m
9	2^{-35}	$2^{-32.87}$	233155	58	22 d 7 h	7 h 42 m
10	2^{-42}	$2^{-38.12}$	1294158	73	32 d 12 h	35 h 18 m
...						

Table 6. The best differentials that we found for various rounds of Midori-64.

r	Differential	$\Pr_{\text{Differential}}$
4	0x0000020000022000 → 0x0020220002022002	$2^{-23.79}$
5	0x0004100000000100 → 0x0222220222222022	$2^{-35.13}$
6	0x0550000000005000 → 0x0000AA0000007707	$2^{-48.36}$
7	0x0AA00500700A0000 → 0x00005AFF0000AAA0	$2^{-57.43}$
8	0x0A000000A0000005 → 0x000000000000A0AA	$2^{-60.87}$
9	0x0000000A050000A0 → 0x770700000AAAA0AA	$2^{-66.52}$
10	0x0500005050000000 → 0xDD7A7D0D25727A7D	$2^{-83.86}$
11	0x0000A00000500500 → 0xAAA0AAA50AAAAA0A	$2^{-98.04}$
12	0xA0A00A0A00007000 → 0x0000DD7A00007077	$2^{-108.59}$
13	0x0000A0070A000AA0 → 0x00000555A5AFAF5F	$2^{-118.70}$
14	0x0000000000000500 → 0x000070777707AAA0	$2^{-131.18}$
15	0x0A0000A00000000A → 0x05550000AA0AAAA0	$2^{-137.07}$
16	0xAA00A0A0AAA00A70 → 0x00007077AA0A7770	$2^{-155.58}$

Table 7. The best differentials that we found for various rounds of Skinny-64.

r	Differential	$\Pr_{\text{Differential}}$
6	0x0041C00001000000 → 0x4044400400404444	$2^{-23.51}$
7	0x002220222B222000 → 0x0444004404004444	$2^{-39.49}$
8	0x0104401000C01C00 → 0x0606060000060666	$2^{-56.55}$
9	0x0020000200020200 → 0x0060000100600160	$2^{-65.36}$
10	0x0008200020000020 → 0x0008808000880088	$2^{-75.98}$
11	0x0002200000000200 → 0x0444004404004444	$2^{-86.63}$
12	0x0004000000000000 → 0x0001000100000001	$2^{-95.00}$
13	0x0200000000002000 → 0x0001001100000001	$2^{-100.06}$
14	0x4000040000400000 → 0x0404040000040444	$2^{-106.71}$
15	0x8008080000800000 → 0x1066100600601666	$2^{-114.65}$
16	0x0020000220000000 → 0x8880088080008888	$2^{-135.41}$
17	0x004C400004000000 → 0x2002022022020022	$2^{-150.07}$
18	0x400C0000C00C0000 → 0x0077001100660077	$2^{-161.64}$
19	0x2200000000002008 → 0x0077001100660077	$2^{-168.27}$
20	0x8800000000008009 → 0x8800080900008800	$2^{-176.74}$
...		

Table 8. The best differentials that we found for various rounds of Sparx-64.

r	Differential	$\Pr_{\text{Differential}}$
1	$(0x0040, 0x8000, 0x0000, 0x0000) \rightarrow (0x0000, 0x0002, 0x0000, 0x0000)$	1
2	$(0x0010, 0x2000, 0x0000, 0x0000) \rightarrow (0x8000, 0x8002, 0x0000, 0x0000)$	2^{-1}
3	$(0x2800, 0x0010, 0x0000, 0x0000) \rightarrow (0x8300, 0x8302, 0x8100, 0x8102)$	2^{-3}
4	$(0x0000, 0x0000, 0x2800, 0x0010) \rightarrow (0x8000, 0x840A, 0x0000, 0x0000)$	$2^{-4.99}$
5	$(0x0000, 0x0000, 0x0211, 0x0A04) \rightarrow (0x8000, 0x840A, 0x0000, 0x0000)$	$2^{-8.99}$
6	$(0x0000, 0x0000, 0x0211, 0x0A04) \rightarrow (0xAF1A, 0xBF30, 0x850A, 0x9520)$	$2^{-12.99}$
7	$(0x0000, 0x0000, 0x7448, 0xB0F8) \rightarrow (0x8004, 0x8C0E, 0x8000, 0x840A)$	$2^{-23.95}$
8	$(0x0000, 0x0000, 0x0050, 0x8402) \rightarrow (0x0040, 0x0542, 0x0040, 0x0542)$	$2^{-28.53}$
9	$(0x2800, 0x0010, 0x2800, 0x0010) \rightarrow (0x5761, 0x1764, 0x5221, 0x1224)$	$2^{-32.87}$
10	$(0x2800, 0x0010, 0x2800, 0x0010) \rightarrow (0x8081, 0x8283, 0x8000, 0x8002)$	$2^{-38.12}$
...		

References

1. Abdelraheem, M.A., Ågren, M., Beelen, P., Leander, G.: On the distribution of linear biases: three instructive examples. In: Safavi-Naini, R., Canetti, R. (eds.) CRYPTO 2012. LNCS, vol. 7417, pp. 50–67. Springer, Heidelberg (2012). https://doi.org/10.1007/978-3-642-32009-5_4

2. Abed, F., List, E., Lucks, S., Wenzel, J.: Differential cryptanalysis of round-reduced SIMON and SPECK. In: Cid, C., Rechberger, C. (eds.) FSE 2014. LNCS, vol. 8540, pp. 525–545. Springer, Heidelberg (2015). https://doi.org/10.1007/978-3-662-46706-0_27

3. Ankele, R., et al.: Related-key impossible-differential attack on reduced-round SKINNY. In: Gollmann, D., Miyaji, A., Kikuchi, H. (eds.) ACNS 2017. LNCS, vol. 10355, pp. 208–228. Springer, Cham (2017). https://doi.org/10.1007/978-3-319-61204-1_11

4. Ankele, R., List, E.: Differential cryptanalysis of round-reduced sparx-64/128. Cryptology ePrint Archive, Report 2018/332 (2018). https://eprint.iacr.org/2018/332

5. Aumasson, J.-P., Jovanovic, P., Neves, S.: Analysis of NORX: investigating differential and rotational properties. In: Aranha, D.F., Menezes, A. (eds.) LATIN-CRYPT 2014. LNCS, vol. 8895, pp. 306–324. Springer, Cham (2015). https://doi.org/10.1007/978-3-319-16295-9_17

6. Banik, S., et al.: Midori: a block cipher for low energy. In: Iwata, T., Cheon, J.H. (eds.) ASIACRYPT 2015. LNCS, vol. 9453, pp. 411–436. Springer, Heidelberg (2015). https://doi.org/10.1007/978-3-662-48800-3_17

7. Beaulieu, R., Shors, D., Smith, J., Treatman-Clark, S., Weeks, B., Wingers, L.: The SIMON and SPECK families of lightweight block ciphers. Cryptology ePrint Archive, Report 2013/404 (2013). http://eprint.iacr.org/2013/404

8. Beierle, C., et al.: The SKINNY family of block ciphers and its low-latency variant MANTIS. In: Robshaw, M., Katz, J. (eds.) CRYPTO 2016. LNCS, vol. 9815, pp. 123–153. Springer, Heidelberg (2016). https://doi.org/10.1007/978-3-662-53008-5_5

9. Biham, E., Shamir, A.: Differential cryptanalysis of DES-like cryptosystems. In: Menezes, A.J., Vanstone, S.A. (eds.) CRYPTO 1990. LNCS, vol. 537, pp. 2–21. Springer, Heidelberg (1991). https://doi.org/10.1007/3-540-38424-3_1

10. Biryukov, A., Derbez, P., Perrin, L.: Differential analysis and meet-in-the-middle attack against round-reduced TWINE. In: Leander, G. (ed.) FSE 2015. LNCS, vol. 9054, pp. 3–27. Springer, Heidelberg (2015). https://doi.org/10.1007/978-3-662-48116-5_1

11. Biryukov, A., Roy, A., Velichkov, V.: Differential analysis of block ciphers SIMON and SPECK. In: Cid, C., Rechberger, C. (eds.) FSE 2014. LNCS, vol. 8540, pp. 546–570. Springer, Heidelberg (2015). https://doi.org/10.1007/978-3-662-46706-0_28

12. Biryukov, A., Velichkov, V.: Automatic search for differential trails in ARX ciphers. In: Benaloh, J. (ed.) CT-RSA 2014. LNCS, vol. 8366, pp. 227–250. Springer, Cham (2014). https://doi.org/10.1007/978-3-319-04852-9_12

13. Blondeau, C., Nyberg, K.: Improved parameter estimates for correlation and capacity deviates in linear cryptanalysis. IACR Trans. Symmetric Cryptol. **2016**(2), 162–191 (2016). https://doi.org/10.13154/tosc.v2016.i2.162-191

14. Bogdanov, A., et al.: PRESENT: an ultra-lightweight block cipher. In: Paillier, P., Verbauwhede, I. (eds.) CHES 2007. LNCS, vol. 4727, pp. 450–466. Springer, Heidelberg (2007). https://doi.org/10.1007/978-3-540-74735-2_31

15. Borghoff, J., et al.: PRINCE – a low-latency block cipher for pervasive computing applications - extended abstract. In: Wang, X., Sako, K. (eds.) ASIACRYPT 2012. LNCS, vol. 7658, pp. 208–225. Springer, Heidelberg (2012). https://doi.org/10.1007/978-3-642-34961-4_14

16. Canteaut, A.: Differential cryptanalysis of Feistel ciphers and differentially uniform mappings. In: Selected Areas on Cryptography, SAC 1997, pp. 172–184 (1997)

17. Canteaut, A., Fuhr, T., Gilbert, H., Naya-Plasencia, M., Reinhard, J.-R.: Multiple differential cryptanalysis of round-reduced PRINCE. In: Cid, C., Rechberger, C. (eds.) FSE 2014. LNCS, vol. 8540, pp. 591–610. Springer, Heidelberg (2015). https://doi.org/10.1007/978-3-662-46706-0_30

18. Daemen, J., Lamberger, M., Pramstaller, N., Rijmen, V., Vercauteren, F.: Computational aspects of the expected differential probability of 4-round AES and AES-like ciphers. Computing **85**(1), 85–104 (2009). https://doi.org/10.1007/s00607-009-0034-y

19. Daemen, J., Rijmen, V.: The wide trail design strategy. In: Honary, B. (ed.) Cryptography and Coding 2001. LNCS, vol. 2260, pp. 222–238. Springer, Heidelberg (2001). https://doi.org/10.1007/3-540-45325-3_20

20. Daemen, J., Rijmen, V.: Plateau characteristics. IET Inf. Secur. **1**(1), 11–17 (2007)

21. Davis, M., Logemann, G., Loveland, D.: A machine program for theorem-proving. Commun. ACM **5**(7), 394–397 (1962). https://doi.org/10.1145/368273.368557

22. Dinu, D., Perrin, L., Udovenko, A., Velichkov, V., Großschädl, J., Biryukov, A.: Private communication

23. Dinu, D., Perrin, L., Udovenko, A., Velichkov, V., Großschädl, J., Biryukov, A.: Design strategies for ARX with provable bounds: SPARX and LAX. In: Cheon, J.H., Takagi, T. (eds.) ASIACRYPT 2016. LNCS, vol. 10031, pp. 484–513. Springer, Heidelberg (2016). https://doi.org/10.1007/978-3-662-53887-6_18

24. Dobraunig, C., Eichlseder, M., Kales, D., Mendel, F.: Practical key-recovery attack on MANTIS5. IACR Trans. Symmetric Cryptol. **2016**(2), 248–260 (2016). https://doi.org/10.13154/tosc.v2016.i2.248-260

25. Eichlseder, M., Kales, D.: Clustering related-tweak characteristics: application to MANTIS-6. IACR Trans. Symmetric Cryptol. **2018**(2), 111–132 (2018). https://doi.org/10.13154/tosc.v2018.i2.111-132

26. Gérault, D., Lafourcade, P.: Related-key cryptanalysis of Midori. In: Dunkelman, O., Sanadhya, S.K. (eds.) INDOCRYPT 2016. LNCS, vol. 10095, pp. 287–304. Springer, Cham (2016). https://doi.org/10.1007/978-3-319-49890-4_16

27. Guo, J., Jean, J., Nikolic, I., Qiao, K., Sasaki, Y., Sim, S.M.: Invariant subspace attack against Midori64 and the resistance criteria for S-box designs. IACR Trans. Symmetric Cryptol. **2016**(1), 33–56 (2016). https://doi.org/10.13154/tosc.v2016.i1.33-56

28. Jean, J., Nikolić, I., Peyrin, T.: Tweaks and keys for block ciphers: the TWEAKEY framework. In: Sarkar, P., Iwata, T. (eds.) ASIACRYPT 2014. LNCS, vol. 8874, pp. 274–288. Springer, Heidelberg (2014). https://doi.org/10.1007/978-3-662-45608-8_15

29. Daemen, J., Peeters, M., Van Assche, G., Rijmen, V.: Nessie Proposal: NOEKEON (2000). http://gro.noekeon.org/Noekeon-spec.pdf

30. Keliher, L., Sui, J.: Exact maximum expected differential and linear probability for two-round advanced encryption standard. IET Inf. Secur. **1**(2), 53–57 (2007). https://doi.org/10.1049/iet-ifs:20060161

31. Kölbl, S., Leander, G., Tiessen, T.: Observations on the SIMON block cipher family. In: Gennaro, R., Robshaw, M. (eds.) CRYPTO 2015. LNCS, vol. 9215, pp. 161–185. Springer, Heidelberg (2015). https://doi.org/10.1007/978-3-662-47989-6_8

32. Kölbl, S., Roy, A.: A brief comparison of SIMON and SIMECK. In: Bogdanov, A. (ed.) LightSec 2016. LNCS, vol. 10098, pp. 69–88. Springer, Cham (2017). https://doi.org/10.1007/978-3-319-55714-4_6

33. Lai, X., Massey, J.L., Murphy, S.: Markov ciphers and differential cryptanalysis. In: Davies, D.W. (ed.) EUROCRYPT 1991. LNCS, vol. 547, pp. 17–38. Springer, Heidelberg (1991). https://doi.org/10.1007/3-540-46416-6_2

34. Leander, G., Abdelraheem, M.A., AlKhzaimi, H., Zenner, E.: A cryptanalysis of PRINTCIPHER: the invariant subspace attack. In: Rogaway, P. (ed.) CRYPTO 2011. LNCS, vol. 6841, pp. 206–221. Springer, Heidelberg (2011). https://doi.org/10.1007/978-3-642-22792-9_12

35. Leurent, G.: Analysis of differential attacks in ARX constructions. In: Wang, X., Sako, K. (eds.) ASIACRYPT 2012. LNCS, vol. 7658, pp. 226–243. Springer, Heidelberg (2012). https://doi.org/10.1007/978-3-642-34961-4_15

36. Lipmaa, H., Moriai, S.: Efficient algorithms for computing differential properties of addition. In: Matsui, M. (ed.) FSE 2001. LNCS, vol. 2355, pp. 336–350. Springer, Heidelberg (2002). https://doi.org/10.1007/3-540-45473-X_28

37. Liu, G.Q., Jin, C.H.: Differential cryptanalysis of PRESENT-like cipher. Des. Codes Cryptogr. **76**(3), 385–408 (2015). https://doi.org/10.1007/s10623-014-9965-1

38. Liu, G., Ghosh, M., Song, L.: Security analysis of SKINNY under related-tweakey settings (long paper). IACR Trans. Symmetric Cryptol. **2017**(3), 37–72 (2017). https://doi.org/10.13154/tosc.v2017.i3.37-72

39. Liu, Z., Li, Y., Wang, M.: Optimal differential trails in SIMON-like ciphers. IACR Trans. Symmetric Cryptol. **2017**(1), 358–379 (2017). https://doi.org/10.13154/tosc.v2017.i1.358-379

40. Mate Soos: CryptoMiniSat SAT solver (2009). https://github.com/msoos/cryptominisat/

41. Mouha, N., Preneel, B.: Towards finding optimal differential characteristics for ARX: application to Salsa20. Cryptology ePrint Archive, Report 2013/328 (2013). http://eprint.iacr.org/2013/328

42. Mouha, N., Wang, Q., Gu, D., Preneel, B.: Differential and linear cryptanalysis using mixed-integer linear programming. In: Wu, C.-K., Yung, M., Lin, D. (eds.) Inscrypt 2011. LNCS, vol. 7537, pp. 57–76. Springer, Heidelberg (2012). https://doi.org/10.1007/978-3-642-34704-7_5

43. Niemetz, A., Preiner, M., Biere, A.: Boolector 20 system description. J. Satisf. Boolean Model. Comput. **9**, 53–58 (2014). (Published 2015)

44. Song, L., Huang, Z., Yang, Q.: Automatic differential analysis of ARX block ciphers with application to SPECK and LEA. In: Liu, J.K., Steinfeld, R. (eds.) ACISP 2016. LNCS, vol. 9723, pp. 379–394. Springer, Cham (2016). https://doi.org/10.1007/978-3-319-40367-0_24

45. Kölbl, S.: CryptoSMT: an easy to use tool for cryptanalysis of symmetric primitives (2015). https://github.com/kste/cryptosmt

46. Sun, S., et al.: Towards finding the best characteristics of some bit-oriented block ciphers and automatic enumeration of (related-key) differential and linear characteristics with predefined properties. Cryptology ePrint Archive, Report 2014/747 (2014). http://eprint.iacr.org/2014/747

47. Suzaki, T., Minematsu, K., Morioka, S., Kobayashi, E.: TWINE: a lightweight block cipher for multiple platforms. In: Knudsen, L.R., Wu, H. (eds.) SAC 2012. LNCS, vol. 7707, pp. 339–354. Springer, Heidelberg (2013). https://doi.org/10.1007/978-3-642-35999-6_22

48. Tezcan, C., Okan, G.O., Şenol, A., Doğan, E., Yücebaş, F., Baykal, N.: Differential attacks on lightweight block ciphers PRESENT, PRIDE, and RECTANGLE revisited. In: Bogdanov, A. (ed.) LightSec 2016. LNCS, vol. 10098, pp. 18–32. Springer, Cham (2017). https://doi.org/10.1007/978-3-319-55714-4_2

49. Todo, Y., Leander, G., Sasaki, Y.: Nonlinear invariant attack - practical attack on full SCREAM, iSCREAM, and Midori64. In: Cheon, J.H., Takagi, T. (eds.) ASIACRYPT 2016. LNCS, vol. 10032, pp. 3–33. Springer, Heidelberg (2016). https://doi.org/10.1007/978-3-662-53890-6_1

50. Ganesh, V., Hansen, T., Soos, M., Liew, D., Govostes, R.: STP constraint solver (2007). https://github.com/stp/stp

51. Wang, M., Sun, Y., Tischhauser, E., Preneel, B.: A model for structure attacks, with applications to PRESENT and Serpent. In: Canteaut, A. (ed.) FSE 2012. LNCS, vol. 7549, pp. 49–68. Springer, Heidelberg (2012). https://doi.org/10.1007/978-3-642-34047-5_4

52. Wang, N., Wang, X., Jia, K.: Improved impossible differential attack on reduced-round LBlock. In: Kwon, S., Yun, A. (eds.) ICISC 2015. LNCS, vol. 9558, pp. 136–152. Springer, Cham (2016). https://doi.org/10.1007/978-3-319-30840-1_9

53. Wang, X., Feng, D., Lai, X., Yu, H.: Collisions for hash functions MD4, MD5, HAVAL-128 and RIPEMD. Cryptology ePrint Archive, Report 2004/199 (2004). http://eprint.iacr.org/2004/199

54. Zhang, W., Bao, Z., Lin, D., Rijmen, V., Yang, B., Verbauwhede, I.: RECTANGLE: a bit-slice lightweight block cipher suitable for multiple platforms. Sci. China Inf. Sci. **58**(12), 1–15 (2015). https://doi.org/10.1007/s11432-015-5459-7

Side Channel and Fault Attacks

Sliding-Window Correlation Attacks Against Encryption Devices with an Unstable Clock

Dor Fledel$^{(\boxtimes)}$ and Avishai Wool$^{(\boxtimes)}$

School of Electrical Engineering, Tel-Aviv University, 69978 Tel-Aviv, Israel
dorfledel@tau.ac.il, yash@eng.tau.ac.il

Abstract. Power analysis side channel attacks rely on aligned traces. As a counter-measure, devices can use a jittered clock to misalign the power traces. In this paper we suggest a way to overcome this counter-measure, using an old method of integrating samples over time followed by a correlation attack (Sliding Window CPA). We theoretically re-analyze this general method with characteristics of jittered clocks and show that it is stronger than previously believed. We show that integration of samples over a suitably chosen window size actually amplifies the correlation both with and without jitter—as long as multiple leakage points are present within the window. We then validate our analysis on a new data-set of traces measured on a board implementing a jittered clock. The data-set we collected is public and accessible online. Our experiments show that the SW-CPA attack with a well-chosen window size is very successful against a jittered clock counter-measure and significantly outperforms previous suggestions, requiring a much smaller set of traces to correctly identify the correct key.

1 Introduction

1.1 Background

The use of encryption in embedded devices is proliferating. Encryption in such devices can be implemented in two ways, either by a hardware (ASIC or FPGA) implementation, or by software. Assuming that reasonable cryptographic algorithms are in use (e.g., AES), a cryptanalyst wanting to break the encryption can use side channel attacks (SCA), exploiting implementation-dependent information leakage captured during the cryptographic operation to find the correct key. A wide range of SCA exist, using leakage sources such as timing [Koc96], electromagnetic radiation [KA98], acoustic emanations [ST04] and even photonics [FH08]. Among these, one of the first and best understood SCA is power analysis. The idea of power analysis attacks is to perform statistical analysis of the CPU power usage, which is influenced by the secret cryptographic keys processed by the device. Some power analysis attacks assume profiling of the board, while others (non-profiling attacks) classify the behavior via a black-box methodology.

© Springer Nature Switzerland AG 2019
C. Cid and M. J. Jacobson, Jr. (Eds.): SAC 2018, LNCS 11349, pp. 193–215, 2019.
https://doi.org/10.1007/978-3-030-10970-7_9

Known non-profiling attacks such as Simple Power Analysis (SPA), traditional difference of means Differential Power Analysis (DPA) [KJJ99] and Correlation Power Analysis (CPA) [BCO04] are described in the literature and can be easily implemented by pre-made kits.

1.2   Power Traces Alignment: Assumptions and Counter-Measures

Alignment Assumption. A crucial property for the success of power SCA is that the power traces are *aligned*. Common power analysis attacks (i.e., DPA and CPA) assume that the information-leaking step (for example an Sbox look-up) will always occur at a fixed sample index. If this assumption does not hold then the leaking information will appear at different offsets, which severely degrades the attack's ability to correlate the power leak to hypothetical key values.

Time Domain Hiding Counter-Measures. One possible SCA counter-measure, originating in the initial days of power SCA (cf. [CCD00]) is "hiding in the time domain". This counter-measure breaks the assumption that traces are aligned. E.g., one variant of time domain hiding (dummy operations insertion) was analyzed by Mangard et al. [MOP08]. They showed that the correlation ratio between the correct key and the power consumption decreases, because not all traces leak in the same sample index.

Alignment problems have two common variants. In the first variant (start point misalignment), the leaking encryption sub-state happens a fixed amount of time after the encryption start, but at a variable sample index within the trace after the measurement start. The second variant of misalignment, more commonly used by defenders, is that the encryption process itself has a variable time duration. Such behavior can be caused in many ways—insertion of random length dummy operations into the machine code execution, Random Process hardware Interrupts (RPIs) or an unstable (jittered) CPU clock. These methods lead to a leaking encryption sub-state happening at an uncertain point in time after the encryption start. Our focus in this paper is dealing with the jittered clock counter-measure.

1.3   Anti-counter-measures Approaches to Trace Misalignment

For the variant of start-point misalignment, several possible solutions were suggested. Homma et al. [HNI+06] suggested a method to align the traces according to trace properties in the frequency domain. Later, Schimmel et al. [SDB+10] suggested Correlation Power Frequency Analysis (CPFA) which is impervious to start-point misalignment because frequency transform magnitude properties are independent of time domain shifting.

Batina et al. [BHvW12] proposed to solve the alignment problem by Principal Component Analysis (PCA). The method changes the possibly correlated linear base of the data-set to another linear uncorrelated base. This transformation may reveal a principal component which stands for the leakage. If such a component is found, there would be a correlation between its values and the correct key hypothesis, while the noise represented in other principal components is reduced. The authors did not suggest a way to predict the number of principal components required for the existence of leakage in these principal components.

The counter-measure variants involving a variable encryption length also have several solutions, typically via a pre-processing step. An early suggestion for time domain hiding was presented by Clavier et al. [CCD00], where the idea of samples integration in the pre-processing stage was introduced. Next, the authors proposed to perform a difference of means attack (traditional DPA), naming this method Sliding Window Differential Power Analysis (SW-DPA). The pre-processing involves aggregating several samples over number of consecutive cycles into one sample. For example, aggregating r out of each n samples for k cycles (creating a "comb-like" transformation). The integration was described as a solution for RPIs, without a specific parameter choosing suggestion. Later, to improve the performance after the pre-processing stage, a more efficient and powerful CPA attack was hinted by Brier et al. [BCO04]. Subsequently, this method was analyzed by Mangard et al. [MOP08]. Their analysis showed that when there is a single leaking sample among the r being aggregated, the correlation coefficient between the correct key hypothesis without jitter and the aggregated trace drops in proportion to $1/\sqrt{r}$; In other words, sliding-window aggregation seems to severely downgrade the performance of CPA.

Another proposed way to overcome the unstable clock counter-measure is to perform a trace alignment pre-processing step. van Woudenberg et al. [vWWB11] suggested using the method of Fast Dynamic Time Warping (FDTW), to align the traces according to one chosen reference trace, by minimizing the disparity. This alignment is done by modifying the aligned trace: inserting, deleting or matching sample points. However, the data-set used to evaluate the algorithm was created synthetically, by duplicating and deleting sampling points, hence the model in use might not be realistic. For example, if the device's power consumption is not constant within an instruction cycle (unstable noise amplitude), or if the clock's jittered frequencies are not divisible by the sampling frequency, then a large difference can be expected between the device's behavior and that of the authors' model. In their evaluation, the FDTW method outperformed two "straw-man" SW-DPA aggregation combinations. The two combinations of window size and number of windows were chosen while considering the instruction cycle length in samples and the "width" of the CPA correlation peaks. The results showed that choosing the window size and number of windows had a major impact on the results. The best results were achieved when the integration consisted of one continuous integration window rather than a "comb" with several distinct "teeth". Later, Muijrers et al. [MvWB11] showed a more computationally efficient way to align the traces using object recognition algorithms (Rapid Alignment Method). The experiments in the article were conducted on a case where random delays are added. This method is considered by the authors to be faster than FDTW but has similar detection results.

Conceptually simpler approaches were suggested in [TH12,HHO15]. Their algorithms were inspired by simple power analysis methods. They used the phenomena of traces' encryption round patterns that are sometimes observable in the traces. Hodgers et al. [HHO15] excluded high jitter traces from the data corpus by identifying peak-to-peak distances, while Tian et al. [TH12] made a specific efficient region alignment by identifying the encryption rounds.

Finally, hardware solutions were proposed for the jittered clock scenario, such as entangling the sampling clock and the board clock [OC15]. In this way, the attack is simple, while the measurement process overcomes the counter-measure. We argue that this idea seems quite difficult to use since the devices' clock is usually much harder to tap than the power supply.

In addition, there are more possible ways to handle alignment if one assumes full board access (profiling). Such approaches include template attacks [CRR02], reducing noise by linear transformations [OP12] and machine learning attacks [CDP17]. Although these methods may have good results, we find their requirements to be challenging, and do not assume full control of the board.

1.4   Contributions and Structure

In this paper we suggest a new flavor of an old sliding-window attack to overcome the counter-measure of an unstable clock and we demonstrate that it works much better than predicted by earlier analysis. Extending the general notion of Clavier et al. [CCD00], we focus on the sliding-window aggregation of consecutive samples, followed by a correlation power analysis (CPA). We start by revisiting the analysis of Mangard et al. [MOP08] and show that SW-CPA actually *amplifies* the correlation between the correct key hypothesis and the aggregated traces, both with and without jitter—as long as multiple leaking sample points are present in the integration window.

Next, we evaluate the jitter introduced by a real commercial board which has a built-in spectrum-spreader. We found it to be a powerful SCA counter-measure—its jittered traces caused severe degradation to standard CPA attacks. We then sampled the power consumption of the board while it executed a software implementation of AES, and collected a new corpus of power traces, both with and without jitter produced by the spectrum-spreader.

Then, we implemented a SW-CPA attack and conducted an extensive evaluation of its performance. The method indeed amplified the correlation and was able to revert the impact of the unstable clock almost completely.

Finally, We compared the performance of SW-CPA to that of several previously suggested SCA on our real-life data corpus: SW-CPA clearly outperformed prior attacks, requiring a vastly smaller number of traces to achieve the same level of secret key detection.

Organization: Section 2 introduces the jittered clock counter-measure and the SW-CPA attack. Section 3 theoretically analyzes the attack and predicts its effectiveness under some mild assumptions on the leakage and the jitter model. Section 4 describes the experiments we conducted with our jittered clock setup and the validation of our analytical model. Section 5 discusses the SW-CPA attack and compares it with other state-of-the-art methods. Section 6 gives final conclusions.

2   The Effect of an Unstable Clock on Standard Attacks

2.1   Unstable CPU Clock and Time Domain Hiding Analysis

An unstable clock (i.e., jittered clock) is a technique in which the CPU does not have a constant clock frequency, but one which can fluctuate in a given frequency domain. In this case, the leaking signal measurements might not occur in the same sample index in the trace. As shown in [MOP08], CMOS circuits have data dependent power consumption called dynamic power consumption which is a dominant factor in the board's total power consumption and:

$$P_{switching} \propto f_{CPU}$$

However, for our analysis we shall assume that the different CPU clock frequencies are relatively close, hence insignificant to the power consumption model.

Following [MOP08], let P, P_{orig} be the random variables representing the board's instantaneous power consumption at sample index t_0, with and without the hiding counter-measure respectively. We use the leaking Hamming weight model commonly used in SCA against software encryption implementations. Let H_{ck} be the random variable representing the hypothetical power consumption of the correct key byte value. Let $\rho(H_{ck}, P)$ denote the Pearson correlation coefficient between these random variables.

Assume that P_{orig} is computed at sample index t_0. When jitter is present, the leak drifts and might be within a range of sample indexes, either before or after t_0. We denote the probability of the leak occurring in a specific sample index t by $\bar{p}(t)$. Let $\hat{p}$ denote $\max_t \bar{p}(t)$. We assume that $\hat{p}$ is achieved at the same sample index t_0 that would likely contain the most of leakage points over the different traces, thus having the highest correlation ratio. For aligned power traces without jitter, $\hat{p} = 1$ because the leakage points all occur in the same sample number. However, for misaligned power traces $\hat{p} \neq 1$, and the maximal correlation ratio between the observed power consumption P and the correct key hypothetical power consumption H_{ck} would be:

$$\rho(H_{ck}, P) = \rho(H_{ck}, P_{orig}) \cdot \hat{p} \tag{1}$$

2.2   Sliding Window CPA Attack on Jittered CPU Clocks

The Sliding Window Differential Power Analysis attack (SW-DPA) was initially proposed in [CCD00]. It was proposed as a way to eliminate RPIs with aggregation parameters similar to a "comb" function transformation. It was performed with traditional difference of means DPA (single bit model attack).

Our attack on jittered CPUs, which we call the Sliding Window Correlation Power Analysis attack (SW-CPA), is inspired by [CCD00]; we use a similar pre-processing idea but then we use a CPA attack (byte model attack). Furthermore, unlike the example in [CCD00], we use only a single continuous integration window with a size of r (aggregating r consecutive samples instead of a sparse "comb" aggregation)—see Algorithm 1. The attack exploits the fact that although each trace's leakage can happen at a different time due to jitter, with a high probability the leakages will occur within some radius $r/2$ of the original leakage sample point (without the counter-measure). If we then apply the CPA attack on the integrated traces, there would be a common trace sample index containing the leakage for many different traces. We chose to aggregate one continuous window (rather than the sparse comb-like integration of [CCD00]) as we cannot assume where the leakage would be.

Algorithm 1. Sliding Window Correlation Power Analysis Attack (SW-CPA)

1: **procedure** PREPROCESSTRACE(Trace, r)
2:   **for** $t \in Trace$ **do**
3:      $SummedTrace(t) = \sum_{i=-r/2}^{r/2} Trace(t + i)$
4:   **return** SummedTrace
5: **procedure** ATTACK(r)
6:   *Acquire set of traces X*
7:   **for** $Trace \in X$ **do**
8:      $Trace \leftarrow PreprocessTrace(Trace, r)$.
9:   *Perform CPA on X.*

2.3   Basic Correlation Analysis of Sliding-Window Integration

To begin with, let us find the Pearson correlation coefficient of a key hypothesis with the pre-processed traces data-set, when no jitter is present and the traces are aligned. Without loss of generality assume that a leakage occurs at sample point 1. Let ρ_1 be coefficient for SW-CPA is: coefficient between the leakage sample P_1 and the correct key hypothesis H_{ck}. Then, by definition we have:

$$\rho_1 \equiv \rho(H_{ck}, P_1) = \frac{Cov(H_{ck}, P_1)}{\sqrt{Var(H_{ck}) \cdot Var(P_1)}} = \frac{E(H_{ck} \cdot P_1) - E(H_{ck}) \cdot E(P_1)}{\sqrt{Var(H_{ck}) \cdot Var(P_1)}} \quad (2)$$

In [MOP08] pp. 210–211, Mangard et al. analyzed the effect of integrating r independent samples, $\{P_{-r/2}, \ldots, P_1, \ldots, P_{r/2}\}$, containing a single leakage sample:

$$\rho(H_{ck}, \sum_{i=-r/2}^{r/2} P_i) = \frac{\rho_1}{\sqrt{r}} \quad (3)$$

Therefore, there is a trade-off on setting the window size r. On the one hand, when we increase r, we increase the likelihood that the leakage sample would be within our aggregation window. Consequently, due to Eq. (1), we would like to increase the window size. On the other hand, Eq. (3) seems to show that integration decreases the correlation by the square root of the window size.

3   A New Analysis of Multiple Leakage Samples Integration

In this section we show that contrary to the degradation predicted by Eq. (3), SW-CPA integration can be an effective technique which actually *amplifies* the correlation with and without jitter. In Sect. 4 we validate that our model assumptions indeed hold on traces collected from a real device with a jittered clock.

3.1   The Correlation Coefficient When Integrating Within a Trace

The leakage model previously mentioned in Sect. 2.3 assumes only a *single* leakage point within the integration window. However, there might be several leakage samples in a trace. This may be caused by several reasons: multiple leakage sources may exist, such as data bus leakage, address bus leakage or different electronic components' glitches which may all happen sequentially. Alternatively, a high sampling frequency of the measurement instrument may cause switching to spread over more than one sample. CPU architecture and software implementation may imply more phenomena creating such behavior. For example, Papagiannopoulos et al. [PV17] showed that the data might be loaded to several registers during the computation. As we shall see, in traces we collected (without jitter), we observed this phenomenon quite clearly: there were multiple leak points, relatively close to each other in time.

We start our analysis with the case of aligned traces: we assume the clock is stable and analyze the effect of SW-CPA with different values of window size r.

Assume that among the r samples $\{P_{-r/2}, \ldots, P_{r/2}\}$, there are $q(r) \geq 1$ leakage points and $r - q(r)$ samples independent of the correct key hypothesis H_{ck} (which we call for short "noise samples"). For the $q(r)$ leakage samples, we assume that the random variables P_i are identically distributed but not independent since they all depend on the leak—but their variability is caused by the noise, which we can reasonably argue to be independent among different sample points. Therefore, they have the same expectation and variance. Without loss of generality, assume that P_1 is a leakage sample point, so for all $q(r)$ leakage samples P_i we have:

$$E(P_i) = E(P_1) \tag{4}$$

Next, we assume that leakage and noise samples have the same variance, since they are all subject to the same noise, i.e.,

$$Var(P_i) = Var(P_1) \text{ for all } i.$$

By definition, for two leakage samples with same variance, using Pearson correlation coefficient $\rho_{i,j}$ between power samples P_i, P_j, we have:

$$Cov(P_i, P_j) \equiv \sqrt{Var(P_i) \cdot Var(P_j)} \cdot \rho_{i,j} = Var(P_1) \cdot \rho_{i,j} \tag{5}$$

For the other $r - q(r)$ noise samples we can assume that they are independent of each other and of the leakage points. Therefore, the samples P_i, P_j where at least one is a noise samples are uncorrelated, i.e., $Cov(P_i, P_j) = 0$. Hence, for all samples' types (leakage or noise), we conclude that for all P_i, P_j:

$$Cov(P_i, P_j) = \begin{cases} Var(P_1) \cdot \rho_{i,j} & i, j \text{ are leakage samples} \\ 0 & \text{Otherwise} \end{cases} \tag{6}$$

Noise samples are also independent of the correct key hypothesis, so for such P_i:

$$E(H_{ck} \cdot P_i) = E(H_{ck}) \cdot E(P_i) \tag{7}$$

Now we return to the correlation coefficient. According to Eq. (2), the correlation coefficient for r integrated samples is:

$$\rho(H_{ck}, \sum_{i=-r/2}^{r/2} P_i) = \frac{E(H_{ck} \cdot (\sum_{i=-r/2}^{r/2} P_i)) - E(H_{ck}) \cdot E(\sum_{i=-r/2}^{r/2} P_i)}{\sqrt{Var(H_{ck}) \cdot Var(\sum_{i=-r/2}^{r/2} P_i)}}$$

$$= \frac{\sum_{i=-r/2}^{r/2} (E(H_{ck} \cdot P_i) - E(H_{ck}) \cdot E(P_i))}{\sqrt{Var(H_{ck})) \cdot Var(\sum_{i=-r/2}^{r/2} P_i)}}$$

Because there are exactly $q(r)$ leakage samples and by Eqs. (4) and (7) and the standard formula for the variance of a sum we get:

$$\rho(H_{ck}, \sum_{i=-r/2}^{r/2} P_i) = \frac{q(r) \cdot (E(H_{ck} \cdot P_1)) - E(H_{ck}) \cdot E(P_1))}{\sqrt{Var(H_{ck})} \cdot \sqrt{\sum_{i=-r/2}^{r/2} Var(P_i) + \sum_{i \neq j} Cov(P_i, P_j)}}$$

By Eq. (6) and plugging in the definition of ρ_1 (non-jittered correlation without integration) from Eq. (2) we can simplify the result to:

$$\rho(H_{ck}, \sum_{i=-r/2}^{r/2} P_i) = \frac{q(r) \cdot (E(H_{ck} \cdot P_1) - E(H_{ck}) \cdot E(P_1))}{\sqrt{Var(H_{ck})} \cdot \sqrt{r + \sum_{i \neq j} \rho_{i,j}} \cdot \sqrt{Var(P_1)}}$$

$$\implies \rho(H_{ck}, \sum_{i=-r/2}^{r/2} P_i) = \frac{q(r)}{\sqrt{r + \sum_{i \neq j, \text{leakage samples}} \rho_{i,j}}} \cdot \rho_1 \tag{8}$$

Let γ denote the normalized sum of correlation coefficients of the leakage points:

$$\gamma \equiv \frac{r + \sum_{i \neq j, \text{leakage samples}} \rho_{i,j}}{r}$$

$$\implies \rho(H_{ck}, \sum_{i=-r/2}^{r/2} P_i) = \frac{q(r)}{\sqrt{r \cdot \gamma}} \cdot \rho_1 \tag{9}$$

If all the leakage points are uncorrelated samples then $\rho_{i,j} = 0 \Rightarrow \gamma = 1$. Conversely, in the worst case the leakage points are fully correlated, with $\rho_{i,j} = 1 \Rightarrow \gamma = r$. Note that γ is derived from the correlation matrix of random variables, which is positive semidefinite and in particular the sum of its items is non-negative, hence also $\gamma \geq 0$. However, γ can be smaller than 1 causing a further amplification. Casting Eq. (9) to also explicitly show the interesting cases we get:

$$\rho(H_{ck}, \sum_{i=-r/2}^{r/2} P_i) = \begin{cases} \frac{q(r)}{\sqrt{r}} \cdot \rho_1 & \text{uncorrelated leakage samples} \\ \frac{q(r)}{\sqrt{r \cdot \gamma}} \cdot \rho_1 & \text{partly correlated leakage samples} \\ \frac{q(r)}{r} \cdot \rho_1 & \text{fully correlated samples} \end{cases} \quad (10)$$

For simplicity, unless mentioned otherwise, in the derivations below we assume leakage samples are uncorrelated, hence:

$$\gamma = 1 \quad (11)$$

In Sect. 4.3, γ is shown to be quite close to 1 and much smaller than r.

We can see that for the special case of $q(r) = 1$ we get exactly Eq. (3), i.e., the result of Mangard et al. [MOP08]. For the most special case, where $r = q = 1$ we obtain the standard CPA attack.

3.2 Correlation Coefficient Amplification

Let P_t be the distribution of trace power values at sample index t. Let

$$\rho_{cpa} = \max_t \rho(H_{ck}, P_t)$$

be the achieved correlation coefficient of a regular CPA attack on the traces. Now, assume we conduct a SW-CPA with a window size of r. Then let

$$\rho_r = \max_t \rho(H_{ck}, \sum_{i=-r/2}^{r/2} P_{t+i})$$

be the correlation achieved by SW-CPA with window size r. Note that $\rho_{cpa} \equiv \rho_1$. We define the correlation coefficient *amplification* to be: $Amplification = \rho_r/\rho_1$.

3.3 The Correlation Coefficient for Specific r and q Relationships

Equation (10) can be made concrete if we have an explicit connection between r and q. We first assume that each key byte has a maximal number of leakage points, q_{max}, which are all temporally close: all located within a distance of r_0 samples from each other. When $r \geq r_0$ we call the window saturated. So we get:

$$q = \begin{cases} q(r) & \text{if } r < r_0 \\ q_{max} & \text{otherwise (saturation)} \end{cases} \quad (12)$$

With this assumption we analyze two important cases:

Constant Number of Leakage Points. In case $r \geq r_0$, our window contains all q_{max} leakage points of the phenomenon. Increasing the window size any further does not change the value of q. According to Eq. (10), the correlation would be:

$$\rho(H_{ck}, \sum_{i=-r/2}^{r/2} P_i) = \frac{q_{max}}{\sqrt{r}} \cdot \rho_1 \tag{13}$$

Hence, when r increases ρ decreases, and for $r > q_{max}^2$ the correlation drops below ρ_1 and eventually $\rho \to 0$. Therefore, r should be selected to be the smallest possible value containing all q_{max} leakage points.

This observation is also valid for the general case. When the number of leakage points $q(r)$ does not change while incrementing r, the correlation decreases by $\sqrt{r}$ until more leakage points are aggregated into the integration window.

Constant Ratio Between r and q. Another important case is when the integration window is not saturated, and increasing r increases the number of leakage points q *linearly* such that $q(r) = r/c$ for some constant c. In this case:

$$\rho(H_{ck}, \sum_{i=-r/2}^{r/2} P_i) = \frac{q(r)}{\sqrt{r}} \cdot \rho_1 = \frac{r/c}{\sqrt{r}} \cdot \rho_1$$

$$\implies \rho(H_{ck}, \sum_{i=-r/2}^{r/2} P_i) = \frac{\sqrt{r}}{c} \cdot \rho_1 \tag{14}$$

The first implication of this equation is that when $\sqrt{r} > c$ we obtain that $\rho > \rho_1$: in other words, without jitter, not only does integration not reduce the correlation coefficient, it can even amplify it. However, as we increase r, eventually the number of leakage points saturates, yielding a non constant ratio between r and $q(r)$ and we fall back to Eq. (13).

Therefore, according to Eqs. (13) and (14), we get that the relationship between ρ, the correlation coefficient of the integrated non-jittered traces; r, the window size; q, the number of leakage points within the window; and c, the ratio between r and q is (Still assuming for simplicity that $\gamma = 1$):

$$\rho(H_{ck}, \sum_{i=-r/2}^{r/2} P_i) = \begin{cases} \frac{\sqrt{r}}{c} \cdot \rho_1 & r < r_0, \text{constant ratio between } q \text{ and } r \\ \frac{q_{max}}{\sqrt{r}} \cdot \rho_1 & r \geq r_0 \text{ (saturated } q) \end{cases} \tag{15}$$

3.4   The Correlation Coefficient with an Unstable Clock

So far, our analysis of SW-CPA assumed a stable clock and aligned traces. When we use an unstable clock, the correlation coefficient is also affected by the probability that the leakage signals happen in the window around the same point in time, as stated in Eq. (1). We denote by $\hat{q}(r)$ the number of leakage points in a window of size r when jitter is present.

Combining $\hat{q}$ leakage points and the case of uncorrelated samples in Eq. (10) yields the general correlation coefficient for the jittered clock:

$$\rho(H_{ck}, \sum_{i=-r/2}^{r/2} P_i) = \frac{\hat{q}(r)}{\sqrt{r}} \cdot \rho_1 \tag{16}$$

Leakage Sample Drift Under a Bounded Jitter. We now assume the clock jitter is bounded and the maximal drift that a logical action in the encryption process can suffer is J sample points (we validate this assumption empirically in Sect. 4.2). We seek to find the relation between $\hat{q}$ and q for different values of r. For simplicity, we assume that the drift of a sample point is uniformly distributed in time around the original non-jittered index, i.e., $Drift \sim U\{\frac{-J}{2}, \frac{J}{2}\}$.

Because the drift is distributed uniformly and $E(Drift) = 0$, the distance between the leakage points might increase as well as decrease, but it's expectation is equal to the non-jittered case.

Hence, with jitter, we take a worst-case scenario in which all q_{max} leakage points are uniformly distributed among the $r_0 + J$ samples. Further, drift causes saturation in a larger window size. Instead of Eq. (12) we get:

$$\hat{q}(r) = \begin{cases} \frac{q_{max}}{r_0 + J} \cdot r & \text{if } r < r_0 + J \\ q_{max} & \text{otherwise (saturation)} \end{cases} \tag{17}$$

The CPA Correlation Coefficient in the Jittered Case. We first calculate $\hat{\rho}_1$, the correlation coefficient for original CPA attack ($r = 1$) with jitter $J > 1$. The leakage signal originally always happens at t_0, but due to the jitter it may occur anywhere within the range $[t_0 - J/2, t_0 + J/2]$.

According to Eq. (17), according to the uniform leakage distribution, the probability that a leakage point appears in sample index t_0 is:

$$\hat{q}(r = 1) = \frac{q_{max}}{r_0 + J} = \frac{r_0}{r_0 + J} \cdot \frac{1}{c} \tag{18}$$

Putting Eqs. (16) and (18) together gives the correlation ratio for the standard CPA ($r = 1$) against jittered traces:

$$\hat{\rho}_1 = \frac{\hat{q}(r = 1)}{\sqrt{r}} \cdot \rho_1 = \frac{r_0}{r_0 + J} \cdot \frac{1}{c} \cdot \rho_1 \tag{19}$$

We can see that according to Eq. (19), when jitter is present the standard CPA attack effectiveness is severely degraded—as we shall see in Sect. 5.2.

The SW-CPA Correlation Coefficient for Different r Values. We now analyze two important cases of r, caused by the different domains of $\hat{q}$ in Eq. (17), under the effect of a bounded jitter.

Constant q/r Ratio: When $r < r_0 + J$ from Eqs. (16) and (17), the correlation coefficient for SW-CPA is:

$$\rho(H_{ck}, \sum_{i=-r/2}^{r/2} P_i) = \hat{q}(r) \cdot \frac{1}{\sqrt{r}} \cdot \rho_1 = \frac{q_{max} \cdot r}{r_0 + J} \cdot \frac{1}{\sqrt{r}} \cdot \frac{r_0 + J}{r_0} \cdot c \cdot \hat{\rho}_1$$

$$\implies \rho(H_{ck}, \sum_{i=-r/2}^{r/2} P_i) = \sqrt{r} \cdot \hat{\rho}_1 \qquad (20)$$

Saturated $\hat{q}$ Values: For $r \geq r_0 + J$, the region around t_0 contains all the leakage points $(\hat{q}(r) = q_{max})$. Combining Eqs. (16) and (19) gives:

$$\rho(H_{ck}, \sum_{i=-r/2}^{r/2} P_i) = \frac{\hat{q}(r)}{\sqrt{r}} \cdot \rho_1 = \frac{q_{max}}{\sqrt{r}} \cdot \frac{r_0 + J}{r_0} \cdot c \cdot \hat{\rho}_1 = \frac{r_0 + J}{\sqrt{r}} \cdot \hat{\rho}_1 \qquad (21)$$

Summarizing Eqs. (20) and (21), we get that the relationship between ρ, the correlation coefficient of the integrated jittered traces; $\hat{\rho}_1$, the correlation coefficient without integration; r, the window size; q, the number of leakage points within the window; c, the ratio between r and q; replugging in the γ factor from Eq. (10); and J, the maximal drift is:

$$\rho(H_{ck}, \sum_{i=-r/2}^{r/2} P_i) = \begin{cases} \frac{\sqrt{r}}{\sqrt{\gamma}} \cdot \hat{\rho}_1 & r < r_0 + J, \text{constant ratio between } q \text{ and } r \\ \frac{r_0 + J}{\sqrt{r}\sqrt{\gamma}} \cdot \hat{\rho}_1 & r \geq r_0 + J \text{ (saturated } q) \end{cases}$$

$$(22)$$

Figure 4 (right) illustrates Eq. (22) theoretically for different γ values and empirically for the data analyzed in Sect. 5.1. For specific parameters SW-CPA can amplify the correlation ratio by factor of 10 for the best r values.

3.5   The Correlation Coefficient with an Unbounded Jitter

While our analysis assumed that the jitter is bounded (and in Sect. 4.2 we demonstrate this is a realistic assumption for our board), we argue that our analysis has merit in more general cases as well. Even if the jitter is unbounded we still expect to observe a randomly changing clock frequency according to some distribution. In such a case, we assume that using a reasonable clock spreading model, it should be possible to build a sample drift model in which with high probability the drift value would be in a specific range, thus making our analysis relevant. We leave the analysis of cases with unbounded jitter to future work.

4   Experiments and Results

4.1   Setup and Measurements

Our experimental setup contains a Rabbit RCM4010 evaluation board which has a 59 MHz processor with a 16-bit architecture [RCM10]. We programmed the board to implement an AES-128 algorithm using open-source code taken from [Con12]. This is a plain-vanilla software implementation, without any side channel counter-measures or software optimizations (i.e., without using T-tables).

The Rabbit processor has a special feature called a spectrum-spreader— designed to reduce electromagnetic interference (EMI). Enabling the spreader introduces jitter into the CPU clock frequency. However, the documentation does not specify precisely how the spectrum-spreader works. Note that the Rabbit has two spreading modes, called Normal and Strong (in addition to no spreading mode), which can be selected by software.

We sampled the board power consumption by a Lecroy WavePro 715Zi oscilloscope. When starting the execution of an encryption, we programmed the board to send a signal to the oscilloscope via one of its I/O pins which can be controlled by the software. This signal sets the trigger for the oscilloscope, which starts sampling at a rate of 500 million samples per second, for 500 μs. This time period contains one round of the full AES encryption. Every encryption process is recorded to a new trace. The voltage of the processor was measured by a shunt resistor soldered to the processor voltage input. The input plaintexts for the program were changed every encryption round, while the key was kept constant during all traces. Two data-sets where captured; one consisted 5,000 traces without jitter and 5,600 traces with Normal spreading, using the same encryption key and plaintexts (for the first 5,000 jittered traces). The second and bigger data-set contains 10,000 traces of each spectrum-spreading mode: no spreading, Normal spreading and Strong spreading. These measurements were done with a different random key than the first data-set, but same plaintexts. The data-sets we collected were uploaded to [FW18] and can be used for side channel attack methods comparison.

Note that while the spectrum-spreader is not an SCA counter-measure by design, we found it to be quite effective as such. E.g., as we shall see in Sect. 5.2, when the spectrum-spreader is turned on, the standard CPA attack is drastically degraded: without jitter the attack correctly discovers all 16 key bytes with as few as 2,500 traces, while with jitter CPA fails to identify more than two key bytes even with all 5,600 traces of the first data-set.

4.2   Jitter Modeling

We explored the jitter injected by the spectrum-spreader to validate the analysis of Sect. 3.4. This part was used for white-box validation of our leakage model only and is not essential for the common adversary. When spectrum-spreading was enabled, frequency analysis revealed several new frequencies that appeared around the original 59 MHz clock frequency, with about 0.15 MHz difference between them. Figure 1(a) shows the spectrum without jitter: notice the peaks at 59 MHz and 60 MHz (the former is the board clock frequency). Figure 1(b) shows the spectrum with Normal jitter: notice how the 59 MHz peak is replaced by some 15–25 separate peaks while the irrelevant 60 MHz peak is unaffected. Figure 1(c) shows the spectrum with Strong jitter: some 15 additional peaks appeared with higher and lower frequencies.

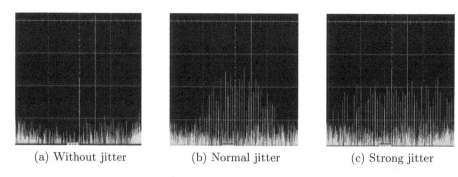

|(a) Without jitter|(b) Normal jitter|(c) Strong jitter|

Fig. 1. FFT magnitude vs. frequency of the power trace from RCM board, computed by the oscilloscope (a) without jitter, (b) with Normal jitter, (c) with Strong jitter, centered around 59 MHz (original clock frequency) and axis between 55–63 MHz

Next, we conducted a set of experiments in order to understand the drift of the jittered clock (Normal jitter). We programmed the board to implement the following steps (see Algorithm 2): send a first signal to the oscilloscope, then perform N times a condition test and a variable assignment, and finally send a second signal when finishing the execution. The time between the two signals (ΔT) was saved and analyzed. We set the execution length N to start at about a quarter of the total AES encryption time ($N = 600 \implies \Delta T = 2 \text{ ms}$), and increased it to more than the encryption time ($N = 3000 \implies \Delta T = 10 \text{ ms}$). We also tested intermediate values of $\Delta T = 4 \text{ ms}$ and $\Delta T = 5 \text{ ms}$. 500 executions were done for each of the N values. When spectrum-spreading was not enabled, ΔT was identical in all executions (per execution length). When Normal spreading was enabled ΔT was not constant per execution length. We denote by D the difference, in number of samples, between the execution length with jitter and the constant execution length without jitter. For different execution lengths, we observed that the magnitude of the drift ($|D|$) was *bounded* by at most 10 samples (20 ns) to each side, regardless of the execution length. Using the terminology of Sect. 3.4, the Rabbit Normal spectrum-spreader has a bound $J = 20, |D| = 10$. Similar experiments with the Strong spectrum-spreader showed that the drift is still bounded but with $J = 40, |D| = 20$. The bounded drift in number of samples is illustrated by a box plot in Fig. 2, for both Normal spreading and Strong spreading (box plots for additional Strong spreading execution lengths omitted).

Algorithm 2. Drift assessment

1: **procedure** PerformInstructions(N)
2:     *Send an initial signal for execution start*
3:     **for** i from 1 to N **do**
4:         **if** *True* **then**
5:             *Var*1 ← 0
6:     *Send a second signal for execution end*

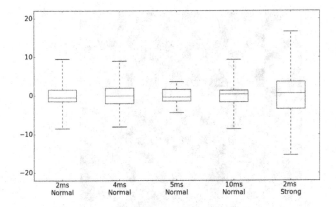

Fig. 2. Drift in number of samples (D) vs. different execution duration (ΔT) with the Normal and Strong spectrum-spreader. The red line is the median, the bottom and top of the boxes represent the first the third quartiles, and the whiskers range from the minimum to the maximum samples drift. Normal spreading is bounded by $|D| = 10$ samples and Strong spreading is bounded by $|D| = 20$ samples. (Color figure online)

We believe that drift is not accumulating beyond $|D| = 10$ for Normal spreading and $|D| = 20$ for Strong spreading because the spreading is probably generated by a fixed cyclic series of clock jitter values, with a cycle time shorter than 2 ms. The bounded drift is consistent with the board design, since even a short cycle of jitter values can achieve the goal of EMI reduction, much more easily than generating true random, or cryptographic pseudo-random, clock jitter.

4.3 Validating Leakage Points' Power Consumption Correlation

We need to validate our assumptions in Eqs. (4), (6) and (10) about the distributions and correlation between leakage points and the value of γ. In Fig. 3 we show a heat-map of the correlation coefficients between 25 leakage sample points of a specific key byte, for 5,000 traces without jitter. These leakage samples form the best window for integration with maximal correlation between the true key byte and the traces as shown in Sect. 5.2. In order to find the leakage points, we set a threshold (of 3 standard deviations above or below the mean) over the correlation coefficient of a sample index to differentiate between leakage and noise samples. Figure 3 shows that the off-diagonal correlations are both negative and positive: these sign alternations in fact help keep the total correlation low, with a total sum of $\gamma = 1.7$ (including diagonal values). Thus, the correlation coefficient in Eq. (10) is divided by $\sqrt{\gamma} = 1.3$, which is still highly amplified in comparison to CPA without integration. This experiment was done for all key bytes, resulting in γ values between 0.5 to 1.7 with average 0.95 and standard deviation of 0.33—supporting our assumption in Eq. (11) that γ is close to 1; hence we can treat the leakages as if they are uncorrelated without a great penalty in the analysis.

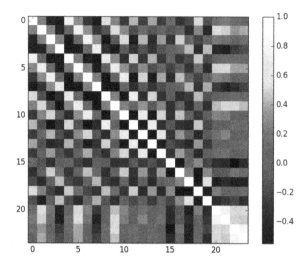

Fig. 3. Correlation matrix heat-map, for 25 leakage sample points for the best leakage window of key byte 7

5 Evaluating the SW-CPA Attack

5.1 Amplification for Different Aggregation Window Sizes

To calibrate the best window size r we examined leaks from the different key bytes in our encryption process. Figure 4 (left) shows the amplification of the correlation coefficient for different window sizes and different correct key bytes when the CPU clock is jittered both theoretically and empirically. Note that these key bytes were not identified correctly by regular CPA due to jitter. For simplicity, we do not show all key bytes. The parameter values of the theoretical Fig. 4 (right) were chosen according to the values found later in our experimental setup. The upper curve models a bounded jitter for uncorrelated leakage ($\gamma = 1$) where $J = 20, r_0 = 70, c = 3$ (leakage in a third of the samples in the window), and q reaches saturation of $q_{max} = 25$ when $r = r_0 + J = 90$. The Figure also illustrates the worst case scenario where the leakage samples are all fully correlated and $\gamma = r$, where we can see no amplification.

The amplification graphs for all key bytes have major similarities. First, they all have an amplification higher than 1 for some window size r, which helps the correct key byte detection and supports SW-CPA as an effective solution for the unstable clock counter-measure. In addition, they all suffer degradation when r grows beyond a certain point and q reaches saturation.

Note that unlike the prediction in Fig. 4 (right), some of the curves do not increase monotonically toward a single peak, and contain a significant peak when r is relatively small, around $5 \leq r \leq 10$, as demonstrated in key byte 10. This is somewhat surprising because as stated in Eq. (22), for a small window size r, the integration might not be as effective as for a large window size. However, the

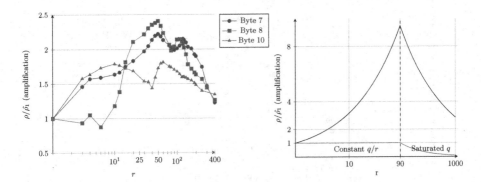

Fig. 4. Amplification of the correlation coefficient vs. window size r (log scale). Amplification above 1 indicates that ρ is amplified beyond the values for $r = 1$. Left: empirical values for three correct key bytes, with the jittered clock data-set of 5,600 traces. Right: theoretical amplification values according to Eq. (22) for $J = 20, r_0 = 70$. The black line is the scenario for uncorrelated leakage samples ($\gamma = 1$). The blue line shows worst case correlated leakage samples ($\gamma = r$). The dashed line at $r = 90$ separates the two regions of the amplification (constant q/r ratio and saturated q). (Color figure online)

analysis leading to Eq. (22) assumed a uniform scatter of the r/c leakages in the window: We speculate that maybe the leaks for some key bytes had leakage points with non-uniform scatter, producing locally-higher densities. Another option is that the leakage samples are correlated in a way that γ is relatively small for this small window of leakage samples.

5.2   Selecting a Window Size r for All Key Bytes

Next, we determine the single, best, r value of all key bytes for our device. Figure 5 shows the overall SW-CPA success rate for different r values together with the results for standard CPA on non-jittered traces (as an ideal) and CPA on the jittered traces (as a worst-case) for Normal spreading.

We've experimentally seen in Fig. 4 that the ρ amplification graphs for separate key bytes had the highest peaks between $25 < r < 75$. We chose the overall value of $r = 75$ experimentally, simply by running the attacks.

Figure 5 shows clearly that SW-CPA is very effective and defeats the clock jitter counter-measure well: for values of $10 \leq r \leq 75$ it finds 12–14 correct key bytes with $\approx$4500 traces—only twice as many traces as needed for an equivalent success rate on non-jittered traces. Further, our attack is not very sensitive to the value of r: values between $10 \leq r \leq 75$ are roughly equally successful. The figure shows that a larger window such as $r = 150$ gives a poor amount of true key byte detections. Windows with $r \leq 10$ have inferior performance (graphs omitted).

We also conducted the same experiment with the larger (10,000 traces) data-set and both Normal and Strong spectrum-spreading. Figure 6 shows the analysis

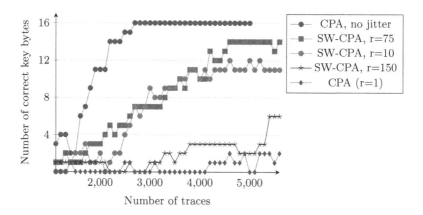

Fig. 5. Number of correct key bytes vs. number of traces, for different values of the integration window size r with Normal spreading.

of Strong spreading and its more noticeable results. The figure shows that SW-CPA is very successful against Strong jitter as well: it correctly finds all key bytes, with about 6,000 traces, for many window sizes, whereas regular CPA cannot find two correct key bytes even with all 10,000 traces. In addition, the higher drift with Strong jitter causes SW-CPA with large window size such as $r = 300$ to be effective and find 15–16 key bytes, whereas with Normal spreading (recall Fig. 5) $r = 150$ was already too high and performance was degraded in comparison to $r = 75$.

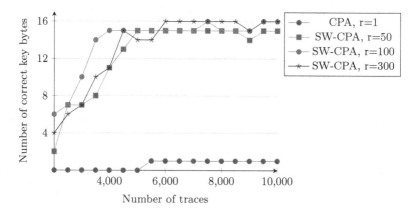

Fig. 6. Number of correct key bytes vs. number of traces, for different values of the integration window size r with Strong spreading and large data-set.

5.3 Correct Key Byte Identification Metric

The metric we used to recognize a correct key byte detection counted a correct key byte when the true key byte was within the highest five correlation possibilities, i.e., the key byte recovery is of the 5th order as stated in [SMY09]. This metric was chosen because a cryptanalyst can iterate (brute force) over the remaining $5^{16} \approx 2^{37}$ options.

To determine the optimal window size r, we suggest choosing its value after analyzing the q/r ratio for all key bytes if possible, or otherwise by trial and error (no profiling). Choosing an imprecise value of r still gives far better results than other state-of-the-art methods as would be shown later: Even for clearly sub-optimal choices of r our method is superior to others (see Fig. 7). In addition, the computational resources for trial and error are low in comparison to other methods. One might also use a different window size for each key byte. We did not explore this possibility since the results with a uniform r were satisfactory.

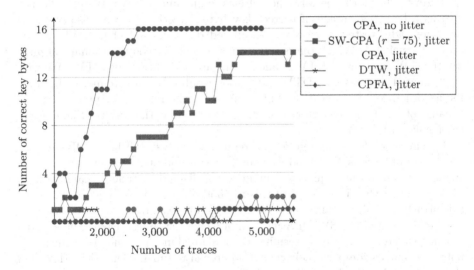

Fig. 7. Number of correct bytes vs. number of traces, for different implemented attacks. Attacks with no successful detections were omitted. We also show the success rate of the standard CPA attack on non-jittered data (as an ideal).

5.4 Comparing SW-CPA with Other Known Methods

We compare the SW-CPA method (with the best integration window size) to previously suggested methods: trace selection pre-processing [HHO15], alignment pre-processing [TH12, vWWB11, BHvW12], and frequency analysis attacks [SDB+10]. Figure 7 summarizes the results.

Applying the methods suggested in [HHO15, TH12] of pre-processing according to simple trace properties was inapplicable to our data-set. These attacks were performed on hardware encryption implementations and assume that the power consumption measurements have clearly visible patterns of the AES rounds. Our data with a software implementation on the Rabbit board exhibited no such patterns. We tried searching for the patterns with different sampling frequencies and different number of samples but the expected 10 spikes marking the 10 AES rounds did not manifest themselves in the traces, possibly because the Rabbit board we used is not idle between the encryption cycles or when waiting for input. Because the attacks rely on visible encryption rounds, we were unable to attack the device by these methods.

Another available solution is using a PCA attack [BHvW12]. This attack works if there exists a principal component representing the leakage. However, in our base transformations, no such principal component was found, even with high numbers of base items and concentrating in the leakage region of the traces.

Therefore, the methods [HHO15, TH12, BHvW12] detected zero key bytes correctly, and were not inserted to the comparison in Fig. 7.

Figure 7 shows the performance of elastic alignment [vWWB11]: it did not give us a high percentage of correct key byte detection (as was also observed by others who tested it with non-simulated data-sets [OP11, GPPT15]). The original article [vWWB11] offers a way to overcome the computational complexity of DTW by using FDTW, which is an approximation for DTW. We first implemented and tested FDTW with poor results. In an attempt to improve its performance, we applied the full DTW (with the relevant alignment margin because of our bounded jitter): this slightly improved the results (Fig. 7 shows the results of full DTW).

The method of Correlation Power Frequency Analysis (CPFA) [SDB+10] was previously offered as a method for handling start-point misalignment, because the magnitude in the frequency domain is not affected by time domain shifting. Figure 7 shows that the results of this method were poor. We tried to optimize this attack as well, by targeting leakage areas, but results stayed the same.

We also tested the SW-DPA method of Clavier et al. [CCD00]. The authors did not suggest a way to determine their algorithm's parameters, hence it is not clear how to compare their general approach to our instantiation. However, their SW-DPA with 1-bit difference of means using our choice of integration parameters gave poor results and was omitted from the comparison figure.

For our SW-CPA attack we chose window size of $r = 75$, as found in Sect. 5.2. Many other choices of r still outperform other methods as well.

Figure 7 clearly shows that SW-CPA yields far better true key byte detection results than the other possible solutions we tried. All the other solutions did not have more than two correct key bytes detections on our small data-set. However, note that the unstable clock still degregates the attack: even our best SW-CPA requires approximately twice the number of traces to achieve an equivalent level of success in comparison to standard CPA against a non-jittered device.

6   Conclusions

In this paper we suggested an attack to overcome the jittered CPU clock counter-measure, proposing a specific parameter setting for the old method of consecutive samples integration followed by a correlation attack (Sliding Window CPA). Former analysis showed that integration of samples degrades the correlation between the correct key hypothesis and the trace. We re-analyzed this method under a new model where multiple leakage points may be present within the window, and we showed that integration of samples over a suitably chosen window size amplifies the correlation significantly. We then validated our analysis on a new data-set of traces measured on a board implementing a jittered clock. Our experiments show that the SW-CPA attack with a well chosen window size is very powerful against a jittered clock counter-measure and significantly outperforms previous state-of-the-art suggestions.

References

[BCO04] Brier, E., Clavier, C., Olivier, F.: Correlation power analysis with a leakage model. In: Joye, M., Quisquater, J.-J. (eds.) CHES 2004. LNCS, vol. 3156, pp. 16–29. Springer, Heidelberg (2004). https://doi.org/10.1007/978-3-540-28632-5_2

[BHvW12] Batina, L., Hogenboom, J., van Woudenberg, J.G.J.: Getting more from PCA: first results of using principal component analysis for extensive power analysis. In: Dunkelman, O. (ed.) CT-RSA 2012. LNCS, vol. 7178, pp. 383–397. Springer, Heidelberg (2012). https://doi.org/10.1007/978-3-642-27954-6_24

[CCD00] Clavier, C., Coron, J.-S., Dabbous, N.: Differential power analysis in the presence of hardware countermeasures. In: Koç, Ç.K., Paar, C. (eds.) CHES 2000. LNCS, vol. 1965, pp. 252–263. Springer, Heidelberg (2000). https://doi.org/10.1007/3-540-44499-8_20

[CDP17] Cagli, E., Dumas, C., Prouff, E.: Convolutional neural networks with data augmentation against jitter-based countermeasures. In: Fischer, W., Homma, N. (eds.) CHES 2017. LNCS, vol. 10529, pp. 45–68. Springer, Cham (2017). https://doi.org/10.1007/978-3-319-66787-4_3

[Con12] Conte, B.: Basic implementations of standard cryptography algorithms, like AES and SHA-1 (2012). https://github.com/B-Con/crypto-algorithms

[CRR02] Chari, S., Rao, J.R., Rohatgi, P.: Template attacks. In: Kaliski, B.S., Koç, K., Paar, C. (eds.) CHES 2002. LNCS, vol. 2523, pp. 13–28. Springer, Heidelberg (2003). https://doi.org/10.1007/3-540-36400-5_3

[FH08]   Ferrigno, J., Hlaváč, M.: When AES blinks: introducing optical side channel. IET Inf. Secur. **2**(3), 94–98 (2008)

[FW18]   Fledel, D., Wool, A.: RCM4010 AES-128 power traces, with and without spectrum-spreading (2018). https://drive.google.com/open?id=1DbcM2Z1RLi1xt8tO7qF5HGCWX8SDAd5BVwxgD7y_bU8

[GPPT15] Genkin, D., Pachmanov, L., Pipman, I., Tromer, E.: Stealing keys from PCs using a radio: cheap electromagnetic attacks on windowed exponentiation. In: Güneysu, T., Handschuh, H. (eds.) CHES 2015. LNCS, vol. 9293, pp. 207–228. Springer, Heidelberg (2015). https://doi.org/10.1007/978-3-662-48324-4_11

[HHO15] Hodgers, P., Hanley, N., O'Neill, M.: Pre-processing power traces to defeat random clocking countermeasures. In: International Symposium on Circuits and Systems (ISCAS), pp. 85–88. IEEE (2015)

[HNI+06] Homma, N., Nagashima, S., Imai, Y., Aoki, T., Satoh, A.: High-resolution side-channel attack using phase-based waveform matching. In: Goubin, L., Matsui, M. (eds.) CHES 2006. LNCS, vol. 4249, pp. 187–200. Springer, Heidelberg (2006). https://doi.org/10.1007/11894063_15

[KA98] Kuhn, M.G., Anderson, R.J.: Soft tempest: hidden data transmission using electromagnetic emanations. In: Aucsmith, D. (ed.) IH 1998. LNCS, vol. 1525, pp. 124–142. Springer, Heidelberg (1998). https://doi.org/10.1007/3-540-49380-8_10

[KJJ99] Kocher, P., Jaffe, J., Jun, B.: Differential power analysis. In: Wiener, M. (ed.) CRYPTO 1999. LNCS, vol. 1666, pp. 388–397. Springer, Heidelberg (1999). https://doi.org/10.1007/3-540-48405-1_25

[Koc96] Kocher, P.C.: Timing attacks on implementations of Diffie-Hellman, RSA, DSS, and other systems. In: Koblitz, N. (ed.) CRYPTO 1996. LNCS, vol. 1109, pp. 104–113. Springer, Heidelberg (1996). https://doi.org/10.1007/3-540-68697-5_9

[MOP08] Mangard, S., Oswald, E., Popp, T.: Power Analysis Attacks: Revealing the Secrets of Smart Cards, vol. 31, pp. 202–211. Springer, Boston (2007). https://doi.org/10.1007/978-0-387-38162-6

[MvWB11] Muijrers, R.A., van Woudenberg, J.G.J., Batina, L.: RAM: rapid alignment method. In: Prouff, E. (ed.) CARDIS 2011. LNCS, vol. 7079, pp. 266–282. Springer, Heidelberg (2011). https://doi.org/10.1007/978-3-642-27257-8_17

[OC15] O'Flynn, C., Chen, Z.: Synchronous sampling and clock recovery of internal oscillators for side channel analysis and fault injection. J. Crypt. Eng. 5(1), 53–69 (2015)

[OP11] Oswald, D., Paar, C.: Breaking Mifare DESFire MF3ICD40: power analysis and templates in the real world. In: Preneel, B., Takagi, T. (eds.) CHES 2011. LNCS, vol. 6917, pp. 207–222. Springer, Heidelberg (2011). https://doi.org/10.1007/978-3-642-23951-9_14

[OP12] Oswald, D., Paar, C.: Improving side-channel analysis with optimal linear transforms. In: Mangard, S. (ed.) CARDIS 2012. LNCS, vol. 7771, pp. 219–233. Springer, Heidelberg (2013). https://doi.org/10.1007/978-3-642-37288-9_15

[PV17] Papagiannopoulos, K., Veshchikov, N.: Mind the gap: towards secure 1st-order masking in software. IACR Cryptology ePrint Archive, p. 345 (2017)

[RCM10] Digi International Inc.: RabbitCore RCM4000 user manual (2010). http://ftp1.digi.com/support/documentation/019-0157_J.pdf

[SDB+10] Schimmel, O., Duplys, P., Boehl, E., Hayek, J., Bosch, R., Rosenstiel, W.: Correlation power analysis in frequency domain. In: COSADE First International Workshop on Constructive Side Channel Analysis and Secure Design (2010)

[SMY09] Standaert, F.-X., Malkin, T.G., Yung, M.: A unified framework for the analysis of side-channel key recovery attacks. In: Joux, A. (ed.) EURO-CRYPT 2009. LNCS, vol. 5479, pp. 443–461. Springer, Heidelberg (2009). https://doi.org/10.1007/978-3-642-01001-9_26

[ST04] Shamir, A., Tromer, E.: Acoustic cryptanalysis (2004). http://www.wisdom.weizmann.ac.il/~tromer

[TH12] Tian, Q., Huss, S.A.: On the attack of misaligned traces by power analysis methods. In: 2012 Seventh International Conference on Computer Engineering and Systems (ICCES), pp. 28–34. IEEE (2012)

[vWWB11] van Woudenberg, J.G.J., Witteman, M.F., Bakker, B.: Improving differential power analysis by elastic alignment. In: Kiayias, A. (ed.) CT-RSA 2011. LNCS, vol. 6558, pp. 104–119. Springer, Heidelberg (2011). https://doi.org/10.1007/978-3-642-19074-2_8

Assessing the Feasibility of Single Trace Power Analysis of Frodo

Joppe W. Bos[1(✉)], Simon Friedberger[1,2], Marco Martinoli[3],
Elisabeth Oswald[3], and Martijn Stam[3]

[1] NXP Semiconductors, Eindhoven, Netherlands
joppe.bos@nxp.com, simon.friedberger@esat.kuleuven.com
[2] KU Leuven - iMinds - COSIC, Leuven, Belgium
[3] University of Bristol, Bristol, UK
{marco.martinoli,elisabeth.oswald,martijn.stam}@bristol.ac.uk

Abstract. Lattice-based schemes are among the most promising post-quantum schemes, yet the effect of both parameter and implementation choices on their side-channel resilience is still poorly understood. Aysu et al. (HOST'18) recently investigated single-trace attacks against the core lattice operation, namely multiplication between a public matrix and a "small" secret vector, in the context of a *hardware* implementation. We complement this work by considering single-trace attacks against software implementations of "ring-less" LWE-based constructions.

Specifically, we target *Frodo*, one of the submissions to the standardisation process of NIST, when implemented on an (emulated) ARM Cortex M0 processor. We confirm Aysu et al.'s observation that a standard divide-and-conquer attack is insufficient and instead we resort to a sequential, extend-and-prune approach. In contrast to Aysu et al. we find that, in our setting where the power model is far from being as clear as theirs, both profiling and less aggressive pruning are needed to obtain reasonable key recovery rates for SNRs of practical relevance. Our work drives home the message that parameter selection for LWE schemes is a double-edged sword: the schemes that are deemed most secure against (black-box) lattice attacks can provide the *least* security when considering side-channels. Finally, we suggest some easy countermeasures that thwart standard extend-and-prune attacks.

Keywords: Side-channel analysis · LWE · Frodo · Template attacks
Lattices

1 Introduction

Recent advances in quantum computing [7,8] have accelerated the research into schemes which can be used as replacements for currently popular public-key encryption, key-exchange and signature schemes, all of which are vulnerable to quantum attacks. The attention of the cryptographic research community in this direction is boosted by the current NIST standardisation process [16].

© Springer Nature Switzerland AG 2019
C. Cid and M. J. Jacobson, Jr. (Eds.): SAC 2018, LNCS 11349, pp. 216–234, 2019.
https://doi.org/10.1007/978-3-030-10970-7_10

Investigating the security of new public-key cryptography proposals in different security settings is an important part of this standardisation process. The current trend, in the era of Internet of Things (IoT), is to connect more and more devices and enable them to transmit sensitive data to other devices or the cloud. These IoT devices can often be physically accessed by potential adversaries, allowing for side-channel attacks. However, the challenges when implementing these novel post-quantum schemes are not as well analysed as for the RSA or ECC-based systems they aim to replace.

Over a third of the submissions to NIST's standardisation process are lattice-based constructions [16]. They come in a number of flavours, of which the two dominant classes are those based on learning with errors (LWE [17]) and its variants (Ring-LWE [11] and Module-LWE [9]). For both scenarios, the key to be recovered is typically a vector of relatively small integers, but the computations involving this vector differ considerably: Ring-LWE and Module-LWE often rely on the Number-Theoretic Transform (NTT) to compute polynomial multiplication, whereas standard LWE depends on textbook matrix–vector multiplication.

One of the standard LWE-based proposals is Frodo. Originally conceived as a key agreement protocol it was expanded to a Key Encapsulation Mechanism (KEM), for the later NIST submission [5,15]. Frodo relies on the equation $\mathbf{B} = \mathbf{AS} + \mathbf{E}$, where $\mathbf{A}, \mathbf{B}, \mathbf{S}$, and $\mathbf{E}$ are all various matrices over $\mathbb{Z}_q$ for q a power of two. The dimensions of these matrices, the modulus q, as well as the distributions from which the error $\mathbf{E}$ and the secret $\mathbf{S}$ are drawn, are all parameters to the scheme. Overall, the Frodo designers proposed six concrete parameter sets, yet the natural resistance of the corresponding matrix multiplication against side-channel analysis is still understood only partially.

Recently, Aysu et al. [2] demonstrated the efficacy of horizontal Correlation Power Analysis (CPA) in a single trace setting against Frodo's matrix multiplication $\mathbf{AS}$ when implemented in hardware. Their attack assumes knowledge of the architecture in order to target specific intermediate registers, as well as that the Hamming distance is a good approximation of *their specific* device's leakage. Even so, for a distinguisher to succeed, knowledge of the algorithm's state so far is required. Aysu et al. cope with this challenge by describing what is known as an extend-and-prune strategy. Seemingly unaware that their method is essentially part of the established methodology of template attacks [6], they do not further explore challenges that may arise in contexts where the device's leakage is too far from Hamming weight/distance for an unprofiled method to work.

Our Contribution. We fill this gap by investigating single-trace attacks against software implementations of "ring-less" LWE-based constructions, as used by Frodo. When Frodo is used as key agreement protocol, the secret $\mathbf{S}$ is ephemeral and the calculation of $\mathbf{AS} + \mathbf{E}$ that we target is only performed once (or twice), resulting in only a single trace. This limited usage implies only a subset of side-channel techniques apply. When Frodo is used as a KEM, the overall private key (of the KEM) *is* used repeatedly for decapsulation and the usual techniques relying on a variable number of traces do apply. However, even then our work provides useful guidance on security, and indeed, we expect our results can be translated to any "small secret" LWE scheme, that is any scheme where the individual entries of **S** are "small" in the space over which the scheme is defined.

Even if only a single trace corresponding to **AS** + **E** is available, each *element* in **S** is still used multiple times in the calculation of **AS**, enabling so called horizontal differential power analysis. Here the single trace belonging to **AS** is cut up into smaller subtraces corresponding to the constituent $\mathbb{Z}_q$ operations. Hence, the number of subtraces available for each targeted $\mathbb{Z}_q$ element (of **S**) is bounded by the dimension of the matrix **A**. For square **A** as given by the suggested parameters, this immediately leads to a situation where high dimensions for **A**, thus **S**, on the one hand imply more elements of **S** need to be recovered (harder), yet on the other hand more subtraces per element are available (easier). To complicate matters, the elements of **S** are chosen to be relatively small in $\mathbb{Z}_q$, with the exact support differing per parameter set. All in all, the effect of parameter selection on the natural side-channel resistance is multi-faceted and potentially counterintuitive; we provide guidance in this respect in Sect. 5.

For our investigation, we opted for the ARM Cortex M0 as platform for Frodo's implementation. The Cortex-M family has high practical relevance in the IoT panorama, where our choice for the M0 is primarily instigated by the availability of the ELMO tool [13], which we use to simulate Frodo's power consumption (see Sect. 2 for details). We believe our results are representative for other 32-bit ARM architectures as well.

Our first research question is how well the unprofiled correlation power analysis, as successfully deployed by Aysu et al. [2] against a hardware implementation of Frodo, works in our software-oriented context. The main operations relevant for Frodo are $\mathbb{Z}_q$ addition and multiplication, which are both known to be poor targets for side-channel attacks [4,10]. This is usually compensated for by employing a larger number of traces and by using a power model sufficiently close to the device's leakage profile. The former is intrinsically not possible in the setting we consider, while the latter necessarily requires a profiling phase in cases where the leakage profile of a device is not well-known (as is the case for registers leaking Hamming distance in Aysu et al.'s case).

Overall, we target up to three points of interest, corresponding to loading of a secret value, the actual $\mathbb{Z}_q$ multiplication, and updating an accumulator with the resulting product. For a classical divide-and-conquer attack, where all positions of the secret matrix **S** are attacked independently, the templates can easily be profiled at the start, but as we find in Sect. 3, the resulting algorithmic variance is too high to allow meaningful key recovery.

Therefore we switch to an extend-and-prune technique (Sect. 4), allowing inclusion of predictions on intermediate variables (such us partial sums stored into an accumulator). This approach drastically reduces the algorithmic variance and hence increases the effective signal strength. We show how different pruning strategies allow for a trade-off between performance and success, concluding that for reasonable levels of success, this type of pruning needs to be less aggressive than that employed by Aysu et al. [2]. We also find that of the two Frodo parameter sets given in the NIST proposal, the one designed for higher security is in fact the most vulnerable against our side-channel cryptanalysis.

We finish with a discussion on possible countermeasures (Sect. 5). In particular, we propose a simple alternative way of evaluating the matrix multiplication that frustrates the extend-and-prune attack, reintroducing the algorithmic variance effectively for free. This deterministic method significantly improves the security of what is otherwise still an unprotected implementation.

2   Preliminaries

Notation. Vectors are denoted by lower case boldface letters and the i-th component of a vector $\mathbf{v}$ is $\mathbf{v}[i]$, where indexing starts at 1. Matrices are denoted by upper case boldface letters and their elements are also indexed using square brackets notation in row major order. The n-dimensional identity matrix is denoted by $\mathbf{I}_n$.

Drawing a random sample x from a distribution $\mathcal{D}$ over a set S is denoted by $x \leftarrow_\$ \mathcal{D}(S)$ or just by $x \leftarrow_\$ \mathcal{D}$ if the set is clear from the context. We denote drawing a random vector of dimension n made of independent and identically distributed random samples by $\mathbf{x} \leftarrow_\$ \mathcal{D}^n(S)$. The support of $\mathcal{D}$, i.e. the values to which $\mathcal{D}$ assigns non-zero probability, is denoted by $\mathrm{Supp}(\mathcal{D})$.

2.1   Frodo: A LWE-Based Key Agreement Protocol/KEM

Originally Frodo was conceived as a key agreement protocol [5]; in the later NIST proposal [15], it was recast as a KEM. It derives its security from a variant of Regev's LWE concept [17], namely the *decisional Matrix-LWE problem with short secrets* (Definition 1), which stipulates secrets and errors as matrices of fixed dimensions, instead of vectors of arbitrary dimension.

Definition 1 ([5, Sect. 5.1]). *Let $n, m, q, \overline{n}$ be positive integers and χ be a distribution over $\mathbb{Z}_q$. Let $\mathbf{A} \leftarrow_\$ \mathcal{U}^{m \times n}(\mathbb{Z}_q)$ where $\mathcal{U}$ is the uniform distribution, $\mathbf{E} \leftarrow_\$ \chi^{m \times \overline{n}}(\mathbb{Z}_q)$ and $\mathbf{S} \leftarrow_\$ \chi^{n \times \overline{n}}(\mathbb{Z}_q)$. Defining $\mathbf{B}$ as $\mathbf{B} = \mathbf{AS} + \mathbf{E}$, the decisional Matrix-LWE problem with short secrets asks to distinguish $(\mathbf{A}, \mathbf{B})$ from $(\mathbf{A}, \mathbf{U})$, where $\mathbf{U} \leftarrow_\$ \mathcal{U}^{m \times \overline{n}}(\mathbb{Z}_q)$.*

Frodo can be instantiated with six different parameter sets, four proposed in the original key agreement protocol [5] and two as part of the NIST submission [15]. Table 1 summarises them all. Matrix dimensions are specified, as well as k, the cardinality of the support of χ. The latter distribution is a discrete Gaussian centred at zero, with range $[-\eta, +\eta]$ for $\eta = (k-1)/2$. This effectively specifies all possibilities for each secret entry.

The core operation of Frodo is the calculation of $\mathbf{B} \leftarrow \mathbf{AS} + \mathbf{E}$. Without loss of generality, we will henceforth concentrate on only a single column of the secret matrix $\mathbf{S}$, which will be denoted by $\mathbf{s}$. Thus we target the operation $\mathbf{b} \leftarrow \mathbf{As} + \mathbf{e}$, where we try to recover the small value $\mathbf{s}$ for known $\mathbf{A}$ and $\mathbf{b}$ based on the leakage from primarily the matrix–vector multiplication $\mathbf{As}$. We note that, given $\mathbf{A}$ and

Table 1. Parameter sets for Frodo where $k = |\mathrm{Supp}(\chi)|$; for all of sets, $m = n$ and $\bar{n} = 8$.

Name	n	q	k
CCS1	352	2^{11}	7
CCS2	592	2^{12}	9
CCS3	752	2^{15}	11
CCS4	864	2^{15}	13
NIST1	640	2^{15}	23
NIST2	976	2^{16}	21

$\mathbf{b}$, it is possible to check whether a guess $\mathbf{s}$ is correct by checking whether $\mathbf{b} - \mathbf{As}$ is in the support of χ. This suffices with very high probability, because a wrong $\mathbf{s}$ would make the result pseudorandom.

Our analysis of a single column recovery $\mathbf{s}$ could easily be extrapolated to the recovery of the full secret matrix $\mathbf{S}$ by taking into account the number of columns $\bar{n}$ and the fact that columns can be attacked independently. Furthermore, for the original Frodo key agreement, a subsequent step in the protocol to arrive at a joint secret, the so-called reconciliation, is component-wise. Consequently, correctly recovering one column of $\mathbf{S}$ immediately translates to recovering part of the eventual session key (between 8 and 32 bits, depending on the selected parameter set). A similar argument applies to the public key encryption scheme on which the KEM variant [15] is based. However, the introduction of hash functions in the final KEM protocol structurally prevents such a threat and full recovery of $\mathbf{S}$ is required.

While we focus on Frodo's operation $\mathbf{As}$, our results apply equally to the transpose operation $\mathbf{s}^{\mathsf{T}}\mathbf{A}$, or indeed to any scenario where a small secret vector is multiplied by a public matrix and there is a method to test (as in the case for LWE) with high probability whether a candidate $\mathbf{s}$ is correct. While we concentrate on the parameter sets relevant to Frodo (which has relatively leak-free modular reductions due to its power-of-two modulus q), the techniques apply to other parameter sets used in different LWE-based schemes as well.

Matrix–Vector Multiplication. Algorithm 1 contains the high level description of textbook matrix–vector multiplication. This is usually deployed as asymptotically faster methods have overhead which makes them unsuitable for the matrix dimensions found in practical lattice-based schemes.

For every iteration of the outer loop, the accumulator sum is initialised to zero and updated n times with as many $\mathbb{Z}_q$ multiplications. This means that for every secret entry $\mathbf{s}[i]$ an adversary can exploit n portions of the power trace, namely each time it is used in Line 5, motivating the use of a horizontal attack.

Note that Line 5 does not include an explicit modular reduction. As the modulus q is a power of two, the accumulator sum is allowed to exceed q and will only be reduced modulo q when it is added to the error in Line 6. The

Algorithm 1. Matrix–vector multiplication as implemented in Frodo.

Input: $\mathbf{A} \in \mathbb{Z}_q^{n \times n}$; $\mathbf{s}, \mathbf{e} \in \mathbb{Z}_q^n$
Output: $\mathbf{b} \leftarrow \mathbf{As} + \mathbf{e}$

1: $\mathbf{b} \leftarrow \mathbf{e}$
2: **for** $r = 1, \ldots, n$ **do**
3:     $sum \leftarrow 0$
4:     **for** $i = 1, \ldots, n$ **do**
5:         $sum \leftarrow sum + \mathbf{A}[r, i] \cdot \mathbf{s}[i]$
6:     $\mathbf{b}[r] \leftarrow (\mathbf{b}[r] + sum) \bmod q$
7: **return** $\mathbf{b}$

modular reduction itself boils down to truncation and similarly, in the earlier Line 5 *sum* will of course be reduced modulo the word size, in our case 32 bits.

2.2 Template Attacks

Template attacks were first introduced by Chari et al. [6]. The idea is that an adversary creates statistical descriptions, called templates, of the device's leakage for specific intermediate values by profiling the target device (or an equivalent one). Subsequently, one can use Bayesian methods (e.g. maximum likelihood estimation) to determine which template best matches the observed leakage, eventually leading to key recovery.

We consider two classes of template attack. For *divide-and-conquer* the secret is split up into many sub-secrets that are recovered *independently* of each other, and subsequently these sub-secrets are recombined. In our case, it would entail recovering the components of the secret vector $\mathbf{s}$ independently of each other. Divide-and-conquer is popular for instance in the context of AES-128 and has the advantage that profiling can easily be done during a preprocessing stage.

Chari et al. already observed that for their use case (RC4), divide-and-conquer was insufficient. Instead they suggested an extend-and-prune approach, where the overall secret is still split up into many sub-secrets, but this time they are recovered *sequentially*. As a result, when recovering the ith sub-secret, it is possible to use knowledge of the preceding $i - 1$ sub-secrets to select more potent templates. The total number of possible templates increases drastically and, while it might still be just about feasible to generate them all as part of preprocessing, it is more common to generate the actually required templates on-the-fly [3].

We analyse both strategies. In Sect. 3 we attack the individual sub-secrets independently using divide-and-conquer. This implies that the templates necessarily cannot rely on the value of the accumulator *sum* as that depends on all the previous sub-secrets. Subsequently, in Sect. 4, we consider the extend-and-prune approach, generating templates on-the-fly, which allows us to profile based on the (likely) correct value of the accumulator.

2.3 Experimental Setup

As target architecture for our experiments we chose the entry level ARM architecture, the Cortex series, because it represents a realistic target and is extremely widely distributed. The Cortex series has several family members, and for the M0 a high quality leakage modelling tool exists. Understanding different attack strategies on different noise levels requires many experiments (we used well over 10^6 full column traces per parameter set), which becomes problematic on real devices. Thus we opted to use simulated *yet realistic* traces which are quicker to generate, modify, and analyse. This allowed us to speed up our analysis, and therefore enable the exploration of a wider noise spectrum.

ELMO. ELMO [12] is a tool to simulate instantaneous power consumption for the ARM Cortex M0 processor. This simulator, created by adapting the open-source instruction set emulator Thumbulator [19], has been designed to enable side-channel analysis without requiring a hardware measurement setup. ELMO takes ARM thumb assembly as input, and its output describes the power consumption, either at instruction or cycle accuracy. The resulting traces are noise free, that is, they are based deterministically on the instructions and their inputs.

ELMO's quality has been established by comparing leakage detection results between simulated and real traces from a STM32F0 Discovery Board [13]. As raw ELMO traces are noise free, the tool is ideal to study the behaviour of template attacks across different noise levels efficiently: both template building and creating noisy traces are straightforward.

We stress that ELMO does capture differential data-dependent effects, such as those caused by neighbouring instructions, as well as higher order leakage terms. Consequently, even though ELMO traces are noise free, the trace for the same machine line of code (same operation with the same operand) will differ depending on the context, leading to *algorithmic variance* (i.e. variation in the trace that deterministically depends on those parts of the input currently not being targeted).

Reference Implementation. We implement the innermost loop of Algorithm 1 in ARM assembly, which for convenience we wrapped in C code for initialization and loop control. This gives us a fine control over the code ELMO simulates the power consumption of and prevents the compiler from inserting redundant instructions which might affect leakage. We refer to Appendix A for the full code, which is then just repeated n times.

Figure 1a plots a partial power trace of our ARM implementation, as simulated by ELMO. After initialisation, a pattern neatly repeats, corresponding to the equivalent of Line 5 in Algorithm 1. After excluding unimportant points (e.g. loop structure), the most relevant instructions responsible for the pattern are given in Fig. 1b.

The index i stored in r4 is used to load values from a row of **A** and s, whose addresses are in r1 and r0 respectively, into r6 and r5. These are then used to

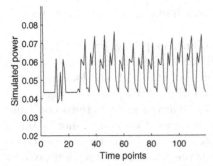

Instruction	Operation
ldrh r5,[r0,r4]	load $s[i]$
ldrh r6,[r1,r4]	load $\mathbf{A}[r,i]$
muls r5,r6	$s[i] \cdot \mathbf{A}[r,i]$
adds r3,r3,r5	$sum + s[i] \cdot \mathbf{A}[r,i]$

(b) Breakdown of instructions forming the repeating pattern.

(a) Power trace as simulated by ELMO of our ARM implementation

Fig. 1. Visual representation and detailed structure of target power traces.

perform one element multiplication, whose result overwrites r5, and finally the accumulator is updated in r3 and eventually returned.

We wrap around negative numbers modulo q. This is in contrast to Frodo's original convention of taking 16-bit cut-off independently on the parameter set. We expect the higher Hamming weights resulting from modulo-2^{16} wraparound to amplify leakage, thus making our decision, motivated by simplicity of analysis, very conservative. Finally, intermediate multiplications and partial sums are truncated only when exceeding 32 bits, being the M0 a 32-bit architecture.

Realistic Noise Estimate. As mentioned before, ELMO traces are noise free. However, when attacking an actual ARM Cortex M0 environmental noise will be introduced. For our experiments, we will artificially add this noise, which we assume independently and identically distributed for all points of interest, according to a normal distribution with mean 0 and variance σ^2.

For the profiling that led to the development of ELMO [13], the observed value[1] of σ was around $4 \cdot 10^{-3}$. We will use this realistic level of environmental noise as benchmark throughout. Furthermore, we will consider a representative range of σ roughly centred around this benchmark. We chose σ in the interval $[10^{-4}, 10^{-2})$ with steps of $5 \cdot 10^{-4}$. Compared to the variance of the signal, our choice implies σ ranges from having essentially no impact to being on the same order of magnitude.

3 Divide-and-Conquer Template Attack

As every entry of $\mathbf{s}$ is an independently and identically distributed sample from χ, we can potentially target each position separately. Thus we first consider a divide-and-conquer template attack. A distinct advantage of this approach is

---

[1] Personal communication with C. Whitnall.

that the total number of templates is fairly small and hence we can preprocess the profiling.

When considering the breakdown of the inner loop (Fig. 1b), we ignore the loading of the public operand (it essentially leaks nothing exploitable), which leaves three potential points of interest. On the one hand, the loading of the secret operand and the multiplication contain direct leakage on the secret, and all relevant inputs appear known. For the accumulator update on the other hand, the leakage is less direct and the value of the accumulator so far cannot be taken into account: it depends on the computation so far, violating the independence requirement for divide-and-conquer. Thus, for the attack in this section we limit ourselves to *two* points of interest, namely the loading of the secret and the $\mathbb{Z}_q$ multiplication.

Of course, one could still generate templates for all *three* points of interest by treating the accumulator as a random variable. However, as the accumulator value is a direct input to the accumulator update and its register is used for the output as well, the resulting algorithmic variance would be overwhelming. Indeed, as we will see below, already for the loading of the secret there is considerable algorithmic variance related to the previous value held by the relevant register. These limitations are intrinsic to a divide-and-conquer approach; in Sect. 4 we show how an extend-and-prune approach bypasses these problems.

Profiling. One feature of LWE instances is that the overall space $\mathbb{Z}_q$ from which elements are drawn is fairly small as q need not be large, certainly compared to classical primitives like ECC or RSA. For Frodo, and in general for "small secret" schemes, the effective space that requires profiling is further reduced as the support of χ (from which secrets are drawn) is even smaller.

For the loading of the secret, we need k templates, whereas for the multiplication $k \cdot q$ templates suffice. We generate these templates as part of the preprocessing, where we are primarily interested in the signal, that is the deterministic part.

Although ELMO is completely deterministic, the power trace it emulates for a given operation still depends on preceding operations, thus introducing algorithmic variance. To profile the loading of secret s, we use the weighted average of k traces, corresponding to the previous value of the register involved, as the deterministic part. For reference, depending on the parameter set, the algorithmic variance is between $1.4 \cdot 10^{-3}$ and $2.9 \cdot 10^{-3}$. For the multiplication, we assumed no algorithmic variance in our profiling and simply performed the operation once for each template.

Estimating Success Rates. For each entry $s[i]$, the distinguisher outputs a distinguishing score vector that can be linked back to a perceived posterior distribution. Selecting the element corresponding to the highest score corresponds to the maximum a posteriori (MAP) estimate and the probability that the correct value is returned this way is referred to as the first-order success rate.

Ultimately, we are more interested in the first order success rate of the full vector **s**. As we assume independence for a divide-and-conquer we can easily extrapolate the success rates for **s** based on those for individual positions as a full vector is recovered correctly iff all its constituent positions are. The advantage of using extrapolated success rates for **s**, rather than using direct sample means, is that it provides us useful estimates even for very small success rates (that would otherwise require an exorbitant number of samples). Thus, analysing the recovery rates of single positions is extremely informative. Additionally, it gives insights on why the extend-and-prune attack in Sect. 4 greatly outperforms divide-and-conquer.

Other metrics, beyond first-order recovery rate, are of course possible to compare distinguishers [18]. However, we regard those alternatives, such as oth-order recovery or more general key ranking, only of interest when first order success rate is low. While for divide-and-conquer this might be the case, for extend-and-prune the first order recovery is sufficiently high to warrant concentrating on that metric only.

Estimating Position Success Rate. Let $\Pr[S]$ be the first order position recovery rate where S is the event that the distinguisher indeed associates the highest score to the actual secret value. We experimentally evaluate $\Pr[S]$ based on the formula

$$\Pr[S] = \sum_{s \in \mathrm{Supp}(\chi)} \Pr[S \mid s] \Pr[s]$$

where $\Pr[s]$ corresponds to the prior distribution χ and the values for $\Pr[S \mid s]$ are estimated by appropriate sample means. To ensure our traces are representative, we range over **A** and **s** (and **e**) for the relevant experiments and generate traces for the *full* computation **b** ← **As**+**e**. This allows us to zoom in on individual positions, highlighting where algorithmic variance occurs. While one could also use direct, position-specific sample means for $\Pr[S]$, our approach links more closely to the confusion matrix and has the advantage that it depends less on the sampling distribution of **s** when running experiments.

Extrapolating Overall Success Rate. If we assume independence of positions, it is easy to express the overall success rate for recovering **s**. If we, temporarily, make the simplifying assumption that $\Pr[S]$ is the same for all n positions, then the first order recovery rate for **s** is $\Pr[S]^n$ (recovery of **s** will be successful if and only if recovery of each of its elements is). Even for extremely high $\Pr[S]$, this value quickly drops, e.g. $0.99^n \approx 5.5 \cdot 10^{-5}$ for NIST2.

Experimental Results. We target each position of **s** individually, but only report on the first and second one. Figure 2 displays the success rate for all parameter sets. Each point in each curve is based on $8 \cdot 10^5$ experiments. The left panel (Fig. 2a) plots the success rate for the first position, whereas the right panel (Fig. 2b) plots it for the second position. The second position is representative for

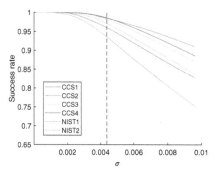
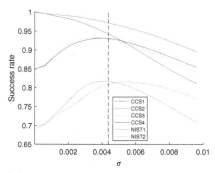

(a) Recovery rate for first position only.   (b) Recovery rate for second position only.

Fig. 2. Comparison of recovery rates between first and second positions. The dashed black line indicates our choice of realistic noise level.

all subsequent positions, but the first position stands out as being significantly easier to tackle due to the lack of algorithmic variance.

The Impact of Algorithmic Variance. The striking difference between Figs. 2a and b, especially in the low environmental noise regime, is due to algorithmic variance. As we mentioned before, algorithmic variance particularly affects the loading of the secret, i.e. the first instruction in Fig. 2b, due to the previous register value contributing to the leakage. This problem only appears from the second position onward; for the first position, no algorithmic variance is present as the initial state is fixed (and profiled for).

With the exception for the two small CCS parameter sets, even with virtually no environmental noise, the success rate for the second position is far from 1. Moreover, when environmental noise is added, the success rate initially goes up. This phenomenon is known as stochastic resonance [14] and has been observed for side-channels before [20]. Even for CCS1 and CCS2, that have the lowest algorithmic variance level, the success rate for the second position is slightly lower than for the first position.

For completeness, our assumption that the noise covariance matrix Σ for our two points of interest is a diagonal matrix $\sigma \cdot \mathbf{I}_2$, is suboptimal in the presence of algorithmic variance. Using a diagonal matrix Σ that incorporates the algorithmic variance would improve the distinguisher while reducing the stochastic resonance. As the extend-and-prune approach from the next section is far more convincing, we refrain from a full analysis.

Full Vector Recovery. The success rates for full vector are more relevant to compare either amongst parameter sets or with other attacks, be it lattice or other side-channel attacks. As a simplification, we assume that the recovery rate for the second position (Fig. 2b) is representative for all positions: we checked this assumption holds for all bar the first position, whose contribution is limited anyway given concrete values of n (the total number of positions).

To ease comparison, for each parameter set we determined the σ for which the divide-and-conquer attack approximately achieves a success rate for recovering s of around 2^{-128} (corresponding to 128-bit security). For the smallest parameter sets, CCS1 and CCS2, all the σ in our range are susceptible (i.e. lead to success rates of at least 2^{-128}), whereas for the NIST parameter sets, none of the σ appear insecure. For the original large sets CCS3 and CCS4, any σ below $7 \cdot 10^{-3}$, which includes our realistic benchmark, leads to a loss of security below the 128-bit level.

As a caveat, a further reduction in residual bit security will be possible by explicitly incorporating algorithmic variance in the templates and by considering key ranking, or possibly even novel lattice reduction algorithms that take into account side-channel information. However, we anticipate none of these approaches will allow straightforward and almost instant key recovery for all parameter sets for realistic levels of noise (as introduced by σ).

4   Extend-and-Prune Template Attack

For the divide-and-conquer approach from the previous section, we assumed that the positions of s are independent of each other. While this assumption is valid for the generation of s, it turned out that for the leakage, it is not. However, Algorithm 1 deals with the elements of s *sequentially*, from position 1 to position n, which we will exploit by a well-known extend-and-prune approach.

In our case, the extend-and-prune algorithm operates as follows. We imagine a k-ary tree of depth n where the nodes at level i in the tree correspond to a partial guess $s[1], \ldots, s[i-1]$ for the secret; for a given node at level i, its k out-going edges are labelled by the k possible values that $s[i]$ can take. This way, each path from the root to one of the k^n possible leaves uniquely corresponds to one of the possible values that the secret vector s can take. A distinguisher can sequentially calculate a score for a vector s by traversing the tree from the root to the leaf representing s where for each edge it encounters it cumulatively updates s's score.

The challenge of an extend-and-prune algorithm is to efficiently traverse a small part of the tree while still ending up with a good overall score. The standard way of doing so is to first calculate the score for all nodes at level 2. For each level-2 node, the score will be that of the edge from the root to that node. Thus the trivial level-1 guess is *extended* to all possible level-2 guesses. The next stage is to *prune* all these guesses to a more reasonable number. For all the remaining level-2 guesses, one then extends to all possible level-3 guesses, and then again these guesses are pruned down. This process repeats until reaching the final level $(n+1)$, where the complete s is guessed.

The advantage of this approach is that, when calculating a score for $s[i]$, the distinguisher already has a guess for $s[1], \ldots, s[i-1]$, which allows it to create templates based on this guess. Our distinguisher will only use the previous secret $s[i-1]$ and the value of the accumulator so far (an inner product of $(s[1], \ldots, s[i-1])$ with the relevant part of $\mathbf{A}$) to create a template. As the total number of possible templates becomes rather unwieldy (around $k^2 \cdot q \cdot 2^{32}$), the

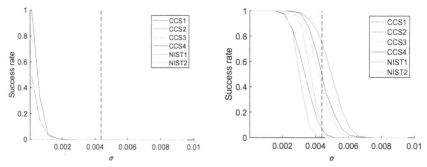

(a) Column recovery rate of divide-and-conquer template attack.

(b) Column recovery rate of extend-and-prune template attack.

Fig. 3. Comparison between column recovery of our two template attacks.

profiling is interleaved with the tree traversal and pruning is used to keep the number of templates manageable.

The success of an extend-and-prune attack depends on the pruning strategy, specifically how many candidates to keep at each step. To the best of our knowledge, there is no comprehensive study comparing different pruning strategies in different scenarios. When Chari et al. [6] introduced template attacks to the cryptanalyst's arsenal, they suggested a pruning strategy that depends on the scores themselves. We instead fix the same number of candidates to keep at each step, which is a classical approach known as *beam search*. The size of the beam, that is the number of candidates to keep after pruning, is denoted by b.

Greedy Pruning Using a Laser Beam ($b = 1$). We start by considering the greediest pruning strategy by restricting the beam size to $b = 1$, meaning that after each step we only keep a single candidate for the secret recovered so far. This "knowledge", provided it is correct, has two very immediate effects. Firstly, the algorithmic variance we observed in the loading of the secret can be reduced as we assume we typically know the previous secret held by the relevant register. Secondly, by recovering s from first to last we can predict the value of the accumulator, which brings into play a third point of interest, namely the update of the accumulator (the last point in Fig. 2b), as here too the algorithmic variance disappears.

Figure 3 presents the vector recovery rates of both last section's divide-and-conquer attack (in the left panel, Fig. 3a), and of extend-and-prune using $b = 1$ (Fig. 3b). Note that the former is extrapolated based on position recovery rates, whereas the latter has been estimated directly, based on $2 \cdot 10^3$ experiments per setting.

The difference between Figs. 3a and b is striking. For the extend-and-prune approach we almost completely removed algorithmic variance and, when virtually no environmental noise is present either ($\sigma \approx 10^{-4}$), this resulted in a vector recovery rate of essentially 1. However, when considering the realistic noise level as indicated by the dashed vertical line, not all parameter sets are as affected and especially for NIST1 there might be still some hope (for the other parameters, recovery rates exceed 5% which translates to less than 5 bits of security, so badly broken).

Increasing the Beam Size ($b > 1$). So far we only considered $b = 1$. Increasing the beam size b will result in a slower key recovery (linear slowdown in b) but should yield higher recovery rates. For $b = 1$ we mentioned two advantages of extend-and-prune, namely reduced algorithmic variance and an additional point of interest. For $b > 1$ a third advantage appears, namely the ability for the distinguisher to self-correct. This self-correcting behaviour has also been observed (for the first position) by Aysu et al. [2], who essentially used a beam size $b > 1$ for the first position and then revert to $b = 1$ for all remaining ones.

Table 2. Minimum values of b to achieve column recovery rate equal to 1, and heuristic column recovery when b is fixed to the listed values.

Name	b_{min}	b								
		2	3	4	5	6	7	8	9	10
CCS1	30709	0	0	0	0	0	0	0	0	0
CCS2	27	0.1	0.13	0.36	0.53	0.68	0.76	0.85	0.90	0.94
CCS3	12	0	0.48	0.77	0.90	0.94	0.96	0.99	0.99	0.99
CCS4	11	0.03	0.63	0.91	0.97	0.97	0.98	0.98	0.99	0.99
NIST1	63	0	0	0.01	0.03	0.13	0.24	0.33	0.41	0.50
NIST2	11	0	0.07	0.63	0.84	0.96	0.99	0.99	0.99	0.99

To assess the effect of the beam size b, we ran two types of experiments. Firstly, for each parameter set and noise level $\sigma = 0.0096$, we ran around 10^3 experiments and looked at the smallest beam b for which all experiments ended with the actual secret **s** part of the final beam (allowing an adversary to identify **s** by a subsequent enumeration of all final beam candidates). The resulting values are reported in the b_{min} column of Table 2. With the exception of CCS1, we notice that b_{min} is at most 2^6, so again only a few bits of security remain. As b_{min} will invariably grow as the number of experiments does, until eventually it is as large as the key space, for our second set of experiment, we estimated final vector recovery rate as a function of the beam size, for $b \leq 10$. The results are again reported in Table 2 and are fairly damning: even for NIST1 a recovery rate of around 50% is achieved.

5   Learning the Lesson: How to Thwart Extend-and-Prune

Choosing Your Parameters. So far, we have compared increasingly effective attack strategies, where we compared different parameter sets purely by name, so without further reference to their actual parameters. We now investigate the

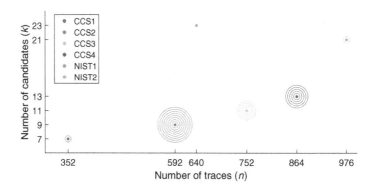

Fig. 4. Visual representation of all parameter sets. For each of them, the x axis lists n, and the y axis lists k. The number of concentric circles around each parameter set encodes how successful our attack is against it.

effect of these parameters on the efficacy and efficiency of the attack. Specifically, we consider the effects of n and k on the natural side-channel vulnerability of the resulting matrix–vector multiplication. We completely ignore the effect on the security of the LWE instance and indeed, leave the combination of side-channel information with lattice reduction methods a tantalizing open problem.

Figure 4 provides a scatter plot of (n, k) for the various parameter sets suggested [5,15]. Furthermore, we encoded the success rate of our extend-and-prune attack with beam $b = 1$ (Sect. 4) and realistic noise level (dashed line in Fig. 3b) with concentric circles around each parameter set. The number of circles is simply the ceil of said success rate times ten, and is helpful in visually quantifying the outcome we achieved in each setting.

The effect that the choice (n, k) has on the hardness of the LWE instance has been well studied [1], but from a side-channel perspective, new meaning emerges: n corresponds both to the number of (sub)traces an adversary obtains on each component of $\mathbf{s}$ *and* to the number of positions to retrieve, whereas k quantifies the keyspace size for individual positions.

Although the divide-and-conquer attack suffers badly when more positions need to be recovered, the extend-and-prune approach is far more robust in this respect. For instance, the main difference between CCS1 and CCS2 is that the latter has a twice as big n, thus providing a much easier target for our attack. Thus increasing n overwhelmingly has the effect of making life easier for an adversary as more leakage will be available. In other words, while increasing the dimension n makes the LWE instance *harder*, it makes the underlying matrix–vector multiplication *easier* to attack in our side-channel scenario. This conclusion does rely on square $\mathbf{A}$, so $n = m$. In case $\mathbf{A}$ is a non-square matrix, then m refers to the number of traces and n to the number of positions to recover. The hardness of LWE appears is mainly governed by n, where increasing n makes both the LWE instance harder and it complicates side-channel cryptanalysis. Similarly, both for LWE and for the side-channel analysis, increasing m makes attacks potentially easier, with the effect for side-channels much, much more pronounced.

The qualitative effect of increasing k is fairly straightforward: a large keyspace means that there are more options to choose from, with corresponding signals that are closer together, making distinguishing harder. This effect is illustrated by comparing the two parameter sets NIST1 and CCS2. These two sets have roughly equal n, but NIST1's k is about thrice that of CCS2: our attacks confirm that CCS2 is a lot easier to attack than NIST1.

Effect of Modifying NIST1. We conducted a final experiment to gain more insights on parameter set selection. We focused our attention on the two NIST parameter sets: they have roughly the same k (it differs by only two) but NIST1 has less than two thirds less traces than NIST2. We therefore increased n in NIST1 to match NIST2's ($n = 976$) and analysed the extend-and-prune attack in two settings: when $b = 1$ and σ is our realistic value, and when $b = 10$ and $\sigma = 0.0096$, i.e. the worst noise level we consider. In the former case the success rate increased from 0.01 to 0.11, almost equating the success rate of 0.12 observed in the NIST2 setting. In the $b = 10$ case, the success rate reported in Table 2 (0.50) skyrocketed to 0.94, again very close to NIST2's. This strongly indicates how having larger matrices, hence more traces per secret element, goes in favour of the adversary. Therefore in general being overpessimistic in the choice of n might prove fatal if side-channel attacks are a concern.

A Simple Countermeasure. Aysu et al. [2] briefly discuss potential countermeasures, including shuffling, based on the observation that randomness is usually introduced to mitigate DPA attacks. However, randomness for countermeasures can be expensive, so we present a much simpler *deterministic* countermeasure that has the effect of re-introducing algorithmic variance in the system even when attempting an extend-and-prune attack.

In order to reduce algorithmic variance, our extend-and-prune attack relies on the sequential manner in which the textbook **As** multiplication processes **s**: for each inner product of a row of **A** with **s**, the elements of the latter are accessed in the same order. However, there is no reason to do so, and we suggest to calculate the rth inner product starting at position r instead. This corresponds to changing Line 5 of Algorithm 1 to

$$sum \leftarrow sum + \mathbf{A}[r, (i + r - 1) \mod n] \cdot \mathbf{s}[(i + r - 1) \mod n].$$

The consequence is that there is no longer a clear ordering of **s**'s elements for an extend-and-prune attack to exploit and, without novel ideas, the attack's success degrades to that of the earlier divide-and-conquer one (Sect. 3).

A natural alternative to frustrate extend-and-prune is to mask the accumulator by setting it to some random value at the beginning, that is only subtracted at the very end. While this alternative would make exploiting the accumulator update hard (as for divide-and-conquer), on its own it would still allow an extend-and-prune attack to reduce algorithmic variance in the loading of the secrets. Thus our first suggestion is preferable.

Acknowledgements

 The research leading to these results has received funding from the European Union's Horizon 2020 research and innovation programme Marie Skłodowska-Curie ITN ECRYPT-NET (Project Reference 643161) and Horizon 2020 project PQCRYPTO (Project Reference 645622). Furthermore, Elisabeth Oswald was partially funded by H2020 grant SEAL (Project Reference 725042). We thank the authors of ELMO for their kind help, comments and feedback.

A   ARM Assembly Code for Inner Product

────────── Assembly ──────────

```
.syntax unified
.text
.thumb

.global Vec_Mult

.func Vec_Mult
Vec_Mult:

push {r1-r7}
 @Load and prepare the data
 @ i->0
 movs r4, #0
 @ number limit->address limit
 lsls r2, #1
loop:
 @Load first[i]
 ldrh r5,[r0,r4]
 @Load second[i]
 ldrh r6,[r1,r4]
 @Multiply
 muls r5,r6
 @Add
 adds r3,r3,r5
 @Update i as address
 adds r4,r4,#2
 @Compare with limit
 cmp r4,r2
bne loop
 @Return Value
 mov r0,r3
 pop {r1-r7}
 bx lr
.endfunc
```

References

1. Albrecht, M.R., Player, R., Scott, S.: On the concrete hardness of learning with errors. J. Math. Cryptol. **9**(3), 169–203 (2015)
2. Aysu, A., Tobah, Y., Tiwari, M., Gerstlauer, A., Orshansky, M.: Horizontal side-channel vulnerabilities of post-quantum key exchange protocols. In: IEEE International Symposium on Hardware Oriented Security and Trust, HOST 2018 (2018, to appear)
3. Batina, L., Chmielewski, L., Papachristodoulou, L., Schwabe, P., Tunstall, M.: Online template attacks. In: Meier, W., Mukhopadhyay, D. (eds.) INDOCRYPT 2014. LNCS, vol. 8885, pp. 21–36. Springer, Cham (2014). https://doi.org/10.1007/978-3-319-13039-2_2
4. Biryukov, A., Dinu, D., Großschädl, J.: Correlation power analysis of lightweight block ciphers: from theory to practice. In: Manulis, M., Sadeghi, A.-R., Schneider, S. (eds.) ACNS 2016. LNCS, vol. 9696, pp. 537–557. Springer, Cham (2016). https://doi.org/10.1007/978-3-319-39555-5_29
5. Bos, J.W., et al.: Frodo: take off the ring! Practical, quantum-secure key exchange from LWE. In: Weippl, E.R., Katzenbeisser, S., Kruegel, C., Myers, A.C., Halevi, S. (eds.) ACM CCS 2016, pp. 1006–1018. ACM Press, Oct. (2016)
6. Chari, S., Rao, J.R., Rohatgi, P.: Template attacks. In: Kaliski, B.S., Koç, K., Paar, C. (eds.) CHES 2002. LNCS, vol. 2523, pp. 13–28. Springer, Heidelberg (2003). https://doi.org/10.1007/3-540-36400-5_3
7. Devoret, M.H., Schoelkopf, R.J.: Superconducting circuits for quantum information: an outlook. Science **339**(6124), 1169–1174 (2013)
8. Kelly, J., et al.: State preservation by repetitive error detection in a superconducting quantum circuit. Nature **519**, 66–69 (2015)
9. Langlois, A., Stehlé, D.: Worst-case to average-case reductions for module lattices. Des. Codes Crypt. **75**(3), 565–599 (2015)
10. Lemke, K., Schramm, K., Paar, C.: DPA on n-bit sized Boolean and arithmetic operations and its application to IDEA, RC6, and the HMAC-construction. In: Joye, M., Quisquater, J.-J. (eds.) CHES 2004. LNCS, vol. 3156, pp. 205–219. Springer, Heidelberg (2004). https://doi.org/10.1007/978-3-540-28632-5_15
11. Lyubashevsky, V., Peikert, C., Regev, O.: On ideal lattices and learning with errors over rings. In: Gilbert, H. (ed.) EUROCRYPT 2010. LNCS, vol. 6110, pp. 1–23. Springer, Heidelberg (2010). https://doi.org/10.1007/978-3-642-13190-5_1
12. McCann, D., Oswald, E., Whitnall, C.: Implementation of ELMO. https://github.com/bristol-sca/ELMO. Accessed 27 Nov 2017
13. McCann, D., Oswald, E., Whitnall, C.: Towards practical tools for side channel aware software engineering: 'grey box' modelling for instruction leakages. In: 26th USENIX Security Symposium, USENIX Security 2017, Vancouver, BC, Canada, 16–18 August 2017, pp. 199–216 (2017)
14. McDonnell, M.D., Stocks, N.G., Pearce, C.E.M., Abbott, D.: Stochastic Resonance - From Suprathreshold Stochastic Resonance to Stochastic Signal Quantization. Cambridge University Press, Cambridge (2008)
15. Naehrig, M., et al.: FrodoKEM. Technical report, National Institute of Standards and Technology (2017). https://frodokem.org/
16. National Institute of Standards and Technology. Post-quantum cryptography standardization. https://csrc.nist.gov/Projects/Post-Quantum-Cryptography/Post-Quantum-Cryptography-Standardization

17. Regev, O.: On lattices, learning with errors, random linear codes, and cryptography. In: Gabow, H.N., Fagin, R. (eds.) 37th ACM STOC, pp. 84–93. ACM Press, May 2005
18. Standaert, F.-X., Malkin, T.G., Yung, M.: A unified framework for the analysis of side-channel key recovery attacks. In: Joux, A. (ed.) EUROCRYPT 2009. LNCS, vol. 5479, pp. 443–461. Springer, Heidelberg (2009). https://doi.org/10.1007/978-3-642-01001-9_26
19. Welch, D.: Thumbulator. https://github.com/dwelch67/thumbulator.git/
20. Whitnall, C., Oswald, E.: A comprehensive evaluation of mutual information analysis using a fair evaluation framework. In: Rogaway, P. (ed.) CRYPTO 2011. LNCS, vol. 6841, pp. 316–334. Springer, Heidelberg (2011). https://doi.org/10.1007/978-3-642-22792-9_18

Cache-Attacks on the ARM TrustZone Implementations of AES-256 and AES-256-GCM via GPU-Based Analysis

Ben Lapid and Avishai Wool[(✉)]

School of Electrical Engineering, Tel Aviv University, Tel Aviv, Israel
ben.lapid@gmail.com, yash@eng.tau.ac.il

Abstract. The ARM TrustZone is a security extension which is used in recent Samsung flagship smartphones to create a Trusted Execution Environment (TEE) called a Secure World, which runs secure processes (Trustlets). The Samsung TEE includes cryptographic key storage and functions inside the Keymaster trustlet. The secret key used by the Keymaster trustlet is derived by a hardware device and is inaccessible to the Android OS. However, the ARM32 AES implementation used by the Keymaster is vulnerable to side channel cache-attacks. The Keymaster trustlet uses AES-256 in GCM mode, which makes mounting a cache attack against this target much harder. In this paper we show that it is possible to perform a successful cache attack against this AES implementation, in AES-256/GCM mode, using widely available hardware. Using a laptop's GPU to parallelize the analysis, we are able to extract a raw AES-256 key with 7 min of measurements and under a minute of analysis time and an AES-256/GCM key with 40 min of measurements and 30 min of analysis.

1 Introduction

1.1 Motivation

The ARM TrustZone [1] is a security extension helping to move the "root of trust" further away from the attacker. TrustZone is a separate environment that can run security dedicated functionality, parallel to the OS and separated from it by a hardware barrier. Recent Samsung flagship smartphones rely on Samsung's Exynos SoC architecture cf. [23]. The ARM cores in Exynos support the TrustZone security extension to create Trusted Execution Environments (TEEs).

In order to support cryptographic modules, the Android OS includes a mechanism for handling cryptographic keys and functions called the Keystore [8]. Keystore is used for several privacy related features such as full disk encryption and password storage. The Keystore depends on a hardware abstraction layer (HAL) module called the Keymaster to implement the underlying key handling and cryptographic functions; and many OEMs, including Samsung, choose to implement the Keymaster as a trustlet in the TrustZone.

© Springer Nature Switzerland AG 2019
C. Cid and M. J. Jacobson, Jr. (Eds.): SAC 2018, LNCS 11349, pp. 235–256, 2019.
https://doi.org/10.1007/978-3-030-10970-7_11

1.2   Related Work

Lipp et al. [14] implemented cache attack techniques to recover secret keys from Java implementation of AES-128 on ARM processors, and exfiltrate additional execution information. In addition they were able to monitor cache activity in the TrustZone.

Zhang et al. [29] demonstrated a successful cache attack on a T-Table implementation of AES-128 that runs inside the TrustZone—however, their target was the C implementation that is part of OpenSSL while we focus on ARM32 assembly implementation found in Samsung Keymaster Trustlet for AES-256 and AES-256/GCM modes. Ryan et al. [15] demonstrated reliable cache side channel techniques that require loading a kernel module into the Normal World—which is disabled or restricted to OEM-verified modules on modern devices. To our knowledge no previous cache attacks on a standard devices' ARM TrustZone AES implementation using publicly available vulnerabilities have been published.

Recently, Green et al. [11] presented AutoLock, an undocumented feature in certain ARM CPUs which prevents eviction of cross-core cache sets. This feature severely reduces the effectiveness of cache side-channel attacks. The authors listed multiple CPUs that include AutoLock, and among them are the A53 and A57 used in the device we used (Samsung Galaxy S6).

Cache side channel attacks on AES were first demonstrated by Bernstein [3] with the target being a remote encryption server with an x86 CPU. Osvik et al. [21,26] demonstrated the *Prime+Probe* technique to attack a T-Table implementation of AES which resides in the Linux kernel on an x86 CPU. Xinjie et al. [28] and Neve et al. [16] presented techniques which improve the effectiveness of cache side channel attacks. Spreitzer et al. [25] demonstrated a specialization of these attacks on misaligned T-Table implementations. Neve et al. [17] discussed the effectiveness of these attacks on AES-256 and demonstrated a successful specialized attack for AES-256.

1.3   Contributions

Our starting point is the observation of [13] that the ARM32 assembly-language AES implementation used by the Keymaster Trustlet uses a T-Table and is vulnerable to cache side-channel attacks. Furthermore, the Keymaster's T-Table is misaligned, which helps the attacker. Unlike prior works, which attacked evaluation boards or AES-128, we successfully demonstrate cache attacks on a real device, against the AES-256 and AES-256/GCM implementation used by the Keymaster trustlet. Beyond the larger keys in AES-256, GCM mode introduces additional challenges, since the cryptanalyst has no control over 4 of the 16 bytes of plaintext in an AES block.

A key aspect of our attack is that we extract the secret key using a *divide and conquer* strategy. In the AES-256/GCM case, rather than analyze all 256 key bits simultaneously, we identify them in 4 phases: we identify 84 bits in phase 1; based on them we identify the next 124 bits in phase 2, and so forth until all 256 bits are discovered.

In addition, we present our approach to implementing the analysis phase of our attacks on a GPU. Such an approach requires careful planning and, when implemented correctly, leads to a significant improvement in analysis speed.

Using a laptop's GPU to parallelize the analysis, we are able to extract a raw AES-256 key with 7 min of measurements and under a minute of analysis time and an AES-256/GCM key with 40 min of measurements and 30 min of analysis.

Organization: Section 2 describes the Keymaster trustlet and its cryptographic functions. Section 3 demonstrates cache side-channel attacks against the AES implementation used by the Keymaster trustlet in isolation. Section 4 describes the use of GPU to mount the attacks and we conclude with Sect. 5. We provide GPU kernel examples in Appendix A.

2 Preliminaries

2.1 ARM TrustZone Overview

ARM TrustZone security extensions [2] enable a processor to run in two states, called Normal World and Secure World. This architecture also extends the concept of "privilege rings" and adds another dimension to it. In the ARMv8 ISA, these rings are called "Exception Levels" (ELs). The most privileged mode is the "Secure Monitor" which runs in EL3 and sits "above" the Secure and Normal Worlds. In the Secure World, the Secure OS kernel runs in EL1 and the Secure userspace runs in EL0. On Samsung devices, the Normal World OS is Android: the Linux kernel runs in EL1 and the user-space programs run in EL0.

The separation of Secure and Normal World allows that certain RAM ranges and bus peripherals may be indicated as "secure" and only be accessed by the Secure World. This means that compromised Normal World code (in userspace or kernel) will not be able to access these memory ranges or devices.

It's important to note that the world separation is completely "virtual". The same cores are used to run both Secure and Normal Worlds and they use the same RAM. Therefore, they use the same cache used by the core to improve memory access times; in [13] we describe how this design decision may be leveraged to mount cache side channel attacks.

In the Samsung ecosystem there are two major players in field of TrustZone implementations. One is Qualcomm, with the QSEE operating system [22] which is compatible with the Snapdragon SoC architecture used on many Samsung devices. The other is Trustonic, with the Kinibi operating system [27] which is used by Samsung in their popular Exynos SoC architecture as a part of the KNOX security system [24]. In this paper we focus on the Trustonic TrustZone.

These Trusted Execution Environments (TEEs) are used for various activities within the smart device: Secure boot, Keymaster implementation (see Sect. 2.2), secure UI, kernel protections, secure payments, digital rights management (DRM) and more.

2.2   Keystore and Keymaster Hardware Abstraction Layer (HAL)

The Android Keystore system [8], which was introduced in Android 4.3, allows applications to create, store and use cryptographic keys while attempting to make the keys themselves hard to extract from the device. The documentation advertises the following security features:

- Extraction Prevention: The keys themselves are never present in the application's memory space. The applications only know of *key-blobs* which cannot be used directly. The *key-blobs* are usually the keys packed with extra metadata and encrypted with a secret key by the Keymaster HAL (Hardware Abstraction Layer).
- Key Use Authorizations: The Keystore system allows the application to place restrictions on the generated keys to mitigate the possibility of unauthorized use.

The Keystore system is implemented in the *keystored* daemon [9], which exposes a binder interface that consists of many key management and cryptographic functions. Under the hood, the *keystored* holds the following responsibilities:

- Expose the binder interface, listen and respond to requests made by applications.
- Manage the application keys. The daemon creates a directory on the filesystem for each application; the key-blobs are stored in files in the application's directory. Each key-blob file is encrypted with a key-blob encryption key (different per application) which is saved as the *masterkey* in the application's directory. The *masterkey* file itself is encrypted when the device is locked, and the encryption employs the user's password and a randomly generated salt to derive the *masterkey* encryption key.
- Relay cryptographic function calls to the Keymaster HAL device (covered below).

The Keymaster hardware abstraction layer (HAL) [7] is an interface between Android's *keystored* and the OEM implementation of a secure-hardware-backed cryptographic module. It requires the OEM to implement several cryptographic functions such as: key generation, init/update/final methods for various cryptographic primitives (public key encryption, symmetric key encryption, and HMAC), key import, public key export and general information requests. The implementation is a library that exports these functions and is implemented by relaying the request to the secure hardware system. The secure system usually encrypts generated keys with some key encryption key (which is usually derived by a hardware-backed mechanism). Therefore, the non-secure system does not know the actual key that is used, but may still save it in the filesystem and subsequently use it through the Keymaster to invoke cryptographic functions with the key. In practice - this is exactly how the *keystored* daemon uses the Keymaster HAL (with the aforementioned addition of an additional encryption of the key blobs).

An example of the usage of the Keymaster HAL is the Android Full Disk Encryption feature, implemented by the userspace daemon *vold* [10], which uses the Keymaster HAL as part of the key derivation.

2.3 Samsung's Keymaster HAL and Trustlet

Samsung's Keymaster HAL library exposes the aforementioned Keymaster interface and implements its functions by making calls to the Keymaster Trustlet. The trustlet itself has UUID: *ffffffff00000000000000000000003e*, and is located in the system partition (*/system/app/mcRegistry/<UUID>.tlbin*). The Trustlet code handles several tasks, of which the following are relevant to our work:

- Key generation of RSA/EC, AES and HMAC keys. Keys are generated using random bytes from the OpenSSL FIPS DRBG module, which seeds its entropy either from *keymaster_add_rng_entropy* calls from the Normal World or from a secure PRNG made available by the Secure World Crypto Driver. Key generation requests receive a list of key characteristics (as defined by the Keymaster HAL), which describe the algorithm, padding, block mode and other restrictions on the key. The generated keys (concatenated with their characteristics) are encrypted by a key-encryption-key (**KEK**) which is unique to the Keymaster trustlet. The trustlet receives this key by making an IPC request along with a *constant* salt to a driver which uses a hardware-based cryptographic function to drive the key. The encryption used for key encryption is AES256-GCM128. The GCM IV and authentication tag are concatenated to the encrypted key before being returned to the user as a key blob. Therefore, an attacker that is able to obtain this KEK is able to decrypt all the key blobs stored in the file system—i.e., the KEK can be viewed as the "key to the kingdom", and it's encryption scheme is the target of our attacks in Sect. 3.
- Execution of cryptographic functions. The trustlet can handle begin/update/-final requests for given keys created by the trustlet. It first decrypts the key-blobs and verifies the authentication tag, then verifies that the key (and the trustlet) supports the requested operation, and then executes it. The cryptographic functions are implemented using the OpenSSL FIPS Object Module [20]. In particular, we discovered that the AES code is a pure ARMv4 assembly implementation that uses a single 1KB T-Table. In general, AES implementations based on T-Tables are vulnerable to cache attacks [21,26]. Our attacks (described in Sect. 3) explore cache side channel attacks on this AES implementation.
- The trustlet handles requests for key characteristics and requests for information on supported algorithms, block modes, padding schemes, digest modes and import/export formats.

2.4   Attack Model

The fundamental reason for the existence of the TrustZone is to provide a hardware-based root of trust for a trusted execution environment (TEE)—that is designed to resist even a compromised Normal World kernel.

Since the Normal World kernel, and all the kernel modules on Samsung's smartphones are signed by Samsung and verified before being loaded, injecting code into the kernel is challenging for the attacker. Our goal in this work is to demonstrate that weaker attacks, that do not require a compromised kernel, are sufficient to exfiltrate Secure World information—in particular secret key material.

Our attack has two stages: a data collection stage and an analysis stage. In the data collection stage we assume an attacker is able to execute code on a Samsung Galaxy S6 device, under **root privileges** and relevant **SELinux permissions**. Note that these privileges are significantly less than kernel privileges, since the attack code runs in EL0.

Root privileges are needed to access the */proc/self/pagemap* to identify cache sets, as described by Lipp et al. [14]. Our attack can theoretically be mounted without access to this file, but it will be substantially more difficult.

To achieve root privileges and the necessary SELinux permissions in our investigation we used the publicly known vulnerability called *dirtycow*. The rooting process is based on Trident [6], which uses *dirtycow*.

The main target of our attack is the Keymaster trustlet. The API to communicate with the trustlet expects a buffer which should hold a key blob. Valid key blobs typically include over 100 bytes of encrypted data, therefore an API call (e.g. to extract some meta-data from a key blob) uses the AES-256 block function at least 9 times (2 for initialization and at least 7 subsequent blocks). If we measure cache access effects only after the trustlet completes its work, the 9 block function invocations will induce too much noise and render our attacks infeasible. Therefore, instead we send *invalid* requests: having the key blob hold just one byte. Such API calls induce the two block function calls for GCM initialization, and a one more call to decrypt the single byte. The request then fails, therefore we do not have access to any ciphertext. Our attacks take this restriction into consideration by focusing on the first AES-256 rounds and knowledge of the plain-text and IV - and avoid relying on the resulting ciphertext.

In the subsequent analysis stage, the collected clock measurement data is analyzed on a separate machine - we utilized a Macbook Pro laptop using a Radeon Pro 460 GPU.

3   Cache Attacks Against the Keymaster AES

3.1   Overview

As stated before, the Keymaster key encryption uses AES256/GCM128, therefore we focused on AES side channel attacks. In this section we present our attack methods.

We begin our work by adapting prior cache attacks on AES to the ARM32 implementation used in the Keymaster trustlet. Our measurements were taken on a stock Samsung Galaxy S6 running original Samsung firmware.

Prior research [16,21,26,28] demonstrated that the use of T-Tables in AES induces cache activity which leads to key leakage. In particular, when the T-Tables are misaligned in memory, better results have been achieved [25]. These methods exploit the fact that the implementation of the AES rounds use memory lookups which may be traced by evicting the T-Table from memory, running the AES encryption and then observing the cache access timing pattern. While the aforementioned methods assume the AES implementation uses four T-Tables, the AES implementation in the Keymaster trustlet uses one T-Table which is misaligned [13]. According to [21,26] this design choice is still vulnerable but requires roughly 3000 times more data and analysis, which is still feasible.

The attack, presented by Osvik et al. [21,26], assumes we can detect cache activity (on the cache sets which hold the T-Table) using the *Prime+Probe* method and focuses on the first round of AES. The *Prime+Probe* method measures cache activity by first *priming* a specific cache set (by writing memory to memory addresses which map to the same set - thereby *evicting* the cache set), then allowing the AES algorithm to run and finally *probing* the cache set (by accessing the *primed* memory addresses and measuring the time it took to fetch them). From the resulting measurements one can infer whether the AES algorithm has evicted a specific set - which would cause the *probing* phase to measure a higher value (due to some of its memory addresses being fetched from memory instead of the cache). In order to differentiate between *probe* measurements of evicted sets and non-evicted sets, a threshold value (denoted T_a below) is used. This value must be calibrated in advance for each hardware (CPU+Cache) that will be used as a target for the attack.

If the *probe* measurement, for the cache set which holds T-Table entry number i, is below T_a, the entry was not accessed and therefore certain k_i values are incorrect. If, in-fact, one of these k_i values was correct, the T-Table entry would have been accessed and eviction would occur for one of our *primed* addresses, resulting in a *probe* measurement above the threshold. More precisely, due to noise in the system, we may only infer that they are more likely, therefore we give each k_i candidate a score based on how many times we deem it likely (0 for each time it is unlikely and 1 otherwise). The k_i values we infer from cache activity depends on p_i, the details of the cache and the alignment (or misalignment) of the T-Table with respect to cache lines.

3.2 Calibrating the Probe Measurement Threshold

We selected the threshold value T_a through analysis of cache access times and eviction strategies as described by Lipp et al. [14]. This method applies *Prime+Probe* to a single address multiple times in two manners: first, it is *primed* and *probed* consecutively and second, memory access is added after the *prime* and before the *probe*. This, essentially, creates statistics on probe measurements for a given eviction strategy on a given CPU and cache. T_a separates between probes on set indexes which were not evicted versus set indexes which at least one address was evicted. Figure 1 shows the results of this method on our Galaxy S6. The strategy we used (using the notation from Lipp et al. [14]) is $N = 5$ (total eviction set size), $A = 5$ (shift offset), $D = 16$ (number of accesses per iteration) which we found to be the best strategy for our device after testing many alternatives; time measurements were made with linux's monotonic clock due to lack of better clock source available under our attack model. Based on the figure we set T_a to be $800[ns]$.

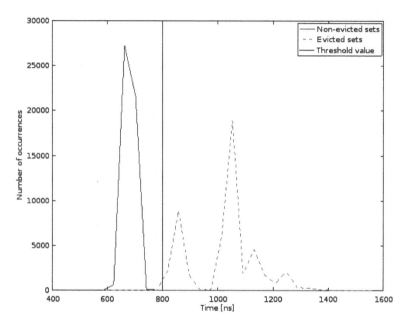

Fig. 1. Histogram of probe timing measurement for $50,000$ probes. Separation between evicted and non-evicted sets is visible at around 800 ns.

3.3   The Analysis Stage for AES-128 Attacks

To begin with, we describe an attack on AES-128 in ECB mode. With the cache activity measurements gathered on the Galaxy S6 in the data collection stage, we implement the analyzing stage on a GPU-equipped laptop. The analysis stage consists of two phases, described below.

Phase 1. In the first round of the AES implementation, each i-th plaintext byte p_i is XOR-ed with the i-th key byte k_i: $x_i^{(0)} = p_i \oplus k_i$. The value of $x_i^{(0)}$ is then used as an index to the T-Table which is accessed subsequently.

Because these calculations only rely on the value of p_i and k_i, it's possible to use a *divide-and-conquer* approach and consider each key byte independently.

Given a probe measurement for T-Table entry x, we iterate through byte index i = 0, ..., 15 and let p_i be the i-th plaintext byte. For all possible values of $k_i = 0$, ..., 255, we check whether k_i is likely based on the method described above and update the scores. We end with a score matrix for the level of likelihood for each candidate value per key byte. We continue the measurements and analysis until for each key byte, one candidate has a z-score above 5 (i.e., 5 standard-deviations above the mean).

On the Samsung Galaxy S6 ARM A53 and A57 CPUs, each cache line is 64 bytes long; therefore each line holds 16 T-Table entries (4 bytes per entry). In the implementation present in the Keymaster trustlet, the T-Table has an 8 byte misalignment with respect to the cache lines, see [13]: So the T-Table actually spans over 17 cache lines, with the first line holding 14 entries and the last line holding 2 entries. This means that our best case resolution is 2: if we use the constraints based on a single AES round we are eventually left with 2 candidates for each key byte which are indistinguishable to us. This means we learn 7 out of 8 bits for each key byte, reducing the unknown key space from 128 bits to 16 bits.

Phase 2. Enumerating through 16 bits is trivial with modern hardware; however, we present the rest of the attack which continues to apply *divide-and-conquer* using analysis of subsequent rounds. It will be useful to understand the next sections in which we attack AES-256 and AES-256/GCM and it may also be of independent interest in cases where the misalignment is less favorable or nonexistent.

To identify the remaining AES128 key bits we focus on the second round of the AES implementation; specifically, the following equations, derived from the Rijndael specification [4], which give 4 of the entries accessed in the second round:

$$
\begin{aligned}
x_2^{(1)} &= s(p_0 \oplus k_0) \oplus s(p_5 \oplus k_5) \\
&\quad \oplus 2 \bullet s(p_{10} \oplus k_{10}) \oplus 3 \bullet s(p_{15} \oplus k_{15}) \oplus s(k_{15}) \oplus k_2 \\
x_5^{(1)} &= s(p_4 \oplus k_4) \oplus 2 \bullet s(p_9 \oplus k_9) \\
&\quad \oplus 3 \bullet s(p_{14} \oplus k_{14}) \oplus s(p_3 \oplus k_3) \oplus s(k_{14}) \oplus k_1 \oplus k_5 \\
x_8^{(1)} &= 2 \bullet (p_8 \oplus k_8) \oplus 3 \bullet s(p_{13} \oplus k_{13}) \\
&\quad \oplus s(p_2 \oplus k_2) \oplus s(p_7 \oplus k_7) \oplus s(k_{13}) \oplus k_0 \oplus k_4 \oplus k_8 \oplus 1 \\
x_{15}^{(1)} &= 3 \bullet s(p_{12} \oplus k_{12}) \oplus s(p_1 \oplus k_1) \\
&\quad \oplus s(p_6 \oplus k_6) \oplus 2 \bullet s(p_{11} \oplus k_{11}) \oplus s(k_{12}) \oplus k_{15} \oplus k_3 \oplus k_7 \oplus k_1
\end{aligned}
\tag{1}
$$

where $s(\cdot)$ denotes Rijndael S-box function and $\bullet$ denotes multiplication over GF(256). There are three properties of these equations which are important to note:

- Each equation refers to 4 "bound" k_i's (that are an input to $s(\cdot)$) and between 1 to 4 "free" k_i's that are simply XOR'ed. In fact, the keen reader may see that if we analyze the equations sequentially, each equation only has 1 "free" k_i: If we solve the equations in sequence then all but one of the "free" k_i is completely discovered by the previous equations.

– Because our measurement resolution is 2 entries, only the 7 most significant bits of the "free" k_i variables are relevant to the index calculations.

– Since the first 7 bits of every k_i are known from phase 1, each equation only has 4 unknown bits - the least significant bit of every "bound" k_i.

These properties allows us to apply *divide-and-conquer* once again, and consider each equation separately. For each pair of plaintext and *probe* measurements, we enumerate the 4 possible key bits, calculate the equation and check whether they are likely based on the cache accesses during the second AES round. Eventually, the most likely candidate of these 4 bits is selected.

Combining the results for the four equations, along with the result of the phase 1, yields the entire 128 bits of the key - full key recovery.

We implemented a cache side-channel attack against the AES-128 implementation used by the Keymaster trustlet after copying it to a user-space sandbox and using AES in ECB mode. We were able to successfully recover the entire 128 bits of the key using the method described above. Our experiment used 100,000 measurements: this amount of data can be collected in under a minute on a Samsung Galaxy S6 and analyzed in less than 15 s on a Radeon Pro 460 GPU. The amount of memory used by the GPU (in phase 1 of the attack) was 1 GB. Details on the GPU analysis implementation in Sect. 4.

3.4   AES-256 Attacks

Phases 1 and 2. As we saw in Sect. 2.3, Samsung's Keymaster trustlet uses AES-256. Attempting to use the attack described in the previous section on AES-256 is not enough for full key recovery. There are relatively few papers discussing the specifics of cache attacks against AES-256. The most relevant seems to be by Neve and Tiri [17]. They proposed an extension of a different attack—one that looks at the *last* two rounds of AES instead of the first, and requires knowing the ciphertext, in contrast to the requirement of knowing the plaintext in the attack we used in Sect. 3.3. As we discussed in Sect. 2.4, relying on last AES round is difficult against the Keymaster trustlet. Therefore, we devised a method which extends the attack of Sect. 3.3 to recover 256 bit keys using the *first* three rounds.

The first part of the attack remains the same as phase 1 of the AES-128 attack (see Sect. 3.3): discover the 7 most significant bits of k_0 through k_{15}. We determine that the first round sieving has ended when for all 16 key bytes the most likely candidate value has a z-score above 5.

In order to recover the missing 16 bits in the lower half of the key, and most of the bits in the upper half, we rely on the second AES round. E.g., consider the first four indexes, which are derived from the Rijndael specifications:

$$
\begin{bmatrix} x_0^{(1)} \\ x_1^{(1)} \\ x_2^{(1)} \\ x_3^{(1)} \end{bmatrix} = \begin{bmatrix} 2\ 3\ 1\ 1 \\ 1\ 2\ 3\ 1 \\ 1\ 1\ 2\ 3 \\ 3\ 1\ 1\ 2 \end{bmatrix} \bullet \begin{bmatrix} s(p_0 \oplus k_0) \\ s(p_5 \oplus k_5) \\ s(p_{10} \oplus k_{10}) \\ s(p_{15} \oplus k_{15}) \end{bmatrix} \oplus \begin{bmatrix} k_{16} \\ k_{17} \\ k_{18} \\ k_{19} \end{bmatrix} \tag{2}
$$

It's important to note that every 4 indexes depend on the same four key bytes from the lower half of the key (k_0, k_5, k_{10}, k_{15} in Eq. (2)) and each index depends on one byte of the upper half of the key. Another important property is that as in Sect. 3.3, from the 8 bits of the bytes from the upper half of the key only the 7 most significant bits affect the measurement of the index. Therefore, each equation has only 11 unknown bits: 1 for each of the 4 lower-half-key bytes and 7 for the single upper-half-key byte.

Once again, we use *divide-and-conquer*; divide the problem to four-equation subproblems, divide each subproblem to it's four equations and on each equation use the same methods described above to select the most likely candidate for the 1 least significant bits of the lower-half key bytes and the 7 most significant bits of the single upper-half key byte. Therefore, for a given equation e and measurement, iterate over all 2^{11} combinations of key-bit values. If the measurement is compatible with key-bit combination c, then increment $score[e][c]$. After this step, we have the entire lower-half key bytes (k_0 through k_{15}) and the 7 most significant bits of every upper-half key byte (k_{16} through k_{31}). Therefore, we have reduced the key space from 256 bits to 16 by using the first two rounds.

Phase 3. While enumeration of 16 bits is feasible, we present a third phase of the attack which may be applied to other misalingment circumstances. We do so by imitating the second phase of the attack on 128 bit AES (Sect. 3.3). Consider Eq. (3) which is derived from the Rijndael specification with 256 key expansion (after substituding the first round indexes $x_i^{(1)}$)

$$
\begin{aligned}
x_2^{(2)} = \ & s(2 \bullet s(p_0 \oplus k_0) \oplus 3 \bullet s(p_5 \oplus k_5) \oplus s(p_{10} \oplus k_{10}) \\
& \oplus s(p_{15} \oplus k_{15}) \oplus \underline{k_{16}}) \\
& \oplus s(s(p_4 \oplus k_4) \oplus 2 \bullet s(p_9 \oplus k_9) \oplus 3 \bullet s(p_{14} \oplus k_{14}) \\
& \oplus s(p_3 \oplus k_3) \oplus \underline{k_{21}}) \\
& \oplus 2 \bullet s(s(p_8 \oplus k_8) \oplus s(p_{13} \oplus k_{13}) \oplus 2 \bullet s(p_2 \oplus k_2) \\
& \oplus 3 \bullet s(p_7 \oplus k_7) \oplus \underline{k_{26}}) \\
& \oplus 3 \bullet s(3 \bullet s(p_{12} \oplus k_{12}) \oplus s(p_{11} \oplus k_{11}) \oplus s(p_6 \oplus k_6) \\
& \oplus 2 \bullet s(p_{11} \oplus k_{11}) \oplus \underline{k_{31}}) \\
& \oplus s(\underline{k_{31}}) \oplus k_2
\end{aligned} \tag{3}
$$

At first sight this equation may seem daunting. However, notice that we, in fact, know p_0 through p_{15}, k_0 through k_{15} and the 7 most significant bits of k_{16}, k_{21}, k_{26}, k_{31}. Therefore, only 4 bits are unknown in this equation. We then use a similar sieving method as in the previous phases to select the most likely candidate for these bits. Then, we apply the same technique to the equations for $x_5^{(2)}$, $x_8^{(2)}$ and $x_{15}^{(2)}$. Eventually, we arrive at full recovery of the AES 256 bit key.

Putting It All Together. We implemented this attack on the AES-256 code used by Samsung's Keymaster trustlet. Figure 2 shows the the number of correct bits in the most likely candidate as a function of the number of measurements used. The horizontal barriers mark the target of each phase of the attack: 112 bits for the 1st phase, 240 bits for the 2nd and 256 bits for the 3rd phase. It's important to note that after we complete the first phase, we reuse the samples for the second phase which explains the sudden increase in known bits between phase one and phase two of the attack. It took 7 min to collect the one million measurements on the Galaxy S6. The sieving process took under a minute to complete (all three stages) and 3.5 GB of memory, using a Radeon Pro 460 GPU on a laptop.

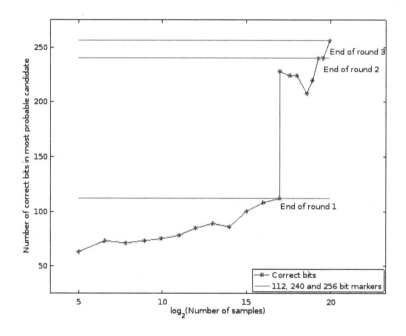

Fig. 2. Number of correct bits in the most likely candidate as function of samples used (log_2 scale).

3.5   Galois Counter Mode (GCM) Attacks

Challenges. A further complication is that Samsung's Keymaster trustlet uses AES-256 in GCM mode [5]. Two factors make cache side channel attacks harder against GCM: the use of the block function in the initialization, and the lack of control over the 4 last bytes of the input to the block function.

According to the GCM specification, the computation of the authentication tag requires two invocations of the block function. When the input *initialization vector* (IV) is 96 bits long, the block function is invoked once with a plaintext of 0^{16} (a 16 zero byte string) and then with a plain-text of $IV||0^4$. Unless it is possible to distinguish between this initialization phase and subsequent encryption phases, the initialization induces substantial cache-access noise.

Furthermore, subsequent block function invocations made by GCM are called with the input $IV||Counter$, where IV is the original 96 bit IV and Counter is a four byte integer counter (starting with the value 2) which is appended (with big endianess) to the IV. This means that we have limited control over the input to the block function. While we control the 96 bits given as IV, the Counter bytes may only be changed by encrypting additional data with the same GCM context. This implies that it is much more difficult to collect enough data to differentiate between key candidates for k_{12} through k_{15}.

Phases 1 and 2. We begin by attempting to apply the same technique used in the previous section to AES-256/GCM. We assume that we can distinguish between cache-access due to the first two block function invocations (initialization calls) and subsequent invocations. However, we continue limit ourselves to scenarios that allow only one encryption call and do not allow knowledge of the resulting ciphertext to allow use against the Keymaster trustlet. Because we do not have control over the last four bytes of the input, the first phase of the technique (Sect. 3.4), which focuses on the first round of AES, only recovers the 7 most significant bits of k_0 through k_{11}, recovering only 84 bits of the key.

The second part of the technique, which focuses on the second round of AES-256, is more difficult under GCM. Instead of the 11 unknown bits we identified in Sect. 3.5, we now face 18 unknown bits: 1 least significant bit for k_0 through k_{11} (three of these per equation), 8 bits for k_{12} through k_{15} (one per equation) and 7 most significant bits for k_{16} through k_{31} (one per equation).

While an enumeration of 18 bits is feasible even with modest resources, another hurdle emerges. Consider the value $t = s(p_{15} \oplus k_{15}) \oplus k_{16}$ in the equation for $x_0^{(1)}$ in Eq. (2). Because p_{15} has a single value (typically $p_{15} = 2$) which we cannot control, t has a constant value. We note that for each key byte candidate value x for k_{15} we can find a key byte candidate k_{16}, which will result in the same value x. More precisely, due to the resolution from the T-Table misalignment, we can find a 7 most significant bit candidate for k_{16}.

By applying the same sieving technique described above, we use pairs of IV (first 12 bytes of plaintext) and *probe* measurements to select the likely value of the 18 unknown bits. Due to the dependency between k_{15} and k_{16} described above, we expect to find 256 likely values: each having the correct least significant bit of k_0, k_5 and k_{10}, one of the 256 candidates for k_{15} and the 7 most significant bits of k_{16} candidate. This method allows us to gain full information on k_0, k_5 and k_{10}, and a constraint on k_{16} depending on k_{15}. This constraint may be visualized as a table indexed by k_{15} and having its value be the constraint on k_{16}.

We apply the same method for the rest of the equations shown in equation set (2) to gain information on the respective constraint between k_{15} and k_{17} through k_{19}. Note that in the case of k_{18} and k_{19} the constrained value t is $t = 3 \bullet s(p_{15} \oplus k_{15}) \oplus k_{18}$ and $t = 2 \bullet s(p_{15} \oplus k_{15}) \oplus k_{19}$ respectively. These four constraints may be grouped into a single table, Table 1 shows an example of such table.

Table 1. The 7 most significant bits of upper key bytes for each possible k_{15} value

k_{15}	k_{16}	k_{17}	k_{18}	k_{19}
0	67	22	60	67
1	69	16	54	79
2	73	38	34	87
...				
253	115	38	109	34
254	32	117	21	9
255	82	7	14	96

The same method may be used to extract similar constraints between the three other bytes k_{12}, k_{13}, k_{14} and their respective four bytes from the upper half of the key.

To summarize, based on 2 rounds of AES in GCM mode we can extract the values of k_0 through k_{11}, and have four table which describe further constraints on the key. The remaining key space is 48 bits: 8 bits per table (32 total) and 1 additional bit per byte in the upper half of the key.

Phase 3. We now shift our focus to round 3 and consider equations $x_{12}^{(2)}$ through $x_{15}^{(2)}$. Equation (4) shows one such equation: These equations are important to us for two reasons:

- The "bound" expressions holding k_{12} to k_{15} in these equations have appeared in our 2nd round analysis and therefore we have a table that constraints the "free" upper-half key bytes to their values (e.g. $s(p_{15} \oplus k_{15}) \oplus k_{17}$ is known up to 1 bit). Thereby reducing the unknown bits in each of these expressions from 8 bits to 1.
- Due to the AES key expansion scheme, each of these equations includes one of the key bytes k_{12} through k_{15} in a "free" manner. Which allows us to receive different measurements for these bytes; note that this is the first round this is possible in.

$$x_{12}^{(2)} =$$

$$2 \bullet s(2 \bullet s(p_{12} \oplus \underline{k_{12}}) \oplus 3 \bullet s(p_1 \oplus k_1) \oplus s(p_6 \oplus k_6) \oplus s(p_{11} \oplus k_{11}) \oplus \underline{k_{28}}) \oplus$$

$$3 \bullet s(s(p_0 \oplus k_0) \oplus 2 \bullet s(p_5 \oplus k_5) \oplus 3 \bullet s(p_{10} \oplus k_{10}) \oplus s(p_{15} \oplus \underline{k_{15}}) \oplus \underline{k_{17}}) \oplus \quad (4)$$

$$s(s(p_4 \oplus k_4) \oplus s(p_9 \oplus k_9) \oplus 2 \bullet s(p_{14} \oplus \underline{k_{14}}) \oplus 3 \bullet s(p_3 \oplus k_3) \oplus \underline{k_{22}}) \oplus$$

$$s(3 \bullet s(p_8 \oplus k_8) \oplus s(p_{13} \oplus \underline{k_{13}}) \oplus s(p_2 \oplus k_2) \oplus 2 \bullet s(p_7 \oplus k_7) \oplus \underline{k_{27}}) \oplus$$

$$\underline{k_{12}} \oplus k_8 \oplus k_4 \oplus k_0 \oplus s(\underline{k_{29}}) \oplus 1$$

While Eq. (4) might seem to have 48 unknown bits, using the knowledge from the previous phases we assert that it only has 13 unknown bits: 8 bits to choose k_{12}, 1 least significant bit for k_{29}, and 1 more least significant bit for each "bound" expression (4 additional bits).

Applying our sieving technique once again for equations $x_{12}^{(2)}$ through $x_{15}^{(2)}$ gives us the most likely value of the 7 most significant bits of k_{12} to k_{15} and the least significant bit of k_{28} through k_{31}. Thereby reducing the amount of unknown bits from 48 to 16: 1 least significant bit of k_{12} to k_{15} and 1 least significant bit of k_{16} to k_{27}. Additional analysis of the 3rd round accesses may reveal the remaining bits, but we chose to apply brute-force enumeration to find them.

It took five million measurements to mount this analysis which took 40 min to collect on the Galaxy S6. The analysis took 30 min and 3.5 GB of memory to complete using a Radeon Pro 460 GPU on a laptop.

In summary, we see that the AES-256 GCM, with a single T-Table implementation used by Samsung's Keymaster trustlet, is vulnerable to cache side-channel attacks when it is used in isolation.

4 Analysis Acceleration Using a GPU with OpenCL

4.1 Overview

Previous work [3, 14, 16, 17, 21, 25, 26, 28, 29] goes into great details about the implementation of the attack phase and candidate sieving; however, little discussion is presented on the implementation of the analysis phase. While designing the attacks described in the previous sections, we found that the amount of data and time required by a sequential implementation of the attack is significant, so, we decided to leverage GPUs to expedite the analysis. The following sections provide detail into our design and implementation of a GPU based cache attack analysis method.

4.2 Programming the GPU

When programming a GPU, one must design a function that will be run in parallel on many data points; such a function is called a *kernel*. We used a

GPGPU (general purpose GPU) programming framework called pyOpenCL [12]. This framework allowed us to write most of the analysis in python while easily deferring the heavy lifting to the GPU. pyOpenCL provides a very convenient way to write OpenCL kernels called "ElementwiseKernel": the programmer only needs to write the calculation for a single element while abstracting away most other details. A kernel is essentially a function, that receives at least one pointer to a GPU memory block (usually containing tables) and an argument i which is used as an index to that memory block. The framework instructs the GPU to run numerous copies of that kernel in parallel, each on a different GPU core, with each copy being allocated a different index i.

A kernel returns output by modifying the memory blocks received as arguments. In order to minimize the need to synchronize the kernels, usually each kernel writes to a separate cell in memory; thus avoiding memory contention and race conditions. The Radeon Pro 460 which we used in our analysis has 1024 cores and 4 GB of internal memory.

4.3 Using the GPU in the Attacks

Previous sections outlined the basic algorithm used for the analysis of the side channel artifacts: Use the plaintext bytes and side channel map to sieve through the possible candidates until one most likely candidate is found. This process can be broken down to the following steps: (i) thresholding the cache access patterns to discern between cache hits and misses, (ii) AES round calculations, (iii) matching between calculation and the cache access pattern and (iv) scoring.

i The thresholding step is straightforward: it receives a cache access timing matrix (rows are different measurements, columns are the relevant cache set indexes which were measured) and a threshold value T_a (recall Sect. 3.2). Each matrix cell is compared against T_a and is set to either 1 if it's above T_a (miss) or 0 otherwise (hit).

ii The AES round calculation varies depending on the step of the analysis (as described in previous sections) but follows the same principles: receive the relevant key candidates and plain-texts used in measurements, apply the relevant round calculation, apply the table misalignment and return the relevant cache set indexes for the given candidates for each given plain-text. Round calculations follow the equations presented in the previous sections, and use lookup tables to calculate the S-Box and GF(256) multiplications. The S-Box and multiplication tables are placed inside the GPU internal memory. The result is a matrix M of the candidates versus the plain-texts where each cell $M_{i,j}$ holds the result of the AES round calculation for i-th plain-text and the j-th key candidate. In other words, if key candidate j is correct, $M_{i,j}$ holds a cache index which we expect to measure as a miss for the i-th plaintext and its cache measurements.

iii The next step takes the thresholded cache access matrix and the cache set index matrix result from the AES round calculation step and returns an array which holds the score for each candidate. This is done in two steps:

matching and summing. The match step takes the two input matrices and outputs a matrix of candidates versus plain-texts in which each cell is 1 if the index predicted by the AES round calculations (for a given plain-text and candidate) was a cache miss in our measurements (which implies the candidate is more likely) and 0 otherwise. The summing phase then sums the result by the plain-text axis, resulting in an array of scores for each key candidate.

iv Finally, we are left with the scores for each key candidate, all we have left is to choose the most likely one. Due to the large key enumeration space in phase 2 and 3 of the AES-256 and AES-256/GCM attacks, the memory on our GPU was not large enough to hold the plain-text over key candidate matrices when trying to analyze all of the plain-texts at once. Instead, we divided the plain-texts into batches, analyzed them separately and combined their result after each batch by simply adding the score array. This allowed us to analyze large amounts of data (over 2^{20} samples) over up to 18 bits of key candidates on a commodity laptop GPU within minutes.

4.4   Kernel Implementation Details

We used several such kernels and provide their code in the appendix:

1. Thresholding: The first kernel is used to reduce the measurements from a matrix of plain-texts over cache indexes which contains the cache timing measurements to a matrix of the same dimensions but with a value 1 if the measurement is above T_a which indicates a cache miss, or 0 otherwise. This is accomplished via a simple ternary operator. See Appendix A.1.

2. Round calculation kernel: The following explanation is relevant to the first four round bytes of the 2nd AES256 round calculations (recall Sect. 3.4), the same principles apply for the rest of the round bytes and the 3rd round as well. This kernel receives our key candidates (5 key bytes serialized as a 64-bit integer), plaintext bytes, S-Box and GF(256) multiplication lookup tables, misalignment parameter and output matrices (plain-texts over candidates). Note that instead of calculating the round for each round byte separately, we optimize this kernel by reusing calculations to calculate four round bytes together (see Eq. (2)). The kernel basically calculates the first AES round (SubBytes, ShiftRows and MixColumns) and then XORs the result with an upper-half key candidate byte. The result is the index of the T-Table which will be access by the 2nd round. It then applies the misalignment parameter and selects the bits which are relevant to the cache index and stores the results in the output matrices. These matrices then hold the cache set which we expect to measure as a miss for a each plain-text, *if* that the key candidates are correct. See Appendix A.2.

3. Round to hit matrix kernel: The previous kernel results in four matrices of cache sets. This kernel performs a pass through those matrices and merges their results with the actual measurements. It receives the four result matrices, and the thresholded measurements matrix. For each cell of the result

matrices (which represent the index which we expect to see as 1 in the measurement matrix for a plain-text and a key candidate, if the candidate is correct), we retrieve the measurement of the relevant plain-text and the relevant cache index. This result will be 1 if the measurements support this candidate (cache index was indeed measured as a miss) and 0 otherwise. We write the result back to the result matrix to save memory. See Appendix A.3.

4. Sum axis kernel: The previous kernel results in four score matrices of plain-text over candidates. Since we are trying to calculate the candidate score, we then need to sum these matrices by the candidates axis. Special care must be taken when summing in GPU code as many cores may access the sum variable concurrently. Several solutions exist, such as: summing in CPU instead, logarithmic reduction kernels or using atomic OpenCL intrinsics. We compared the CPU solution (using the Python Numpy package sum by axis function) with a kernel which uses the "atomic_add" intrinsic and found that the kernel is about twice as fast. That being said, both solutions took negligible time compared to the other operations. We did not attempt to implement a more optimized sum kernel. See Appendix A.4.

5   Conclusions

The ARM TrustZone is a security extension which is used in recent Samsung flagship smartphones to create a Trusted Execution Environment (TEE) called a Secure World, which runs secure processes called Trustlets. The Samsung TEE includes cryptographic key storage and functions inside the Keymaster trustlet. The secret key material used by the Keymaster trustlet is derived by a hardware device and is inaccessible to the Android OS. However, the ARM32 AES implementation used by the Keymaster is vulnerable to side channel cache-attacks. The Keymaster trustlet uses AES-256 in GCM mode, which makes mounting a cache attack against this target much harder. In this paper we show that it is possible to perform a successful cache attack against this AES implementation, in AES-256/GCM mode using widely available hardware. Using a laptop's GPU to parallelize the analysis, we are able to extract a raw AES-256 key with 7 min of measurements and under a minute of analysis time and an AES-256/GCM key with 40 min of measurements and 30 min of analysis.

We conclude that cache side-channel effects are a serious threat to the current AES implementation inside the Keymaster trustlet. However, side-channel-resistant implementations, that do not use memory accesses for round calculations, do exist for the ARM platform, such as a bit-sliced implementation [19] or one using ARMv8 cryptographic extensions [18]. Using such an implementation would render most cache attacks, including ours, ineffective.

A   OpenCL Kernels Code

A.1   Thresholding Kernel

```
threshold_kernel = ElementwiseKernel(ctx,
''',
uint* in,  uint thresh,  uint* out
''',
''',
out[i] = (in[i] > thresh) ? (1) : (0)
''',
''threshold_kernel''
)
```

A.2   Round Two Kernel

```
round_kernel = ElementwiseKernel(ctx,
'''
uint *x0, uint *x1, uint *x2, uint *x3, ulong *candidates, uint *p0,
uint *p5, uint *p10, uint *p15, uint row_size,  uint *sbox,
uint *mult2, uint *mult3, uint disalignment
''',
'''
// Extract key byte candidate from serialized candidate, apply SubBytes
uint t0 = SHIFT_RIGHT(candidates[i % row_size], 0) ^p0[i/row_size];
uint t5 = SHIFT_RIGHT(candidates[i % row_size], 8) ^p5[i/row_size];
uint t10 = SHIFT_RIGHT(candidates[i % row_size], 16)^p10[i/row_size];
uint t15 = SHIFT_RIGHT(candidates[i % row_size], 24)^p15[i/row_size];
uint k_e = SHIFT_LEFT(SHIFT_RIGHT(candidates[i % row_size], 32), 1);

t0  = sbox[t0 ];
t5  = sbox[t5 ];
t10 = sbox[t10];
t15 = sbox[t15];

// apply ShiftRows and MixColumns
// also XOR with the upper key byte candidate
x0[i] = mult2[t0]^mult3[t5]^      t10 ^       t15 ^k_e;
x1[i] =       t0 ^mult2[t5]^mult3[t10]^       t15 ^k_e;
x2[i] =       t0 ^      t5 ^mult2[t10]^ mult3[t15]^k_e;
x3[i] = mult3[t0]^      t5 ^      t10 ^ mult2[t15]^k_e;

// apply disalignment
x0[i] = (x0[i] + disalignment) & 0xff;
x1[i] = (x1[i] + disalignment) & 0xff;
x2[i] = (x2[i] + disalignment) & 0xff;
x3[i] = (x3[i] + disalignment) & 0xff;
```

```
// select bits which affect cache index
x0[i] = SHIFT_RIGHT(x0[i], 4);
x1[i] = SHIFT_RIGHT(x1[i], 4);
x2[i] = SHIFT_RIGHT(x2[i], 4);
x3[i] = SHIFT_RIGHT(x3[i], 4);
''',
"round_kernel",
preamble='''
#define SHIFT_RIGHT(X, Y) ((X >> Y) & 0xff)
#define SHIFT_LEFT(X, Y) ((X << Y) & 0xff)
'''
)
```

A.3   Round to Hit Matrix Kernel

```
round_to_hits_kernel = ElementwiseKernel(ctx,
'''
uint *x0, uint *x1, uint *x2, uint *x3 ,
uint row_size, uint * sets_data_thresh ,
uint  sets_data_thresh_row_size
''',
'''
// (i/row_size) provides an index to the measurement row,
// x[i] provides the offset to the cache set we wish to check
x0[i]=sets_data_thresh [(i/row_size)* sets_data_thresh_row_size +x0[i]];
x1[i]=sets_data_thresh [(i/row_size)* sets_data_thresh_row_size +x1[i]];
x2[i]=sets_data_thresh [(i/row_size)* sets_data_thresh_row_size +x2[i]];
x3[i]=sets_data_thresh [(i/row_size)* sets_data_thresh_row_size +x3[i]];
''',
" round_to_hits_kernel"
)
```

A.4   Axis Sum Kernel

```
sum_axis_column_kernel = ElementwiseKernel(ctx,
'''
uint *tmp, uint tmp_row_size, uint *out
''',
'''
// Use atomic_add to avoid data races,
// not the fastest approach but the time it takes is negligible anyway
atomic_add(&out[i % tmp_row_size], tmp[i]);
''',
"sum_axis_column_kernel"
)
```

References

1. ARM. Building a secure System using TrustZone Technology. http://infocenter.arm.com/help/topic/com.arm.doc.prd29-genc-009492c/PRD29-GENC-009492C_trustzone_security_whitepaper.pdf
2. ARM. ARM trustzone (2018). https://www.arm.com/products/security-on-arm/trustzone
3. Bernstein, D.J.: Cache-timing attacks on AES (2005). https://cr.yp.to/antiforgery/cachetiming-20050414.pdf
4. Daemen, J., Rijmen, V.: AES proposal: Rijndael. In: AES submission document (1999). http://csrc.nist.gov/CryptoToolkit/aes/rijndael/Rijndael-ammended.pdf
5. Dworkin, M.J.: SP 800–38D: recommendation for block cipher modes of operation: Galois/counter mode GCM and GMAC. National Institute of Standards & Technology (2007)
6. freddierice. Trident - temporary root for the Galaxy S7 active. https://github.com/freddierice/trident
7. Google. Android keymaster HAL. https://source.android.com/security/keystore/implementer-ref
8. Google. Android keystore. https://developer.android.com/training/articles/keystore.html
9. Google. Android keystore - source code. http://androidxref.com/6.0.0_r1/xref/system/security/keystore/keystore.cpp
10. Google. Android vold cryptfs. http://androidxref.com/6.0.0_r1/xref/system/vold/cryptfs.c
11. Green, M., Rodrigues-Lima, L., Zankl, A., Irazoqui, G., Heyszl, J., Eisenbarth, T.: AutoLock: why cache attacks on ARM are harder than you think. In: 26th USENIX Security Symposium, pp. 1075–1091 (2017)
12. Klöckner, A., Pinto, N., Lee, Y., Catanzaro, B., Ivanov, P., Fasih, A.: PyCUDA and PyOpenCL: a scripting-based approach to GPU run-time code generation. Parallel Comput. **38**(3), 157–174 (2012)
13. Lapid, B., Wool, A.: Navigating the Samsung TrustZone with applications to cache-attacks on AES-256 in the Keymaster trustlet. In: Proceedings of 23rd European Symposium on Research in Computer Security (ESORICS), Barcelona, September 2018, to appear
14. Lipp, M., Gruss, D., Spreitzer, R., Maurice, C., Mangard, S.: ARMageddon: cache attacks on mobile devices. In: USENIX Security Conference, pp. 549–564 (2016). https://www.usenix.org/system/files/conference/usenixsecurity16/sec16_paper_lipp.pdf
15. nccgroup. Cachegrab. https://github.com/nccgroup/cachegrab
16. Neve, M., Seifert, J.-P.: Advances on access-driven cache attacks on AES. In: Biham, E., Youssef, A.M. (eds.) SAC 2006. LNCS, vol. 4356, pp. 147–162. Springer, Heidelberg (2007). https://doi.org/10.1007/978-3-540-74462-7_11
17. Neve, M., Tiri, K.: On the complexity of side-channel attacks on AES-256 - methodology and quantitative results on cache attacks. Technical report (2007). https://eprint.iacr.org/2007/318
18. OpenSSL. ARM AES implementation using cryptographic extensions. https://github.com/openssl/openssl/blob/master/crypto/aes/asm/aesv8-armx.pl
19. OpenSSL. ARMv7 AES bit sliced implementation. https://github.com/openssl/openssl/blob/master/crypto/aes/asm/bsaes-armv7.pl
20. OpenSSL. OpenSSL FIPS. https://www.openssl.org/docs/fips.html

21. Osvik, D.A., Shamir, A., Tromer, E.: Cache attacks and countermeasures: the case of AES. In: Pointcheval, D. (ed.) CT-RSA 2006. LNCS, vol. 3860, pp. 1–20. Springer, Heidelberg (2006). https://doi.org/10.1007/11605805_1

22. Qualcomm. Snapdragon security (2018). https://www.qualcomm.com/solutions/mobile-computing/features/security

23. Samsung. Mobile processor: Exynos 7 Octa (7420) (2018). http://www.samsung.com/semiconductor/minisite/exynos/products/mobileprocessor/exynos-7-octa-7420/

24. Samsung. Platform security (2018). http://developer.samsung.com/tech-insights/knox/platform-security

25. Spreitzer, R., Plos, T.: Cache-access pattern attack on disaligned AES T-tables. In: Prouff, E. (ed.) COSADE 2013. LNCS, vol. 7864, pp. 200–214. Springer, Heidelberg (2013). https://doi.org/10.1007/978-3-642-40026-1_13

26. Tromer, E., Osvik, D.A., Shamir, A.: Efficient cache attacks on AES, and countermeasures. J. Cryptol. **23**(1), 37–71 (2010)

27. Trustonic. Trustonic Kinibi technology. https://developer.trustonic.com/discover/technology

28. Xinjie, Z., Tao, W., Dong, M., Yuanyuan, Z., Zhaoyang, L.: Robust first two rounds access driven cache timing attack on AES. In: 2008 International Conference on Computer Science and Software Engineering, vol. 3, pp. 785–788. IEEE (2008)

29. Zhang, N., Sun, K., Shands, D., Lou, W., Hou, Y.T.: TruSpy: cache side-channel information leakage from the secure world on ARM devices. IACR Cryptology ePrint Archive, 2016/980 (2016)

Fault Attacks on Nonce-Based Authenticated Encryption: Application to Keyak and Ketje

Christoph Dobraunig[1], Stefan Mangard[1], Florian Mendel[2], and Robert Primas[1(✉)]

[1] Graz University of Technology, Graz, Austria
{christoph.dobraunig,stefan.mangard,robert.primas}@iaik.tugraz.at
[2] Infineon Technologies AG, Neubiberg, Germany
florian.mendel@infineon.com

Abstract. In the context of fault attacks on nonce-based authenticated encryption, an attacker faces two restrictions. The first is the uniqueness of the nonce for each new encryption that prevents the attacker from collecting pairs of correct and faulty outputs to perform, e.g., differential fault attacks. The second restriction concerns the verification/decryption, which releases only verified plaintext. While many recent works either exploit misuse scenarios (e.g. nonce-reuse, release of unverified plaintext), we turn the fact that the decryption/verification gives us information on the effect of a fault (whether a fault changed a value or not) against it.

In particular, we extend the idea of statistical ineffective fault attacks (SIFA) to target the initialization performed in nonce-based authenticated encryption schemes. By targeting the initialization performed during decryption/verification, most nonce-based authenticated encryption schemes provide the attacker with an oracle whether a fault was ineffective or not. This information is all the attacker needs to mount statistical ineffective fault attacks. To demonstrate the practical threat of the attack, we target software implementations of the authenticated encryption schemes Keyak and Ketje. The presented fault attacks can be carried out without the need of sophisticated equipment. In our practical evaluation the inputs corresponding to 24 ineffective fault inductions were required to reveal large parts of the secret key in both scenarios.

Keywords: Fault attack · Statistical ineffective fault attack · SIFA · Authenticated encryption · Keyak · Ketje

1 Introduction

With the rise of the Internet of Things (IoT), devices implementing authenticated encryption schemes will become ubiquitous. A trend, NIST is planning to address with standardization efforts in the area of lightweight authenticated

© Springer Nature Switzerland AG 2019
C. Cid and M. J. Jacobson, Jr. (Eds.): SAC 2018, LNCS 11349, pp. 257–277, 2019.
https://doi.org/10.1007/978-3-030-10970-7_12

encryption schemes [21,24]. As a consequence, authenticated encryption schemes will be more and more applied on devices in areas, where the physical access of malicious entities is unavoidable. Hence, implementation attacks like side-channel attacks and fault attacks, are a major concern for such devices as demonstrated, e.g., by Ronen, Shamir, Weingarten, and O'Flynn [28] in their attack on smart lamps. To identify and protect against the potential threats raised by implementation attacks, research in the practicability and applicability of implementation attacks on authenticated encryption schemes is needed.

As observed by many publications [29–31], the uniqueness of the nonce in authenticated encryption schemes prohibits the straight-forward application of prominent fault attacks like differential fault analysis (DFA) [10] to the authenticated encryption. In the case of authenticated decryption, the built-in validation of the authenticity of the processed data often provides an implicit detection of induced faults. Therefore, a lot of attacks published so far assume scenarios, where the uniqueness of the nonce is not ensured [31] or unverified plaintext is released [29], or even require a precise induction of faults at multiple locations during one execution of the authenticated encryption scheme [30]. Recently, statistical fault attacks (SFA) that are applicable to a wide-range of AES-based authenticated encryption schemes including popular schemes like GCM, CCM and OCB have been published [15]. However, the presented attacks face some limitations. In particular, they are only applicable to schemes where the secret key is processed right before the data is output. Thus, it is typically not applicable to sponge or stream cipher-based constructions. Moreover, they only work in the case of authenticated encryption, leaving fault attacks targeting authenticated decryption (assuming that the unverified plaintext is not released) as an open problem.

Our Contribution. In this work, we close the aforementioned gaps. We present the—to the best of our knowledge—first fault attacks targeting authenticated decryption/verification that are applicable to a broad range of nonce-based authenticated encryption schemes. In particular, the presented attacks are applicable whenever the nonce is mixed with the secret key during the initialization as it is the case in a wide range of authenticated encryption schemes. This includes sponge and stream cipher-based authenticated encryption schemes for which most of the existing fault attacks are not applicable.

We focus our analysis on Keyak and Ketje designed by Bertoni, Daemen, Peeters, Van Assche, and Van Keer [6,7]. Both designs are based on the Keccak-f permutation [4], which also underlies Keccak/SHA-3 [23]. Please note that the presented attacks do not exploit a weakness inherent in the design of Keyak and Ketje, these two primitives just serve as an example to show the applicability of fault attacks on sponge and stream cipher-based authenticated encryption schemes.

Our attacks are based on statistical ineffective fault attacks [14,16] and do not require an extensive profiling or characterization of the attacked device. Additionally, they are resilient against "errors" induced by miss-located faults, or in general fault inductions that do not behave as intended. As a consequence, they can be easily applied in practice as demonstrated by our attack targeting 8-bit software implementations of Keyak and Ketje running on an AVR Xmega 128D4. After inducing faults during authenticated decryptions and filtering for the inputs of 24 unaffected computations, we can recover large parts of the secret keys. The remaining unknown key bits can then either be brute-forced or further reduced by repeating the attack and inducing the fault at a different point in time.

Outline. In Sect. 2, we cover the required background of our attack. After, we describe the state-of-the-art of fault attacks, we give a short overview of authenticated encryption schemes. We provide a more detailed description of Keyak and Ketje, the two authenticated encryption schemes that are the main target of our practical attack evaluation, in Sect. 3. In Sect. 4, we discuss the idea and working principle of the attack. Section 5 describes the practical evaluation of our fault attack on a real microprocessor. We conclude the paper in Sect. 6.

2   Background

In this section, we give a brief introduction to fault attacks in general and state the idea behind Statistical Ineffective Fault Attacks (SIFA), recently proposed by Dobraunig, Eichlseder, Korak, Mangard, Mendel, and Primas [16], in more detail. We then recall the concept of nonce-based authenticated encryption with associated data.

2.1   Fault Attacks

The threat of fault attacks was demonstrated by Boneh, DeMillo, and Lipton [11] in 1997 when they showed the vulnerability of several asymmetric primitives like RSA to erroneous computations. Since then, fault attacks have been demonstrated targeting many other cryptographic schemes [9], including symmetric ones [10,25].

   The way in which faults can be induced into a cryptographic computation is manifold. Originally, the most popular fault attacks were based on clock glitches or variations on the supply voltage. However, by the time, more and more sophisticated fault induction methods were presented like attacks based on lasers [32], EM-pulses [19], or even X-rays [1].

   While the induction of (more or less) precise faults into a cryptographic computation is an essential prerequisite for the attack, the exploitation of the observed erroneous behavior is equally important. Biham and Shamir [10] proposed Differential Fault Analysis (DFA) as an effective key recovery method for DES. DFA requires the collection of pairs of valid and faulty ciphertexts where a fault was induced in the last few rounds of the encryption. The difference between valid and faulty ciphertexts together with knowledge about the faulted operation can then be used to recover the used secret key. Later it has been shown that DFA is not limited to DES and can be applied to broad range of block ciphers.

One immediate consequence of fault attacks was the evaluation of possible countermeasures that can prevent such attacks. One commonly used countermeasure is the detection of the induced fault by means of redundancy like double-execution [2]. Here, the cryptographic computation is performed twice and the output is only released, if the results of both computations match up. While double-execution does prevent the attacks presented so far, a more powerful attacker can still succeed by either inducing a fault that skips the final comparison or by inducing a fault with equivalent effect during both computations. On top of that, Safe Error Attacks (SEA) [34] or Ineffective Fault Analysis (IFA) [13] solely rely on valid outputs of faulted cryptographic computations and hence are unaffected by double-execution.

So far, most fault attacks require the attacker to send specific inputs multiple times to the attacked cryptographic implementation. This raises the question whether or not such attacks also apply to nonce-based authenticated encryption schemes where unique nonces prevent attackers from doing so. Indeed, the feasibility of fault attacks has been shown by Dobraunig, Eichlseder, Korak, Lomné, and Mendel [15] for various block cipher-based authenticated encryption schemes by using Statistical Fault Attacks (SFA) [18]. However, their attacks face some limitations. For instance, they are not applicable in a straight-forward manner to most sponge-based and stream cipher-based authenticated encryption schemes. In our attack, we make use of Statistical Ineffective Fault Attacks (SIFA) [16] that build upon the concepts of both SFA [18] and IFA [13].

2.2 Statistical Ineffective Fault Attacks

The Statistical Ineffective Fault Attack (SIFA) [16] is a technique that exploits distributions of faults that have been induced, but do not affect the outcome of a computation (ineffective faults). Concretely, the effect of an induced fault depends on the values that are currently processed by a device. As a consequence, the distribution of the values where an induced fault does not change the processed value is often biased in practice. This distribution can then be exploited in attacks, which cannot be precluded by popular detection/infection countermeasures [16]. As shown in [14], even additional masking does not preclude such attacks.

To discuss the basic working principles of SIFA, let us consider an encryption where an attacker is able to force (using fault inductions) one specific intermediate value to follow an unknown but non-uniform distribution during the computation. Such fault inductions are rather easy to achieve in practice as it has been shown, e.g. in [16] by using clock glitches for various microprocessors, or in [15] by using lasers on a hardware AES co-processor. If we continuously perform such faulted encryptions we will probably observe plain- or ciphertexts where the fault was ineffective. In those cases, the distribution of the targeted value, where the fault has been ineffective, might also follow a biased/non-uniform distribution.

Once the attacker has collected a sufficiently large set of unaffected plain- or ciphertexts, key recovery can be performed as follows. First, the attacker needs to identify all key bits that are involved in the calculation of the targeted value. Clearly, the time frame of the fault induction has to be either towards the beginning or the end of the encryption such that the targeted value only depends on parts of the key. Hence, when attacking sponge or stream cipher-based authenticated encryption schemes, the usual location for fault inductions is the initialization phase. Next, she calculates the targeted value for each collected unaffected plain- or ciphertext and every possible key candidate. The targeted value should, when calculated using the correct key candidate, follow a non-uniform distribution (which is usually not known to the attacker). In contrast, the calculated distribution for a wrong key guess is typically unrelated to the event that there has been an ineffective fault and hence, is expected to be closer to uniform. As a consequence, we are able to distinguish wrong key guesses from a right key guess. For a detailed description of the working principles of the attack including statistical background and on the effects of faults we refer to [14, 16].

2.3 Authenticated Encryption

An authenticated encryption scheme provides confidentiality and authenticity for a given plaintext. It is usually modeled as a function of four input parameters: a secret key K, unique nonce N, associated data A and plaintext P [26]. The output of authenticated encryption is a tuple that consists of a ciphertext C and tag T:

$$\mathcal{E}(K, N, A, P) = (C, T)$$

The corresponding authenticated decryption takes the following five inputs: a secret key K, unique nonce N, associated data A, ciphertext C and tag T. During decryption T is used to verify the authenticity of A and C. If they are not authentic the original plaintext P is not released and the special error symbol $\perp$ is returned instead:

$$\mathcal{D}(K, N, A, C, T) \in \{P, \perp\}$$

The concrete implementation of authenticated encryption schemes can differ significantly. Currently, many of the popular schemes like GCM [20], CCM [33], EAX [3], and OCB [27] are all based on block ciphers like AES. However, since the announcement of CAESAR [12], we can also see an increasing number of stream cipher-based and sponge-based authenticated encryption schemes. In the next section, we will present two such sponge-based designs: Keyak and Ketje, in more detail, since we will use them to describe the attack and for the practical evaluation.

3 Keyak and Ketje

Keyak [7] and Ketje [6] are sponge-based authenticated encryption schemes. Their design is heavily inspired by the hash function Keccak [4], the winner of the SHA-3 competition. While both schemes make use of variants of the permutation in Keccak, their modes of operation are slightly different. At first, we give a short description of Keccak and its underlying permutation. We then describe how Keyak and Ketje make use of the Keccak permutation in order to build an authenticated encryption scheme.

3.1 Keccak

Keccak is a sponge-based hash function that was selected as the winner of the SHA-3 competition. It is parameterized by the permutation Keccak-f, rate r, and capacity c.

Keccak-f, more precisely denoted by Keccak-$f[b]$, is an iterated permutation that operates on a b-bit state that is organized in 5×5 lanes of 2^l bits where l ranges from 0 to 6. The number of rounds n_r is determined by the width of the permutation and is equal to $12 + 2l$. Keccak-f consists of the 5 operations: $\theta, \rho, \pi, \chi, \iota$ that are applied to the state in the presented order in every round. From these 5 operations χ is the only non-linear transformation. The purpose of θ, π and ρ is to cause diffusion while ι breaks any symmetries.

In the case of Keccak, the lane size l equals 6, thus the state has a size of $5 \times 5 \times 64 = 1600$ bits and the number of rounds n_r is $12 + 2 \times 6 = 24$. Depending on the desired security, c is chosen as twice the desired preimage resistance in bits and $r = 1600 - c$. Following the sponge construction design principle, Keccak can be divided into two phases: an initial absorbing phase and a subsequent squeezing phase. During the absorbing phase input chunks of r bits are repeatedly XOR-ed into the state and subsequently processed by Keccak-f. Once all input chunks have been absorbed, a chunk of the desired hash bit-size can be extracted from the state (squeezing phase).

Besides Keccak-f, a variety of similar permutations Keccak-$p[b, n_r]$ were proposed by the Keccak designers. In contrast to Keccak-f, in Keccak-p the number of rounds n_r does not depend on the state size b anymore and can be set to any positive integer. This allows for more flexibility in the design of Keccak-based cryptographic primitives. The state size b is however still restricted to the same values. Next, we give basic descriptions of the authenticated encryption schemes Keyak and Ketje.

3.2 Keyak

Keyak is an authenticated encryption scheme that uses the *Motorist* mode of operation and is based on the Keccak-p permutation. Even though Keyak supports a parameterized degree of parallelism we limit our description to the (recommended) Lake Keyak variant that does not support parallelization and thus can be used even on constrained devices. Lake Keyak utilizes a 1600-bit state, uses the 12-round Keccak-$p[1600, 12]$ permutation and performs authenticated encryption with 128 to 256 bits of secret key, up to 150 bytes of nonce and 128-bit tags. In the following, we describe the *Motorist* mode of operation, as used in Keyak. Whenever we refer to Keyak we mean Lake Keyak.

Motorist Mode. The *Motorist* mode defines how incoming messages are processed together with key, nonce, associated data and tag in Keyak. It is closely related to the duplex construction [5], with the main difference being the size of the input blocks. While the original duplex construction only allows input blocks as large as the outer part (rate r) of the underlying permutation, *Motorist* uses full-state keyed duplexes [22] that can make use of the full width of the permutation and thus allow higher throughput as shown in Fig. 1.

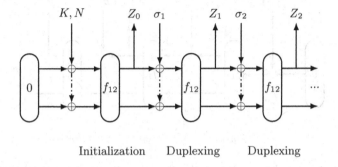

Initialization   Duplexing   Duplexing

Fig. 1. Lake Keyak. f_{12} denotes the Keccak-$p[1600,12]$ permutation, σ denotes the input string, and Z denotes the key stream.

3.3 Ketje

Ketje is an authenticated encryption scheme that consists of 2 parts: The mode of operation *MonkeyWrap* and the Keccak-p permutation. While 4 different versions of Ketje have been proposed by the designers for the 4 different permutation sizes of 200, 400, 800 and 1600 bits, our practical evaluation is performed on Ketje Jr. The main use case of Ketje Jr is lightweight authenticated encryption for constraint devices. Hence, the permutation is based on Keccak-$p[200, n_r]$, meaning that only a rather small 200-bit state is used and the number of permutation rounds n_r is variable. Ketje Jr performs authenticated encryption with a 96-bit secret key and up to 86-bits of nonce. Different to Keccak and Keyak, in Ketje every call of the permutation is slightly twisted. The twisted permutation Keccak-p^* is an extended version of the standard permutation Keccak-p. It always starts with an additional call of π^{-1} and ends with an additional call to π. In the following we describe the *MonkeyWrap* mode of operation, as used in Ketje. Whenever we refer to Ketje we mean Ketje Jr.

Monkey Wrap. The *MonkeyWrap* mode defines how incoming messages are processed together with key, nonce, associated data and tag in Ketje.

The initialization of *MonkeyWrap* is called *Start* which is similar to the *Motorist* mode. First, key K and nonce N are XOR-ed into the zero-initialized state. Then 12 rounds of twisted Keccak-p^* permutation are performed.

The key stream generation *Step* is accomplished by performing duplexing calls, yet this time not the full width of the permutation is utilized, as illustrated in Fig. 2. Since the rate r of the permutation in Ketje is very small only a 1-round twisted Keccak-p^* permutation is needed in between *Step* calls.

Before the extraction of the tag starts, a 6-round twisted Keccak-p^* permutation is performed.

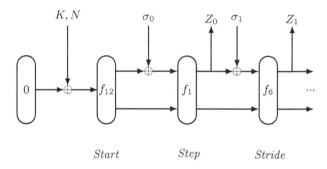

Fig. 2. Ketje Jr. f_{n_r} denotes the application of a n_r-round twisted Keccak-$p^*[200]$ permutation, σ denotes the input string, and Z denotes the key stream.

4 Attack Strategy

In our attack, we target the decryption/verification of Lake Keyak and Ketje Jr ($\mathcal{D}(K, N, A, C, T)$). To be precise, we observe the behavior of the authenticated decryption of valid messages (N, A, C, T) in the presence of faults that are induced during the initialization phase. For both schemes, the initialization is the application of variants of Keccak-f to a state, which is composed out of the secret key K and a publicly known nonce N. If the fault induction affects and changes the outcome of this computation, also the value of the afterwards computed tag T will change compared to the value of the transmitted tag T and thus, the verification will fail. If the induced fault does not change the outcome of the initialization, the verification will succeed and the authenticated decryption will return a plaintext. Please note that the actual plaintext is not needed for the attack, we solely assume that the attacker is able to distinguish a failed verification from a successful one.

As shown in [16], inducing faults in multiple runs of the same computation with differing inputs, followed by a subsequent filtering for unaffected computations, most likely leads to biased distribution in the targeted intermediate value. In our case, unaffected computations (and thus ineffective faults) can be deduced from the condition that the verification succeeds. Hence, we assume that the attacker is able to affect one or multiple bits ($A_{\chi_2}[x, y, z]$) of the internal state before the application of χ in the 2$^\text{nd}$ round of the initialization, so that the distribution of these bits is non-uniform for the filtered inputs (N, A, C, T). More concretely, we assume that the attacker is able to collect several nonces N, which lead to one or multiple biased bits before the 2nd round χ-layer of the initialization.

Out of this knowledge, the attacker is able to extract information about the secret key. In the following section, we give a detailed description of how key recovery is achieved for Keyak. A very similar approach can then be used to perform key recovery for Ketje Jr. The major difference is the fact that a 200-bit permutation is used and hence bits of the equivalent key directly before the application of the 1st round χ-layer are guessed in the attack.

4.1   Involved Bits in Keyak

Information about the secret can be deduced by identifying key bits that are involved in the calculation of $A_{\chi_2}[x, y, z]$ and evaluate the value of $A_{\chi_2}[x, y, z]$ under every possible assignment of the key bits for every previously collected value of the nonce N. For the right key guess, we expect to observe the highest bias in the values of $A_{\chi_2}[x, y, z]$. But at first, we have to identify the involved bits.

First, we need to determine the bits at the input of the linear layer of the 2nd round, which are involved in the calculation of $A_{\chi_2}[x, y, z]$. The linear layer of one round of Keccak-$p[1600, 12]$ consists of the application of the single round functions θ, ρ, and π. The function π just swaps the words, so that

$$A_{\chi_2}[x, y, z] = A_{\pi_2}[(x + 3y) \bmod 5, x, z] .$$

The function ρ rotates each lane by a different offset $R[x, y]$. Hence,

$$A_{\chi_2}[x, y, z] = A_{\rho_2}[(x + 3y) \bmod 5, x, (z - R[(x + 3y) \bmod 5, x]) \bmod 64] .$$

Finally, θ computes its output by XOR-ing each bit with the parity of two columns in the array, thus, one bit $A_{\chi_2}[x, y, z]$ is the sum of 11 input bits to θ.

$$A_{\chi_2}[x, y, z] = A_{\theta_2}[(x + 3y) \bmod 5, x, (z - R[(x + 3y) \bmod 5, x]) \bmod 64]$$
$$\oplus \bigoplus_{y'=0}^{4} A_{\theta_2}[(x + 3y - 1) \bmod 5, y', (z - R[(x + 3y) \bmod 5, x]) \bmod 64]$$
$$\oplus \bigoplus_{y'=0}^{4} A_{\theta_2}[(x + 3y + 1) \bmod 5, y', (z - R[(x + 3y) \bmod 5, x] - 1) \bmod 64]$$

Each of the 11 bits $A_{\theta_2}[x_i, y_i, z_i]$ can be calculated using three input bits to χ. Therefore,

$$A_{\theta_2}[x_i, y_i, z_i] = A_{\chi_1}[x_i, y_i, z_i] \oplus$$
$$((A_{\chi_1}[(x_i + 1) \bmod 5, y_i, z_i] \oplus 1) \cdot A_{\chi_1}[(x_i + 2) \bmod 5, y_i, z_i]) .$$

Note that two bits at the input of θ in the 2nd round needed in the calculation of $A_{\chi_2}[x, y, z]$ are adjacent bits of the same S-box, namely

$$A_{\theta_2}[(x + 3y) \bmod 5, x, (z - R[(x + 3y) \bmod 5, x]) \bmod 64]$$
$$A_{\theta_2}[(x - 3y - 1) \bmod 5, y, (z - R[(x + 3y) \bmod 5, x]) \bmod 64] \ .$$

As a consequence, only 31 bits of $A_{\chi_1}[x_j, y_j, z_j]$ are involved in the calculation of $A_{\chi_2}[x, y, z]$.

The bits at the input to the 1st round that are needed to compute the 31 bits $A_{\chi_1}[x_j, y_j, z_j]$ can be determined in a similar manner as done for the second round. However, doing so for general values of x and y gets a bit clumsy, hence, we focus on the restricted case of calculating $A_{\chi_2}[0, 0, 0]$. The necessary equations are given in Appendix B.

Determining the necessary bits to calculate $A_{\chi_2}[x, y, z]$ by hand is quite time consuming and also error prone. Thus, we have used a search tool [17], which has been developed to search for linear characteristics to identify the bits at the input of the Keccak-f permutation that are involved in the calculation of a certain $A_{\chi_2}[x, y, z]$. In Fig. 3, we give the involved bits for calculating $A_{\chi_2}[0, 0, 0]$. The figure represents one lane as hexadecimal value, where bits that are set to 1 are needed in the calculation of $A_{\chi_2}[0, 0, 0]$. A corresponding figure for Ketje Jr is given in Appendix A.

```
Input to                                    Bit positions

         C-62-C1---9C---1 8E-12----C3----C 6-1--45----E-3-4 -384983--118---6 1---4228184--181
         C-62-C1-189C---- 8E-12----C38---C 661-45----E-3-- -384982--118-3-6 1---5A28184--181
  θ₁     D-62-C1---9C---- 8E212----C3----C 6-1--45----3E-3-- -384986--118---6 1---4228194--181
         C-62-C1---DC---- 8E-12----C34---C 6-11-45----E-3--- -3849C2--118---6 1-8-4228184--181
         C-624C1---9C---- CE-12----C3----C 6-1--45----E-3-8 -384982--118-1-6 1--44228184--181


         ---------------1 8---------------1 8---------------1 8--------------- ---------------1
         ---------------1 8---------------1 8--------------- 8--------------- ---------------1
  χ₁     ---------------1 8---------------1 8--------------- 8--------------- ---------------1
         ---------------1 8---------------1 8--------------- 8--------------- ---------------1
         ---------------1 8---------------1 8--------------- 8--------------- ---------------1


         ---------------1 8--------------- --------------- --------------- ---------------1
         --------------- 8--------------- --------------- --------------- ---------------1
  θ₂     --------------- 8--------------- --------------- --------------- ---------------1
         --------------- 8--------------- --------------- --------------- ---------------1
         --------------- 8--------------- --------------- --------------- ---------------1


         ---------------1 --------------- --------------- --------------- ---------------
         --------------- --------------- --------------- --------------- ---------------
  χ₂     --------------- --------------- --------------- --------------- ---------------
         --------------- --------------- --------------- --------------- ---------------
         --------------- --------------- --------------- --------------- ---------------
```

Fig. 3. Bits involved in calculation of $A_{\chi_2}[0, 0, 0]$. The position of the 128-bit key is highlighted in gray. Zeros are replaced by - to improve readability.

4.2   Recovered Bits

In this section, we will discuss how much information on the key bits can be recovered by exploiting a bias in $A_{\chi_2}[x, y, z]$. For the sake of simplicity, we will stick to the example of $A_{\chi_2}[0, 0, 0]$. Bits having a gray background in Fig. 3 are bits that represent the 128 key bits. Hence, to compute $A_{\chi_2}[0, 0, 0]$, 25 bits of the key have to be guessed. However, from the equation given in Appendix B, we can see that only the 17 bits:

$$A_{\theta_1}[0, 0, 0] \,, A_{\theta_1}[0, 0, 18], A_{\theta_1}[0, 0, 20], A_{\theta_1}[0, 0, 23], A_{\theta_1}[0, 0, 36], A_{\theta_1}[0, 0, 43],$$
$$A_{\theta_1}[0, 0, 53], A_{\theta_1}[0, 0, 54], A_{\theta_1}[1, 0, 2] \,, A_{\theta_1}[1, 0, 20], A_{\theta_1}[1, 0, 21], A_{\theta_1}[1, 0, 27],$$
$$A_{\theta_1}[1, 0, 48], A_{\theta_1}[1, 0, 58], A_{\theta_1}[1, 0, 59], A_{\theta_1}[1, 0, 63], A_{\theta_1}[2, 0, 62]$$

can influence $A_{\chi_2}[0, 0, 0]$ in a non-linear manner, while the 8 bits:

$$A_{\theta_1}[0, 0, 19], A_{\theta_1}[0, 0, 42], A_{\theta_1}[0, 0, 49], A_{\theta_1}[1, 0, 3] \,, A_{\theta_1}[1, 0, 26], A_{\theta_1}[1, 0, 45],$$
$$A_{\theta_1}[1, 0, 57], A_{\theta_1}[2, 0, 61]$$

only have a linear influence.

As a consequence, we can at most uniquely identify the 17 bits that influence $A_{\chi_2}[0, 0, 0]$ in a non-linear way. For the 8-bits that influence $A_{\chi_2}[0, 0, 0]$ in a linear way, only their XOR-sum (parity) effects the value of $A_{\chi_2}[0, 0, 0]$. Since for 8 bits, half of the possible assignments have parity 0 and the other half has parity one, we get at least 2^7 key candidates that always lead to the same result. Please note that this is a rather simplistic evaluation and does not consider the dependencies of the non-linear bits and also the bits, which are used as nonce and constants. In fact, the key recovery depends on the value of these bits, since an unfortunate choices for the nonce can, for instance, lead to situations, where some S-boxes are linearized for some key bits, or some key bits are always blocked, so that they do not influence $A_{\chi_2}[0, 0, 0]$. For instance, let us have a look at the results of one of our concrete experiments given in Sect. 5. Instead of recovering 17 out of the 25 bits uniquely from 2^7 key candidates scoring best, we are able to recover 15 of the 25 bits uniquely out of 2^9 key candidates that score best.

5   Practical Evaluation

We now describe the practical evaluation of our attack on a microprocessor implementation. Although we have performed attacks on both Lake Keyak and Ketje Jr, we limit our description to Lake Keyak, since the attack procedure is similar for both schemes. We do, however, state the results for both schemes at the end of this section. We start this section by giving a quick overview of the attack procedure in Sect. 5.1. We then describe the hardware/software that we have used to perform our attack evaluation in Sect. 5.2. After that, we state requirements on a fault setup more generally in Sect. 5.3. Finally, we present the results of our fault attacks on Lake Keyak and Ketje Jr in Sect. 5.4.

5.1   Attack Procedure

As described in Sect. 4, our key recovery exploits the input of specific Keyak decryptions. We are interested in decryptions that have a bias in one or multiple bits of the Keccak state before χ in the 2nd round. To achieve the required filtering of inputs we use statistical ineffective fault attacks (SIFA), as proposed in [16].

Before the attack we set the secret key of the microprocessor Keyak implementation to a constant and unknown value. During the attack we send inputs, consisting of random nonce and tag, to the microprocessor, induce a clock glitch with constant offset during the computation and observe the behavior. The tag verification is used to detect whether or not an induced fault was ineffective.

5.2   Attack Setup

The practical evaluation of our fault attack was done on an 8-bit Xmega 128D4 microprocessor. The attacked software implementation of Lake Keyak consists of two parts. The first part is a C implementation of the *Motorist* mode of operation. The second part is a fast 8-bit AVR optimized assembler implementation of the Keccak permutation. Both implementations are taken from the Keccak Code Package [8] and therefore represent a good target software implementation for our practical evaluation. The clock signal of the microprocessor is generated by a Spartan-6 FPGA running at 12 MHz. We additionally use this FPGA for the insertion of glitches onto the clock signal. The insertion of clock glitches is achieved by XOR-ing an additional fast clock edge onto the original clock signal at a specified time. By doing so, we can violate the critical path to force undefined behavior of the microprocessor.

In our practical evaluation we can force strong biases in virtually every state bit that is affected by χ, however only in blocks of 8 bits at a time (which is not surprising on a 8-bit architecture). We suspect that our glitch does skip one of the XOR instructions in the bit-sliced χ implementation, but we cannot say for sure though.

5.3   Attack Setup - Requirements

As we use SIFA [16], the requirements we have on the locality and especially the effect of the fault are quite relaxed. Basically, we only need some sort of bias in any bit at the input of χ in the 2nd round. This can be achieved by e.g. faulting instructions in χ, slightly before χ, or by directly faulting registers using lasers. In the case of AES, such fault inductions have already been demonstrated for multiple microprocessors and even for hardware co-processors [15,16]. One way to find a suitable glitch location in practice would be to estimate the clock cycles until the targeted operation is executed. Hence, in our scenario, one can estimate the time frame of the 2nd round and try to induce a glitch in several different clock cycles towards the end of that round.

5.4 Results

Keyak. As already mentioned in Sect. 4.2, when getting a bias in the bit $A_{\chi_2}[0,0,0]$ located at the input of the 2nd round χ-layer, 25 bits of the key are involved in its calculation. In our attack, we guess these 25 bits and evaluate the bias in $A_{\chi_2}[0,0,0]$ for each key guess. Since some of the guessed key bits only influence $A_{\chi_2}[0,0,0]$ in a linear manner, we get several equivalent key candidates having the same bias. As a consequence, Fig. 4 shows the advantage in bits the attacker gets from guessing key candidates down to a bias which also the correct key guess over just randomly guessing the key, which is $\log_2(\#\text{total keys}) - \log_2(\#\text{candidate keys})$.

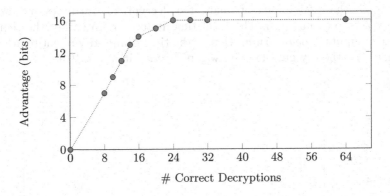

Fig. 4. Attack on Keyak. Advantage in bits when targeting $A_{\chi_2}[0,0,0]$ and guessing the associated 25 bits of the 128-bit key.

As shown in Fig. 4, 24 inputs of such unaffected decryptions are necessary to get a maximum advantage of 16 bits. In our case, we get 2^9 keys ranked top that have the same bias (not considering its sign). From those 2^9 keys, the values of 15 key bits can be uniquely determined. Due to the architecture of the implementation, we do not only get a bias in one bit, but one byte. By combining this information, we can uniquely determine 82-bits of the key.

In our attack setup, we are able to perform about 20 faulted decryptions per second. According to the practical evaluation, in about 1 out of 250 decryptions the induced fault is ineffective. The total time it took us to gather the required amount of inputs is roughly 5 min.

Ketje. In the attack on Ketje Jr we use the same fault location as in the attack on Lake Keyak. This is however not strictly necessary. Even though both schemes use variants of Keccak-f during initialization, the influence of key bits on one of our biased bits before χ in the 2nd round is quite different, mainly due to the fact that the lane sizes are different (see Fig. 6). In contrast to Lake Keyak, in Ketje Jr nearly all key bits influence each of our biased bits, most of the time in a linear way.

Hence, for Ketje Jr we instead guess the 200-bit equivalent key before χ in the 1st round (i.e. after the first linear layer). By doing so we can reduce the dependency on the equivalent key to 31 bit and guessing becomes feasible in practice.

In our attack setup we can recover about 19 bits of the equivalent key that correspond to one biased bit in about 10 h using a single thread on an Intel Xeon CPU. Note that this time can be significantly improved, since we used for our evaluation purposes just the unoptimized reference implementation. Furthermore, the task of key guessing can be parallelized trivially. If we parallelize the computations for e.g. the 8 bits that were affected by our fault induction we can recover 152 bits of the equivalent key in the same amount of time. The remaining bits can be determined either by brute-force or repeating key recovery for a different fault location.

In total, again 24 inputs of unaffected decryptions are necessary for key recovery as shown in Fig. 5. The total time it took us to gather the required amount of inputs is below 5 min. Hence, the time complexity of entire attack is dominated by the key guessing and was performed in about 10 h.

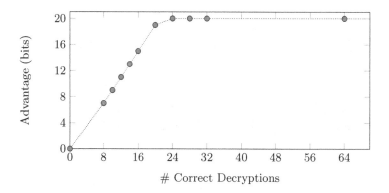

Fig. 5. Attack on Ketje. Advantage in bits when targeting $A_{\chi_2}[0, 0, 0]$ and guessing the associated 31 bits of the 200-bit equivalent key.

6   Conclusion

In this work, we present the first fault attacks targeting a broad range of nonce-based authenticated encryption schemes. While fault attacks on authenticated encryption have already been shown at Asiacrypt 2016 [15], this attack is mostly limited to schemes that additionally feature a final key addition and thus, is not directly applicable to most sponge-based or stream cipher-based constructions. We close this gap and show attacks based on SIFA [16], which are in principle applicable to most nonce-based authenticated encryption schemes that perform some sort of initialization where the nonce (or an other publicly known input) is mixed with the secret key. Since we only need to know whether a fault induction was ineffective or not, attacking the decryption function of authenticated encryption schemes gives us a perfect oracle. Our attack evaluation is focused on

Keyak and Ketje, however, we conjecture that our attack can also be adopted to other schemes like the CAESAR finalists ACORN, AEGIS, Ascon, MORUS, etc. in a rather straight-forward way.

SIFA is resistant to popular fault countermeasures like double-execution and infection-based countermeasures as shown in [16]. Even additional masking does not preclude this attack vector [14]. The key recovery is capable of dealing with an arbitrary amount of noise (however requiring more faulted decryptions) that might arise due to possibly imperfect fault inductions. The effort required to perform our attack is rather low. We neither require perfectly timed faults nor precise knowledge about the effect of the induced fault. In our fault setup we are able to collect enough material for key recovery within 5 min. The actual key recovery for Keyak and Ketje is easily parallelizable and takes about 30 min and 10 h, respectively. The hardware cost of the attack setup does not exceed 300$.

Acknowledgments. This project has received funding in part from the European Research Council (ERC) under the European Union's Horizon 2020 research and innovation programme (grant agreement No 681402) and by the Austrian Research Promotion Agency (FFG) via the project ESPRESSO, which is funded by the province of Styria and the Business Promotion Agencies of Styria and Carinthia.

A   Bits Involved in the Calculation for Ketje Jr

Input to	Bit positions
θ_1	ff bf 7f bf fb
	fe bf 7f bf fb
	fe bf 7f ff fb
	fe bf 7f bf fb
	fe ff 7f bf ff
χ_1	-1 81 81 8- -1
	-1 81 8- 8- -1
	-1 81 8- 8- -1
	-1 81 8- 8- -1
	-1 81 8- 8- -1
θ_2	-1 8- -- -- -1
	-- 8- -- -- -1
	-- 8- -- -- -1
	-- 8- -- -- -1
	-- 8- -- -- -1
χ_2	-1 -- -- -- --
	-- -- -- -- --
	-- -- -- -- --
	-- -- -- -- --
	-- -- -- -- --

Fig. 6. Bits involved in calculation of $A_{\chi_2}[0,0,0]$. Zeros are replaced by - to improve readability.

B   Equations to Calculate $A_{\chi_2}[0,0,0]$

$$A_{\chi_2}[0,0,0] = A_{\theta_2}[0,0,0] \oplus \bigoplus_{y'=0}^{4} A_{\theta_2}[4,y',0] \oplus \bigoplus_{y'=0}^{4} A_{\theta_2}[1,y',63]$$

$$A_{\theta_2}[0,0,0] = A_{\chi_1}[0,0,0] \oplus ((A_{\chi_1}[1,0,0] \oplus 1) \cdot A_{\chi_1}[2,0,0])$$
$$A_{\theta_2}[4,0,0] = A_{\chi_1}[4,0,0] \oplus ((A_{\chi_1}[0,0,0] \oplus 1) \cdot A_{\chi_1}[1,0,0])$$
$$A_{\theta_2}[4,1,0] = A_{\chi_1}[4,1,0] \oplus ((A_{\chi_1}[0,1,0] \oplus 1) \cdot A_{\chi_1}[1,1,0])$$
$$A_{\theta_2}[4,2,0] = A_{\chi_1}[4,2,0] \oplus ((A_{\chi_1}[0,2,0] \oplus 1) \cdot A_{\chi_1}[1,2,0])$$
$$A_{\theta_2}[4,3,0] = A_{\chi_1}[4,3,0] \oplus ((A_{\chi_1}[0,3,0] \oplus 1) \cdot A_{\chi_1}[1,3,0])$$
$$A_{\theta_2}[4,4,0] = A_{\chi_1}[4,4,0] \oplus ((A_{\chi_1}[0,4,0] \oplus 1) \cdot A_{\chi_1}[1,4,0])$$

$$A_{\theta_2}[1,0,63] = A_{\chi_1}[1,0,63] \oplus ((A_{\chi_1}[2,0,63] \oplus 1) \cdot A_{\chi_1}[3,0,63])$$
$$A_{\theta_2}[1,1,63] = A_{\chi_1}[1,1,63] \oplus ((A_{\chi_1}[2,1,63] \oplus 1) \cdot A_{\chi_1}[3,1,63])$$
$$A_{\theta_2}[1,2,63] = A_{\chi_1}[1,2,63] \oplus ((A_{\chi_1}[2,2,63] \oplus 1) \cdot A_{\chi_1}[3,2,63])$$
$$A_{\theta_2}[1,3,63] = A_{\chi_1}[1,3,63] \oplus ((A_{\chi_1}[2,3,63] \oplus 1) \cdot A_{\chi_1}[3,3,63])$$
$$A_{\theta_2}[1,4,63] = A_{\chi_1}[1,4,63] \oplus ((A_{\chi_1}[2,4,63] \oplus 1) \cdot A_{\chi_1}[3,4,63])$$

$$A_{\chi_1}[0,0,0] = A_{\theta_1}[0,0,0] \oplus \bigoplus_{y'=0}^{4} A_{\theta_1}[4,y',0] \oplus \bigoplus_{y'=0}^{4} A_{\theta_1}[1,y',63]$$

$$A_{\chi_1}[1,0,0] = A_{\theta_1}[1,1,20] \oplus \bigoplus_{y'=0}^{4} A_{\theta_1}[0,y',20] \oplus \bigoplus_{y'=0}^{4} A_{\theta_1}[2,y',19]$$

$$A_{\chi_1}[2,0,0] = A_{\theta_1}[2,2,21] \oplus \bigoplus_{y'=0}^{4} A_{\theta_1}[1,y',21] \oplus \bigoplus_{y'=0}^{4} A_{\theta_1}[3,y',20]$$

$$A_{\chi_1}[4,0,0] = A_{\theta_1}[4,4,50] \oplus \bigoplus_{y'=0}^{4} A_{\theta_1}[3,y',50] \oplus \bigoplus_{y'=0}^{4} A_{\theta_1}[0,y',49]$$

$$A_{\chi_1}[4,1,0] = A_{\theta_1}[2,4,3] \oplus \bigoplus_{y'=0}^{4} A_{\theta_1}[1,y',3] \oplus \bigoplus_{y'=0}^{4} A_{\theta_1}[3,y',2]$$

$$A_{\chi_1}[0,1,0] = A_{\theta_1}[3,0,36] \oplus \bigoplus_{y'=0}^{4} A_{\theta_1}[2,y',36] \oplus \bigoplus_{y'=0}^{4} A_{\theta_1}[4,y',35]$$

$$A_{\chi_1}[1,1,0] = A_{\theta_1}[4,1,44] \oplus \bigoplus_{y'=0}^{4} A_{\theta_1}[3,y',44] \oplus \bigoplus_{y'=0}^{4} A_{\theta_1}[0,y',43]$$

$$A_{\chi_1}[4,2,0] = A_{\theta_1}[0,4,46] \oplus \bigoplus_{y'=0}^{4} A_{\theta_1}[4,y',46] \oplus \bigoplus_{y'=0}^{4} A_{\theta_1}[1,y',45]$$

$$A_{\chi_1}[0,2,0] = A_{\theta_1}[1,0,63] \oplus \bigoplus_{y'=0}^{4} A_{\theta_1}[0,y',63] \oplus \bigoplus_{y'=0}^{4} A_{\theta_1}[2,y',62]$$

$$A_{\chi_1}[1,2,0] = A_{\theta_1}[2,1,58] \oplus \bigoplus_{y'=0}^{4} A_{\theta_1}[1,y',58] \oplus \bigoplus_{y'=0}^{4} A_{\theta_1}[3,y',57]$$

$$A_{\chi_1}[4,3,0] = A_{\theta_1}[3,4,8] \oplus \bigoplus_{y'=0}^{4} A_{\theta_1}[2,y',8] \oplus \bigoplus_{y'=0}^{4} A_{\theta_1}[4,y',7]$$

$$A_{\chi_1}[0,3,0] = A_{\theta_1}[4,0,37] \oplus \bigoplus_{y'=0}^{4} A_{\theta_1}[3,y',37] \oplus \bigoplus_{y'=0}^{4} A_{\theta_1}[0,y',36]$$

$$A_{\chi_1}[1,3,0] = A_{\theta_1}[0,1,28] \oplus \bigoplus_{y'=0}^{4} A_{\theta_1}[4,y',28] \oplus \bigoplus_{y'=0}^{4} A_{\theta_1}[1,y',27]$$

$$A_{\chi_1}[4,4,0] = A_{\theta_1}[1,4,62] \oplus \bigoplus_{y'=0}^{4} A_{\theta_1}[0,y',62] \oplus \bigoplus_{y'=0}^{4} A_{\theta_1}[2,y',61]$$

$$A_{\chi_1}[0,4,0] = A_{\theta_1}[2,0,2] \oplus \bigoplus_{y'=0}^{4} A_{\theta_1}[1,y',2] \oplus \bigoplus_{y'=0}^{4} A_{\theta_1}[3,y',1]$$

$$A_{\chi_1}[1,4,0] = A_{\theta_1}[3,1,9] \oplus \bigoplus_{y'=0}^{4} A_{\theta_1}[2,y',9] \oplus \bigoplus_{y'=0}^{4} A_{\theta_1}[4,y',8]$$

$$A_{\chi_1}[1, 0, 63] = A_{\theta_1}[1, 1, 19] \oplus \bigoplus_{y'=0}^{4} A_{\theta_1}[0, y', 19] \oplus \bigoplus_{y'=0}^{4} A_{\theta_1}[2, y', 18]$$

$$A_{\chi_1}[2, 0, 63] = A_{\theta_1}[2, 2, 20] \oplus \bigoplus_{y'=0}^{4} A_{\theta_1}[1, y', 20] \oplus \bigoplus_{y'=0}^{4} A_{\theta_1}[3, y', 19]$$

$$A_{\chi_1}[3, 0, 63] = A_{\theta_1}[3, 3, 42] \oplus \bigoplus_{y'=0}^{4} A_{\theta_1}[2, y', 42] \oplus \bigoplus_{y'=0}^{4} A_{\theta_1}[4, y', 41]$$

$$A_{\chi_1}[1, 1, 63] = A_{\theta_1}[4, 1, 43] \oplus \bigoplus_{y'=0}^{4} A_{\theta_1}[3, y', 43] \oplus \bigoplus_{y'=0}^{4} A_{\theta_1}[0, y', 42]$$

$$A_{\chi_1}[2, 1, 63] = A_{\theta_1}[0, 2, 60] \oplus \bigoplus_{y'=0}^{4} A_{\theta_1}[4, y', 60] \oplus \bigoplus_{y'=0}^{4} A_{\theta_1}[1, y', 59]$$

$$A_{\chi_1}[3, 1, 63] = A_{\theta_1}[1, 3, 18] \oplus \bigoplus_{y'=0}^{4} A_{\theta_1}[0, y', 18] \oplus \bigoplus_{y'=0}^{4} A_{\theta_1}[2, y', 17]$$

$$A_{\chi_1}[1, 2, 63] = A_{\theta_1}[2, 1, 57] \oplus \bigoplus_{y'=0}^{4} A_{\theta_1}[1, y', 57] \oplus \bigoplus_{y'=0}^{4} A_{\theta_1}[3, y', 56]$$

$$A_{\chi_1}[2, 2, 63] = A_{\theta_1}[3, 2, 38] \oplus \bigoplus_{y'=0}^{4} A_{\theta_1}[2, y', 38] \oplus \bigoplus_{y'=0}^{4} A_{\theta_1}[4, y', 37]$$

$$A_{\chi_1}[3, 2, 63] = A_{\theta_1}[4, 3, 55] \oplus \bigoplus_{y'=0}^{4} A_{\theta_1}[3, y', 55] \oplus \bigoplus_{y'=0}^{4} A_{\theta_1}[0, y', 54]$$

$$A_{\chi_1}[1, 3, 63] = A_{\theta_1}[0, 1, 27] \oplus \bigoplus_{y'=0}^{4} A_{\theta_1}[4, y', 27] \oplus \bigoplus_{y'=0}^{4} A_{\theta_1}[1, y', 26]$$

$$A_{\chi_1}[2, 3, 63] = A_{\theta_1}[1, 2, 53] \oplus \bigoplus_{y'=0}^{4} A_{\theta_1}[0, y', 53] \oplus \bigoplus_{y'=0}^{4} A_{\theta_1}[2, y', 52]$$

$$A_{\chi_1}[3, 3, 63] = A_{\theta_1}[2, 3, 48] \oplus \bigoplus_{y'=0}^{4} A_{\theta_1}[1, y', 48] \oplus \bigoplus_{y'=0}^{4} A_{\theta_1}[3, y', 47]$$

$$A_{\chi_1}[1,4,63] = A_{\theta_1}[3,1,8] \oplus \bigoplus_{y'=0}^{4} A_{\theta_1}[2,y',8] \oplus \bigoplus_{y'=0}^{4} A_{\theta_1}[4,y',7]$$

$$A_{\chi_1}[2,4,63] = A_{\theta_1}[4,2,24] \oplus \bigoplus_{y'=0}^{4} A_{\theta_1}[3,y',24] \oplus \bigoplus_{y'=0}^{4} A_{\theta_1}[0,y',23]$$

$$A_{\chi_1}[3,4,63] = A_{\theta_1}[0,3,22] \oplus \bigoplus_{y'=0}^{4} A_{\theta_1}[4,y',22] \oplus \bigoplus_{y'=0}^{4} A_{\theta_1}[1,y',21]$$

References

1. Anceau, S., Bleuet, P., Clédière, J., Maingault, L., Rainard, J., Tucoulou, R.: Nanofocused X-ray beam to reprogram secure circuits. In: Fischer, W., Homma, N. (eds.) CHES 2017. LNCS, vol. 10529, pp. 175–188. Springer, Cham (2017). https://doi.org/10.1007/978-3-319-66787-4_9

2. Bar-El, H., Choukri, H., Naccache, D., Tunstall, M., Whelan, C.: The sorcerer's apprentice guide to fault attacks. Proc. IEEE **94**(2), 370–382 (2006). https://doi.org/10.1109/JPROC.2005.862424

3. Bellare, M., Rogaway, P., Wagner, D.A.: EAX: a conventional authenticated-encryption mode. Cryptology ePrint Archive, Report 2003/069 (2003). http://eprint.iacr.org/2003/069

4. Bertoni, G., Daemen, J., Peeters, M., Van Assche, G.: The Keccak SHA-3 submission (Version 3.0) (2011). http://keccak.noekeon.org/Keccak-submission-3.pdf

5. Bertoni, G., Daemen, J., Peeters, M., Van Assche, G.: Duplexing the sponge: single-pass authenticated encryption and other applications. In: Miri, A., Vaudenay, S. (eds.) SAC 2011. LNCS, vol. 7118, pp. 320–337. Springer, Heidelberg (2012). https://doi.org/10.1007/978-3-642-28496-0_19

6. Bertoni, G., Daemen, J., Peeters, M., Van Assche, G., Van Keer, R.: CAESAR submission: Ketje v2. https://keccak.team/files/Ketjev2-doc2.0.pdf

7. Bertoni, G., Daemen, J., Peeters, M., Van Assche, G., Van Keer, R.: CAESAR submission: Keyak v2. https://keccak.team/files/Keyakv2-doc2.2.pdf

8. Bertoni, G., Daemen, J., Peeters, M., Van Assche, G., Van Keer, R.: Keccak code package. https://github.com/gvanas/KeccakCodePackage. Accessed 05 Dec 2017

9. Biehl, I., Meyer, B., Müller, V.: Differential fault attacks on elliptic curve cryptosystems. In: Bellare, M. (ed.) CRYPTO 2000. LNCS, vol. 1880, pp. 131–146. Springer, Heidelberg (2000). https://doi.org/10.1007/3-540-44598-6_8

10. Biham, E., Shamir, A.: Differential fault analysis of secret key cryptosystems. In: Kaliski, B.S. (ed.) CRYPTO 1997. LNCS, vol. 1294, pp. 513–525. Springer, Heidelberg (1997). https://doi.org/10.1007/BFb0052259

11. Boneh, D., DeMillo, R.A., Lipton, R.J.: On the importance of checking cryptographic protocols for faults. In: Fumy, W. (ed.) EUROCRYPT 1997. LNCS, vol. 1233, pp. 37–51. Springer, Heidelberg (1997). https://doi.org/10.1007/3-540-69053-0_4

12. CAESAR committee: CAESAR: Competition for authenticated encryption: Security, applicability, and robustness (2014). http://competitions.cr.yp.to/caesar.html

13. Clavier, C.: Secret external encodings do not prevent transient fault analysis. In: Paillier, P., Verbauwhede, I. (eds.) CHES 2007. LNCS, vol. 4727, pp. 181–194. Springer, Heidelberg (2007). https://doi.org/10.1007/978-3-540-74735-2_13

14. Dobraunig, C., Eichlseder, M., Gross, H., Mangard, S., Mendel, F., Primas, R.: Statistical ineffective fault attacks on masked AES with fault countermeasures. Cryptology ePrint Archive, Report 2018/357 (2018). https://eprint.iacr.org/2018/357

15. Dobraunig, C., Eichlseder, M., Korak, T., Lomné, V., Mendel, F.: Statistical fault attacks on nonce-based authenticated encryption schemes. In: Cheon, J.H., Takagi, T. (eds.) ASIACRYPT 2016. LNCS, vol. 10031, pp. 369–395. Springer, Heidelberg (2016). https://doi.org/10.1007/978-3-662-53887-6_14

16. Dobraunig, C., Eichlseder, M., Korak, T., Mangard, S., Mendel, F., Primas, R.: SIFA: exploiting ineffective fault inductions on symmetric cryptography. In: IACR Transactions on Cryptographic Hardware and Embedded Systems, vol. 2018, no. 3, pp. 547–572, August 2018. https://tches.iacr.org/index.php/TCHES/article/view/7286

17. Dobraunig, C., Eichlseder, M., Mendel, F.: Heuristic tool for linear cryptanalysis with applications to CAESAR candidates. In: Iwata, T., Cheon, J.H. (eds.) ASIACRYPT 2015. LNCS, vol. 9453, pp. 490–509. Springer, Heidelberg (2015). https://doi.org/10.1007/978-3-662-48800-3_20

18. Fuhr, T., Jaulmes, É., Lomné, V., Thillard, A.: Fault attacks on AES with faulty ciphertexts only. In: Fischer, W., Schmidt, J.M. (eds.) FDTC 2013, pp. 108–118. IEEE Computer Society (2013)

19. Maurine, P.: Techniques for EM fault injection: equipments and experimental results. In: Bertoni, G., Gierlichs, B. (eds.) FDTC 2012, pp. 3–4. IEEE Computer Society (2012)

20. McGrew, D.A., Viega, J.: The security and performance of the Galois/Counter mode (GCM) of operation. In: Canteaut, A., Viswanathan, K. (eds.) INDOCRYPT 2004. LNCS, vol. 3348, pp. 343–355. Springer, Heidelberg (2004). https://doi.org/10.1007/978-3-540-30556-9_27

21. McKay, K.A., Bassham, L., Turan, M.S., Mouha, N.: NISTIR 8114: report on lightweight cryptography (2017). https://doi.org/10.6028/NIST.IR.8114

22. Mennink, B., Reyhanitabar, R., Vizár, D.: Security of full-state keyed sponge and duplex: applications to authenticated encryption. In: Iwata, T., Cheon, J.H. (eds.) ASIACRYPT 2015. LNCS, vol. 9453, pp. 465–489. Springer, Heidelberg (2015). https://doi.org/10.1007/978-3-662-48800-3_19

23. National Institute of Standards and Technology: FIPS PUB 202: SHA-3 Standard: Permutation-Based Hash and Extendable-Output Functions. Federal Information Processing Standards Publication 202, U.S. Department of Commerce, August 2015. http://nvlpubs.nist.gov/nistpubs/FIPS/NIST.FIPS.202.pdf

24. National Institute of Standards and Technology: DRAFT submissionrequirements and evaluation criteria for the lightweight cryptographystandardization process (2018). https://csrc.nist.gov/CSRC/media/Projects/Lightweight-Cryptography/documents/Draft-LWC-Submission-Requirements-April2018.pdf

25. Piret, G., Quisquater, J.-J.: A differential fault attack technique against SPN structures, with application to the AES and KHAZAD. In: Walter, C.D., Koç, Ç.K., Paar, C. (eds.) CHES 2003. LNCS, vol. 2779, pp. 77–88. Springer, Heidelberg (2003). https://doi.org/10.1007/978-3-540-45238-6_7

26. Rogaway, P.: Authenticated-encryption with associated-data. In: CCS 2002, pp. 98–107. ACM (2002)

27. Rogaway, P., Bellare, M., Black, J., Krovetz, T.: OCB: a block-cipher mode of operation for efficient authenticated encryption. In: Reiter, M.K., Samarati, P. (eds.) CCS 2001, pp. 196–205. ACM (2001)

28. Ronen, E., Shamir, A., Weingarten, A.O., O'Flynn, C.: IoT goes nuclear: creating a ZigBee chain reaction. In: SP 2017, pp. 195–212. IEEE Computer Society (2017)
29. Saha, D., Chowdhury, D.R.: SCOPE: on the side channel vulnerability of releasing unverified plaintexts. In: Dunkelman, O., Keliher, L. (eds.) SAC 2015. LNCS, vol. 9566, pp. 417–438. Springer, Cham (2016). https://doi.org/10.1007/978-3-319-31301-6_24
30. Saha, D., Chowdhury, D.R.: ENCOUNTER: on breaking the nonce barrier in differential fault analysis with a case-study on PAEQ. In: Gierlichs, B., Poschmann, A. (eds.) CHES 2016. LNCS, vol. 9813, pp. 581–601. Springer, Heidelberg (2016). https://doi.org/10.1007/978-3-662-53140-2_28
31. Saha, D., Kuila, S., Roy Chowdhury, D.: EscAPE: diagonal fault analysis of APE. In: Meier, W., Mukhopadhyay, D. (eds.) INDOCRYPT 2014. LNCS, vol. 8885, pp. 197–216. Springer, Cham (2014). https://doi.org/10.1007/978-3-319-13039-2_12
32. Skorobogatov, S.P., Anderson, R.J.: Optical fault induction attacks. In: Kaliski, B.S., Koç, K., Paar, C. (eds.) CHES 2002. LNCS, vol. 2523, pp. 2–12. Springer, Heidelberg (2003). https://doi.org/10.1007/3-540-36400-5_2
33. Whiting, D., Housley, R., Ferguson, N.: Counter with CBC-MAC (CCM) (2003)
34. Yen, S.M., Joye, M.: Checking before output may not be enough against fault-based cryptanalysis. IEEE Trans. Comput. **49**(9), 967–970 (2000)

Post-Quantum Cryptography

EFLASH: A New Multivariate Encryption Scheme

Ryann Cartor[1]([✉]) and Daniel Smith-Tone[1,2]

[1] Department of Mathematics, University of Louisville, Louisville, KY, USA
`ryann.cartor@louisville.edu`
[2] National Institute of Standards and Technology, Gaithersburg, MD, USA
`daniel.smith@nist.gov`

Abstract. Multivariate Public Key Cryptography is a leading option for security in a post quantum society. In this paper we propose a new encryption scheme, EFLASH, and analyze its efficiency and security.

Keywords: Multivariate cryptography · HFE · PFLASH
Discrete differential · MinRank

1 Introduction

In the 1990's, Peter Shor developed a polynomial time algorithm to factor and compute discrete logarithms using a quantum computer. This discovery has changed the focus of the future of cryptography. With large scale quantum computing increasingly being viewed as an inevitability, as opposed to a mere possibility, research in the field of post-quantum cryptography is more important than ever.

A plethora of possible post-quantum cryptosystems have been proposed at this time, including (but not limited to) lattice-based cryptosystems, code-based cryptosystems, multivariate cryptosystems, and hash-based signatures. Each of these areas rely on mathematical problems for which there is no obvious quantum advantage. In this article, we focus on the application of multivariate cryptography to secure encryption.

1.1 Recent History of Multivariate Encryption

Multivariate encryption has had a complicated history, with an increase in activity in the recent past. These schemes are composed of systems of multivariate quadratic polynomials over a finite field $\mathbb{F}$. The security of these schemes is based on the MQ-problem, the problem of solving systems of quadratic equations over a field, which is known to be NP-hard. This fact suggests that the problem remains hard even for quantum computers.

Recently we have seen new candidates and strategies emerge for multivariate encryption. Previously, multivariate schemes centered around bijective functions

© Springer Nature Switzerland AG 2019
C. Cid and M. J. Jacobson, Jr. (Eds.): SAC 2018, LNCS 11349, pp. 281–299, 2019.
https://doi.org/10.1007/978-3-030-10970-7_13

that map from vector spaces of size n back into a vector space of size n. The problem with this strategy is that there are not many bijective quadratic maps. Furthermore, of the maps that do exist, many of these functions were either too hard to invert, or too easy to invert. The common practice to try to overcome this downfall was to try to hide an easily invertible function by composing the bijective function with affine maps.

In 2013, Tao et al. proposed relaxing the *bijective* condition for the central function and replacing it with an *injective* map with a much larger codomain in [1]. In theory, this would make hiding the structure of the map while maintaining efficient inversion easier to accomplish. The recent resurgence of multivariate encryption is due primarily to this change in philosophy. Many schemes have been proposed along these apparently promising lines.

Some notable schemes that increase the codomain size of the central mappings include the ABC Simple Matrix scheme, see [1], which utilizes a large matrix algebra structure; ZHFE, see [2], which is similar to a high degree version of HFE with a single variable over the extension; and SRP, see [3], which combines the Square encryption scheme, Rainbow signature scheme, and Plus method. Although these schemes appear promising, many of these schemes have subsequently been the victims of surprising (if not disabling) cryptanalysis. The attacks on ABC from [4–6] work well if the base field is small, and both ZHFE and SRP were broken in [7] and [8], respectively.

1.2   Our Contribution

We propose a new encryption system, EFLASH, based on a primitive with strong security results. The scheme is a projected C^{*-} scheme with a parameterization effective for encryption. This scheme also follows the philosophy of increasing the size of the codomain to avoid ciphertext collision. We accomplish this increase in codomain size by replacing the traditional projection with an embedding into a larger space.

This construction introduces challenges that a projected C^{*-} signature scheme does not have to address. Since valid decryption requires a unique preimage, it is a requirement that there is a single assignment of the missing coordinates of the output of the central map corresponding to a valid input. Thus, for constant time implementations, every such assignment of coordinates must be computed. We introduce a new method of decryption satisfying these constraints in realistic amounts of time.

1.3   Organization of Paper

The paper is organized as follows. The section following the introduction introduces the idea of big field schemes and describes relevant big field schemes, namely C^{*}, PFLASH and HFE. Then the subsequent section outlines the cryptanalytic techniques that have had the most success attacking big field schemes. After that, we introduce the algebraic structure of our scheme in Sect. 4 where we discuss the algebraic aspects of EFLASH and methods for encryption and decryption. Finally, we discuss the resistance to relevant attacks and parameter selection for EFLASH.

2   Big Field Schemes

EFLASH belongs to a family of multivariate cryptosystems known as "big field" schemes. These schemes rely on the multiplicative structure of a degree d extension $\mathbb{F}_{q^d}$ of the finite field $\mathbb{F}_q$. Let $\phi : \mathbb{F}_q^d \to \mathbb{F}_{q^d}$ be a vector space isomorphism (we will also denote $\mathbb{F}_{q^d}$ as $\mathbb{K}$). Notice that univariate monomials of the form $X^{q^i+q^j}$ in $\mathbb{F}_{q^d}[X]$ are the product of two Frobenius automorphisms over $\mathbb{F}_q$, and hence are the product of two $\mathbb{F}_q$-linear functions. Thus $\phi^{-1} \circ X^{q^i+q^j} \circ \phi$ is coordinate-wise quadratic when expressed over $\mathbb{F}_q$. Thus functions of the form

$$\sum_{0 \le i,j < d} \alpha_{ij} X^{q^i+q^j}$$

are said to be $\mathbb{F}_q$-quadratic.

To disguise the structure of the central map of such schemes one applies a morphism of polynomials, essentially choosing random linear maps mixing the input and output spaces of the central map. Formally, we define these morphisms as follows.

Definition 1. *A* polynomial morphism *is a map between two systems of polynomials,* $F : \mathbb{F}_q^d \to \mathbb{F}_q^d$ *and* $P : \mathbb{F}_q^n \to \mathbb{F}_q^m$ *defined by a pair of affine maps* $T : \mathbb{F}_q^d \to \mathbb{F}_q^m$ *and* $U : \mathbb{F}_q^n \to \mathbb{F}_q^d$ *such that* $P = T \circ F \circ U$. *If both* T *and* U *are invertible, then the morphism is said to be an isomorphism and* F *and* P *are said to be isomorphic.*

The following diagram illustrates the entire construction utilizing the big field.

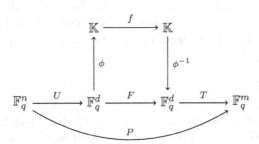

2.1 C^*

Matsumoto and Imai introduced the C^* scheme in [9] at Eurocrypt '88, effectively introducing the world to massively multivariate cryptography. The scheme uses a big field construction where the quadratic monomial map $f : \mathbb{K} \to \mathbb{K}$ is defined by $f(x) = x^{q^\theta+1}$ and is hidden by a polynomial isomorphism. The public key for the scheme is given by $P = T \circ \phi^{-1} \circ f \circ \phi \circ U$.

Encryption of a plaintext is accomplished by evaluating the public polynomials P at an encoding of the plaintext x, and is thus very efficient. Decryption is accomplished by inverting each of the three component maps individually. The inversion of $v = f(u)$ is performed by solving $h(q^\theta + 1) = 1 (\mod q^n - 1)$, and calculating $u = v^h$. This process can be cumbersome, depending on the degree of extension and the exponent θ.

2.2   PFLASH

Following the break of C^*, efforts to modify the scheme to add security lead to the discovery of PFLASH, introduced in [10]. The PFLASH scheme is a specific parametrization of a projected C^{*-} scheme. Both the projection and minus modifiers were initially proposed in relation to C^* in [11]. The purpose of the projection modifier is to change the simplicity of the central map by fixing the value of d input variables. The composition of the projection and an affine map U create a projection onto a codimension d hyperplane. The minus modifier eliminates r equations from the public key. Note that the composition of the minus projection with the affine map T has corank r. The public key of PFLASH(q, n, r, d) is given by $P(\overline{x}) = \pi_r \circ T \circ \phi^{-1} \circ f \circ \phi \circ U \circ \pi_d(\overline{x})$.

The scheme works as a digital signature primitive. To verify a signature, an individual evaluates the public polynomials at the given signature. To create a signature, the signer finds a preimage of each of the private maps. In order to find a preimage of $\pi_r \circ T \circ \phi^{-1}$, randomly append r values to the message, then apply T^{-1} and ϕ. After inverting f, an element that is in the preimage of $\phi \circ U$ and in the image of π_d is selected as the signature.

PFLASH has strong security arguments, including a proof of security against differential attacks that can be found in [12]. Due to the modifications of the scheme, the public key is not isomorphic to the private monomial function, but rather only a polynomial morphism exists between the central map and the public key. As shown in [13], the morphism of polynomials problem is NP-hard, which gives hope that the information lost to the public key may secure the scheme.

2.3   HFE

Another descendent of the C^* scheme is the Hidden Field Equation (HFE) scheme of [14]. HFE replaces the monomial map of the C^* scheme with a more general polynomial with a degree bound D.

Given $\mathbb{K}$, the degree n extension of $\mathbb{F}$, a quadratic polynomial $f : \mathbb{K} \to \mathbb{K}$ with degree bound D is chosen. The function f has the following form:

$$f(x) = \sum_{\substack{i \le j \\ q^i+q^j \le D}} \alpha_{i,j} x^{q^i+q^j} + \sum_{\substack{i \\ q^i \le D}} \beta_i x^{q^i} + \gamma,$$

where $\alpha_{i,j}, \beta_i, \gamma \in \mathbb{K}$. The public key is then constructed via the isomorphism:

$$P = T \circ \phi^{-1} \circ f \circ \phi \circ U.$$

Inversion for this scheme is achieved by taking a ciphertext $\overline{y} = P(\overline{x})$ and computing $v = \phi \circ T^{-1}(\overline{y})$. The next step is to solve $v = f(u)$ for u via the Berlekamp algorithm, see [15], and finally recovering $\overline{x} = U^{-1} \circ \phi^{-1}(u)$.

3 Cryptanalyses of Big Field Schemes

There are three main cryptanalytic techniques that are applicable to big field multivariate cryptosystems. In a sense, all of these techniques are related to Q-rank. The MinRank key recovery attack has a complexity directly dependent on the Q-rank of the central map. The differential symmetry attack is relevant when the Q-rank of the central map is minimal in the relevant algebra. The direct algebraic attack has a complexity dependent on the degree of regularity of the public key which is usually a linear function of the Q-rank. We review each of these techniques.

3.1 MinRank

The first effective attack on HFE was presented in [16] and is now commonly called the Kipnis-Shamir (KS) attack. Their idea is to express the central polynomial as a single quadratic form on an a large representation of the extension field. Specifically, choose a representation $\psi : \mathbb{K} \to \mathbb{A}$ of the form $\psi(X) = (X, X^q, \ldots, X^{d-1})$. Then one can choose a matrix representation $\mathbf{F}$ of the central map f such that

$$f(X) = \begin{bmatrix} X & X^q & \cdots & X^{q^{d-1}} \end{bmatrix} \mathbf{F} \begin{bmatrix} X & X^q & \cdots & X^{q^{d-1}} \end{bmatrix}^{\top}.$$

As the reader easily notices, the degree bound on f implies that $\mathbf{F}$ has only a small block of nonzero values and thus has low rank. We call the rank of this quadratic form the Q-rank of f.

The attack in [16] exploits this low Q-rank property by using interpolation to find a formula for the public key over the extension field, computing the matrix forms of all of the Frobenius powers of this map, and then finding a low rank linear combination of these matrices with coefficients chosen from $\mathbb{K}$. The attack can be effective, but all of the algebra takes place in $\mathbb{K}$ which can be cumbersome.

The KS attack was significantly improved for determined or slightly over-determined schemes in [17], where the authors introduce minors modeling. Whereas the modeling of the low rank property in the KS attack requires structures defined over $\mathbb{K}$, the authors of [17] noticed that a $\mathbb{K}$-linear combination of the *public* quadratic forms defined over $\mathbb{F}_q$ has low rank. Thus one may construct a system of equations over the small field, resolve this system via Gröbner bases over the small field, and finally recover the variety over the big field. Requiring the most intensive calculations to be performed over the base field provided a significant advantage.

PFLASH is algebraically equivalent to an HFE- scheme (with a more efficient inversion process), and though the MinRank problem is less over-defined, the technique can, in principle, still be applied. To see this equivalence, note that the removal of equations can be modeled as a projection whose minimal polynomial, see [18, Definition 1], has low degree. Thus, there is a basis in which one can compose a low degree linear map with the low Q-rank central map of PFLASH producing a low Q-rank composition. As shown in [12, 19], the Q-rank of the PFLASH public key is too large for this attack to be effective.

3.2 Differential Techniques

A second class of attacks that has proven effective against big field schemes is the family of differential attacks involving the recovery of a symmetric relation to remove the minus modifier, or as a tool for accessing a low Q-rank. The discrete differential of a function $f : \mathbb{K} \to \mathbb{K}$ is the bivariate function

$$Df(a, x) = f(a + x) - f(a) - f(x) + f(0).$$

The differential operation D is linear and acts in many ways like a derivative; e.g. the differential of a $\mathbb{F}_q$-quadratic map is $\mathbb{F}_q$-bilinear, the differential of a $\mathbb{F}_q$-cubic function is $\mathbb{F}_q$-bi-quadratic, etc. The operators D^2, D_x, and so on all work analogously as do $\frac{d^2}{dx^2}$, $\frac{\partial}{\partial x}$, etc.

Differential attacks have been the basis of several cryptanalyses, see [4–6, 8, 20, 21]. The two basic techniques are linear differential symmetry attacks and differential invariant attacks.

Linear differential symmetry attacks attempt to find linear maps L that "factor through" the differential of the central map in an interesting way. Specifically, the goal is to find maps L satisfying

$$Df(La, x) + Df(a, Lx) = \Lambda_L Df(a, x).$$

If such a map can be found, it allows one to "remove" the minus modifier by discovering new linear combinations of the central maps that are linearly independent of the public key.

Such an attack is what broke SFLASH in [21]. If L represents multiplication by an element $\sigma \in \mathbb{K}$, then one can factor out σ from the differential due to the fact that the central map f is multiplicative. This vulnerability is provably removed via projection as shown in [12]. Thus PFLASH is invulnerable to this attack.

The other differential attack model, the invariant attacks, use the low rank structure on a large subspace of the public key to enhance the linear algebra search version of MinRank. Specifically, if a large subspace of the public key have the property that the matrices representing the functions as quadratic form map a particular subspace V simultaneously into another subspace W of the same dimension, then any projection producing two full rank differentials Df_1 and Df_2 allow one an advantage in recovering V, since V is left invariant by $Df_1 Df_2^{-1}$. This attack has been applied to undermine some of the proposed parameters for ABC and cubic ABC in [4–6] and was used to break the balanced oil-vinegar scheme in [22]. This attack was shown to be useless against PFLASH in [19].

3.3 Algebraic Attacks

The most straightforward attack is to try to directly invert the public key via Gröbner bases. The complexity of solving such systems relies on the degree of regularity of the system, which can be defined as the smallest degree at which a nontrivial syzygy producing a degree fall is generated in the Gröbner basis algorithm.

As shown in [23], the degree of regularity for HFE- systems, with a equations removed, satisfies the bound

$$d_{reg} \leq (q-1) \left\lfloor \frac{\lceil log_q(D) \rceil + a}{2} \right\rfloor + 2.$$

This upper bound is fairly tight for small fields and provides a fair estimate of the complexity of the direct algebraic attack on HFE-.

4   Description of EFLASH

Our scheme may be considered an atypical parameterization of a projected C^{*-}, which introduces new challenges. The major difference major difference between our scheme and the previously studied PFLASH, is the size of the projection. The size of our projection π will be much larger. This modification produces a significantly different scheme with different security properties.

4.1   Algebraic Structure

We will let n be the number of variables and $d > n$ be the degree of the extension field over $\mathbb{F}_q$. We will let $m \geq n$ be the number of equations ($m < d$) and denote the number of equations removed by $a = d - m$. We will compose our central map $f(x) = x^{q^\theta + 1}$ with affine maps S and T from $(\mathbb{F}_q)^d$ to $(\mathbb{F}_q)^d$. We let ϕ be a vector space isomorphism from $(\mathbb{F}_q)^d$ to $\mathbb{F}_{q^d}$, π be a linear embedding from $(\mathbb{F}_q)^n$ to $(\mathbb{F}_q)^d$, and τ be a linear projection from $(\mathbb{F}_q)^d$ to $(\mathbb{F}_q)^m$.

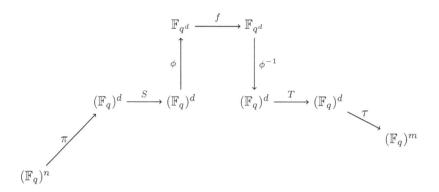

Our public equations P can be found by computing $P = \tau \circ T \circ \phi^{-1} \circ f \circ \phi \circ S \circ \pi$, where $f(x) = x^{q^\theta + 1}$.

4.2   Encryption and Decryption

To encrypt a message $\overline{x}$, the sender would just compute $P(\overline{x}) = \tau \circ T \circ \phi^{-1} \circ f \circ \phi \circ S \circ \pi(\overline{x}) = \overline{y}$ to get ciphertext $\overline{y}$. To decrypt the message we will take advantage of some of the weaknesses that an unmodified C^* scheme possesses.

To decrypt, we exploit the more efficient method of inversion Patarin developed in his linearization equations attack from [24]. Specifically, if $\overline{v} = (\phi^{-1} \circ f \circ \phi)(\overline{u})$ then there is a system of d polynomials of the form

$$\sum_{0 \le i,j < d} \alpha_{i,j,\ell} u_i v_j + \sum_{0 \le i < d} \beta_{i,\ell} u_i + \sum_{0 \le i < d} \gamma_{i,\ell} v_i + \delta_\ell$$

in the coefficients of $\overline{u}$ and $\overline{v}$ which are simultaneously zero. Composing the right inverse of $S \circ \pi$ and T with $\overline{u}$ and $\overline{v}$, respectively, we obtain a bilinear relation between the plaintext $\overline{x}$ and $\overline{y}' = T \circ \phi^{-1} \circ f \circ \phi \circ S \circ \pi(\overline{x})$. Given access to the private key (which includes the linearization equations) the calculation of this bilinear relation is immediate. Adding the linearization equations to the private key can be considered a drawback as it increases the private key size, but is an important aspect for our algorithm.

Inversion, given the ciphertext $\overline{y}$, is then accomplished by concatenating every possible suffix $\overline{y}_a$ to discover $\overline{y}' = \overline{y} \| \overline{y}_a$. Success is determined by solving the affine system in $\overline{x}$ induced from the linearization equations upon input $\overline{y}'$. If the affine system has a solution, $\overline{x}$, we can be assured that $P(\overline{x}) = \overline{y}$.

4.3   Decryption Failure Rate

We want to find the probability that there are multiple preimages of y under τ, which would result in a decryption failure. Specifically, we want to compute the probability that $x_1, x_2, y \in \mathbb{F}_q$ exists such that $P(x_1) = P(x_2) = y$, given that $P(x_1) = y$. Given our function $P(x) = \tau \circ T \circ \phi^{-1} \circ f \circ \phi \circ S \circ \pi(x)$, it is clear that the only part of this function that is not injective is τ, and that π is the only additional map that is not bijective. Thus we compute the probability of decryption failure under the simplifying heuristic that the central map $\hat{P}(x) = T \circ \phi^{-1} \circ f \circ \phi \circ S(x)$ is a random bijection. This assumption is obviously false as f is a quadratic map, but we believe this heuristic to be statistically useful. Let $A = \text{image}(\pi)$, $|A| = q^n$. We can consider B to be the preimage of y under τ, so under our simplifying heuristic B is a random set of q^a elements from $\mathbb{F}_q^d$.

We will use Bernoulli trials to estimate the probability that y is the image of at least two distinct elements of $\mathbb{F}_q^n$, given that it is the image of at least one. If $Pr(\hat{P}(x) \in B : \hat{P}(x) \in A) = p$, then the probability of k elements in A being in B is $\binom{q^n}{k}(1-p)^{(q^n-k)}p^k$.

The probability of $\hat{P}(x) \in B$ is $\frac{q^a}{q^d} = q^{-m}$, and the probability that $\hat{P}(x)$ is not in B is $1 - q^{-m}$. Thus we compute:

$$Pr(|A \cap B| \geq 2 \mid |A \cap B| \geq 1) = \frac{Pr(|A \cap B| \geq 2)}{Pr(|A \cap B| \geq 1)}$$

$$= \frac{1 - \left(Pr(|\mathbf{G} \cap \tau^{-1}(y)| = 0) + Pr(|\mathbf{G} \cap \tau^{-1}(y)| = 1) \right)}{1 - Pr(|\mathbf{G} \cap \tau^{-1}(y)| = 0)}$$

Therefore we find $Pr(|A \cap B| \geq 2 \mid |A \cap B| \geq 1)$ to be

$$p = \frac{1 - (1 - q^{-m})^{q^n} - q^{n-m}(1 - q^{-m})^{q^n - 1}}{1 - (1 - q^{-m})^{q^n}}.$$

To find an upper bound for the probability p, we find an upper bound for the numerator, and a lower bound for the denominator.

Claim 1. $\binom{a}{i+1}(q^{-(i+1)m}) < \binom{a}{i}(q^{-im})$ when $a < q^m$

Proof. Notice that $\binom{a}{i+1}(q^{-(i+1)m}) = \frac{a!}{(i+1)!(a-i-1)!q^{(i+1)m}}$ has the same numerator as $\binom{a}{i}(q^{-im}) = \frac{a!}{(i)!(a-i)!q^{im}}$, so we will prove the claim by showing the denominator of the left hand side is larger than the denominator of the right hand side.

Clearly $(i+1)! > i!$, and $q^{(i+1)m} > q^{im}$ by a factor of q^m. We see that $(a-i-1)! < (a-i)!$ by a factor of $a - i$, but we know that $a - i < a < q^m$. Thus we can conclude $(i+1)!(a-i-1)!q^{(i+1)m} > (i)!(a-i)!q^{im}$ and therefore $\binom{a}{i+1}(q^{-(i+1)m}) < \binom{a}{i}(q^{-im})$ when $a < q^m$. □

Bounding the numerator: $1 - (1 - q^{-m})^{q^n} - q^{n-m}(1 - q^{-m})^{q^n - 1}$.

Using binomial coefficients and the above claim, we see that:

$$(1 - q^{-m})^{q^n} = (1 - \binom{q^n}{1}q^{-m} + \binom{q^n}{2}q^{-2m} - \cdots) \geq 1 - q^n q^{-m}.$$

Thus $1 - (1 - q^{-m})^{q^n} \leq 1 - (1 - q^{n-m})$.

By the same argument, we are given:

$$(1 - q^{-m})^{q^n-1} = (1 - \binom{q^n-1}{1}q^{-m} + \binom{q^n-1}{2}q^{-2m} - \cdots) \geq 1 - (q^n - 1)q^{-m}.$$

Therefore, $-q^{n-m}(1-q^{-m})^{q^n-1} \leq -q^{n-m}(1-(q^n-1)q^{-m})$. Thus the numerator is bounded above by $1 - (1 - q^{n-m}) - q^{n-m}(1 - (q^n - 1)q^{-m})$.

Bounding the denominator: $1 - (1 - q^{-m})^{q^n}$ Similar to our argument for bounding the numerator, we will use binomial coefficients and claim 1 to find:

$$(1 - q^{-m})^{q^n} = (1 - \binom{q^n}{1}q^{-m} + \binom{q^n}{2}q^{-2m} - \cdots) \leq 1 - \binom{q^n}{1}q^{-m} + \binom{q^n}{2}q^{-2m}$$

Hence the denominator is bounded below by $1 - (1 - q^{n-m} + \frac{q^n q^n - 1}{2}q^{-2m})$.

Finding a bound for the probability, p

$$\begin{aligned}
p &= \frac{1 - (1 - q^{-m})^{q^n} - q^{n-m}(1 - q^{-m})^{q^n-1}}{1 - (1 - q^{-m})^{q^n}} \\
&\leq \frac{1 - (1 - q^{n-m}) - q^{n-m}(1 - (q^n - 1)q^{-m})}{1 - (1 - q^{n-m} + \frac{q^n q^n - 1}{2}q^{-2m})} \\
&= \frac{1 - 1 + q^{n-m} - q^{n-m} + q^{n-m}(q^n - 1)q^{-m}}{1 - 1 + q^{n-m} - \frac{q^n(q^n-1)}{2}q^{-2m}} = \frac{q^{n-m}(q^n - 1)q^{-m}}{q^{n-m} - \frac{q^n(q^n-1)}{2}q^{-2m}} \\
&= \frac{q^{n-m}(q^n - 1)q^{-m}}{q^{n-m} - q^{n-m}(q^{-(n-m)}\frac{q^n(q^n-1)}{2}q^{-2m})} \\
&= \frac{q^{n-m} - q^{-m}}{1 - (\frac{q^{n-m}-q^{-m}}{2})}
\end{aligned}$$

When $q = 2$, empirical evidence shows we can approximate this by 2^{n-m-1}. The data to support this claim are shown in Table 1.

Table 1. Probability of decryption failure for specific parameters of EFLASH.

q	n	d	a	m	$n - m$	Decrypt fail rate
2	14	34	8	26	-12	$2^{-13.13}$
2	14	35	8	27	-13	$2^{-13.94}$
2	14	36	8	28	-14	$2^{-14.94}$
2	14	37	8	29	-15	$2^{-15.94}$
2	14	38	8	30	-16	$2^{-17.64}$

5   Resistance to Known Attacks

The security analysis of EFLASH is quite related to that of PFLASH because of the similar algebraic structure. There are three attack methods that must be considered. Since the scheme requires more equations than variables to ensure a low probability of decryption failure, we require a careful analysis of the direct algebraic attack to ensure that the degree of regularity of the scheme is not too low. Second, in light of the attack on HFE- schemes, see [25], we require a MinRank analysis. Finally, given the history of the lineage of the C^* family, we require an analysis of symmetric differential methods.

5.1   Algebraic Attack

The first relevant attack for EFLASH is the direct algebraic attack. Algebraically, EFLASH is a high degree projected HFE- scheme, in the sense that EFLASH has a low Q-rank like HFE. Applying a projection to the input variables cannot increase the Q-rank, so we analyze the Q-rank of the central map composed with the minus modifier.

The key observation is that, unlike the case of HFE in which removing one equation in general increases the Q-rank by one, since the quadratic form associated with the central map is so sparse, the removal of one equation in general increases the rank by *two*. To see this, note that the coefficients of the quadratic form associated with HFE are restricted to a square submatrix whose size is typically the Q-rank of the map. A codimension one projection allows these coefficients to bleed into another row and column, which increases the size of the square by one. In contrast, the size of the smallest square containing the nonzero values in the quadratic form of the EFLASH central map is usually much larger than the Q-rank of EFLASH; in fact, the codimension one projection can produce two elements in original rows and columns, see Fig. 1.

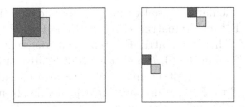

Fig. 1. The shape of the matrices representing the central maps of HFE- and C^{*-}. The darkly shaded regions represent nonzero values of the central map without the minus modifier, the lightly shaded regions represent new nonzero values introduced by the removal of one equation. Unshaded areas have coefficients of zero.

Thus, the central map of EFLASH has Q-rank $2+2a$. By the formula provided in [23], we compute an upper bound on the degree of regularity,

$$d_{reg} \leq (q-1)(a+1) + 2. \tag{1}$$

When q is small this bound is known to be fairly tight. The complexity of the algebraic attack on EFLASH is therefore estimated to be $\mathcal{O}\left(\binom{n+d_{reg}}{d_{reg}}^{\omega}\right)$, where $2 \leq \omega \leq 3$ is the linear algebra constant.

We conducted experiments on some small scale instances of EFLASH to study the behavior of the degree of regularity for values of n and $m = d - a$ of a similar ratio to a full sized scheme with a low decryption failure rate. The results are shown in Table 2.

Table 2. The degree of regularity of small scale EFLASH parameters in comparison to that of random systems of the same size.

n	d	a	m	d_{reg}	d_{reg} (RANDOM)
16	28	9	19	4	4
24	37	9	28	4	5
32	47	9	38	5	6
40	56	9	47	≥ 6	7

The data show that the degree of regularity grows with the size of the system when a is fixed. Until our resource permissions were limited on the machine, each sufficiently large system exhibited a degree of regularity at most one less than that of a random system. We do not have a solid theoretical argument for why the degree of regularity should be bounded thusly; however, for the sizes of schemes necessary to achieve security, the upper bound provided by (1) is already strictly less than the degree of regularity of random systems of the same size.

5.2   MinRank Attack

We can denote the calculations used to find our public equations P as matrix multiplications. Let $\mathbf{F^{*i}}$ be the matrix representation of the ith Frobenius power of the central map f. Then the matrix $\mathbf{F^{*0}}$ represents our central map f, and is the $d \times d$ matrix with 1's in the $(0, \theta)$ and $(\theta, 0)$ coordinates and zeros elsewhere. Matrices $\mathbf{S}$ and $\mathbf{T}$ are $d \times d$ affine maps. We can also consider π as a linear embedding from $(\mathbb{F}_q)^n$ to $(\mathbb{F}_q)^d$, and τ as a linear projection from $(\mathbb{F}_q)^d$ to $(\mathbb{F}_q)^m$. Let σ be a primitive element of the extension, and thus $\{1, \sigma, \sigma^2, \ldots, \sigma^{d-1}\}$ is a basis vector over $\mathbb{F}_q$. Then mappings of ϕ and ϕ^{-1} can be represented as multiplication of M_d and M_d^{-1}, respectively, where

$$M_d = \begin{pmatrix} 1 & 1 & \cdots & 1 \\ \sigma & \sigma^q & \cdots & \sigma^{q^{d-1}} \\ \vdots & \vdots & \cdots & \vdots \\ \sigma^{d-1} & \sigma^{(d-1)q} & \cdots & \sigma^{(d-1)q^{d-1}} \end{pmatrix}$$

We can express the actions of τ by the following $d \times d$ matrix,

$$\tau^* = \begin{bmatrix} I_m & 0_{m \times a} \\ 0_{a \times m} & 0_{a \times a} \end{bmatrix}.$$

Notice that $\tau^* : (\mathbb{F}_q)^d \to (\mathbb{F}_q)^d$. We will call $P^* := \tau^* \circ T \circ \phi^{-1} \circ f \circ \phi \circ S \circ \pi$. P and P^* will be comprised of the same m public equations, but P^* will then have a rows of 0 appended to it.

Consider $R = \phi \circ \tau^* \circ T \circ \phi^{-1}$. Then $R : \mathbb{F}_{q^d} \to \mathbb{F}_{q^d}$ is $\mathbb{F}_q$-linear. If we let $\tilde{\tau}(x) = \Pi_{r \in \ker(R)}(x - r)$, then we know by proposition 2 in [18], there exists a nonsingular linear map $\tilde{R}$ from $\mathbb{F}_{q^d}$ to $\mathbb{F}_{q^d}$ such that $Rx = \tilde{R}\tilde{\tau}x$. Let $\tilde{T} = \phi^{-1} \circ \tilde{R} \circ \tilde{\tau} \circ \phi$. This brings us to the following claim.

Claim 2. $P^*(x) = \tau^* \circ T \circ \phi^{-1} \circ f \circ \phi \circ S \circ \pi x = \tilde{T} \circ \phi^{-1} \circ f \circ \phi \circ S \circ \pi x$

Proof.

$$\begin{aligned}
\tilde{T} \circ \phi^{-1} \circ f \circ \phi \circ S \circ \pi &= \phi^{-1} \circ \tilde{R} \circ \tilde{\tau} \circ \phi \circ \phi^{-1} \circ f \circ \phi \circ S \circ \pi \\
&= \phi^{-1} \circ \tilde{R} \circ \tilde{\tau} \circ f \circ \phi \circ S \circ \pi \qquad (*) \\
&= \phi^{-1} \circ R \circ f \circ \phi \circ S \circ \pi \\
&= \phi^{-1} \circ \phi \circ \tau^* \circ T \circ \phi^{-1} \circ f \circ \phi \circ S \circ \pi \\
&= \tau^* \circ T \circ \phi^{-1} \circ f \circ \phi \circ S \circ \pi \\
&= P^*
\end{aligned}$$

$\square$

Now, let us reconsider $(*)$. We know that our public key is equivalent to $(*)$, so we see that

$$\begin{aligned}
P^* &= \phi^{-1} \circ \tilde{R} \circ \tilde{\tau} \circ f \circ \phi \circ S \circ \pi \\
&= \phi^{-1} \circ \tilde{R} \circ \phi \circ \phi^{-1} \circ \tilde{\tau} \circ f \circ \phi \circ S \circ \pi \\
&= \hat{T} \circ \phi^{-1} \circ \hat{f} \circ \phi \circ S \circ \pi
\end{aligned}$$

where $\hat{f}$ is our new central map and $\hat{f} = \tilde{\tau} \circ f$ and $\hat{T} = \phi^{-1} \circ \tilde{R} \circ \phi$. We now consider $\widehat{\mathbf{F}}^{*i}$ to be the ith Frobenius power of the new central map $\hat{f} = \tilde{\tau} \circ f$. If we denote $h = \phi^{-1} \circ \hat{f} \circ \phi$, then we can find symmetric matrices $(\mathbf{H}_1, \ldots, \mathbf{H}_d) \in (\mathbb{F}_q)^d$ such that $h_i = \bar{x} \mathbf{H}_i \bar{x}^\top$. As shown in [17] we see,

$$(\mathbf{H}_1, \ldots, \mathbf{H}_d) = (\mathbf{M}_d \widehat{\mathbf{F}}^{*0} \mathbf{M}_d^\top, \ldots, \mathbf{M}_d \widehat{\mathbf{F}}^{*(d-1)} \mathbf{M}_d^\top) \mathbf{M}_d^{-1}. \qquad (2)$$

If we denote the public key by $P = (g_1, g_2, \ldots, g_m)^\top$, then we can consider the symmetric matrices $(\mathbf{G}_1, \mathbf{G}_2, \ldots, \mathbf{G}_m)$ that correspond to the public polynomials, such that $g_i = \bar{x} \mathbf{G}_i \bar{x}$. By analysis in [17] we find,

$$(\mathbf{G}_1, \ldots, \mathbf{G}_m) = (\pi \mathbf{S} \mathbf{M}_d \widehat{\mathbf{F}}^{*0} \mathbf{M}_d^\top \mathbf{S}^\top \pi^\top, \ldots, \pi \mathbf{S} \mathbf{M}_d \tilde{\mathbf{F}}^{*(d-1)} \mathbf{M}_d^\top \mathbf{S}^\top \pi^\top) \mathbf{M}_d^{-1} \tilde{\mathbf{T}} \qquad (3)$$

When we consider our original central map, we saw that $\mathbf{F}^{*0}$ has rank 2. Looking at our new central map $\hat{f}$, we see that $\tilde{\tau}$ increases the rank. If we insist that θ is between $a + 1$ and $d - a - 1$, then $\widehat{\mathbf{F}}^{*0}$ has rank $2(a + 1)$, as discussed in Sect. 5.1.

Notice that the embedding $\pi : (\mathbb{F}_q)^n \to (\mathbb{F}_q)^d$, and the affine map S will not *increase* the rank of the right hand side of (3), so it will not affect our MinRank attack. Applying $\widehat{T}$ normally does increase the rank, but it does not increase the min-Q-rank because it just produces new linear combinations of these matrices.

Using these facts and the analysis from [17] we find that we are solving the MinRank problem:

$$\text{rank}\left(\sum_{k=0}^{m-1} \lambda_i \mathbf{G_i} \right) \leq 2(a + 1)$$

By the analysis in [26] and [27], the complexity of solving MinRank with the given parameters is $\mathcal{O}\left(\binom{m+d_{reg}}{d_{reg}}^{\omega} \right)$, where d_{reg} is the degree of regularity of the minors system and ω is the linear algebra constant. Treating EFLASH as a special case of HFE-, we may derive the degree of regularity of the minors system from [25, Conjecture 2] by using the Q-rank in place of the sum of the logarithm of the degree bound and the number of equations removed. Then we may estimate that the degree of regularity of the minors system is $d_{reg} = 2a + 3$.

5.3   Discrete Differential Attack

In [12], it is shown that almost all parameters of PFLASH are secure against differential adversaries. The proof relies on the fact that the corank of the projection is relatively small. Since EFLASH uses a corank $d - n$ projection, the security proof does not apply and so we must use other arguments.

By the symmetric argument to that in [25], we can express π under the appropriate basis as a polynomial in $\mathbb{K}$ of degree q^{d-n}. Thus, the central quadratic form can be considered a quadratic form in the $d - n$ "variables" $\pi(x)^{q^i}$, for $0 \leq i \leq d - n$. In characteristic two, there are at least as many linearly independent quadratic monomials as in GF(2); thus, there are at least $\binom{d-n+1}{2}$ linearly independent quadratic monomials in $\pi(x)^{q^i}$, for $0 \leq i \leq d - n$ over $\mathbb{K}$.

We expect that the locus of stabilizing pairs of matrices is zero-dimensional over $\mathbb{K}$, though it is necessarily positive dimensional over $\mathbb{F}_q$ since scalar multiples induce symmetry for any map. We performed experiments and found that the solution space was zero-dimensional over $\mathbb{K}$ in all cases. We conclude that the space of linear maps inducing symmetry on EFLASH is too small to be exploited like in the attack on SFLASH of [21].

6   Parameter Selection

In choosing parameters for EFLASH, we need to consider security against the direct algebraic attack, the MinRank attack, and fault attacks exploiting decryption failure. We address the constraints each of these attacks places on parameters, as well as efficiency concerns.

The complexity of both the direct attack and the MinRank attack is directly related to the Q-rank of the public key. In the case of very small fields, such as $GF(2)$, the degree of regularity is little larger than the Q-rank, $2a + 2$; thus, several equations must be removed to achieve security. Over $GF(2)$, each increase in a doubles decryption time while making the direct attack approximately n times harder and the MinRank attack approximately $2m$ times harder.

To address decryption failures, we note that the probability estimate of Sect. 4 is approximately q^{n-m}. We set an reasonable bound 2^{-B} on the probability of decryption failure and may set $m = n + \frac{B}{lg(q)}$ to achieve this bound.

For larger q, the MinRank attack seems to be the most concerning. For efficiency reasons, it is impractical to have a large a; therefore, an instance with large q is vulnerable to MinRank. For this reason, we recommend the choice $q = 2$ with a and n sufficiently large to resist the algebraic attack. Our specific parameter selections for classical security levels are summarized in Table 3. It is important to note that our implementation is a proof of concept, and not at all optimized. This is a magma implementation, and we are only using one core.

Table 3. Parameters and unoptimized performance of $EFLASH(q, n, d, a)$ at the 80-bit and 128-bit classical security levels.

Scheme	Security	Public key B	Enc. (ms)	Dec. (ms)	Dec. failure
$EFLASH(2, 80, 101, 5)$	80-bit	38892	0.7	194	2^{-17}
$EFLASH(2, 134, 159, 9)$	128-bit	169613	1.3	12758	2^{-17}

In principle, Grover search should affect the security of these schemes, but at this time we are not aware of a result that indicates a Grover search would be feasible for such large parameters. It is possible that Grover search could halve the dimension of the preimage search space. Thus, we may have to roughly double the size of the plaintext. To protect against the possible threat of Grover search we consider the parameter selections shown in Table 4.

On the other hand, we may consider the possibility of the cryptosystem being implemented on a quantum device so that the search step in decryption may be Groverized. Therefore Grover's algorithm may, in fact, improve efficiency.

Table 4. Parameters and unoptomized performance of $EFLASH(q, n, d, a)$ at the 80-bit and 128-bit quantum security levels.

Scheme	Security	Public key B	Enc. (ms)	Dec. (ms)	Dec. failure
$EFLASH(2, 160, 181, 5)$	80-bit	141691	1.9	1140	2^{-17}
$EFLASH(2, 256, 279, 7)$	128-bit	559249	5.3	16177	2^{-17}

7 Conclusion

In this paper we propose a new multivariate encryption scheme, EFLASH, derived from the lineage of PFLASH. One can view EFLASH as a parameterized projected C^{*-} scheme, where the projection π may be viewed as an embedding that maps from a smaller field to a much larger field. Thus, EFLASH follows the recent trend of achieving encryption with injective expansion maps. A possible direction to improve this result is to handle decryption failures in a more clever way. It may be possible to handle decryption failures in a generic way, as in [28]. it may also be interesting to consider reaction attacks against the scheme. Our algorithm implementation is a proof-of-concept implementation and is not optimized. Some possible optimizations may include making it not constant time, which should halve the decryption time.

EFLASH inherits some of the solid security justification from its digital signature forebear, PFLASH, though some of the security arguments are weakened by the massive cokernel of the projection. Still, the analysis of the security of EFLASH against each of the primary modes of attack on big field schemes is straightforward and encouraging. In this sense, it makes sense to consider the scheme as a sort of "standard candle" for the advancement of big field multivariate cryptanalysis. If EFLASH is to be broken, it seems that a new technique will need to be discovered.

A Toy Example

We illustrate our scheme by presenting a toy example. We generate public and private keys and perform a decryption. For simplicity, we consider linear, as opposed to affine, transformations.

A.1 Public and Private Key Generation

Let $q = 2$, $n = 4$, $d = 8$ and $a = 2$. We construct the degree d extension field $\mathbb{K} = \mathbb{F}_2[x]/\langle x^8 + x^4 + x^3 + x^2 + 1 \rangle$. We randomly select the C^* monomial $f(x) = x^{2^5+1}$. We next randomly select the invertible transformation T and embedding U:

$$T = \begin{bmatrix} 1 & 0 & 1 & 1 & 1 & 0 & 1 & 0 \\ 0 & 1 & 0 & 1 & 0 & 1 & 1 & 1 \\ 1 & 1 & 0 & 0 & 0 & 0 & 0 & 0 \\ 0 & 1 & 1 & 0 & 1 & 1 & 0 & 1 \\ 0 & 0 & 1 & 0 & 0 & 1 & 0 & 1 \\ 1 & 0 & 0 & 0 & 0 & 1 & 1 & 1 \\ 0 & 1 & 0 & 1 & 1 & 1 & 1 & 0 \\ 0 & 0 & 0 & 1 & 1 & 1 & 0 & 1 \end{bmatrix}, \text{ and } U = \begin{bmatrix} 0 & 1 & 1 & 0 & 1 & 0 & 0 & 1 \\ 0 & 1 & 1 & 1 & 0 & 0 & 0 & 0 \\ 1 & 0 & 1 & 1 & 1 & 0 & 1 & 0 \\ 1 & 0 & 1 & 0 & 0 & 0 & 0 & 0 \end{bmatrix}.$$

We fix $\Pi : \mathbb{F}_q^8 \rightarrow \mathbb{F}_q^6$, the projection onto the first 6 coordinates. Then the public key $P = \Pi \circ T \circ \phi^{-1} \circ f \circ \phi \circ U$ in symmetric matrix form over $\mathbb{F}_q$ is given by:

$$\mathbf{P_1} = \begin{bmatrix} 0 & 0 & 0 & 1 \\ 0 & 0 & 0 & 0 \\ 0 & 0 & 0 & 1 \\ 1 & 0 & 1 & 0 \end{bmatrix}, \mathbf{P_2} = \begin{bmatrix} 1 & 0 & 1 & 1 \\ 0 & 1 & 0 & 0 \\ 1 & 0 & 0 & 1 \\ 1 & 0 & 1 & 0 \end{bmatrix}, \mathbf{P_3} = \begin{bmatrix} 0 & 1 & 0 & 1 \\ 1 & 0 & 1 & 0 \\ 0 & 1 & 1 & 0 \\ 1 & 0 & 0 & 1 \end{bmatrix},$$

$$\mathbf{P_4} = \begin{bmatrix} 0 & 1 & 1 & 1 \\ 1 & 0 & 0 & 1 \\ 1 & 0 & 0 & 0 \\ 1 & 1 & 0 & 0 \end{bmatrix}, \mathbf{P_5} = \begin{bmatrix} 0 & 0 & 1 & 1 \\ 0 & 1 & 0 & 1 \\ 1 & 0 & 0 & 1 \\ 1 & 1 & 1 & 0 \end{bmatrix}, \mathbf{P_6} = \begin{bmatrix} 0 & 0 & 1 & 1 \\ 0 & 0 & 1 & 0 \\ 1 & 1 & 0 & 0 \\ 1 & 0 & 0 & 0 \end{bmatrix}.$$

We note that P has some linear terms that can be found on the diagonals of the public matrices.

The plaintext $\overline{x}$ and the output, $\overline{v}$, of $\phi^{-1} \circ f \circ \phi \circ U$ are related by linearization equations due to the relation

$$uv^{2^5} + u^{2^{10}}v = 0,$$

where $u = \phi(U\overline{x})$ and $v = \phi(\overline{v})$. Given our choice of basis we generate these linearization equations, written here in matrix form:

$$\mathbf{L_1} = \begin{bmatrix} 1 & 1 & 1 & 1 & 1 & 1 & 0 & 1 \\ 0 & 0 & 0 & 0 & 1 & 0 & 1 & 0 \\ 1 & 0 & 1 & 1 & 0 & 1 & 1 & 0 \\ 1 & 1 & 0 & 0 & 0 & 0 & 1 & 0 \end{bmatrix}, \mathbf{L_2} = \begin{bmatrix} 1 & 1 & 1 & 1 & 0 & 0 & 1 & 1 \\ 1 & 0 & 1 & 0 & 0 & 0 & 0 & 0 \\ 1 & 1 & 1 & 0 & 0 & 0 & 1 & 1 \\ 0 & 1 & 1 & 1 & 1 & 1 & 1 & 0 \end{bmatrix}, \mathbf{L_3} = \begin{bmatrix} 1 & 0 & 0 & 1 & 0 & 0 & 1 & 1 \\ 1 & 0 & 0 & 1 & 0 & 0 & 1 & 0 \\ 1 & 0 & 0 & 0 & 0 & 0 & 1 & 1 \\ 0 & 0 & 1 & 0 & 1 & 1 & 0 & 1 \end{bmatrix},$$

$$\mathbf{L_4} = \begin{bmatrix} 1 & 1 & 0 & 0 & 0 & 1 & 1 & 1 \\ 1 & 1 & 1 & 0 & 1 & 1 & 0 & 1 \\ 1 & 1 & 0 & 1 & 1 & 1 & 1 & 0 \\ 1 & 1 & 1 & 1 & 0 & 1 & 0 & 1 \end{bmatrix}, \mathbf{L_5} = \begin{bmatrix} 0 & 0 & 0 & 0 & 1 & 1 & 0 & 1 \\ 0 & 0 & 1 & 0 & 1 & 0 & 0 & 0 \\ 0 & 1 & 1 & 1 & 0 & 1 & 0 & 1 \\ 1 & 0 & 0 & 1 & 1 & 1 & 0 & 1 \end{bmatrix}, \mathbf{L_6} = \begin{bmatrix} 0 & 0 & 0 & 0 & 0 & 1 & 1 & 1 \\ 0 & 1 & 1 & 1 & 1 & 1 & 0 & 1 \\ 0 & 0 & 1 & 0 & 1 & 1 & 1 & 1 \\ 0 & 1 & 0 & 1 & 1 & 1 & 1 & 1 \end{bmatrix},$$

$$\mathbf{L_7} = \begin{bmatrix} 1 & 1 & 0 & 0 & 1 & 0 & 0 & 0 \\ 1 & 1 & 0 & 0 & 1 & 0 & 0 & 1 \\ 1 & 1 & 1 & 1 & 0 & 0 & 1 & 1 \\ 0 & 1 & 1 & 1 & 1 & 1 & 1 & 1 \end{bmatrix}, \mathbf{L_8} = \begin{bmatrix} 1 & 1 & 1 & 1 & 1 & 0 & 0 & 0 \\ 1 & 0 & 0 & 0 & 1 & 0 & 0 & 0 \\ 0 & 1 & 0 & 1 & 1 & 1 & 1 & 1 \\ 0 & 1 & 0 & 0 & 0 & 1 & 1 & 1 \end{bmatrix}.$$

A.2   Encryption and Decryption

Encryption is accomplished by evaluating the public key at the ciphertext. We randomly choose a plaintext

$$\overline{x} = \begin{bmatrix} 0 & 1 & 1 & 0 \end{bmatrix}.$$

Computing $(\overline{x})P$ we obtain the ciphertext

$$\overline{y} = \begin{bmatrix} 0 & 1 & 0 & 0 & 1 & 1 \end{bmatrix}.$$

Decryption is accomplished by first appending a random suffix on $\overline{y}$ to form $\overline{y_0}$, applying T^{-1}, and then solving the linear system defined by the linearization equations. Our first attempt appends the zero vector to $\overline{y}$. Thus

$$\overline{y}_0 = \begin{bmatrix} 0 & 1 & 0 & 0 & 1 & 1 & 0 & 0 \end{bmatrix}.$$

We then solve the system $\mathbf{0} = \overline{x}\mathbf{L}_i\overline{y}_0^\top$, where $1 \leq i \leq 8$, for $\overline{x}$. We immediately obtain the valid plaintext $\begin{bmatrix} 0 & 1 & 1 & 0 \end{bmatrix}$. We subsequently append all remaining possible suffixes on $\overline{y}$ and attempt to invert. Each of these linear systems is inconsistent, however; thus $\overline{x}$ is the unique preimage and so the valid plaintext.

References

1. Tao, C., Diene, A., Tang, S., Ding, J.: Simple matrix scheme for encryption. In: Gaborit, P. (ed.) PQCrypto 2013. LNCS, vol. 7932, pp. 231–242. Springer, Heidelberg (2013). https://doi.org/10.1007/978-3-642-38616-9_16
2. Porras, J., Baena, J., Ding, J.: ZHFE, a new multivariate public key encryption scheme. [30], pp. 229–245 (2014)
3. Yasuda, T., Sakurai, K.: A multivariate encryption scheme with rainbow. In: Qing, S., Okamoto, E., Kim, K., Liu, D. (eds.) ICICS 2015. LNCS, vol. 9543, pp. 236–251. Springer, Cham (2016). https://doi.org/10.1007/978-3-319-29814-6_19
4. Moody, D., Perlner, R., Smith-Tone, D.: An asymptotically optimal structural attack on the ABC multivariate encryption scheme. [30], pp. 180–196 (2014)
5. Moody, D., Perlner, R., Smith-Tone, D.: Key recovery attack on the cubic ABC simple matrix multivariate encryption scheme. In: Avanzi, R., Heys, H. (eds.) SAC 2016. LNCS, vol. 10532, pp. 543–558. Springer, Cham (2017). https://doi.org/10.1007/978-3-319-69453-5_29
6. Moody, D., Perlner, R., Smith-Tone, D.: Improved attacks for characteristic-2 parameters of the cubic ABC simple matrix encryption scheme. [29], pp. 255–271 (2017)
7. Cabarcas, D., Smith-Tone, D., Verbel, J.A.: Key recovery attack for ZHFE. [29], pp. 289–308 (2017)
8. Perlner, R., Petzoldt, A., Smith-Tone, D.: Total break of the SRP encryption scheme. In: Adams, C., Camenisch, J. (eds.) SAC 2017. LNCS, vol. 10719, pp. 355–373. Springer, Cham (2018). https://doi.org/10.1007/978-3-319-72565-9_18
9. Matsumoto, T., Imai, H.: Public quadratic polynomial-tuples for efficient signature-verification and message-encryption. In: Barstow, D., et al. (eds.) EUROCRYPT 1988. LNCS, vol. 330, pp. 419–453. Springer, Heidelberg (1988). https://doi.org/10.1007/3-540-45961-8_39
10. Ding, J., Dubois, V., Yang, B.-Y., Chen, O.C.-H., Cheng, C.-M.: Could SFLASH be repaired? In: Aceto, L., Damgård, I., Goldberg, L.A., Halldórsson, M.M., Ingólfsdóttir, A., Walukiewicz, I. (eds.) ICALP 2008. LNCS, vol. 5126, pp. 691–701. Springer, Heidelberg (2008). https://doi.org/10.1007/978-3-540-70583-3_56
11. Patarin, J., Goubin, L., Courtois, N.: C^*_{-+} and HM: variations around two schemes of T. Matsumoto and H. Imai. In: Ohta, K., Pei, D. (eds.) ASIACRYPT 1998. LNCS, vol. 1514, pp. 35–50. Springer, Heidelberg (1998). https://doi.org/10.1007/3-540-49649-1_4
12. Cartor, R., Smith-Tone, D.: An updated security analysis of PFLASH. [29], pp. 241–254 (2017)

13. Patarin, J., Goubin, L., Courtois, N.: Improved algorithms for isomorphisms of polynomials. In: Nyberg, K. (ed.) EUROCRYPT 1998. LNCS, vol. 1403, pp. 184–200. Springer, Heidelberg (1998). https://doi.org/10.1007/BFb0054126

14. Patarin, J.: Hidden fields equations (HFE) and isomorphisms of polynomials (IP): two new families of asymmetric algorithms. In: Maurer, U. (ed.) EUROCRYPT 1996. LNCS, vol. 1070, pp. 33–48. Springer, Heidelberg (1996). https://doi.org/10.1007/3-540-68339-9_4

15. Berlekamp, E.R.: Factoring polynomials over large finite fields. Math. Comput. **24**, 713–735 (1970)

16. Kipnis, A., Shamir, A.: Cryptanalysis of the HFE public key cryptosystem by relinearization. In: Wiener, M. (ed.) CRYPTO 1999. LNCS, vol. 1666, pp. 19–30. Springer, Heidelberg (1999). https://doi.org/10.1007/3-540-48405-1_2

17. Bettale, L., Faugère, J., Perret, L.: Cryptanalysis of HFE, multi-HFE and variants for odd and even characteristic. Des. Codes Cryptogr. **69**, 1–52 (2013)

18. Daniels, T., Smith-Tone, D.: Differential properties of the *HFE* cryptosystem. [30], pp. 59–75 (2014)

19. Chen, M.S., Yang, B.Y., Smith-Tone, D.: PFLASH - secure asymmetric signatures on smart cards. In: Lightweight Cryptography Workshop 2015 (2015). http://csrc.nist.gov/groups/ST/lwc-workshop2015/papers/session3-smith-tone-paper.pdf

20. Hashimoto, Y.: Cryptanalysis of multi-HFE. IACR Cryptol. ePrint Arch. **2015**, 1160 (2015)

21. Dubois, V., Fouque, P.-A., Shamir, A., Stern, J.: Practical cryptanalysis of SFLASH. In: Menezes, A. (ed.) CRYPTO 2007. LNCS, vol. 4622, pp. 1–12. Springer, Heidelberg (2007). https://doi.org/10.1007/978-3-540-74143-5_1

22. Kipnis, A., Shamir, A.: Cryptanalysis of the oil and vinegar signature scheme. In: Krawczyk, H. (ed.) CRYPTO 1998. LNCS, vol. 1462, pp. 257–266. Springer, Heidelberg (1998). https://doi.org/10.1007/BFb0055733

23. Ding, J., Kleinjung, T.: Degree of regularity for HFE-. IACR Cryptol. ePrint Arch. **2011**, 570 (2011)

24. Patarin, J.: Cryptanalysis of the Matsumoto and Imai Public Key Scheme of Eurocrypt'88. In: Coppersmith, D. (ed.) CRYPTO 1995. LNCS, vol. 963, pp. 248–261. Springer, Heidelberg (1995). https://doi.org/10.1007/3-540-44750-4_20

25. Vates, J., Smith-Tone, D.: Key recovery attack for all parameters of HFE-. [29], pp. 272–288 (2017)

26. Bardet, M., Faugere, J.C., Salvy, B.: On the complexity of Gröbner basis computation of semi-regular overdetermined algebraic equations. In: Proceedings of the International Conference on Polynomial System Solving (2004)

27. Bardet, M., Faugére, J., Salvy, B., Yang, B.: Asymptotic behaviour of the degree of regularity of semi-regular polynomial systems. In: Proceedings of MEGA 2005, Eighth International Symposium on Effective Methods in Algebraic Geometry (2005)

28. Smart, N.P., Albrecht, M.R., Orsini, E., Paterson, K.G., Peer, G.: LIMA: a PQC encryption scheme

29. Lange, T., Takagi, T. (eds.): PQCrypto 2017. LNCS, vol. 10346. Springer, Cham (2017). https://doi.org/10.1007/978-3-319-59879-6

30. Mosca, M. (ed.): PQCrypto 2014. LNCS, vol. 8772. Springer, Cham (2014). https://doi.org/10.1007/978-3-319-11659-4

Public Key Compression for Constrained Linear Signature Schemes

Ward Beullens, Bart Preneel, and Alan Szepieniec[(✉)]

imec-COSIC KU Leuven, Leuven, Belgium
{ward.beullens,bart.preneel,alan.szepieniec}@esat.kuleuven.be

Abstract. We formalize the notion of a *constrained linear trapdoor* as an abstract strategy for the generation of signature schemes, concrete instantiations of which can be found in MQ-based, code-based, and lattice-based cryptography. Moreover, we revisit and expand on a transformation by Szepieniec *et al.* [39] to shrink the public key at the cost of a larger signature while reducing their combined size. This transformation can be used in a way that is provably secure in the random oracle model, and in a more aggressive variant whose security remained unproven. In this paper we show that this transformation applies to any constrained linear trapdoor signature scheme, and prove the security of the first mode in the quantum random oracle model. Moreover, we identify a property of constrained linear trapdoors that is sufficient (and necessary) for the more aggressive variant to be secure in the quantum random oracle model. We apply the transformation to an MQ-based scheme, a code-based scheme and a lattice-based scheme targeting 128-bits of post quantum security, and we show that in some cases the combined size of a signature and a public key can be reduced by more than a factor 300.

Keywords: Digital signatures · Post-quantum
Quantum random oracle model · Key size reduction

1 Introduction

Trapdoor functions are an important tool in public key cryptography due to the computational asymmetry they bring about. On the one hand, the function is a proper cryptographic one-way function to anyone who is ignorant of the secret trapdoor information; but on the other hand, anyone who does know this trapdoor information can use it to find inverse images quickly.

The case of *surjective* trapdoor functions is especially interesting for generating *digital signature schemes*. A cryptographic hash function maps a message of any size to a random point in the trapdoor function's output space. An inverse of this point under the trapdoor function, or *signature*, testifies to the involvement of the trapdoor information, or *secret key*, in its generation. This testimony ensures the target property of *non-repudiation of origin*: the secret key holder cannot deny generating the signature at a later date.

© Springer Nature Switzerland AG 2019
C. Cid and M. J. Jacobson, Jr. (Eds.): SAC 2018, LNCS 11349, pp. 300–321, 2019.
https://doi.org/10.1007/978-3-030-10970-7_14

Since their inception in the seminal paper by Diffie and Hellman [10], various digital signature schemes have been deployed whose security is based on the hardness of integer factorization [35] and the discrete logarithm problem [30, 36]. However, the advent of quantum computers threatens the security of these signature schemes because both hard problems are solved efficiently by Shor's quantum algorithm [37]. This ultimatum drives the need to design, develop and deploy so-called *post-quantum* cryptosystems, *i.e.*, cryptography that can be run on classical hardware but promises to resist attacks by quantum computers.

Even though the RSA trapdoor is broken by quantum computers, the hash-and-sign construction that RSA signatures are based on seems to survive the transition to post-quantum cryptography. To achieve post-quantum secure signature schemes it suffices to exchange the underlying trapdoor for one that has the desired security against quantum adversaries. There is no shortage of trapdoor-based signature schemes based on the MQ problem [11,21,34], coding theory [8,9], or lattices [3,15,27].

Unfortunately, the public keys in these schemes are prohibitively large, measurable in hundreds of kilobytes if not megabytes. In contrast, post-quantum signature schemes derived from zero-knowledge proofs require only a one-way function whose selection can be random or might as well be determined by a short seed and an implicit pseudorandom generator. Signature schemes based on zero-knowledge proofs tend to exchange tiny public keys for prohibitively large signatures [7,18,23,38], and moreover require complicated and expansive non-interactivity transforms to retain security against quantum attackers [40]. Although provable security in the case of hash-based signature schemes is much more straightforward, this family of constructions follows the same pattern: tiny public keys but huge signatures [4,5].

Szepieniec, Beullens and Preneel offer an alternative to the dilemma between large public keys or large signatures [39], motivated by the desire to minimize the combined size of public key and signature. This minimization is particularly important in the context of public key infrastructure (PKI) where a chain of signatures and public keys is transmitted in order to authenticate a message with respect to a pre-shared root public key. The construction of Szepieniec *et al.* applies specifically to MQ trapdoors and relies on the observation that verifying a couple of random linear combinations of the public key's polynomial equations can be as good as verifying all of them. The coefficients of this linear combination are determined as a function of the produced signature, and the combination itself is transmitted along with this signature in addition to information authenticating its link to the public key. This transformation reduces the size of public key plus that of the signature by roughly a factor three whilst provably retaining security in the random oracle model; and by a much larger factor at the expense of a heuristic security argument.

This article expands on the paper of Szepieniec *et al.* in several ways. We observe that this transformation also applies to other post-quantum trapdoor signature schemes, most notably code-based and lattice-based trapdoors. From a general perspective, these three hard problems are variations on a common theme, which we call *constrained linear signature schemes*. This commonality allows a generic presentation of the transformation. The security proofs of Szepieniec *et al.* only work in the classical random oracle model. However, security proofs that purport to defend against quantum adversaries should additionally hold in the *quantum random oracle model*, which our proof does. Moreover, we identify a necessary and sufficient security property, called (σ, r)-hash-and-sign-security $((\sigma, r)$-HSS), that a constrained linear signature scheme must have in order for the more aggressive parameter choices of Szepieniec *et al.* to be provably secure. This leads to an improved understanding of the security of instantiations of this construction, which includes the DualModeMS submission of Faugère *et al.* [12] to the NIST PQC standardization project [29]. To showcase the key size improvements that can be achieved with the transformation, we apply the transformation to a lattice-based, code-based and multivariate constrained linear signature scheme with parameters targeting 128 bits of security against quantum computers.

2   Preliminaries

Random Oracle Model. We use a hash function in our construction. For the purpose of proving security we model it by a *random oracle*, which is a random function $\mathsf{H} : \{0,1\}^* \rightarrow \{0,1\}^\kappa$ with a fixed output length, typically equal to the security parameter. If necessary, the random oracle's output space can be lifted to any finite set X. We use subscripts to differentiate the random oracles associated with different output spaces. A security proof relying on the modelling of hash function as random oracles is said to hold in the *random oracle model*. When quantum adversaries are considered, the security proofs should allow for superposition queries to the random oracle [6]; a security proof with this property is said to hold in the *quantum random oracle model*.

Trapdoor Functions. A trapdoor function is a function that can be efficiently computed in one direction, but for which it is hard to compute preimages *unless* by someone who knows a secret piece of information called the *trapdoor*. We associate three algorithms to a trapdoor function family:

- GenTrapdoor takes a security parameter as input and outputs a trapdoor function f and a trapdoor t.
- Evaluate takes a description of the trapdoor function f and an argument x as input, and returns the evaluation of f at x. In the rest of the paper, we simply write this as $f(x)$.
- Invert takes the function f, the trapdoor t and an image y as input, and outputs a value x such that $f(x) = y$.

Signature Scheme. A public key signature scheme is defined as a triple of polynomial-time algorithms (KeyGen, Sign, Verify). The probabilistic key generation algorithm takes the security level κ (in unary notation) and produces a secret and public key: $\mathsf{KeyGen}(1^\kappa) = (sk, pk)$; the signature generation algorithm produces a signature: $s = \mathsf{Sign}(\mathsf{sk}, m) \in \{0,1\}^*$. The verification algorithm takes the public key, the message and the signature and decides if the signature is valid: $\mathsf{Verify}(pk, m, s) \in \{0,1\}$; we refer to these outputs as "reject" and "accept", respectively. The signature scheme is *correct* if signing a message with the secret key produces a valid signature under the matching public key:

$$(\mathsf{KeyGen}(1^\kappa) \Rightarrow (sk, pk)) \quad \Longrightarrow \quad \forall m \in \{0,1\}^* \,.\, \mathsf{Verify}\,(pk, m, \mathsf{Sign}(sk, m)) = 1.$$

Here and elsewhere we use $\Rightarrow$ to denote the event of the probabilistic algorithm on the left hand producing the output on the right hand, and $\Longrightarrow$ to denote logical implication.

Security is defined with respect to the Existential Unforgeability under Chosen Message Attack (EUF-CMA) game of Goldwasser *et al.* [17]. The adversary A is allowed to make a polynomial number of queries $m_i, i \in \{1, \ldots, q\}, q \leq \kappa^c$ for some c, which the challenger signs using the secret key and sends back: $s_i \leftarrow \mathsf{Sign}(\mathsf{sk}, m_i)$. At the end of the game, the adversary must produce a pair of values (m', s') where m' was not queried before: $m' \notin \{m_i\}_{i=1}^q$. The adversary wins if $\mathsf{Verify}(pk, m', s') = 1$. In the game below, the Iverson brackets $[\![\cdot]\!]$ return 0 if the expression is False or 1 if it is True.

Game EUF-CMA

1:   $sk, pk \leftarrow \mathsf{KeyGen}(1^\kappa)$
2:   $\mathcal{M} \leftarrow \varnothing$
3:   **define** S(m) **as**
4:     | $\mathcal{M} \leftarrow \mathcal{M} \cup \{m\}$
5:     | **return** $\mathsf{Sign}(sk, m)$
6:   **end definition**
7:   $(m, s) \leftarrow \mathsf{A}^{\mathsf{S}}(pk)$
8:   **return** $[\![\mathsf{Verify}(pk, m, s) = \mathsf{True} \wedge m \notin \mathcal{M}]\!]$

We define the insecurity function $\mathsf{InSec}_{\mathsf{scheme}}^{\mathrm{EUF\text{-}CMA}}(Q_{\mathsf{S}}; t)$ as the maximum winning probability across all quantum adversaries that run in time t and that make at most Q_{S} signature queries.

Hash-and-Sign Signature Schemes. Given a trapdoor function family and a hash function H that hashes arbitrary messages to elements in the range of the trapdoor functions we can use the hash-and-sign construction to build a (not necessarily secure) signature scheme. The key generation algorithm simply calls the GenTrapdoor function to get (f, t). The public key is then the description of f, and the trapdoor t is the private key. To sign a message m, the signer uses his trapdoor t to produce a preimage s for $\mathsf{H}(m)$. This preimage is the signature for m. Lastly, to verify the validity of a signature the verifier computes $\mathsf{H}(m)$, uses the public key to evaluate f at s and checks if $f(s) = \mathsf{H}(m)$.

Merkle Tree. A Merkle tree [26] is a balanced binary tree whose root authenticates a list of data items which are contained in the leaves. Every non-leaf node, including the root, has a value equal to the hash of the concatenation of its two children. A leaf can be proven to be a member of the tree by tracing a path from the leaf to the root and listing all siblings of nodes on that path: every step can be verified by computing one hash. We associate three algorithms with a Merkle tree:

- CalculateMerkleRoot takes a list of leaf items, computes the entire Merkle tree, and returns its root.
- OpenMerklePath takes a list of leaf nodes and an index, and outputs its authentication path: the list of all siblings of nodes on the path from the indicated leaf node to the root.
- VerifyMerklePath takes an index, a leaf node, a Merkle path, and a root, and decides whether the leaf node is a member of the tree with the given root.

3   Trapdoor-Based Signature Schemes

3.1   MQ Trapdoors

Multivariate quadratic (MQ) trapdoor functions date back to the C^* scheme of Matsumoto and Imai [25], which has since given rise to a number of viable candidates including HFE_v^- [32], UOV [21] and Rainbow [11]. The idea is to compose a special quadratic map $\mathcal{F} : \mathbb{F}_q^n \to \mathbb{F}_q^m$ with two linear transforms, $T \in \mathsf{GL}_m(\mathbb{F}_q)$ and $S \in \mathsf{GL}_n(\mathbb{F}_q)$ to obtain the public key $\mathcal{P} = T \circ \mathcal{F} \circ S$. A vector $\mathbf{s} \in \mathbb{F}_q^n$ that represents an assignment to the variables, is a valid signature for the document $d \in \{0, 1\}^*$ whenever

$$\mathcal{P}(\mathbf{s}) = \mathsf{H}(d). \tag{1}$$

In order to find $\mathbf{s}$, the signer computes $\mathbf{z} = \mathsf{H}(d)$, $\mathbf{y} = T^{-1}\mathbf{z}$, uses the special structure of $\mathcal{F}$ to sample an inverse $\mathbf{x}$ such that $\mathcal{F}(\mathbf{x}) = \mathbf{y}$, and then computes $\mathbf{s} = S^{-1}\mathbf{x}$.

We focus on the Rainbow submission to the NIST PQC project [29], where the parameter set ($q = 256, v = 68, o_1 = 36, o_2 = 36$) is proposed. In this case, $n = v + o_1 + o_2 = 140$ and $m = o_1 + o_2 = 72$. While the proposal does not employ Petzoldt's compression trick [33] we note that it is possible in principle, in which case $v(v + 1)/2 + vo_1$ columns of the public Macaulay matrix are set as the output of a PRG expanding a seed of 32 bytes.[1] Allocating five bits per field element, we obtain signatures of 140 bytes and public keys of 356.9 kB. Without Petzoldt's compression trick the public key is 694.0 kB.

3.2   Code-Based Trapdoors

The first code-based signature scheme was proposed by Courtois, Finiasz and Sendrier (CFS) [8]; it relies on the difficulty of finding a low Hamming weight

---

[1] In fact, Petzoldt manages to fix more elements of the public key's Macaulay matrix, but as these elements are not arranged into columns they are incompatible with our compression technique.

word associated with a given syndrome. The public key in such a signature scheme is a parity check matrix $H \in \mathbb{F}_2^{(n-k) \times n}$. A signature $(\mathbf{s}, i) \in \mathbb{F}_2^{1 \times n} \times \mathbb{Z}$ on a document $d \in \{0,1\}^*$ consists of an error vector and an index; it is valid when the error vector has Hamming weight at most t and syndrome equal to the hash of the document concatenated with the index i. The index i can be thought of as selecting a different hash function. Formulaically:

$$H\mathbf{s}^\mathsf{T} = \mathsf{H}(d\|i) \quad \text{and} \quad \mathsf{HW}(\mathbf{s}) \le t. \tag{2}$$

By our calculations, a 128-bit post-quantum security level is achieved with the parameter set $m = 26$, $t = 15$ and thus $n = 2^m = 2^{26}$ and $n - k = tm = 390$. At this point the public key is 3.05 GB but the signatures are 390 bits. We refer to Appendix A for a derivation of these parameters. We choose not to consider the question whether the cryptosystem is practically usable with these parameters and instead focus on the obtained compression factor. The CFS scheme is used as a generic stand-in for code-based signature schemes using the hash-and-sign paradigm and relying on the hardness of syndrome decoding.

3.3 Lattice-Based Trapdoors

A first trapdoor-based signature schemes from lattices was proposed by Goldreich, Goldwasser and Halevi (GGH) at Crypto'97 [16]. The signatures of this scheme leak information about the private key, and the scheme was broken by Nguyen and Regev [31]. Gentry et al. [15] showed how to sample signatures that do not leak information and constructed a provably secure signature scheme. Later improvements by Alwen and Peikert [3] and by Micciancio and Peikert [27] make the scheme more efficient. The main idea is the same in all schemes: the public key is a matrix $A \in \mathbb{F}_q^{n \times m}$ with large coefficients but such that there exists another matrix $S \in \mathbb{Z}^{m \times m}$ with small coefficients with $AS = 0 \bmod q$. In order to generate a signature for a document $d \in \{0,1\}^*$, the signer uses the secret key S to obtain a small-coefficient vector $\mathbf{z} \in \mathbb{Z}^m$. It is a valid signature whenever

$$A\mathbf{z} = \mathsf{H}(d) \bmod q \quad \text{and} \quad \|\mathbf{z}\|_2 \le \beta, \tag{3}$$

for some length bound $\beta \in \mathbb{R}_{>0}$.

Using the methodology of [28], and the estimator for the concrete hardness of the SIS problem of Albrecht et al. [1], we choose parameters for the scheme of [27] that achieves 128 bits of security. This results in the parameters $n = 321, q = 2^{26} - 5, m = 16692$ and $\beta = 112296$, a public key of $n \times m \times 26$ bits $= 16.6$ MB, and signatures of $\lceil \log_2(\beta) \rceil \times m$ bits $= 34.6$ KB. We chose q to be prime as this is required for our security proof to work. The first half of the matrix A can be chosen randomly, so we can fix this part with a PRG to cut the size of the public key in half.

3.4 A Unifying View

The above three signature schemes can be thought of as variations on a common theme. These schemes are all hash-and-sign signature schemes with a linear trapdoor function $f : \mathbb{F}_q^\ell \to \mathbb{F}_q^k$, but with f restricted to a domain defined by a nonlinear constraint function $\mathsf{nc} : \mathbb{F}_q^\ell \to \{\mathsf{True}, \mathsf{False}\}$. We call these trapdoor functions **constrained linear trapdoor functions**, and if they are used in a hash-and-sign construction, we call the resulting signature scheme a **constrained linear signature scheme**.

For all the constrained linear signature schemes the public key is a matrix $M \in \mathbb{F}_q^{k \times \ell}$ with $k < \ell$ which represents the trapdoor function f and a signature is represented by a vector $\mathbf{s} \in \mathbb{F}_q^\ell$. A signature is valid if $M\mathbf{s}$ is equal to a target $\mathbf{t} \in \mathbb{F}_q^k$, which is the evaluation of a hash function at a document, and if the vector $\mathbf{s}$ also satisfies the constraint nc. Symbolically:

$$\mathsf{Verify}(sk, m, \mathbf{s}) = 1 \quad \Longleftrightarrow \quad M\mathbf{s} = \mathbf{t} = \mathsf{H}(m) \wedge \mathsf{nc}(\mathbf{s}) = \mathsf{True}.$$

In the case of lattice-based trapdoors, the signature is valid only if $\mathbf{s}$ is a short vector. In the case of code-based trapdoors, it is valid only if the Hamming weight of $\mathbf{s}$ is low. And in the case of MQ trapdoors, the matrix M is the coefficient matrix (or Macaulay matrix) of the quadratic polynomial map $\mathcal{P}$ and the signature $\mathbf{s}$ must be factorizable as a vector of products of n variables: $\mathbf{s}^\mathsf{T} = (x_1^2, x_1 x_2, \dots, x_n^2)$. Formally, we capture this difference between MQ, code-based, and lattice-based trapdoors with the nonlinear constraint nc, namely by defining for

- code-based trapdoors: $\mathsf{nc}(\mathbf{s}) = \mathsf{True} \Leftrightarrow \mathrm{HW}(\mathbf{s}) \leq t$;
- lattice-based trapdoors: $\mathsf{nc}(\mathbf{s}) = \mathsf{True} \Leftrightarrow \|\mathbf{s}\|_2 \leq \beta$;
- MQ trapdoors: $\mathsf{nc}(\mathbf{s}) = \mathsf{True} \Leftrightarrow \exists x_1, \dots, x_n \in \mathbb{F}_q . \mathbf{s}^\mathsf{T} = (x_1^2, x_1 x_2, \dots, x_n^2)$.

3.5 Additional Security Properties

We say that a surjective trapdoor function f is one-way (OW) if it is hard to find a preimage for a randomly chosen output, and we say that f is hash-and-sign secure (HSS) if using the trapdoor function f in the hash-and-sign construction leads to a signature scheme that is EUF-CMA secure. If f is a constrained linear trapdoor function we can define stronger versions of the OW and HSS security properties that will be useful for the security analysis of the transformation (Fig. 1).

(σ, r)-One-Wayness. For any two non-negative integers $\sigma > r$ we define (σ, r)-one-wayness and (σ, r)-hash-and-sign security. To break (σ, r)-one-wayness, an adversary has to find σ preimages $\mathbf{x}_1, \dots, \mathbf{x}_\sigma \in \mathbb{F}_q^\ell$ for σ vectors $\mathbf{y}_1, \dots, \mathbf{y}_\sigma \in \mathbb{F}_q^k$. However, the adversary is allowed to make mistakes in each of the σ preimages it produces, as long as the errors $f(\mathbf{x}_i) - \mathbf{y_i}$ are contained in a vector space of dimension r. The $(1, 0)$-one-wayness property is identical to the one-wayness property, because the adversary only needs to find a preimage for one target and it is not allowed to make any mistakes.

The (σ, r)-OW property is a generalization of the AMQ problem introduced in [39]; an MQ trapdoor $\mathcal{P}$ is (σ, r)-one-way precisely if the Approximate MQ problem with σ targets and rank r is hard for the map $\mathcal{P}$.

(σ, r)-**Hash-and-Sign Security.** We also define a (σ, r)-variant of the HSS property. The security game behind this property is similar to the EUF-CMA game of the hash-and-sign signature scheme induced by f. To break this property, an adversary has to come up with a message m and σ 'signatures' $s_1, \cdots, s_\sigma$ such that the errors $f(s_i) - H(m||i)$ are contained in a a subspace of dimension r. The adversary can query a signing oracle S any (polynomially bounded) number of times. When given a message m', this signing oracle uses the trapdoor to produce preimages for $H(m'||1), \cdots, H(m'||\sigma)$ and returns these σ preimages. The adversary loses the game if it returns a message m for which it has queried the signing oracle, as is the case for the familiar EUF-CMA game.

We define the insecurity function $\mathsf{InSec}_f^{(\sigma,r)-\mathsf{HSS}}(Q_S, Q_H; t)$ as the maximal winning probability of an adversary that plays the (σ, r)-HSS game of f, that makes Q_S queries to the signing oracle, Q_H queries to the random oracle and that runs in time t. The $(1, 0)$-HSS property is equivalent to the HSS property.

Remark 1. If f is a *collision-resistant preimage-sampleable trapdoor function* (as is the case for some lattice-based trapdoor functions), the one-wayness of f can be reduced tightly to its hash-and-sign security and so OW and HSS are equivalent [15, Proposition 6.1]. Under the same assumption on f, the security proof of [15] can be modified to prove that (σ, r)-OW and (σ, r)-HSS are equivalent for all $\sigma > r \geq 0$ (Fig. 2).

4 Construction

4.1 Description

This section describes the transform of Szepieniec *et al.* but adapted to apply generically to constrained linear signature schemes. The parameters for the transformation are:

- (KeyGen, Sign, Verify), the constrained linear signature scheme to start from. We denote the hash function used in the verification algorithm by H_1 and the nonlinear constraint by nc.
- τ, the number of leaves in the Merkle tree.
- e, the extension degree of $\mathbb{F}_{q^e}$, which is the field over which the error-correcting code is defined. This value is constrained by $q^e \geq \tau$.
- ϑ, the number of Merkle paths that are opened with each new signature.
- σ, the number of signatures of the original signature scheme that is included in each signature of the new scheme.
- H_2, a hash function that outputs a α-by-k matrix over $\mathbb{F}_q$.
- H_3, a hash function that outputs a set of ϑ numbers between 1 and τ.
- H_4, a hash function used for building a Merkle tree.

The transformation outputs a new signature scheme (NEW.KeyGen, NEW.Sign, NEW.Verify) with a smaller public key but larger signatures.

```
┌─────── Game (σ,r)-OW ───────┐    ┌─────── Game (σ,r)-HSS ───────┐
│ 1: (f,t) ← GenTrapdoor(1^κ)  │    │ 1: (f,t) ← GenTrapdoor(1^κ)       │
│ 2: y₁,...,y_σ ←$ F_q^k       │    │ 2: M ← ∅                          │
│ 3: x₁,...,x_σ ← A(f,y₁,...,y_σ)│  │ 3: define S(m) as                 │
│ 4: return [dim(⟨f(x_i) − y_i⟩_i) ≤ r] │ 4:    M ← M ∪ {m}             │
│                              │    │ 5:    for i from 1 to σ do        │
│                              │    │ 6:       s_i ← Invert(f,t,H(m‖i)) │
│                              │    │ 7:    end for                     │
│                              │    │ 8:    return s₁,...,s_σ           │
│                              │    │ 9: end definition                 │
│                              │    │ 10: m,s₁,...,s_σ ← A^{H,S(·)}(f)  │
│                              │    │ 11: d = dim(⟨f(s_i) − H(m‖i)⟩_i)  │
│                              │    │ 12: return [(d ≤ r) ∧ (m ∉ M)]    │
└──────────────────────────────┘    └──────────────────────────────────┘
```

Fig. 1. The security game of the $(\sigma, r) - \mathsf{OW}$ property (left) and of the $(\sigma, r) - \mathsf{HSS}$ property (right).

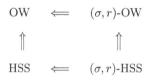

Fig. 2. Security properties of constrained linear trapdoor functions, and implications between them.

Random Linear Combinations. A signature of the new signature scheme consists of σ signatures of the original signature scheme, along with some information to verify them. The ith signature is obtained by using the signature generation algorithm of the original contrained-linear signature scheme to sign $d\|i$. It is not necessary to communicate the entire public key $M \in \mathbb{F}_q^{k\times\ell}$. Rather, it suffices to transmit a few random linear combinations of its rows. Therefore, part of the new signature consists of a matrix T that is equal to RM, where R is drawn uniformly at random from the space of $\alpha \times k$ matrices. Instead of checking whether $Ms_i = \mathsf{H}_1(d\|i)$, the verifier can now check wheter $Ts_i = R\mathsf{H}_1(d\|i)$. Obviously, if all signatures are valid, then the latter equations will also be satisfied for any matrix R. Conversely, if at least one signature is invalid, *i.e.*, $Ms_i \neq \mathsf{H}_1(d\|i)$ for some i, then the probability that $RMs = R\mathsf{H}_1(d\|i)$ is at most $q^{-\alpha}$. By choosing α large enough, the probability of accepting an invalid signature can be made arbitrarily small.

Determining R. In order for the above argument to work, R must be chosen independently from $s = s_1\|\cdots\|s_\sigma$. Therefore, we determine R with a hash function as $R = \mathsf{H}_2(d\|s_1\|\cdots\|s_\sigma)$ to ensure that a forger cannot use knowledge about R in his choice of the s_i.

Verifying T. An attacker can present the verifier with a signature containing a matrix T which is totally unrelated to the matrix M. How can the verifier be sure that the matrix T that is included in the signature, is really equal to RM with $R = \mathsf{H}_2(d\|\mathbf{s}_1\|\cdots\|\mathbf{s}_\sigma)$? We solve this problem with a probabilistic test based on an $\mathbb{F}_q$-linear error correcting code. This is a code whose alphabet consists of the elements of a finite field $\mathbb{F}_q$, with the property that any $\mathbb{F}_q$-linear combination of codewords is again a codeword. We work with Reed-Solomon Codes[2] over $\mathbb{F}_{q^e}$ with message length $L = \lceil \ell/e \rceil$ (we pack e elements of $\mathbb{F}_q$ into each symbol), codeword length τ and minimal codeword distance $D = \tau - L$. We use $\mathsf{Enc} : \mathbb{F}_{q^e}^{a \times L} \rightarrow \mathbb{F}_{q^e}^{a \times \tau}$ to denote the operation of encoding the rows of a matrix.

In the key generation phase, we compute $E = \mathsf{Enc}(M)$. Then we commit to this matrix E by building a Merkle tree whose leaves contain the columns of E, which are denoted by e_i for $i \in \{1,\ldots,\tau\}$. The new public key is the root of this tree. If $T = RM$, then by $\mathbb{F}_q$-linearity of the error correcting code, we have that $\mathsf{Enc}(T)$ is equal to $R\mathsf{Enc}(M) = RE$. Conversely, if $T \neq RM$, then $\mathsf{Enc}(T)$ and RE differ in at least one row. These rows are different codewords, so they differ in at least D of the τ symbols. To verify that $T = RM$, we now select ϑ columns $e_{b_1},\cdots,e_{b_\vartheta}$ of E with the hash function H_3 and we check whether the b_i-th column of T agrees with Re_{b_i} for all i in $1,\cdots,\vartheta$. If T is not equal to RM, this will go undetected with a probability of at most $(\frac{L}{\tau})^\vartheta$.

Pseudocode. Algorithms 1, 2 and 3 present pseudocode for the new signature scheme (NEW.KeyGen, NEW.Sign, NEW.Verify) obtained from transforming the old constrained-linear signature scheme (KeyGen, Sign, Verify).

Algorithm NEW.KeyGen

input: 1^κ — security level (in unary)
      random coins
output: *root* — A public key
      (sk, M) — A corresponding secret key

1: $(sk, M) \leftarrow \mathsf{KeyGen}(1^\kappa)$
2: $E \leftarrow \mathsf{Enc}(M)$ ▷ Encode M row by row.
3: *root* $\leftarrow \mathsf{CalculateMerkleRoot}(e_1,\cdots,e_\tau)$ ▷ Build tree on columns of E
4: **return** $(root, (sk, M))$

Algorithm 1. The key generation algorithm

---

[2] While the original description of the transformation used MAC-polynomials, we think it is better to describe the same transformation it in the language of Reed-Solomon error correcting codes.

Key and Signature Sizes. For a post-quantum security level of κ bits, the new public key is 2κ bits in size, as it represents the Merkle root. The new signature consists of σ old signatures, α linear combinations of the rows of M (each one of which consists of ℓ field elements of size $\lceil \log_2 q \rceil$ bits), ϑ columns of $\mathsf{Enc}(M)$ (each one of which consists of k field elements of $e \times \lceil \log_2 q \rceil$ bits), and ϑ Merkle paths of consisting of $\log_2 \tau$ hash images of 2κ bits each. Put all together, we have

$$|\mathsf{NEW}.signature| = \sigma|\mathsf{OLD}.signature| + (\alpha\ell + \vartheta ke) \times \lceil \log_2 q \rceil + 2\vartheta\kappa \times \log_2 \tau. \quad (4)$$

The old signatures can be represented as ℓ field elements but in some cases a more concise encoding is possible. For instance, CFS signatures require only the positions of the 1-bits, and MQ signatures require only an assignment to the variables from which the vector of quadratic monomials can be derived.

4.2   Security

Before we present the security statement and its proof, we need to introduce a pair of security games that will be important for our security analysis. In particular, we need hash functions that are one-way and second-preimage resistant, in both cases with respect to multiple targets. Both games are formalized with respect to a hash function H that is randomly selected from a hash function family $\mathcal{H}$. We follow the formalisms of Hülsing $et\ al.$ [20].

Algorithm NEW.Sign

input: d — A document to sign
$\quad\quad\quad (sk, M)$ — A private key

output: $(\mathbf{s}_1, \cdots, \mathbf{s}_\sigma, T, v_{b_1}, \cdots, v_{b_\vartheta}, paths)$ — A signature for d

```
 1: for i from 1 to σ do
 2: |   s_i ← Sign(d‖i, sk)
 3: end for
 4: R ← H_2(d‖s_1‖ ⋯ ‖s_σ)
 5: T ← RM
 6: E ← Enc(M)                                    ▷ Encode M row by row.
 7: b_1, ⋯, b_ϑ ← H_3(d‖s_1‖ ⋯ ‖s_σ‖T)
 8: paths ← empty list
 9: for i from 1 to ϑ do
10: |   paths.append(OpenMerklePath(e_1, ⋯, e_τ, b_i))
11: end for
12: return (s_1, ⋯, s_ϑ, T, e_{b_1}, ⋯, e_{b_ϑ}, paths)
```

Algorithm 2. The signature generation algorithm.

- In the *single-function, multiple-target one-wayness* (SM-OW) game, the adversary is given a list of target outputs and it wins if it can produce a single input that maps to any one of the outputs. We write $\mathsf{InSec}_{\mathsf{H},P}^{\mathrm{SM\text{-}OW}}(Q)$ to denote the maximum success probability across all adversaries that make at most Q queries and with respect to the hash function family $\mathcal{H}$ and where P is the number of target outputs.

- In the *single-function, multiple-target second-preimage resistance* (SM-SPR) game, the adversary is given a list of inputs and it wins if it can produce a second preimage that maps to the same output as any one of the input preimages. We write $\mathsf{InSec}_{\mathsf{H},P}^{\mathrm{SM\text{-}SPR}}(Q)$ to denote the maximum success probability across all adversaries that make at most Q queries and with respect to the hash function family $\mathcal{H}$ and where P is the number of input preimages.

Game SM-OW	Game SM-SPR
1: $\mathsf{H} \xleftarrow{\$} \mathcal{H}$	1: $\mathsf{H} \xleftarrow{\$} \mathcal{H}$
2: **for** i from 1 to P **do**	2: **for** i from 1 to P **do**
3: $\quad M_i \xleftarrow{\$} \{0,1\}^m$	3: $\quad M_i \xleftarrow{\$} \{0,1\}^m$
4: $\quad Y_i \leftarrow \mathsf{H}(M_i)$	4: **end for**
5: **end for**	5: $M' \leftarrow \mathsf{A}^{\mathsf{H}}(M_1,\ldots,M_P)$
6: $M' \leftarrow \mathsf{A}^{\mathsf{H}}(Y_1,\ldots,Y_P)$	6: **return** $[\![\exists i . \mathsf{H}(M') = Y_i \wedge$
7: **return** $[\![\exists i . \mathsf{H}(M') = Y_i]\!]$	$\quad M' \neq M_i]\!]$

Hülsing *et al.* obtain values for these insecurity functions in the random oracle model, *i.e.* where H is drawn uniformly at random from the set of all functions from the given input space to the given output space. In the classical random oracle model we have

$$\mathsf{InSec}_{\mathsf{H},P}^{\mathrm{SM\text{-}OW}}(Q) = \mathsf{InSec}_{\mathsf{H},P}^{\mathrm{SM\text{-}SPR}}(Q) = \frac{(Q+1)P}{|\mathsf{range}(\mathsf{H})|}. \tag{5}$$

In the quantum random oracle model, where the adversary is allowed $\hat{Q}$ quantum queries, we have

$$\mathsf{InSec}_{\mathsf{H},P}^{\mathrm{SM\text{-}OW}}(\hat{Q}) = \mathsf{InSec}_{\mathsf{H},P}^{\mathrm{SM\text{-}SPR}}(\hat{Q}) = \Theta\left(\frac{(\hat{Q}+1)^2 P}{|\mathsf{range}(\mathsf{H})|}\right). \tag{6}$$

The SM-OW game does not quite capture one of the transitions in our security proof. The reason for this is that the adversary cannot be given a definite list of target output images because whether an output of the hash function is suitable for the adversary depends on the input of the hash function. We model this task by a new game, *marked element search* (MES), in which the adversary does not have a list of target outputs but a marking function $\mathsf{mark} : \mathsf{domain}(\mathsf{H}) \times \mathsf{range}(\mathsf{H}) \rightarrow \{0,1\}$ that determines whether the pair

---

Algorithm NEW.Verify

input: d — document
$(\mathbf{s}_1, \cdots, \mathbf{s}_\vartheta, T, v_{b_1}, \cdots, v_{b_\vartheta}, paths)$ — signature
$root$ — public key

output: 1 if the signature is valid, 0 otherwise

1: $R \leftarrow \mathsf{H}_2(d\|\mathbf{s}_1\|\cdots\|\mathbf{s}_\sigma)$
2: **for** i from 1 to σ **do**
3:     **if** $T\mathbf{s}_i \neq R\mathsf{H}_1(d\|i)$ or $\mathsf{nc}(\mathbf{s}_i) = \mathsf{False}$ **then**
4:         **return** 0
5:     **end if**
6: **end for**
7: $b_1, \cdots, b_\vartheta \leftarrow \mathsf{H}_3(d\|\mathbf{s}_1\|\cdots\|\mathbf{s}_\sigma\|T)$
8: **for** i from 1 to ϑ **do**
9:     **if** $\mathsf{Enc}(T)_{*,b_i} \neq Re_{b_i}$ **then**
10:         **return** 0
11:     **end if**
12:     **if** $\mathsf{VerifyMerklePath}(b_i, e_{b_i}, paths[i], root) = \mathsf{Fail}$ **then**
13:         **return** 0
14:     **end if**
15: **end for**
16: **return** 1

Algorithm 3. The signature verification algorithm.

($input$, $output$) is suitable. We write $\mathsf{InSec}^{\mathrm{MES}}_{\mathsf{H,mark}}(Q)$ to denote the maximum success probability across all adversaries that make at most Q queries to the hash oracle in the MES game. In the quantum random oracle model this notion is reducible to SM-OW.

Game MES

1: $\mathsf{H} \xleftarrow{\$} \mathcal{H}$
2: $M \leftarrow \mathsf{A}^{\mathsf{H}}()$
3: **return** $\mathsf{mark}(M, \mathsf{H}(M))$

Proposition 1 (SM-OW $\leq$ MES). *In the (quantum) random oracle model, we have that for any marking function mark with $P = \max_X |\{Y \mid \mathsf{mark}(X, Y) = 1\}|$,*

$$\mathsf{InSec}^{\mathrm{MES}}_{\mathsf{H,mark}}(Q) \leq \mathsf{InSec}^{\mathrm{SM\text{-}OW}}_{\mathsf{H},P}(Q). \tag{7}$$

Proof. We show an algorithm, $\mathsf{B}_{\mathsf{SM\text{-}OW}}$ in the SM-OW game, that simulates a given algorithm $\mathsf{A}_{\mathsf{MES}}$ for the MES game with marking function mark, and wins with at least the same probability. The input of $\mathsf{B}_{\mathsf{SM\text{-}OW}}$ is a list of P images $\{Y_1, \ldots, Y_P\}$ and access to a random oracle H. The algorithm $\mathsf{B}_{\mathsf{SM\text{-}OW}}$ programs a random oracle H' that on input X returns $\sigma_X^{-1}(\mathsf{H}(X))$, where σ_X is a permutation (chosen deterministically) with the property that the elements Y that satisfy $\mathsf{mark}(X,Y) = 1$ are mapped into the set $\{Y_1, \ldots, Y_P\}$. By assumption, $|\{Y \mid \mathsf{mark}(X,Y) = 1\}| \leq P$, so such a permutation always exists. Note that $\mathsf{B}_{\mathsf{SM\text{-}OW}}$ is bounded in the number of queries it can make to H, but not bounded in time or memory. Therefore it will be able to choose such a permutation σ_X. Then, $\mathsf{B}_{\mathsf{SM\text{-}OW}}$ invokes $\mathsf{A}_{\mathsf{MES}}$ with the programmed random oracle H'. Since H' only applies a permutation to the ouput of H, the ouputs of H' will be independent and uniformly distributed. Hence, H' is itself a perfect random oracle. Pseudocode for $\mathsf{B}_{\mathsf{SM\text{-}OW}}$ is given below.

Algorithm $\mathsf{B}_{\mathsf{SM\text{-}OW}}$

1: **define** $\mathsf{H}'(X)$ **as**
2:   pick σ_X s.t. $\sigma_X(\{Y \mid \mathsf{mark}(X,Y) = 1\}) \subset \{Y_1, \cdots, Y_P\}$
3:   **return** $\sigma_X^{-1} \circ \mathsf{H}(X)$
4: **end definition**
5: **return** $\mathsf{A}_{\mathsf{MES}}^{\mathsf{H}'(\cdot)}()$

Clearly, the number of queries that $\mathsf{B}_{\mathsf{SM\text{-}OW}}$ makes to H is identical to the number of queries made by the simulated algorithm $\mathsf{A}_{\mathsf{MES}}$. Eventually, $\mathsf{A}_{\mathsf{MES}}$ returns a preimage X. $\mathsf{A}_{\mathsf{MES}}$ wins the MES game if $\mathsf{mark}(X, \sigma_X^{-1}(\mathsf{H}(X))) = \mathsf{True}$. By our choice of σ_X this implies that $\sigma_X(\sigma_X^{-1}(\mathsf{H}(X))) = \mathsf{H}(X) \in \{Y_1, \cdots, Y_P\}$, which shows that $\mathsf{B}_{\mathsf{SM\text{-}OW}}$ wins his SM-OW game in this case. So $\mathsf{InSec}_{\mathsf{H},\mathsf{mark}}^{\mathsf{MES}}(Q) \leq \mathsf{InSec}_{\mathsf{H},P}^{\mathsf{SM\text{-}OW}}(Q)$. $\qquad\square$

We are now in a position to state and prove our security claim.

Theorem 1. *Let* NEW *be the signature scheme derived from applying the transformation to a constrained linear scheme* OLD. *The maximum winning probability across all time-t adversaries in the EUF-CMA game against* NEW *that make* Q_s *signature queries and* Q_1, Q_2, Q_3, Q_4 *queries to the random oracles* $\mathsf{H}_1, \mathsf{H}_2, \mathsf{H}_3, \mathsf{H}_4$ *respectively is bounded by*

$$\mathsf{InSec}_{\mathsf{NEW}}^{\mathrm{EUF\text{-}CMA}}(Q_s, Q_1, Q_2, Q_3, Q_4; t) \leq \mathsf{InSec}_f^{(\sigma,\mathrm{r})\text{-}\mathrm{HSS}}(Q_s, Q_1; O(t)) + \mathsf{InSec}_{\mathsf{H}_4, 2\tau-1}^{\mathrm{SM\text{-}SPR}}(Q_4)$$
$$+ \mathsf{InSec}_{\mathsf{H}_3, L^\vartheta}^{\mathrm{SM\text{-}OW}}(Q_3) + \mathsf{InSec}_{\mathsf{H}_2, q^{\alpha \times (k-r+1)}}^{\mathrm{SM\text{-}OW}}(Q_2). \quad (8)$$

Proof. We show through a sequence of four games how an adversary for the EUF-CMA game against NEW can be transformed into an adversary for the (σ, r)-HSS property of the underlying constrained linear trapdoor function f that wins with the same probability conditional on each of the transitions being successful. By bounding the failure probability of each transition and summing the terms we obtain a bound on the winning probability of the adversary against NEW. The sequence of games is as follows:

- The first game G_1 is the EUF-CMA game against NEW.
- The second game G_2 drops the Merkle tree. Instead, the public key consists of all the τ columns of E, and the verifier checks directly if the columns that are included in the signature are correct.
- The game G_3 drops the codeword identity testing. Instead, the public key is now the original public key (*i.e.*, M), and the verifier tests directly if the matrix T, which is included in the signature is equal to RM.
- The last game G_4 drops the random linear combinations for signature validity testing, instead G_4 is won if the errors $f(\mathbf{s}_i) - H_1(m||i)$ are contained in a subspace of dimension r. G_4 is thus the (σ, r)-HSS game for the constrained linear trapdoor function f.

In games G_2, G_3 and G_4, the adversary B simulates the previous game's adversary A in order to win his own game. In particular, this means that B must answer the signing queries that A makes. This is not a problem, because in all cases B can just forward the queries that A makes to its own signing oracle, remove some information that is not required for the game that A is playing from the signature and pass the response back to A. In each case, we define the transition's failure probability as the probability that A wins but B does not. In all cases the adversary A has unbridled access (perhaps even quantum access) to the hash functions H_1, H_2, H_3 and H_4.

The event that A wins G_1 but B does not win G_2 occurs only if the signature outputted by A passes the Merkle root test, but the columns included in this signature do not agree with the columns in $E = \mathsf{Enc}(M)$. This event requires finding a second preimage for one of the $2\tau - 1$ nodes of the Merkle tree, so the failure probability is bounded by

$$\mathsf{InSec}^{\text{SM-SPR}}_{H_4, 2\tau-1}(Q_4).$$

Likewise, the event that A wins the G_2 game, but B does not win the G_3 game occurs only if the columns $e_{b_1}, \cdots, e_{b_\vartheta}$ of E in the signature outputted by A are correct, but still T is not equal to RM. This implies that $\mathsf{Enc}(T)$ differs from RE in at least $\tau - L$ columns (since the rows are codewords from a code with minimal distance $\tau - L$), but that none of these columns were not chosen by the random oracle H_3. Finding $m||\mathbf{s}_1|| \cdots ||\mathbf{s}_\sigma||T$, such that this happens is a marked element search with marking function

$$\mathsf{mark}_1(m||\mathbf{s}_1|| \cdots ||\mathbf{s}_\sigma||T, b_1|| \cdots ||b_\vartheta) = \begin{cases} \text{False} & \text{if } T = RM \\ \text{False} & Re_{b_i} \neq \mathsf{Enc}(T)_{\star, b_i} \text{ for some } i \\ \text{True} & \text{otherwise} \end{cases}.$$

Since there are at most L indices for which the columns of $\mathsf{Enc}(T)$ and $R\mathsf{Enc}(E)$ are identical, there are at most $\binom{L}{\vartheta} \leq L^\vartheta$ marked elements for a given input. The failure probability is therefore bounded by

$$\mathsf{InSec}^{\text{MES}}_{H_3, \mathsf{mark}_1}(Q_3) \leq \mathsf{InSec}^{\text{SM-OW}}_{H_3, L^\vartheta}(Q_3).$$

Finally, the event that A wins game G_3 but that B does not win G_4 happens when the errors span a vector space of dimension strictly larger than r (B does not win), but that all these error lie in the kernel of $R = H_2(m||s_1||\cdots||s_\sigma)$ (otherwise A does not win). Finding $m||s_1||\cdots||s_\sigma$ such that this happens is a marked element search for the marking function

$$\mathsf{mark}_2(m||s_1||\cdots||s_\sigma, R) = \begin{cases} \mathsf{False} & \text{if } R(f(s_i) - H_1(m||i)) \neq 0 \text{ for some } i \\ \mathsf{False} & \text{if } \dim(\langle f(s_i) - H_1(m||i)\rangle_{i=0,\cdots,\sigma}) > r \\ \mathsf{True} & \text{otherwise} \end{cases}.$$

For a choice of $m||s_1||\cdots||s_\sigma$ there are only good matrices R if the space spanned by the errors $f(s_i) - H_1(m||i)$ has dimension at least $r+1$. If this is the case then the good matrices R are precisely the α-by-k matrices whose kernel contains the error space. Therefore there are at most $q^{\alpha(k-r+1)}$ good matrices for each choice of $m||s_1||\cdots||s_\sigma$. Therefore the failure probability of the last step is bounded by

$$\mathsf{InSec}_{H_2,\mathsf{mark}_2}^{\mathsf{MES}}(Q_2) \leq \mathsf{InSec}_{H_2,q^{\alpha\times(k-r+1)}}^{\mathsf{SM-OW}}(Q_2). \qquad \square$$

Joining Theorem 1 with Eqs. (5) and (6) gives the following corollaries.

Corollary 1. *In the classical random oracle model,*

$$\mathsf{InSec}_{\mathsf{NEW}}^{\mathsf{EUF\text{-}CMA}}(Q_s, Q_1, Q_2, Q_3, Q_4; t) \leq \mathsf{InSec}_f^{(\sigma,r)\text{-}\mathsf{HSS}}(Q_s, Q_1; t) + (Q_2 + 1)q^{-\alpha(r+1)}$$
$$+(Q_3 + 1)(\ell/\tau)^\vartheta + (Q_4 + 1)(2\tau - 1)/2^\kappa.$$

Corollary 2. *In the quantum random oracle model,*

$$\mathsf{InSec}_{\mathsf{NEW}}^{\mathsf{EUF\text{-}CMA}}(Q_s, \hat{Q}_1, \hat{Q}_2, \hat{Q}_3, \hat{Q}_4; t) \leq \mathsf{InSec}_f^{(\sigma,r)\text{-}\mathsf{HSS}}(Q_s, \hat{Q}_1; t) + \Theta\left((\hat{Q}_2 + 1)^2 q^{-\alpha(r+1)}\right)$$
$$+\Theta\left((\hat{Q}_3 + 1)^2(\ell/\tau)^\vartheta\right) + \Theta\left((\hat{Q}_4 + 1)^2(2\tau - 1)/2^\kappa\right).$$

There are two ways to use the transformation. One can choose $\sigma = 1$ and α large enough such that $q^{\alpha/2}$ reaches the required post-quantum security level, i.e., $q^{\alpha/2} > 2^\kappa$. Corollary 2 with $r = 0$ then guarantees that the resulting signature scheme is EUF-CMA secure, provided that the constrained linear trapdoor function f that we started from is $(1, 0)$-HSS. This assumption is equivalent to the EUF-CMA security of the original signature scheme OLD. We also note that in this case the security proof is tight, meaning that no security is lost (in the QROM) by applying the transformation in this way.

One can also use the transformation with $\sigma > r$, and a lower value of α such that $q^{\alpha\cdot(r+1)/2}$ reaches the required security level. This reduces the size of the public keys even further, but this comes at the cost of a stronger security assumption on the constrained linear trapdoor function f. In this case Corollary 2 says that the resulting signature scheme is EUF-CMA secure, if the underlying constrained linear trapdoor function is (σ, r)-HSS.

4.3 Applying the Transformation

Table 1 presents a comparison of the transformation applied to the three constrained linear trapdoor signature schemes treated in Sect. 3. For the Rainbow and Micciancio-Peikert schemes part of the public key can be generated with a PRNG to reduce the size of the public key. This trick is compatible with our construction, so we have taken this into account. In all cases, 128 bits of security against quantum computers was targeted for an apples-to-apples comparison.

Table 1. Comparison of constrained linear signature schemes before and after public key compression. Legend: NC = no compression; PS = our provably secure technique based on the assumption that the original hash-and-sign signature scheme is secure; SA = the approach relying on stronger assumptions.

Scheme	q	Other parameters	α	σ	ϑ	τ	e	$|pk|$	$|sig|$
Rainbow NC	256	$v = 68, o_1 = 36, o_2 = 36$	-	-	-	-	-	0.35 MB	0.14 kB
Rainbow PS			32	1	25	2^{20}	3	64 bytes	0.18 MB
Rainbow SA			2	16	25	2^{20}	3	64 bytes	35.51 kB
CFS NC	2	$m = 26, t = 15$	-	-	-	-	-	3.05 GB	59 bytes
CFS PS			256	1	71	2^{25}	25	32 bytes	2.00 GB
CFS SA			1	256	71	2^{25}	25	32 bytes	8.15 MB
Micciancio-Peikert NC	$2^{26} - 5$	$n = 321, m = 16692, \beta = 112296$	-	-	-	-	-	8.30 MB	34.64 kB
Micciancio-Peikert PS			10	1	37	2^{20}	1	64 bytes	0.35 MB
Micciancio-Peikert SA			5	2	37	2^{20}	1	64 bytes	0.26 MB

The shrinkage is the most striking when $k \gg \alpha \cdot \sigma$, because this is when the largest part of the matrix M is omitted. The mediocre shrinkage of $|pk| + |sig|$ for the provably secure case ($\sigma = 1$) suggests that for the trapdoors considered, k is already quite close to the lower bound $k \geq \kappa/\log_2 q$ needed for κ bits of security. The greater compression factor attained when $\sigma > 1$ is due mostly to the representation of the old signatures in far less than $\ell \cdot \log_2 q$ bits.

5   Conclusion

This paper generalizes the construction of Szepieniec *et al.* [39] to a wide class of signature schemes called constrained linear signature schemes. This construction transforms a constrained linear signature scheme into a new signature scheme with tiny public keys, at the cost of larger signatures and while reducing their combined size. We prove the EUF-CMA security of the resulting signature scheme in the quantum random oracle model, and for a more aggressive parametrization we identify the (σ, r)-hash-and-sign security notion as a sufficient property for security. This improves the understanding of the security of instantiations of this construction, which includes the DualModeMS submission to the NIST PQC standardization project [12,29]. Finally, to showcase the generality and facilitate comparison, the construction is applied to an $\mathcal{MQ}$-based, a code-based and a lattice-based signature scheme, all targeting the same security level. In some cases the combined size of a signature and a public key can be reduced by more than a factor 300.

We close with some notes on the practicality of the transformation. From Table 1 we see that our transformation improves the practicality of state of the art multivariate and code-based signature schemes for applications such as public key infrastructure (PKI), where the metric $|\mathsf{sig}| + |\mathsf{pk}|$ is important and the performance of signing a message is less critical (most signatures in a PKI chain are long-lived and need not be created often). Code-based signature schemes remain not very practical, despite the improvements our construction makes. For example, applying the construction to the CFS scheme results in signatures of 8.15 MB. Still, if better code based signature schemes are developed, the construction will likely to be able to improve the quantity $|\mathsf{sig}|+|\mathsf{pk}|$. For example, even though the pqsigRM [22] proposal to the NIST PQC project does not have a completely unstructured matrix as public key, our construction can still reduce $|\mathsf{sig}| + |\mathsf{pk}|$ by a factor 6 from 329 kB to 60 kB in this case (with $\alpha = 4, \sigma = 64$). Unfortunately, comments on the NIST forum indicate that the pqsigRM proposal might not be secure [2].

State of the art hash-and-sign lattice-based signature schemes are built on structured lattices to achieve smaller public keys (e.g. Falcon relies on NTRU lattices [14]). Therefore, our construction does not improve on state of the art lattice-based schemes. Rather, our construction can be seen as an alternative to using structured lattices that provably does not deteriorate the security of the original schemes. In contrast, it is possible that switching to structured lattices has a negative impact on security.

Acknowledgements. This work was supported in part by the Research Council KU Leuven: C16/15/058. In addition, this work was supported by the European Commission through the EC H2020 FENTEC under grant agreement No 780108. In addition, this work was supported by imec through ICON Diskman and by FWO through SBO SPITE S002417N. Ward Beullens is funded by an FWO SB fellowship. Alan Szepieniec is being supported by a doctoral grant from the Flemish Agency for Innovation and Entrepreneurship (VLAIO, formerly IWT).

A   CFS Parameters

Perhaps surprisingly, the most efficient attack on the CFS cryptosystem is not information set decoding (as is the case for the closely related Niederreiter cryptosystem) but a generalized birthday algorithm credited to Bleichenbacher by Finiasz and Sendrier [13]. The offline phase of this attack consists of building three lists L_0, L_1, L_2 containing sums of respectively w_0, w_1, w_2 columns from H, where $t = w_0 + w_1 + w_2$. Next, L_0 and L_1 are merged and pruned by taking the sum of each pair and keeping it only if it starts with λ zeros; the result of this operation is stored in L'_0. In the online phase a random counter i is appended to the document and the sum of $\mathsf{H}(d\|i)$ with every element of L_2 that agrees on the first λ positions is looked up in L'_0—because if this sum is present then that means that $\mathsf{H}(d\|i)$ equals the sum of $w_1 + w_2 + w_3 = t$ columns of H which can be identified by tracing the origins of the elements from L'_0, L_2, L_0, L_1 that were used. Let L'_1 denote the list obtained from pruning the sums of elements of L_2 and $\mathsf{H}(d\|i)$.

A single trial is successful if there is a collision between L'_0 and L'_1. This is essentially a generalized birthday problem as described by Wendl [41], and the same result shows that the much more easily computed binomial distribution approximates the probability of zero collisions very well when this quantity is overwhelming. The number of pairs to consider is $\#L'_0 \times \#L'_1$ and the proportion of pairs representing a collision is $1/2^{k-\lambda}$. All considered pairs fail to collide with probability $(1-2^{\lambda-k})^{\#L_0 \times \#L_1}$. By approximating $\#L'_0 \approx \mathrm{E}[\#L'_0] = 2^{-\lambda}\binom{n}{w_0+w_1}$ and $\#L'_1 \approx \mathrm{E}[\#L'_1] = 2^{-\lambda}\binom{n}{w_2}$ we have a probability of success of

$$P_s = 1 - \left(1 - 2^{\lambda-k}\right)^{2^{-2\lambda}\binom{n}{w_0+w_1}\binom{n}{w_2}} \tag{9}$$

$$\approx 2^{-\lambda-k}\binom{n}{w_0+w_1}\binom{n}{w_2} + O(2^{2(\lambda-k)}). \tag{10}$$

The online complexity is $O(\mathrm{C} \cdot P_s^{-1})$. The offline complexity is dominated by sorting the largest list of L_0, L_1 and L_2, as merging L_0 and L_1 can be done in linear time. Therefore, the offline complexity is $O\left(\binom{n}{\lceil w/3\rceil}\log_2\binom{n}{\lceil w/3\rceil}\right)$.

Quantumly, there is no speed-up for sorting, and so the offline phase might as well remain classical. The online phase can be improved by applying Grover's algorithm to the "random" guess for the counter i. While sorted list lookup requires only $\frac{1}{\pi}(\ln(n) - 1)$ operations [19], this speed-up factor is hidden by the big-O. The λ that minimizes the online quantum complexity $O(\mathrm{C} \cdot P_s^{-1/2})$ is small enough to make the offline complexity the algorithm's bottleneck. All complexities are larger than 2^{128} for the parameter set $m = 26, t = 15$, with $\lambda = 31$ being the smallest such value for which the offline complexity is larger than the quantum online complexity. At this point the public key is a bit matrix of $(15 \cdot 26) \times 2^{26}$ elements, or roughly 3.05 GB. In contrast, a signature represents a bitstring of length 2^{26} and of Hamming weight 15, which can be straightforwardly represented as 15 integers of 26 bits each, by 390 bits in total.

References

1. Albrecht, M.R., Player, R., Scott, S.: On the concrete hardness of learning with errors. J. Math. Cryptol. **9**(3), 169–203 (2015)
2. Alperin-Sheriff, J., Lee, Y., Perlner, R., Lee, W., Moody, D.: Official comments on pqsigRM (2018). https://csrc.nist.gov/CSRC/media/Projects/Post-Quantum-Cryptography/documents/round-1/official-comments/pqsigRM-official-comment.pdf
3. Alwen, J., Peikert, C.: Generating shorter bases for hard random lattices. In: Albers, S., Marion, J. (eds.) 26th International Symposium on Theoretical Aspects of Computer Science, STACS 2009. Proceedings of LIPIcs, Freiburg, Germany, 26–28 February 2009, vol. 3, pp. 75–86. Schloss Dagstuhl - Leibniz-Zentrum fuer Informatik, Germany (2009). https://doi.org/10.4230/LIPIcs.STACS.2009.1832

4. Aumasson, J.P., Endignoux, G.: Improving stateless hash-based signatures. Cryptology ePrint Archive, Report 2017/933 (2017). http://eprint.iacr.org/2017/933
5. Bernstein, D., et al.: SPHINCS: Practical Stateless Hash-Based Signatures. In: Oswald, E., Fischlin, M. (eds.) EUROCRYPT 2015. LNCS, vol. 9056, pp. 368–397. Springer, Heidelberg (2015). https://doi.org/10.1007/978-3-662-46800-5_15
6. Boneh, D., Dagdelen, Ö., Fischlin, M., Lehmann, A., Schaffner, C., Zhandry, M.: Random oracles in a quantum world. In: Lee, D.H., Wang, X. (eds.) ASIACRYPT 2011. LNCS, vol. 7073, pp. 41–69. Springer, Heidelberg (2011). https://doi.org/10.1007/978-3-642-25385-0_3
7. Chen, M.-S., Hülsing, A., Rijneveld, J., Samardjiska, S., Schwabe, P.: From 5-pass $\mathcal{MQ}$-based identification to $\mathcal{MQ}$-based signatures. In: Cheon, J.H., Takagi, T. (eds.) ASIACRYPT 2016. LNCS, vol. 10032, pp. 135–165. Springer, Heidelberg (2016). https://doi.org/10.1007/978-3-662-53890-6_5
8. Courtois, N.T., Finiasz, M., Sendrier, N.: How to achieve a McEliece-based digital signature scheme. In: Boyd, C. (ed.) ASIACRYPT 2001. LNCS, vol. 2248, pp. 157–174. Springer, Heidelberg (2001). https://doi.org/10.1007/3-540-45682-1_10
9. Debris-Alazard, T., Sendrier, N., Tillich, J.: A new signature scheme based on (U|U+V) codes. IACR Cryptology ePrint Archive 2017/662 (2017). http://eprint.iacr.org/2017/662
10. Diffie, W., Hellman, M.E.: New directions in cryptography. IEEE Trans. Inf. Theor. 22(6), 644–654 (1976). https://doi.org/10.1109/TIT.1976.1055638
11. Ding, J., Schmidt, D.: Rainbow, a new multivariable polynomial signature scheme. In: Ioannidis, J., Keromytis, A., Yung, M. (eds.) ACNS 2005. LNCS, vol. 3531, pp. 164–175. Springer, Heidelberg (2005). https://doi.org/10.1007/11496137_12
12. Faugère, J.C., Perret, L., Ryckeghem, J.: DualModeMS: a dual mode for Multivariate-based signature 20170918 draft. UPMC-Paris 6 Sorbonne Universités; INRIA Paris; CNRS (2017)
13. Finiasz, M., Sendrier, N.: Security bounds for the design of code-based cryptosystems. In: Matsui, M. (ed.) ASIACRYPT 2009. LNCS, vol. 5912, pp. 88–105. Springer, Heidelberg (2009). https://doi.org/10.1007/978-3-642-10366-7_6. [24]
14. Fouque, P.A., et al.: Falcon (2017). submission to the NIST PQC project
15. Gentry, C., Peikert, C., Vaikuntanathan, V.: Trapdoors for hard lattices and new cryptographic constructions. In: Dwork, C. (ed.) Proceedings of the 40th Annual ACM Symposium on Theory of Computing, Victoria, British Columbia, Canada, 17–20 May 2008, pp. 197–206. ACM (2008). https://doi.org/10.1145/1374376.1374407
16. Goldreich, O., Goldwasser, S., Halevi, S.: Public-key cryptosystems from lattice reduction problems. In: Kaliski, B.S. (ed.) CRYPTO 1997. LNCS, vol. 1294, pp. 112–131. Springer, Heidelberg (1997). https://doi.org/10.1007/BFb0052231
17. Goldwasser, S., Micali, S., Rivest, R.L.: A digital signature scheme secure against adaptive chosen-message attacks. SIAM J. Comput. 17(2), 281–308 (1988). https://doi.org/10.1137/0217017
18. Güneysu, T., Lyubashevsky, V., Pöppelmann, T.: Practical lattice-based cryptography: a signature scheme for embedded systems. In: Prouff, E., Schaumont, P. (eds.) CHES 2012. LNCS, vol. 7428, pp. 530–547. Springer, Heidelberg (2012). https://doi.org/10.1007/978-3-642-33027-8_31
19. Høyer, P., Neerbek, J., Shi, Y.: Quantum complexities of ordered searching, sorting, and element distinctness. In: Orejas, F., Spirakis, P.G., van Leeuwen, J. (eds.) ICALP 2001. LNCS, vol. 2076, pp. 346–357. Springer, Heidelberg (2001). https://doi.org/10.1007/3-540-48224-5_29

20. Hülsing, A., Rijneveld, J., Song, F.: Mitigating multi-target attacks in hash-based signatures. In: Cheng, C.-M., Chung, K.-M., Persiano, G., Yang, B.-Y. (eds.) PKC 2016. LNCS, vol. 9614, pp. 387–416. Springer, Heidelberg (2016). https://doi.org/10.1007/978-3-662-49384-7_15

21. Kipnis, A., Patarin, J., Goubin, L.: Unbalanced oil and vinegar signature schemes. In: Stern, J. (ed.) EUROCRYPT 1999. LNCS, vol. 1592, pp. 206–222. Springer, Heidelberg (1999). https://doi.org/10.1007/3-540-48910-X_15

22. Lee, W., Kim, Y.S., Lee, Y.W., Kim, J.-S.: pqsigRM (2017). submission to the NIST PQC project

23. Lyubashevsky, V.: Fiat-Shamir with aborts: applications to lattice and factoring-based signatures. In: Matsui, M. (ed.) ASIACRYPT 2009. LNCS, vol. 5912, pp. 598–616. Springer, Heidelberg (2009). https://doi.org/10.1007/978-3-642-10366-7_35. [24]

24. Matsui, M. (ed.): ASIACRYPT 2009. LNCS, vol. 5912. Springer, Heidelberg (2009). https://doi.org/10.1007/978-3-642-10366-7

25. Matsumoto, T., Imai, H.: Public quadratic polynomial-tuples for efficient signature-verification and message-encryption. In: Barstow, D., et al. (eds.) EUROCRYPT 1988. LNCS, vol. 330, pp. 419–453. Springer, Heidelberg (1988). https://doi.org/10.1007/3-540-45961-8_39

26. Merkle, R.C., Charles, R., et al.: Secrecy, authentication, and public key systems (1979)

27. Micciancio, D., Peikert, C.: Trapdoors for lattices: simpler, tighter, faster, smaller. IACR Cryptology ePrint Archive 2011/501 (2011). http://eprint.iacr.org/2011/501

28. Micciancio, D., Regev, O.: Lattice-based cryptography. In: Bernstein, D.J., Buchmann, J., Dahmen, E. (eds.) Post-Quantum Cryptography. Springer, Heidelberg (2009). https://doi.org/10.1007/978-3-540-88702-7_5

29. National Institute for Standards and Technology (NIST): post-quantum crypto standardization (2018). http://csrc.nist.gov/groups/ST/post-quantum-crypto/

30. National Institute of Standards and Technology: FIPS PUB 186–4: Digital Signature Standard (DSS) (2013). http://nvlpubs.nist.gov/nistpubs/FIPS/NIST.FIPS.186-4.pdf

31. Nguyen, P.Q., Regev, O.: Learning a parallelepiped: cryptanalysis of GGH and NTRU signatures. In: Vaudenay, S. (ed.) EUROCRYPT 2006. LNCS, vol. 4004, pp. 271–288. Springer, Heidelberg (2006). https://doi.org/10.1007/11761679_17

32. Patarin, J.: Hidden fields equations (HFE) and Isomorphisms of polynomials (IP): two new families of asymmetric algorithms. In: Maurer, U. (ed.) EUROCRYPT 1996. LNCS, vol. 1070, pp. 33–48. Springer, Heidelberg (1996). https://doi.org/10.1007/3-540-68339-9_4

33. Petzoldt, A., Bulygin, S., Buchmann, J.: CyclicRainbow – a multivariate signature scheme with a partially cyclic public key. In: Gong, G., Gupta, K.C. (eds.) INDOCRYPT 2010. LNCS, vol. 6498, pp. 33–48. Springer, Heidelberg (2010). https://doi.org/10.1007/978-3-642-17401-8_4

34. Petzoldt, A., Chen, M.-S., Yang, B.-Y., Tao, C., Ding, J.: Design principles for HFEv- based multivariate signature schemes. In: Iwata, T., Cheon, J.H. (eds.) ASIACRYPT 2015. LNCS, vol. 9452, pp. 311–334. Springer, Heidelberg (2015). https://doi.org/10.1007/978-3-662-48797-6_14

35. Rivest, R.L., Shamir, A., Adleman, L.M.: A method for obtaining digital signatures and public-key cryptosystems. Commun. ACM **21**(2), 120–126 (1978). https://doi.org/10.1145/359340.359342

36. Schnorr, C.P.: Efficient identification and signatures for smart cards. In: Brassard, G. (ed.) CRYPTO 1989. LNCS, vol. 435, pp. 239–252. Springer, New York (1990). https://doi.org/10.1007/0-387-34805-0_22

37. Shor, P.W.: Algorithms for quantum computation: discrete logarithms and factoring. In: 35th Annual Symposium on Foundations of Computer Science, Santa Fe, New Mexico, USA, 20–22 November 1994, pp. 124–134. IEEE Computer Society (1994). https://doi.org/10.1109/SFCS.1994.365700

38. Stern, J.: A new paradigm for public key identification. IEEE Trans. Inf. Theor. **42**(6), 1757–1768 (1996). https://doi.org/10.1109/18.556672

39. Szepieniec, A., Beullens, W., Preneel, B.: MQ signatures for PKI. In: Lange, T., Takagi, T. (eds.) PQCrypto 2017. LNCS, vol. 10346, pp. 224–240. Springer, Cham (2017). https://doi.org/10.1007/978-3-319-59879-6_13

40. Unruh, D.: Non-interactive zero-knowledge proofs in the quantum random oracle model. In: Oswald, E., Fischlin, M. (eds.) EUROCRYPT 2015. LNCS, vol. 9057, pp. 755–784. Springer, Heidelberg (2015). https://doi.org/10.1007/978-3-662-46803-6_25

41. Wendl, M.C.: Collision probability between sets of random variables. Stat. Probab. Lett. **64**(3), 249–254 (2003)

On the Cost of Computing Isogenies Between Supersingular Elliptic Curves

Gora Adj[1], Daniel Cervantes-Vázquez[2], Jesús-Javier Chi-Domínguez[2], Alfred Menezes[1], and Francisco Rodríguez-Henríquez[2(✉)]

[1] Department of Combinatorics and Optimization, University of Waterloo, Waterloo, Canada
gora.adj@gmail.com, ajmeneze@uwaterloo.ca
[2] Computer Science Department, CINVESTAV-IPN, Mexico City, Mexico
{dcervantes,jjchi}@computacion.cs.cinvestav.mx, francisco@cs.cinvestav.mx

Abstract. The security of the Jao-De Feo Supersingular Isogeny Diffie-Hellman (SIDH) key agreement scheme is based on the intractability of the Computational Supersingular Isogeny (CSSI) problem—computing $\mathbb{F}_{p^2}$-rational isogenies of degrees 2^e and 3^e between certain supersingular elliptic curves defined over $\mathbb{F}_{p^2}$. The classical meet-in-the-middle attack on CSSI has an expected running time of $O(p^{1/4})$, but also has $O(p^{1/4})$ storage requirements. In this paper, we demonstrate that the van Oorschot-Wiener golden collision finding algorithm has a lower cost (but higher running time) for solving CSSI, and thus should be used instead of the meet-in-the-middle attack to assess the security of SIDH against classical attacks. The smaller parameter p brings significantly improved performance for SIDH.

1 Introduction

The Supersingular Isogeny Diffie-Hellman (SIDH) key agreement scheme was proposed by Jao and De Feo [14] (see also [9]). It is one of 69 candidates being considered by the U.S. government's National Institute of Standards and Technology (NIST) for inclusion in a forthcoming standard for quantum-safe cryptography [13]. The security of SIDH is based on the difficulty of the Computational Supersingular Isogeny (CSSI) problem, which was first defined by Charles, Goren and Lauter [4] in their paper that introduced an isogeny-based hash function. The CSSI problem is also the basis for the security of isogeny-based signature schemes [11,30] and an undeniable signature scheme [15].

Let p be a prime, let ℓ be a small prime (e.g., $\ell \in \{2,3\}$), and let E and E' be two supersingular elliptic curves defined over $\mathbb{F}_{p^2}$ for which a (separable) degree-ℓ^e isogeny $\phi : E \to E'$ defined over $\mathbb{F}_{p^2}$ exists. The CSSI problem is that of constructing such an isogeny. In [9], the CSSI problem is assessed as having a complexity of $O(p^{1/4})$ and $O(p^{1/6})$ against classical and quantum attacks [26], respectively. The classical attack is a meet-in-the-middle attack (MITM) that has time complexity $O(p^{1/4})$ and space complexity $O(p^{1/4})$. We observe that the (classical) van Oorschot-Wiener golden collision finding algorithm [19,20] can be employed to construct ϕ. Whereas the time complexity of the van Oorschot-Wiener algorithm is higher than that of the meet-in-the-middle attack, its space requirements are smaller. Our cost analysis of these two CSSI attacks leads to the conclusion that, despite its higher running time, the golden collision finding CSSI attack has a lower cost than the meet-in-the-middle attack, and thus should be used to assess the security of SIDH against (known) classical attacks.

The remainder of this paper is organized as follows. The CSSI problem and relevant mathematics background are presented in Sect. 2. In Sects. 3 and 4, we report on our implementation of the meet-in-the-middle and golden collision search methods for solving CSSI. Our implementations confirm that the heuristic analysis of these CSSI attacks accurately predicts their performance in practice. Our cost models and cost comparisons are presented in Sect. 5. Finally, in Sect. 6 we make some concluding remarks.

2 Computational Supersingular Isogeny Problem

2.1 Mathematical Prerequisites

Let $p = \ell_A^{e_A} \ell_B^{e_B} - 1$ be a prime[1], where ℓ_A and ℓ_B are distinct small primes and $\ell_A^{e_A} \approx \ell_B^{e_B} \approx p^{1/2}$. Let E be a (supersingular) elliptic curve defined over $\mathbb{F}_{p^2}$ with $\#E(\mathbb{F}_{p^2}) = (p+1)^2$. Then $E(\mathbb{F}_{p^2}) \cong \mathbb{Z}_{p+1} \oplus \mathbb{Z}_{p+1}$, whence the torsion groups $E[\ell_A^{e_A}]$ and $E[\ell_B^{e_B}]$ are contained in $E(\mathbb{F}_{p^2})$.

In the following, we write (ℓ, e) to mean either (ℓ_A, e_A) or (ℓ_B, e_B). All isogenies ϕ considered in this paper are separable, whereby $\deg \phi = \#\mathrm{Ker}(\phi)$.

Let S be an order-ℓ^e subgroup of $E[\ell^e]$. Then there exists an isogeny $\phi : E \to E'$ (with both ϕ and E' defined over $\mathbb{F}_{p^2}$) with kernel S. The isogeny ϕ is unique up to isomorphism in the sense that if $\tilde{\phi} : E \to \tilde{E}$ is another isogeny defined over $\mathbb{F}_{p^2}$ with kernel S, then there exists an $\mathbb{F}_{p^2}$-isomorphism $\psi : E' \to \tilde{E}$ with $\tilde{\phi} = \psi \circ \phi$.

Given E and S, an isogeny ϕ with kernel S and the equation of E' can be computed using Vélu's formula [27]. The running time of Vélu's formula is polynomial in $\#S$ and $\log p$. Since $\#S \approx p^{1/2}$, a direct application of Vélu's formula does not yield a polynomial-time algorithm for computing ϕ and E'. However, since $\#S$ is a power of a small prime, one can compute ϕ and E' in time that is polynomial in $\log p$ by using Vélu's formula to compute a sequence of e degree-ℓ isogenies (see Sect. 2.2).

We will denote the elliptic curve that Vélu's formula yields by E/S and the (Vélu) isogeny by $\phi_S : E \to E/S$. As noted above, ϕ_S is unique up to isomorphism. Thus, for any fixed E, there is a one-to-one correspondence between order-ℓ^e subgroups of $E[\ell^e]$ and degree-ℓ^e isogenies $\phi : E \to E'$ defined over $\mathbb{F}_{p^2}$. It follows that the number of degree-ℓ^e isogenies $\phi : E \to E'$ is $\ell^e + \ell^{e-1} = (\ell+1)\ell^{e-1}$.

---

[1] More generally, one can take $p = \ell_A^{e_A} \ell_B^{e_B} d \pm 1$ where d is a small cofactor.

2.2 Vélu's Formula

Vélu's formula (see [4]) can be used to compute degree-ℓ isogenies. We present Vélu's formula for $\ell = 2$ and $\ell = 3$.

Consider the elliptic curve $E/\mathbb{F}_{p^2} : Y^2 = X^3 + aX + b$, and let $P = (X_P, Y_P) \in E(\mathbb{F}_{p^2})$ be a point of order two. Let $v = 3X_P^2 + a$, $a' = a - 5v$, $b' = b - 7vX_P$, and define the elliptic curve $E'/\mathbb{F}_{p^2} : Y^2 = X^3 + a'X + b'$. Then the map

$$(X, Y) \mapsto \left(X + \frac{v}{X - X_P}, \ Y - \frac{vY}{(X - X_P)^2} \right)$$

is a degree-2 isogeny from E to E' with kernel $\langle P \rangle$.

Let $P = (X_P, Y_P) \in E(\mathbb{F}_{p^2})$ be a point of order three. Let $v = 6X_P^2 + 2a$, $u = 4Y_P^2$, $a' = a - 5v$, $b' = b - 7(u + vX_P)$, and define the elliptic curve $E'/\mathbb{F}_{p^2} : Y^2 = X^3 + a'X + b'$. Then the map

$$(X, Y) \mapsto \left(X + \frac{v}{X - X_P} + \frac{u}{(X - X_P)^2}, \ Y \left(1 - \frac{v}{(X - X_P)^2} - \frac{2u}{(X - X_P)^3} \right) \right)$$

is a degree-3 isogeny from E to E' with kernel $\langle P \rangle$.

Suppose now that $R \in E(\mathbb{F}_{p^2})$ has order ℓ^e where $\ell \in \{2, 3\}$ and $e \geq 1$. Then the isogeny $\phi : E \to E/\langle R \rangle$ can be efficiently computed as follows. Define $E_0 = E$ and $R_0 = R$. For $i = 0, 1, \ldots, e-1$, let $\phi_i : E_i \to E_{i+1}$ be the degree-ℓ isogeny obtained using Vélu's formula with kernel $\langle \ell^{e-1-i}R_i \rangle$, and let $R_{i+1} = \phi_i(R_i)$. Then $\phi = \phi_{e-1} \circ \cdots \circ \phi_0$.

Remark 1 (cost of computing an ℓ^e-isogeny). As shown in [9], a 'balanced strategy' for computing a degree-ℓ^e isogeny requires approximately $\frac{e}{2} \log_2 e$ point multiplications by ℓ, $\frac{e}{2} \log_2 e$ degree-ℓ isogeny evaluations, and e constructions of degree-ℓ isogenous curves. Also presented in [9] is a slightly faster 'optimal strategy' that accounts for the relative costs of a point multiplication and a degree-ℓ isogeny evaluation.

2.3 SIDH

In SIDH, the parameters ℓ_A, ℓ_B, e_A, e_B, p and E are fixed and public, as are bases $\{P_A, Q_A\}$ and $\{P_B, Q_B\}$ for the torsion groups $E[\ell_A^{e_A}]$ and $E[\ell_B^{e_B}]$.

In (unauthenticated) SIDH, Alice selects $m_A, n_A \in_R [0, \ell_A^{e_A} - 1]$, not both divisible by ℓ_A, and sets $R_A = m_A P_A + n_A Q_A$ and $A = \langle R_A \rangle$; note that A is an order-$\ell_A^{e_A}$ subgroup of $E[\ell_A^{e_A}]$. Alice then computes the isogeny $\phi_A : E \to E/A$ while keeping A and ϕ_A secret. She transmits

$$E/A, \quad \phi_A(P_B), \quad \phi_A(Q_B)$$

to Bob. Similarly, Bob selects $m_B, n_B \in_R [0, \ell_B^{e_B} - 1]$, not both divisible by ℓ_B, and sets $R_B = m_B P_B + n_B Q_B$ and $B = \langle R_B \rangle$. Bob then computes the isogeny $\phi_B : E \to E/B$. He keeps B and ϕ_B secret and transmits

$$E/B, \quad \phi_B(P_A), \quad \phi_B(Q_A)$$

to Alice. Thereafter, Alice computes $\phi_B(R_A) = m_A\phi_B(P_A) + n_A\phi_B(Q_A)$ and

$$(E/B)/\langle\phi_B(R_A)\rangle,$$

whereas Bob computes $\phi_A(R_B) = m_B\phi_A(P_B) + n_B\phi_A(Q_B)$ and

$$(E/A)/\langle\phi_A(R_B)\rangle.$$

The compositions of isogenies

$$E \to E/A \to (E/A)/\langle\phi_A(R_B)\rangle$$

and

$$E \to E/B \to (E/B)/\langle\phi_B(R_A)\rangle$$

both have kernel $\langle R_A, R_B\rangle$. Hence the elliptic curves computed by Alice and Bob are isomorphic over $\mathbb{F}_{p^2}$, and their shared secret k is the j-invariant of these curves.

Remark 2 (SIDH vs. SIKE). SIDH is an unauthenticated key agreement protocol. The NIST submission [13] specifies a variant of SIDH that is a key encapsulation mechanism (KEM) called SIKE (Supersingular Isogeny Key Encapsulation). In SIKE, Alice's long-term public key is $(E/A, \phi_A(P_B), \phi_A(Q_B))$. Bob sends Alice an ephemeral public key $(E/B, \phi_B(P_A), \phi_B(Q_A))$ where B is derived from Alice's public key and a random string, and then computes a session key from the j-invariant of the elliptic curve $(E/A)/\langle\phi_A(R_B)\rangle$, the aforementioned random string, and the ephemeral public key. One technical difference between the original SIDH specification in [9,14] and the SIKE specification in [13] (and also the SIDH implementation in [5]) is that in the latter the secret R_A is of the form $P_A + n_A Q_A$ where n_A is selected (almost) uniformly at random from the interval $[0, \ell_A^{e_A} - 1]$ (and similarly for R_B). Thus, R_A is selected uniformly at random from a subset of size approximately ℓ^{e_A} of the set of all order-$\ell_A^{e_A}$ subgroups (which has cardinality $\ell_A^{e_A} + \ell_A^{e_A-1}$).

2.4   CSSI

The challenge faced by a passive adversary is to compute k given the public parameters, E/A, E/B, $\phi_A(P_B)$, $\phi_A(Q_B)$, $\phi_B(P_A)$ and $\phi_B(Q_A)$. A necessary condition for hardness of this problem is the intractability of the Computational Supersingular Isogeny (CSSI) problem: Given the public parameters ℓ_A, ℓ_B, e_A, e_B, p, E, P_A, Q_A, P_B, Q_B, the elliptic curve E/A, and the auxiliary points $\phi_A(P_B)$ and $\phi_A(Q_B)$, compute the Vélu isogeny $\phi_A : E \to E/A$ (or, equivalently, determine a generator of A).

An assumption one makes (e.g., see [9]) is that the auxiliary points $\phi_A(P_B)$ and $\phi_A(Q_B)$ are of no use in solving CSSI. Thus, we can simplify the statement of the CSSI problem to the following:

Problem 1 (CSSI). Given the public parameters ℓ_A, ℓ_B, e_A, e_B, p, E, P_A, Q_A, and the elliptic curve E/A, compute a degree-$\ell_A^{e_A}$ isogeny $\phi_A : E \to E/A$.

3   Meet-in-the-Middle

For the sake of simplicity, we will suppose that e is even. We denote the number of order-$\ell^{e/2}$ subgroups of $E[\ell^e]$ by $N = (\ell+1)\ell^{e/2-1} \approx p^{1/4}$.

Let $E_1 = E$ and $E_2 = E/A$. Let R denote the set of all j-invariants of elliptic curves that are isogenous to E_1; then $\#R \approx p/12$ [23]. Let R_1 denote the set of all j-invariants of elliptic curves over $\mathbb{F}_{p^2}$ that are $\ell^{e/2}$-isogenous to E_1. Since $\#R \gg N$, one expects that the number of pairs of distinct order-$\ell^{e/2}$ subgroups (A_1, A_2) of $E_1[\ell^e]$ with $j(E_1/A_1) = j(E_1/A_2)$ is very small. Thus, we shall assume for the sake of simplicity that $\#R_1 = N$. Similarly, we let R_2 denote the set of all j-invariants of elliptic curves that are $\ell^{e/2}$-isogenous to E_2, and assume that $\#R_2 = N$. Since E_1 is ℓ^e-isogenous to E_2, we know that $R_1 \cap R_2 \neq \emptyset$. Moreover, since $\#R_1 \ll \#R$ and $\#R_2 \ll \#R$, it is reasonable to assume that $\#(R_1 \cap R_2) = 1$; in other words, we can assume that there is a unique degree-ℓ^e isogeny $\phi : E_1 \to E_2$.

3.1   Basic Method

The meet-in-the-middle attack on CSSI [9], which we denote by MITM-basic, proceeds by building a (sorted) table with entries $(j(E_1/A_1), A_1)$, where A_1 ranges over all order-$\ell^{e/2}$ subgroups of $E_1[\ell^e]$. Next, for each order-$\ell^{e/2}$ subgroup A_2 of $E_2[\ell^e]$, one computes E_2/A_2 and searches for $j(E_2/A_2)$ in the table (see Fig. 1). If $j(E_2/A_2) = j(E_1/A_1)$, then the composition of isogenies

$$\phi_{A_1} : E_1 \to E_1/A_1, \quad \psi : E_1/A_1 \to E_2/A_2, \quad \hat{\phi}_{A_2} : E_2/A_2 \to E_2,$$

where ψ is an $\mathbb{F}_{p^2}$-isomorphism and $\hat{\phi}_{A_2}$ denotes the dual of ϕ_{A_2}, is the desired degree-ℓ^e isogeny from E_1 to E_2. The worst-case time complexity of MITM-basic is $T_1 = 2N$, where a unit of time is a degree-$\ell^{e/2}$ Vélu isogeny computation (cf. Remark 1). The average-case time complexity is $1.5N$. The attack has space complexity N.

3.2   Depth-First Search

The set of pairs $(j(E/A), A)$, with A ranging over all order-$\ell^{e/2}$ subgroups of $E[\ell^e]$, can also be generated by using a depth-first search (DFS) to traverse the tree in the left of Fig. 1 (and also the tree in the right of Fig. 1). We denote this variant of the meet-in-the-middle attack by MITM-DFS. We describe the depth-first search for $\ell = 2$.[2]

Let $\{P, Q\}$ be a basis for $E[2^{e/2}]$. Let $R_0 = 2^{e/2-1}P$, $R_1 = 2^{e/2-1}Q$, $R_2 = R_0 + R_1$ be the order-2 points on E. For $i = 0, 1, 2$, the degree-2 isogenies $\phi_i : E \to E_i = E/\langle R_i \rangle$ are computed, as are bases $\{P_0 = \phi_0(P), Q_0 = \phi_0(2Q)\}$,

---

[2] For the sake of concreteness, all implementation reports of CSSI attacks in this paper are for the case $\ell = 2$. However, all conclusions about the relative efficiencies of classical and quantum CSSI attacks for $\ell = 2$ are also valid for the $\ell = 3$ case.

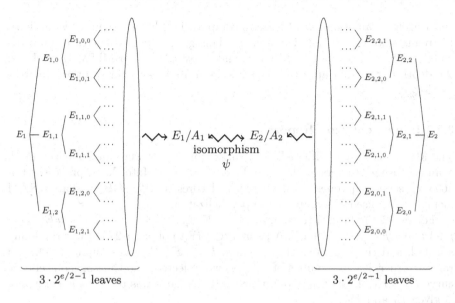

Fig. 1. Meet-in-the-middle attack for degree-2 isogeny trees.

$\{P_1 = \phi_1(Q), Q_1 = \phi_1(2P)\}$, $\{P_2 = \phi_2(P+Q), Q_2 = \phi_2(2P)\}$ for $E_0[2^{e/2-1}]$, $E_1[2^{e/2-1}]$, $E_2[2^{e/2-1}]$, respectively. A memory stack is initialized with the tuples $(E_0, 0, P_0, Q_0)$, $(E_1, 1, P_1, Q_1)$, $(E_2, 2, P_2, Q_2)$, and the tuple on the top of the stack is processed recursively as described next.

Suppose that we have to process (E_x, x, P_x, Q_x), where $x \in \{0,1,2\} \times \{0,1\}^{n-1}$ and $1 \le n \le e/2 - 1$. Let $B_0 = 2^{e/2-n-1}P_x$, $B_1 = 2^{e/2-n-1}Q_x$ and $B_2 = B_0 + B_1$ be the order-2 points on E_x. Let $R_{x0} = B_0$ and $R_{x1} = B_2$ (B_1 is the backtracking point), and compute the degree-2 isogenies $\phi_{xi} : E_x \to E_{xi} = E_x/\langle R_{xi}\rangle$ for $i = 0, 1$. Then two cases arise:

(i) If $n < e/2 - 1$, then let $P_{x0} = \phi_{x0}(P_x)$, $Q_{x0} = \phi_{x0}(2(P_x + Q_x))$, $P_{x1} = \phi_{x1}(P_x + Q_x)$, $Q_{x1} = \phi_{x1}(2P_x)$; one can check that $\{P_{xi}, Q_{xi}\}$ is a basis for $E_{xi}[2^{e/2-n-1}]$ for $i = 0, 1$. Then, $(E_{x1}, x1, P_{x1}, Q_{x1})$ is added to the stack and $(E_{x0}, x0, P_{x0}, Q_{x0})$ is processed next.

(ii) If $n = e/2 - 1$, the leaves $(j(E_{x0}), x0)$ and $(j(E_{x1}), x1)$ of the tree are stored in the table. If the stack is non-empty, then its topmost entry is processed next; otherwise the computation terminates.

The cost of building each of the two depth-first search trees is approximately $2N$ degree-2 isogeny computations, $2N$ degree-2 isogeny evaluations, $N/2$ point additions, and $2N$ point doublings (where $N = 3 \cdot 2^{e/2-1}$).

In contrast, the cost of building the table in MITM-basic (with $\ell = 2$) is approximately $\frac{Ne}{2}$ 2-isogeny computations, $\frac{Ne}{4}\log_2\frac{e}{2}$ 2-isogeny evaluations, and $\frac{Ne}{4}\log_2\frac{e}{2}$ point doublings (cf. Remark 1). A count of $\mathbb{F}_{p^2}$ multiplications and squarings yields the following costs for the core operations when Jacobian coordinates are used for elliptic curve arithmetic, isogeny computations, and isogeny evaluations: 8 (2-isogeny computation), 12 (2-isogeny evaluation), 14 (point addition), 9 (point doubling). This gives a per-table cost of approximately $5.25Ne \log_2 e$ for MITM-basic, and a cost of $65N$ for MITM-DFS. Thus, the depth-first search approach yields a speedup by a factor of approximately $\frac{e}{12.4} \log_2 e$.

3.3 Implementation Report

The MITM-basic and MITM-DFS attacks (for $\ell = 2$) were implemented in C, compiled using gcc version 4.7.2, and executed on an Intel Xeon processor E5-2658 v2 server equipped with 20 physical cores and 256 GB of shared RAM memory.[3] We used fopenmp for the parallelization.

For $p = 2^{e_A} 3^{e_B} d - 1$, the elliptic curve $E/\mathbb{F}_p : Y^2 = X^3 + X$ has $\#E(\mathbb{F}_p) = p+1$ and $\#E(\mathbb{F}_{p^2}) = (p+1)^2$. A point $P \in E(\mathbb{F}_{p^2})$ of order $2 \cdot 3 \cdot d$ was randomly selected, and the isogenous elliptic curve $E_1 = E/\langle P \rangle$ was computed. Then, a random order-2^{e_A} subgroup A of $E_1(\mathbb{F}_{p^2})$ was selected, and the isogenous elliptic curve $E_2 = E_1/A$ was computed. Our CSSI challenge was to find a generator of A given E_1 and E_2.

We used Jacobian coordinates for elliptic curve arithmetic, isogeny computations, and isogeny evaluations. For MITM-basic, the leaves of the E_1-rooted tree shown in Fig. 1 were generated as follows. Let $\{P, Q\}$ be a basis for $E_1[2^{e/2}]$. Then for each pair $(b, k) \in \{0, 1, 2\} \times \{0, 1, \ldots, 2^{e/2-1} - 1\}$, triples

$$\left(j(E_1/\langle P + (b2^{e/2-1} + k)Q \rangle), \; b, \; b2^{e/2-1} + k \right), \text{ for } b = 0, 1,$$

$$(j(E_1/\langle (2k)P + Q \rangle), \; b, \; k), \text{ for } b = 2,$$

Table 1. Meet-in-the-middle attacks for finding a 2^{e_A}-isogeny between two supersingular elliptic curves over $\mathbb{F}_{p^2}$ with $p = 2^{e_A} \cdot 3^{e_B} \cdot d - 1$. For each p, 25 randomly generated CSSI instances were solved and the average of the results are reported. The 'expected time' and 'measured time' columns give the expected number and the actual number of degree-$2^{e_A/2}$ isogeny computations for MITM-basic. The space is measured in bytes.

e_A	e_B	d	MITM-basic				MITM-DFS
			Expected time	Space	Measured time	Clock cycles	Clock cycles
32	20	23	$2^{17.17}$	$2^{20.72}$	$2^{17.26}$	$2^{34.50}$	$2^{31.73}$
34	21	109	$2^{18.17}$	$2^{21.83}$	$2^{18.24}$	$2^{35.49}$	$2^{32.71}$
36	22	31	$2^{19.17}$	$2^{22.87}$	$2^{19.14}$	$2^{36.43}$	$2^{33.67}$
38	23	271	$2^{20.17}$	$2^{23.99}$	$2^{20.20}$	$2^{37.59}$	$2^{34.60}$
40	25	71	$2^{21.17}$	$2^{25.04}$	$2^{21.15}$	$2^{38.63}$	$2^{35.71}$
42	26	37	$2^{22.17}$	$2^{26.09}$	$2^{22.11}$	$2^{39.83}$	$2^{36.78}$
44	27	37	$2^{23.17}$	$2^{27.14}$	$2^{23.25}$	$2^{41.07}$	$2^{37.87}$

---

[3] Our code for the MITM-basic, MITM-DFS and VW golden collision search CSSI attacks is available at https://github.com/JJChiDguez/CSSI.

were computed and stored in 20 tables sorted by j-invariant (each of the 20 cores was responsible for generating a portion of the leaves). The 20 tables were stored in shared RAM memory.

MITM-DFS was executed using 12 cores. Each core was responsible for generating a portion of the leaves, and the 12 sets of leaves were stored in shared RAM memory. Table 1 shows the time expended for finding 2^e-isogenies for $e \in \{32, 34, 36, 38, 40, 42, 44\}$ with the MITM-basic and MITM-DFS attacks. These experimental results confirm the accuracy of the attacks' heuristic analysis.

4   Golden Collision Search

4.1   Van Oorschot-Wiener Parallel Collision Search

Let S be a finite set of cardinality M, and let $f : S \to S$ be an efficiently-computable function which we shall heuristically assume is a random function. The van Oorschot-Wiener (VW) method [20] finds a collision for f, i.e., a pair $x, x' \in S$ with $f(x) = f(x')$ and $x \neq x'$.

Define an element x of S to be *distinguished* if it has some easily-testable distinguishing property. Suppose that the proportion of elements of S that are distinguished is θ. For $i = 1, 2, \ldots$, the VW method repeatedly selects $x_{i,0} \in_R S$, and iteratively computes a sequence $x_{i,j} = f(x_{i,j-1})$ for $j = 1, 2, 3, \ldots$ until a distinguished element $x_{i,a}$ is encountered. In that event, the triple $(x_{i,a}, a, x_{i,0})$ is stored in a table sorted by first entry. If $x_{i,a}$ was already in the table, say $x_{i,a} = x_{i',b}$ with $i \neq i'$, then a collision has been *detected* (see Fig. 2). The two colliding table entries $(x_{i,a}, a, x_{i,0})$, $(x_{i',b}, b, x_{i',0})$ can then be used to find a collision for f by iterating the longer sequence (say the ith sequence) beginning at $x_{i,0}$ until it is the same distance from $x_{i,a}$ as $x_{i',0}$ is from $x_{i',b}$, and then stepping both sequences in unison until they collide (see Fig. 3).

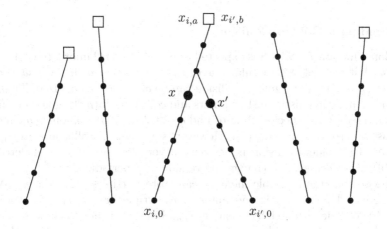

Fig. 2. VW method: detecting a collision (x, x').

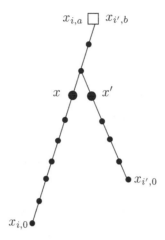

Fig. 3. VW method: finding a collision (x, x').

By the birthday paradox, the expected time before a collision occurs is $\sqrt{\pi M/2}$, where a unit of time is an f evaluation. After a collision has occurred, the expected time before it is detected is $1/\theta$, and thereafter the expected time to find the collision is approximately $3/\theta$. Thus, the expected time complexity of the VW method is approximately $\sqrt{\pi M/2} + 4/\theta$. The expected storage complexity is $\theta\sqrt{\pi M/2}$. The parameter θ can be selected to control the storage requirements.

The collision detecting stage of the VW method can be effectively parallelized. Each of the available m processors computes its own sequences, and the distinguished elements are stored in shared memory. The expected time complexity of parallelized VW is then $\frac{1}{m}\sqrt{\pi M/2} + \frac{2.5}{\theta}$. The space complexity is $\theta\sqrt{\pi M/2}$.

4.2 Finding a Golden Collision

A random function $f : S \to S$ is expected to have $(M-1)/2$ unordered collisions. Suppose that we seek a particular one of these collisions, called a *golden collision*; we assume that the golden collision can be efficiently recognized. Thus one continues generating distinguished points and collisions until the golden collision is encountered. The expected time to find q collisions is only about $\sqrt{q}$ times as much as that to find one collision. However, since not all collisions are equally likely and the golden collision might have a very low probability of detection (see [19]), it is necessary to change the version of f periodically.

Suppose that the available memory can store w triples $(x_{i,a}, a, x_{i,0})$. When a distinguished point $x_{i,a}$ is encountered, the triple $(x_{i,a}, a, x_{i,0})$ is stored in a memory cell determined by hashing $x_{i,a}$. If that memory cell was already occupied with a triple holding a distinguished point $x_{i',b} = x_{i,a}$, then the two triples are used to locate a collision.

Van Oorschot and Wiener proposed setting

$$\theta = \alpha\sqrt{w/M} \tag{1}$$

and using each version of f to produce βw distinguished points. Experimental data presented in [20] suggested that the total running time to find the golden collision is minimized by setting $\alpha = 2.25$ and $\beta = 10$. Then, for $2^{10} \le w \le M/2^{10}$, the expected running time to find the golden collisions when m processors are employed is slightly overestimated as

$$\frac{1}{m}(2.5\sqrt{M^3/w}). \tag{2}$$

Remark 3 (verifying the VW heuristic analysis). The running time estimate (2) relies on several heuristics, the most significant of which is that when $2^{10} \le w \le M/2^{10}$ then each version of f generates approximately $1.3w$ collisions, of which approximately $1.1w$ are distinct. The numbers $1.3w$ and $1.1w$ were determined experimentally in [20]. Then the probability that a particular version of f yields the golden collision is approximately $1.1w/(M/2)$, whence the expected number of function versions needed to locate the golden collision is approximately $0.45M/w$, and the expected total time is

$$0.45\frac{M}{w} \times 10w \times \frac{1}{2.25}\sqrt{M/w} \approx 2\sqrt{M^3/w}.$$

Table 2. Observed number $c_1 w$ of collisions and number $c_2 w$ of distinct collisions per version v of the MD5-based random function $f_{n,v} : \{0,1\}^n \to \{0,1\}^n$. The numbers are averages for 20 function versions when $w \le 2^8$ and 10 function versions when $w \ge 2^9$.

w	2^2	2^3	2^4	2^5	2^6	2^7	2^8	2^9	2^{10}	2^{11}	2^{12}	2^{13}	2^{14}	2^{15}	2^{16}	2^{17}	2^{18}
$M = 2^{20}$																	
c_1	1.66	1.30	1.48	1.30	1.48	1.38	1.28	1.27	1.29	1.27	1.28	1.27	1.24	1.18	1.08	—	—
c_2	1.31	1.14	1.26	1.11	1.22	1.15	1.08	1.05	1.03	1.02	1.03	1.00	0.94	0.83	0.61	—	—
$M = 2^{24}$																	
c_1	1.38	1.36	1.38	1.37	1.33	1.31	1.31	1.36	1.32	1.33	1.31	1.30	1.30	1.29	1.29	1.27	1.24
c_2	1.21	1.14	1.16	1.16	1.12	1.10	1.11	1.13	1.11	1.11	1.09	1.06	1.06	1.05	1.04	1.00	0.95
$M = 2^{28}$																	
c_1	1.09	1.21	1.33	1.35	1.36	1.35	1.30	1.34	1.32	1.34	1.33	1.34	1.33	1.32	1.31	1.31	1.30
c_2	0.98	1.06	1.10	1.15	1.15	1.12	1.09	1.12	1.12	1.13	1.12	1.13	1.12	1.10	1.08	1.07	1.07
$M = 2^{32}$																	
c_1	1.21	1.44	1.35	1.35	1.35	1.31	1.30	1.32	1.33	1.35	1.33	1.34	1.33	1.34	1.33	1.33	1.32
c_2	1.00	1.18	1.17	1.12	1.16	1.10	1.10	1.11	1.13	1.13	1.13	1.13	1.12	1.13	1.12	1.12	1.11
$M = 2^{36}$																	
c_1	1.34	1.31	1.29	1.32	1.38	1.34	1.31	1.32	1.35	1.32	1.33	1.34	1.33	1.33	1.33	1.33	1.33
c_2	1.10	1.10	1.08	1.13	1.16	1.13	1.11	1.10	1.13	1.12	1.12	1.13	1.13	1.13	1.13	1.13	1.13

To verify these numbers, we ran some experiments using a "random" function $f_{n,v} : \{0,1\}^n \to \{0,1\}^n$ (so $M = 2^n$), where v is a string identifying the function version, and $f_{n,v}(X)$ is defined to be the n most significant bits of MD5(v, X). Table 2 lists the numbers of collisions and distinct collisions that were found for different values of (n, w), confirming the $1.3w$ and $1.1w$ numbers reported in [20].

4.3   The Attack

Let $I = \{1, 2, \ldots, N\}$ and $S = \{1, 2\} \times I$. For $i = 1, 2$, let $\mathcal{A}_i$ denote the set of all order-$\ell^{e/2}$ subgroups of $E_i[\ell^e]$, define $f_i : \mathcal{A}_i \to R_i$ by $f_i(A_i) = j(E_i/A_i)$, and let $h_i : I \to \mathcal{A}_i$ be bijections. Let $g : R \to S$ be a random function. Finally, define $f : S \to S$ by

$$f : (i, x) \mapsto g(f_i(h_i(x))).$$

Then one can view f as a "random" function from S to S.

Recall that one expects there are unique order-$\ell^{e/2}$ subgroups A_1, A_2 of $E_1[\ell^e]$, $E_2[\ell^e]$, respectively, with $j(E_1/A_1) = j(E_2/A_2)$. Let $y_1 = h_1^{-1}(A_1)$ and $y_2 = h_2^{-1}(A_2)$. Then the collision for f that we seek is the golden collision $(1, y_1)$, $(2, y_2)$. Using m processors and w cells of memory, the VW method can be used to find this golden collision in expected time

$$\frac{1}{m}(2.5\sqrt{8N^3/w}) \approx 7.1p^{3/8}/(w^{1/2}m).$$

Remark 4 (finding any collision vs. finding a golden collision). The problem of finding a collision for a hash function $H : \{0,1\}^* \to \{0,1\}^n$ and the problem of computing discrete logarithms in a cyclic group $\mathcal{G}$ can be formulated as problems of finding a collision for a random function $f : S \to S$, where $\#S = 2^n$ for the first problem and $\#S = \#\mathcal{G}$ for the second problem (see [20]). For both formulations, *any* collision for f yields a solution to the original problem. Thus, letting $N = 2^n$ or $N = \#\mathcal{G}$, the problems can be solved using van Oorschot-Wiener collision search in time approximately

$$\frac{1}{m}N^{1/2}.$$

In contrast, the only formulation of CSSI as a collision search problem for $f : S \to S$ that we know requires one to find a *golden* collision for f. For this problem, the van Oorschot-Wiener algorithm has running time approximately

$$N^{3/2}/(w^{1/2}m).$$

4.4   Implementation Report

The VW attack (for $\ell = 2$) was implemented in C, compiled using gcc version 4.7.2, and executed on an Intel Xeon processor E5-2658 v2 server equipped with 20 physical cores and 256 GB of shared RAM memory. We used fopenmp for the parallelization and openssl's MD5 implementation. The CSSI challenges were the same as the ones in Sect. 3.3.

Let $\{P_1, Q_1\}$, $\{P_2, Q_2\}$ be bases for $E_1[2^{e/2}]$, $E_2[2^{e/2}]$, respectively. Noting that $N = 3 \cdot 2^{e/2-1}$, we identify the elements of $I = \{1, 2, \ldots, N\}$ with elements of $I_1 \times I_2$ where $I_1 = \{0, 1, 2\}$ and $I_2 = \{0, 1, \ldots, 2^{e/2-1} - 1\}$. The bijections $h_i : I_1 \times I_2 \to \mathcal{A}_i$ for $i = 1, 2$ are defined by

$$h_i : (b, k) \mapsto \begin{cases} P_i + (b2^{e/2-1} + k)Q_i, & \text{if } b = 0, 1, \\ (2k)P_i + Q_i, & \text{if } b = 2. \end{cases}$$

Let $S = \{1, 2\} \times I_1 \times I_2$. For $n \in \{0, 1\}^{64}$, we let $g_n : R \to S$ be the function computed using Algorithm 1. We then define the version $f_n : S \to S$ of f by $(i, x) \mapsto g_n(f_i(h_i(x)))$.

Algorithm 1. The "random" function g_n

Require: $n \in \{0, 1\}^{64}$ and $j \in \mathbb{F}_{p^2}$.
Ensure: Output $c \in \{1, 2\}$, $b \in I_1$, $k \in I_2$.
1: *counter* := 0.
2: **repeat**
3:    $h :=$ MD5$(1, j, n, counter)$.
4:    Let h' be the $e/2 + 2$ least significant bits of h, and parse h' as (k, c, b), where k, c, b have bitlengths $e/2 - 1$, 1, and 2, respectively.
5:    *counter* := *counter* + 1.
6: **until** $b \neq 11$
7: **return** $(c + 1, b, k)$.

We set $\theta = 2.25\sqrt{w/2N}$, where $w = 2^t$, and declare an element $X \in S$ to be distinguished if the integer formed from the 32 least significant bits of MD5$(2, X)$ is $\leq 2^{32}\theta$. If X is distinguished, then it is placed in memory cell s, where s is the integer determined by the t least significant bits of MD5$(3, X)$. If a distinguished point is not encountered after $10/\theta$ iterations, then that trail is abandoned and a new trail is formed.

Table 3 shows the time expended for finding 2^e-isogenies for $e \in \{32, 34, 36, 38, 40, 42, 44\}$ with the VW attack. These experimental results confirm the accuracy of the VW attack's heuristic analysis.

To gain further confidence that the VW attack's heuristic analysis is accurate for cryptographically-interesting CSSI parameters (e.g., $e = 256$), we ran some experiments to estimate the number of collisions and distinct collisions for functions f_n when $e = 50, 60, 70, 80$. The results, listed in Table 4, confirm the $1.3w$ and $1.1w$ estimates in [20].

Table 3. Van Oorschot-Wiener golden collision search for finding a 2^{e_A}-isogeny between two supersingular elliptic curves over $\mathbb{F}_{p^2}$ with $p = 2^{e_A} \cdot 3^{e_B} \cdot d - 1$. For each p, the listed number of CSSI instances were solved and the median and average of the results are reported. The $\#f_n$'s column indicates the number of random functions f_n that were tested before the golden collision was found. The expected and measured times list the number of degree-$2^{e_A/2}$ isogeny computations.

e_A	e_B	d	w	Expected time	Number of runs	Median			Average		
						$\#f_n$'s	Measured time	Clock cycles	$\#f_n$'s	Measured time	clock cycles
32	20	23	2^9	$2^{23.20}$	25	180	$2^{23.55}$	$2^{40.79}$	319	$2^{24.38}$	$2^{41.62}$
34	21	109	2^9	$2^{24.70}$	25	256	$2^{24.54}$	$2^{41.89}$	714	$2^{26.02}$	$2^{43.37}$
36	22	31	2^{10}	$2^{25.70}$	25	369	$2^{26.06}$	$2^{43.51}$	838	$2^{27.25}$	$2^{44.70}$
38	23	271	2^{11}	$2^{26.70}$	25	196	$2^{26.15}$	$2^{43.70}$	567	$2^{27.69}$	$2^{45.23}$
40	25	71	2^{11}	$2^{28.20}$	25	162	$2^{26.36}$	$2^{43.99}$	1015	$2^{29.01}$	$2^{46.64}$
42	26	37	2^{12}	$2^{29.20}$	25	477	$2^{28.92}$	$2^{46.52}$	1940	$2^{30.95}$	$2^{48.55}$
44	27	37	2^{13}	$2^{30.20}$	25	431	$2^{29.78}$	$2^{47.46}$	942	$2^{30.91}$	$2^{48.58}$

Table 4. Observed number $c_1 w$ of collisions and number $c_2 w$ of distinct collisions per CSSI-based random function f_n. The numbers are averages for 25 function versions (except for $(e, w) \in \{(80, 2^{12}), (80, 2^{14}), (80, 2^{16})\}$ for which 5 function versions were used).

e	p	w	2^8	2^{10}	2^{12}	2^{14}	2^{16}
50	$2^{50} 3^{31} 179 - 1$	c_1	1.37	1.36	1.37	1.41	1.49
		c_2	1.14	1.12	1.12	1.11	1.09
60	$2^{60} 3^{37} 31 - 1$	c_1	1.37	1.34	1.34	1.35	1.36
		c_2	1.15	1.13	1.13	1.12	1.12
70	$2^{70} 3^{32} 127 - 1$	c_1	1.33	1.34	1.34	1.34	1.34
		c_2	1.13	1.14	1.13	1.13	1.13
80	$2^{80} 3^{25} 71 - 1$	c_1	1.35	1.32	1.33	1.34	1.33
		c_2	1.14	1.12	1.13	1.13	1.13

5   Comparisons

There are many factors that can affect the efficacy of an algorithm.

1. *Time*: the worst-case or average-case number of basic arithmetic operations performed by the algorithm.
2. *Space*: the amount of storage (RAM, hard disk, etc.) required.
3. *Parallelizability*: the speedup achievable when running the algorithm on multiple processors. Ideally, the speedup is by a factor equal to the number of

processors, and the processors do not need to communicate with each other; if this is the case then the parallelization is said to be *perfect*[4].

4. *Communication costs*: the time taken for communication between processors, and the memory access time for retrieving data from large storage devices. Memory access time can be a dominant cost factor when using extremely large storage devices [2].

5. *Custom-designed devices*: the possible speedups that can be achieved by executing the algorithm on custom-designed hardware. Examples of such devices are TWINKLE [24] and TWIRL [25] that were designed for the number field sieve integer factorization algorithm.

In this section we analyze and compare the efficacy of the meet-in-the-middle algorithm, VW golden collision search, and a mesh sorting algorithm for solving CSSI. We make two assumptions:

1. The number m of processors available is at most 2^{64}.
2. The total amount of storage w available is at most 2^{80} units.

Our analysis will ignore communication costs, and thus our running time estimates can be considered to be lower bounds on the "actual" running time.

Remark 5 (feasible amount of storage and number of processors). The Sunway TaihuLight supercomputer, the most powerful in the world as of March 2018, has $2^{23.3}$ CPU cores [29]. In 2013, it was estimated that Google's data centres have a total storage capacity of about a dozen exabytes[5] [29]. Thus it is reasonable to argue that acquiring 2^{64} processors and a storage capacity (with low access times) of several dozen yottabytes[6] for the purpose of solving a CSSI problem will be prohibitively costly for the foreseeable future.

5.1 Meet-in-the-Middle

As stated in Sect. 3, the running time of MITM-basic and MITM-DFS is approximately $2N$ and the storage requirements are N, where $N \approx p^{1/4}$. Since for $N \geq 2^{80}$ the storage requirements are infeasible, we deem the meet-in-the-middle attacks to be prohibitively expensive when $N \gg 2^{80}$.

Of course, one can trade space for time. One possible time-memory tradeoff is to store a table with entries $(j(E_1/A_1), A_1)$, where A_1 ranges over a w-subset of order-$\ell^{e/2}$ subgroups of $E_1[\ell^e]$. Next, for each order-$\ell^{e/2}$ subgroup A_2 of $E_2[\ell^e]$, E_2/A_2 is computed and $j(E_2/A_2)$ is searched in the table. If no match is found, then the algorithm is repeated for a disjoint w-subset of order-$\ell^{e/2}$ subgroups of $E_1[\ell^e]$, and so on. The running time of this time-memory tradeoff is approximately

$$(w + N)\frac{N}{w} \approx N^2/w.$$

---

[4] If the processors share the same storage space, then frequent storage accesses might decrease the parallelizability of the algorithm.

[5] An exabyte is 2^{60} bytes.

[6] A yottabyte is 2^{80} bytes.

For MITM-basic, the unit of time is an $\ell^{e/2}$-isogeny computation. For MITM-DFS, the running time (for $\ell = 2$) can be scaled to $\ell^{e/2}$-isogeny computations by dividing by $\frac{e}{12.4} \log_2 e$ (cf. Sect. 3.2). One can see that this time-memory-tradeoff can be parallelized perfectly.

Another possible time-memory tradeoff is to store $(j(E_1/A_1), A_1)$, where A_1 ranges over all order-ℓ^c subgroups of $E_1[\ell^e]$ and $c \approx \log_\ell w$. Let $d = e - c$. Then, for each order-ℓ^d subgroup A_2 of $E_2[\ell^e]$, E_2/A_2 is computed and $j(E_2/A_2)$ is searched in the table. One can check that the running time of this time-memory tradeoff is approximately N^2/w, and that it can be parallelized perfectly. Note that the unit of time here is an ℓ^d-isogeny computation instead of an $\ell^{e/2}$-isogeny computation. The larger tree of ℓ^d-isogenies can be traversed using a depth-first search; the running time is then the same as that of the MITM-DFS variant described in the previous paragraph.

5.2   Golden Collision Search

As stated in Sect. 4.3, the running time of van Oorschot-Wiener golden collision search is approximately

$$N^{3/2}/w^{1/2}.$$

The algorithm parallelizes perfectly.

5.3   Mesh Sorting

The mesh sorting attack is analogous to the one described by Bernstein [2] for finding hash collisions. Suppose that one has m processors arranged in a two-dimensional grid. Each processor only communicates with its neighbours in the grid. In one unit of time, each processor computes and stores pairs $(j(E_1/A_1), A_1)$, where A_1 is an order-$\ell^{e/2}$ subgroup of $E_1[\ell^e]$. Next, these stored pairs are sorted in time $\approx m^{1/2}$ (e.g., see [22]). In the next stage, a second two-dimensional grid of m processors computes and stores pairs $(j(E_2/A_2), A_2)$, where A_2 is an order-$\ell^{e/2}$ subgroup of $E_2[\ell^e]$, and the two sorted lists are compared for a match. This is repeated for a disjoint m-subset of order-$\ell^{e/2}$ subgroups A_2 until all order-$\ell^{e/2}$ subgroups of $E_2[\ell^e]$ have been tested. Then, the process is repeated for a disjoint subset of order-$\ell^{e/2}$ subgroups A_1 of $E_1[\ell^e]$ until a match is found. One can check that the calendar running time[7] is approximately

$$\left(m^{1/2} + m^{1/2}\frac{N}{m}\right)\frac{N}{m} \approx N^2/m^{3/2}.$$

5.4   Targetting the 128-Bit Security Level

The CSSI problem is said to have a 128-bit security level if the fastest known attack has total time complexity at least 2^{128} and feasible space and hardware costs.

---

[7] *Calendar time* is the elapsed time taken for a computation, whereas *total time* is the sum of the time expended by all m processors.

Suppose that $p \approx 2^{512}$, whereby $N \approx 2^{128}$; this would be a reasonable choice for the bitlength of p if the meet-in-the-middle attacks were assessed to be the fastest (classical) algorithm for solving CSSI. However, as noted above, the storage costs for the attacks are prohibitive. Instead, one should consider the time complexity of the time-memory tradeoffs, VW golden collision search, and mesh sorting under realistic constraints on the storage space w and the number m of processors. Table 5 lists the calendar time and the total time of these CSSI attacks for $(m, w) \in \{(2^{48}, 2^{64}), (2^{48}, 2^{80}), (2^{64}, 2^{80})\}$. One sees that in all cases the total time complexity is significantly greater than 2^{128}, even though we have ignored communication costs.

Table 5. Time complexity estimates of CSSI attacks for $p \approx 2^{512}$ and $p \approx 2^{448}$, and $\ell = 2$. All numbers are expressed in their base-2 logarithms. The unit of time is a $2^{e/2}$-isogeny computation.

	# processors m	space w	$p \approx 2^{512}$		$p \approx 2^{448}$	
			calendar time	total time	calendar time	total time
Meet-in-the-middle (DFS) time-memory tradeoff	48	64	138	186	106	154
	48	80	122	170	90	138
	64	80	106	170	74	138
Van Oorschot-Wiener golden collision search	48	64	112	160	88	136
	48	80	104	152	80	128
	64	80	88	152	64	128
Mesh sorting	48	—	184	232	152	200
	64	—	160	224	128	192

Since the total times for $p \approx 2^{512}$ in Table 5 are all significantly greater than 2^{128}, one can consider using smaller primes p while still achieving the 128-bit security level. Table 5 also lists the calendar time and the total time of these CSSI attacks for $(m, w) \in \{(2^{48}, 2^{64}), (2^{48}, 2^{80}), (2^{64}, 2^{80})\}$ when $p \approx 2^{448}$ and $N \approx 2^{112}$. One sees that all attacks have total time complexity at least 2^{128}, even though we have ignored communication costs. We can conclude that selecting SIDH parameters with $p \approx 2^{448}$ provides 128 bits of security against known classical attacks. For example, one could select the 434-bit prime

$$p_{434} = 2^{216} 3^{137} - 1;$$

this prime is balanced in the sense that $3^{137} \approx 2^{217}$, thus providing maximal resistance to Petit's SIDH attack [21].

Remark 6 (communication costs). Consider the case $p \approx 2^{448}$, $e = 224$, $m = 2^{64}$, $w = 2^{80}$. From (1) and (2) we obtain $\theta \approx 1/2^{15.62}$ and an expected running time of $2^{131.7}$. For each function version, the 2^{64} processors will generate approximately $2^{48.4}$ distinguished points per unit of time (i.e., a 2^{112}-isogeny computation). So, on average, the 2^{80} storage device will be accessed $2^{48.4}$ times during each unit of time. The cost of these accesses will certainly dominate the computational costs. Thus our security estimates, which ignore communication costs, should be regarded as being conservative.

5.5 Targetting the 160-Bit Security Level

Using similar arguments as in Sect. 5.4, one surmises that SIDH parameters with $p \approx 2^{536}$ offer at least 160 bits of CSSI security against known classical (see Table 6). For example, one could select the 546-bit prime

$$p_{546} = 2^{273}3^{172} - 1;$$

this prime is nicely balanced since $3^{172} \approx 2^{273}$.

Table 6. Time complexity estimates of CSSI attacks for $p \approx 2^{536}$ and $p \approx 2^{614}$, and $\ell = 2$. All numbers are expressed in their base-2 logarithms. The unit of time is a $2^{e/2}$-isogeny computation.

	# processors m	space w	$p \approx 2^{536}$		$p \approx 2^{614}$	
			calendar time	total time	calendar time	total time
Meet-in-the-middle (DFS) time-memory tradeoff	48	64	150	198	188	236
	48	80	134	182	172	220
	64	80	118	182	156	220
Van Oorschot-Wiener golden collision search	48	64	121	169	149	197
	48	80	113	161	141	189
	64	80	97	161	125	189
Mesh sorting	48	—	196	244	234	282
	64	—	172	236	210	274

5.6 Targetting the 192-Bit Security Level

Using similar arguments as in Sect. 5.4, one surmises that SIDH parameters with $p \approx 2^{614}$ offer at least 192 bits of CSSI security against known classical (see Table 6). For example, one could select the 610-bit prime

$$p_{610} = 2^{305}3^{192} - 1;$$

this prime is nicely balanced since $3^{192} \approx 2^{304}$.

5.7   Resistance to Quantum Attacks

The appeal of SIDH is its apparent resistance to attacks by quantum computers. What remains to be determined then is the security of CSSI against quantum attacks.

The fastest known quantum attack on CSSI is Tani's algorithm [26]. Given two generic functions $g_1 : X_1 \to Y$ and $g_2 : X_2 \to Y$, where $\#X_1 \approx \#X_2 \approx N$ and $\#Y \gg N$, Tani's quantum algorithm finds a claw, i.e., a pair $(x_1, x_2) \in X_1 \times X_2$ such that $g_1(x_1) = g_2(x_2)$ in time $O(N^{2/3})$. The CSSI problem can be recast as a claw-finding problem by defining X_i to be the set of all degree-$\ell^{e/2}$ isogenies originating at E_i, g_i to be the function that maps a degree-$\ell^{e/2}$ isogeny originating at E_i to the j-invariant of its image curve, and $Y = R$. Since $\#X_1 = \#X_2 = N \approx p^{1/4}$, this yields an $O(p^{1/6})$-time CSSI attack.

CSSI can also be solved by an application of Grover's quantum search [12]. Recall that if $g : X \to \{0, 1\}$ is a generic function such that $g(x) = 1$ for exactly one $x \in X$, then Grover's algorithm can determine the x with $g(x) = 1$ in quantum time $O(\sqrt{\#X})$. The CSSI problem can be recast as a Grover search problem by defining X to be the set of all ordered pairs (ϕ_1, ϕ_2) of degree-$\ell^{e/2}$ isogenies originating at E_1, E_2, respectively, and defining $g(\phi_1, \phi_2)$ to be equal to 1 if and only if the j-invariants of the image curves of ϕ_1 and ϕ_2 are equal. Since $\#X = N^2 \approx p^{1/2}$, this yields an $O(p^{1/4})$-time quantum attack on CSSI.

The Jao-De Feo paper [14] that introduced SIDH identified Tani's claw-finding algorithm as the fastest known attack, whether classical or quantum, on CSSI. The subsequent literature on SIDH used the simplified running time $p^{1/6}$ of Tani's algorithm (i.e., ignoring the implied constant in its $O(p^{1/6})$ running time expression) to select SIDH primes p for a desired level of security. In other words, in order to achieve a b-bit security level against known classical and quantum attacks, one selects an SIDH prime p of bitlength approximately $6b$. For example, the 751-bit prime $p = 2^{372}3^{239} - 1$ was proposed in [8] for the 128-bit security level, and this prime has been used in many subsequent works, e.g., [6,7,13,17,32]. Also, the 964-bit prime $p = 2^{486}3^{301} - 1$ was proposed in [13] for the 160-bit security level.

However, this assessment of SIDH security does not account for the cost of the $O(p^{1/6})$ quantum space requirements of Tani's algorithm, nor for the fact that Grover's search does not parallelize well—using m quantum circuits only yields a speedup by a factor of $\sqrt{m}$ and this speedup has been proven to be optimal [31]. Some recent work [1,16] suggests that Tani's and Grover's attacks on CSSI are costlier than the van Oorschot-Wiener golden collision search algorithm. If this is indeed the case, then one can be justified in selecting SIDH primes p_{434} (instead of p_{751}), p_{546} (instead of p_{964}) and p_{610} in order to achieve the 128-, 160- and 192-bit security levels, respectively, against both classical and quantum attacks. Furthermore, SIDH parameters with p_{434} could be deemed to meet the security requirements in NIST's Category 2 [18] (classical and quantum security comparable or greater than that of SHA-256 with respect to collision resistance), and p_{610} could be deemed to meet the security requirements in NIST's Category 4 [18] (classical and quantum security comparable to that of SHA-384).

5.8 SIDH Performance

A significant benefit of using smaller SIDH primes is increased performance. The reasons for the boost in SIDH performance are twofold. First, since the computation of the ground field $\mathbb{F}_p$ multiplication operation has a quadratic complexity, any reduction in the size of p will result in significant savings. Since high-end processors have a word size of 64 bits, the primes p_{751}, p_{546} and p_{434} can be accommodated using twelve, nine and seven 64-bit words, respectively. Hence, if $\mathbb{F}_p$ multiplication using p_{751} can be computed in T clock cycles, then a rough estimation of the computational costs for $\mathbb{F}_p$ multiplication using p_{434} and p_{546} is as low as $0.34T$ and $0.56T$, respectively. Second, since the exponents of the primes 2 and 3 in p_{434} and p_{546} are smaller than the ones in p_{751}, the computation of the isogeny chain described in Sect. 2.2 (see Remark 1) is faster.

Table 7 lists timings for SIDH operations for p_{434}, p_{546} and p_{751} using the SIDH library of Costello et al. [5]. The timings show that SIDH operations are about 4.8 times faster when p_{434} is used instead of p_{751}.

Table 7. Performance of the SIDH protocol. All timings are reported in 10^6 clock cycles, measured on an Intel Core i7-6700 supporting a Skylake micro-architecture. The "CLN + enhancements" columns are for our implementation that incorporates improved formulas for degree-2 and degree-3 isogenies from [6] and Montgomery ladders from [10] into the CLN library.

Protocol phase		CLN library [8]			CLN + enhancements		
		p_{751}	p_{434}	p_{546}	p_{751}	p_{434}	p_{546}
Key Gen.	Alice	35.7	7.51	13.20	26.9	5.3	10.5
	Bob	39.9	8.32	14.84	30.5	6.0	11.7
Shared secret	Alice	33.6	7.01	12.56	24.9	5.0	10.0
	Bob	38.4	7.94	14.35	28.6	5.8	11.5

6 Concluding Remarks

Our implementations of the MITM and golden collision search CSSI attacks are, to the best of our knowledge, the first ones reported in the literature. The implementations confirm that the performance of these attacks is accurately predicted by their heuristic analysis.

Our concrete cost analysis of the attacks leads to the conclusion that golden collision search is more effective that the meet-in-the-middle attack. Thus one can use 448-bit primes and 536-bit primes p in SIDH to achieve the 128-bit and 160-bit security levels against *known* classical attacks on the CSSI problem. We emphasize that these conclusions are based on our understanding of how to best implement these algorithms, and on assumptions on the amount of storage and the number of processors that an adversary might possess. On the other hand, our conclusions are somewhat conservative in that the analysis does not account for communication costs. Moreover, whereas it is generally accepted that the AES-128 and AES-256 block ciphers attain the 128-bit security level in the classical and quantum settings, the time it takes to compute a degree-2^{112} isogeny (which is the unit of time for the golden collision search CSSI attack with balanced 448-bit prime p) is considerably greater than the time for one application of AES-128 or AES-256.

Acknowledgements. We thank Steven Galbraith for the suggestion to traverse the MITM trees using depth-first search. We also thank Sam Jaques for the many discussions on Grover's and Tani's algorithms.

References

1. Adj, G., Cervantes-Vázquez, D., Chi-Domínguez, J., Menezes, A., Rodríguez-Henríquez, F.: On the cost or computing isogenies between supersingular elliptic curves. Cryptology ePrint Archive: Report 2018/313. http://eprint.iacr.org/2018/313
2. Bernstein, D.: Cost analysis of hash collisions: will quantum computers make SHARCS obsolete? In: Workshop Record of SHARCS 2009: Special-purpose Hardware for Attacking Cryptographic Systems (2009). https://cr.yp.to/papers.html#collisioncost
3. Brassard, G., Høyer, P., Tapp, A.: Quantum cryptanalysis of hash and claw-free functions. In: Lucchesi, C.L., Moura, A.V. (eds.) LATIN 1998. LNCS, vol. 1380, pp. 163–169. Springer, Heidelberg (1998). https://doi.org/10.1007/BFb0054319
4. Charles, D., Goren, E., Lauter, K.: Cryptographic hash functions from expander graphs. J. Cryptol. **22**, 93–113 (2009)
5. Costello, C., et al.: SIDH Library. https://www.microsoft.com/en-us/research/project/sidh-library/
6. Costello, C., Hisil, H.: A simple and compact algorithm for SIDH with arbitrary degree isogenies. In: Takagi, T., Peyrin, T. (eds.) ASIACRYPT 2017. LNCS, vol. 10625, pp. 303–329. Springer, Cham (2017). https://doi.org/10.1007/978-3-319-70697-9_11
7. Costello, C., Jao, D., Longa, P., Naehrig, M., Renes, J., Urbanik, D.: Efficient compression of SIDH public keys. In: Coron, J.-S., Nielsen, J.B. (eds.) EUROCRYPT 2017. LNCS, vol. 10210, pp. 679–706. Springer, Cham (2017). https://doi.org/10.1007/978-3-319-56620-7_24
8. Costello, C., Longa, P., Naehrig, M.: Efficient algorithms for supersingular isogeny Diffie-Hellman. In: Robshaw, M., Katz, J. (eds.) CRYPTO 2016. LNCS, vol. 9814, pp. 572–601. Springer, Heidelberg (2016). https://doi.org/10.1007/978-3-662-53018-4_21
9. De Feo, L., Jao, D., Plût, J.: Towards quantum-resistant cryptosystems from supersingular elliptic curve isogenies. J. Math. Cryptol. **8**, 209–247 (2014)
10. Faz-Hernández, A., López, J., Ochoa-Jiménez, E., Rodríguez-Henríquez, F.: A faster software implementation of the supersingular isogeny Diffie-Hellman key exchange protocol. IEEE Trans. Comput. **67**, 1622–1636 (2018)

11. Galbraith, S.D., Petit, C., Silva, J.: Identification protocols and signature schemes based on supersingular isogeny problems. In: Takagi, T., Peyrin, T. (eds.) ASIACRYPT 2017. LNCS, vol. 10624, pp. 3–33. Springer, Cham (2017). https://doi.org/10.1007/978-3-319-70694-8_1

12. Grover, L.: A fast quantum mechanical algorithm for database search. In: Proceedings of the Twenty-Eighth Annual Symposium on Theory of Computing – STOC 1996. ACM Press, pp. 212–219 (1996)

13. Jao, D., et al.: Supersingular isogeny key encapsulation. Round 1 submission, NIST Post-Quantum Cryptography Standardization, 30 November 2017

14. Jao, D., De Feo, L.: Towards quantum-resistant cryptosystems from supersingular elliptic curve isogenies. In: Yang, B.-Y. (ed.) PQCrypto 2011. LNCS, vol. 7071, pp. 19–34. Springer, Heidelberg (2011). https://doi.org/10.1007/978-3-642-25405-5_2

15. Jao, D., Soukharev, V.: Isogeny-based quantum-resistant undeniable signatures. In: Mosca, M. (ed.) PQCrypto 2014. LNCS, vol. 8772, pp. 160–179. Springer, Cham (2014). https://doi.org/10.1007/978-3-319-11659-4_10

16. Jaques, S., Schanck, J.: Quantum cryptanalysis in the RAM model. Preprint (2018)

17. Koziel, B., Azarderakhsh, R., Mozaffari-Kermani, M.: Fast hardware architectures for supersingular isogeny Diffie-Hellman key exchange on FPGA. In: Dunkelman, O., Sanadhya, S.K. (eds.) INDOCRYPT 2016. LNCS, vol. 10095, pp. 191–206. Springer, Cham (2016). https://doi.org/10.1007/978-3-319-49890-4_11

18. National Institute of Standards and Technology: Submission requirements and evaluation criteria for the post-quantum cryptography standardization process, December 2016. https://csrc.nist.gov/csrc/media/projects/post-quantum-cryptography/documents/call-for-proposals-final-dec-2016.pdf

19. van Oorschot, P.C., Wiener, M.J.: Improving implementable meet-in-the-middle attacks by orders of magnitude. In: Koblitz, N. (ed.) CRYPTO 1996. LNCS, vol. 1109, pp. 229–236. Springer, Heidelberg (1996). https://doi.org/10.1007/3-540-68697-5_18

20. van Oorschot, P., Wiener, M.: Parallel collision search with cryptanalytic applications. J. Cryptol. **12**, 1–28 (1999)

21. Petit, C.: Faster algorithms for isogeny problems using torsion point images. In: Takagi, T., Peyrin, T. (eds.) ASIACRYPT 2017. LNCS, vol. 10625, pp. 330–353. Springer, Cham (2017). https://doi.org/10.1007/978-3-319-70697-9_12

22. Schnorr, C., Shamir, A.: An optimal sorting algorithm for mesh connected computers. In: Proceedings of the Eighteenth Annual Symposium on Theory of Computing – STOC 1986. ACM Press, pp. 255–263 (1986)

23. Schoof, R.: Nonsingular plane cubic curves over finite fields. J. Comb. Theory Ser. A **46**, 183–211 (1987)

24. Shamir, A.: Factoring large numbers with the TWINKLE device. In: Koç, Ç.K., Paar, C. (eds.) CHES 1999. LNCS, vol. 1717, pp. 2–12. Springer, Heidelberg (1999). https://doi.org/10.1007/3-540-48059-5_2

25. Shamir, A., Tromer, E.: Factoring large numbers with the TWIRL device. In: Boneh, D. (ed.) CRYPTO 2003. LNCS, vol. 2729, pp. 1–26. Springer, Heidelberg (2003). https://doi.org/10.1007/978-3-540-45146-4_1

26. Tani, S.: Claw finding algorithms using quantum walk. Theor. Comput. Sci **410**, 5285–5297 (2009)

27. Vélu, J.: Isogénies entre courbes elliptiques. C. R. Acad. Sc. Paris **273**, 238–241 (1971)

28. Wikipedia: Sunway TaihuLight. https://en.wikipedia.org/wiki/Sunway_TaihuLight

29. Wikipedia: Exabyte. https://en.wikipedia.org/wiki/Exabyte#Google

30. Yoo, Y., Azarderakhsh, R., Jalali, A., Jao, D., Soukharev, V.: A post-quantum digital signature scheme based on supersingular isogenies. In: Kiayias, A. (ed.) FC 2017. LNCS, vol. 10322, pp. 163–181. Springer, Cham (2017). https://doi.org/10.1007/978-3-319-70972-7_9

31. Zalka, C.: Grover's quantum searching algorithm is optimal. Phys. Rev. A **60**, 2746–2751 (1999)

32. Zanon, G.H.M., Simplicio, M.A., Pereira, G.C.C.F., Doliskani, J., Barreto, P.S.L.M.: Faster isogeny-based compressed key agreement. In: Lange, T., Steinwandt, R. (eds.) PQCrypto 2018. LNCS, vol. 10786, pp. 248–268. Springer, Cham (2018). https://doi.org/10.1007/978-3-319-79063-3_12

Lattice-Based Cryptography

A Full RNS Variant of Approximate Homomorphic Encryption

Jung Hee Cheon[1], Kyoohyung Han[1], Andrey Kim[1], Miran Kim[2],
and Yongsoo Song[3(✉)]

[1] Seoul National University, Seoul, South Korea
{jhcheon,satanigh,kimandrik}@snu.ac.kr
[2] University of Texas, Health Science Center at Houston, Houston, USA
miran.kim@uth.tmc.edu
[3] University of California, San Diego, La Jolla, USA
yongsoosong@ucsd.edu

Abstract. The technology of Homomorphic Encryption (HE) has improved rapidly in a few years. The newest HE libraries are efficient enough to use in practical applications. For example, Cheon et al. (ASIACRYPT'17) proposed an HE scheme with support for arithmetic of approximate numbers. An implementation of this scheme shows the best performance in computation over the real numbers. However, its implementation could not employ a core optimization technique based on the Residue Number System (RNS) decomposition and the Number Theoretic Transformation (NTT).

In this paper, we present a variant of approximate homomorphic encryption which is optimal for implementation on standard computer system. We first introduce a new structure of ciphertext modulus which allows us to use both the RNS decomposition of cyclotomic polynomials and the NTT conversion on each of the RNS components. We also suggest new approximate modulus switching procedures without any RNS composition. Compared to previous exact algorithms requiring multi-precision arithmetic, our algorithms can be performed by using only word size (64-bit) operations.

Our scheme achieves a significant performance gain from its full RNS implementation. For example, compared to the earlier implementation, our implementation showed speed-ups 17.3, 6.4, and 8.3 times for decryption, constant multiplication, and homomorphic multiplication, respectively, when the dimension of a cyclotomic ring is 32768. We also give experimental result for evaluations of some advanced circuits used in machine learning or statistical analysis. Finally, we demonstrate the practicability of our library by applying to machine learning algorithm. For example, our single core implementation takes 1.8 min to build a logistic regression model from encrypted data when the dataset consists of 575 samples, compared to the previous best result 3.5 min using four cores.

Keywords: Homomorphic encryption · Approximate arithmetic
Residue number system

© Springer Nature Switzerland AG 2019
C. Cid and M. J. Jacobson, Jr. (Eds.): SAC 2018, LNCS 11349, pp. 347–368, 2019.
https://doi.org/10.1007/978-3-030-10970-7_16

1   Introduction

As the growth of big data analysis have led to many concerns about security and privacy of data, researches on secure computation have been highlighted in cryptographic community. Homomorphic Encryption (HE) is a cryptosystem that allows an arbitrary circuit to be evaluated on encrypted data without decryption. It has been one of the most promising solutions that make it possible to outsource computation and securely aggregate sensitive information of individuals. After the first construction of fully homomorphic encryption by Gentry [20], several researches [7,11,16–18] have improved the efficiency of HE schemes.

There are a few software implementations of HE schemes based on the Ring Learning with Errors (RLWE) problem such as HElib [25] of the BGV scheme [7] and SEAL [32] of the BFV scheme [6,18]. These HE schemes are constructed over the residue ring of a cyclotomic ring (with a huge characteristic) so they manipulate modulo operations between high-degree polynomials, resulting in a performance degradation. For an efficient implementation of polynomial arithmetic, Gentry et al. [21] suggested a representation of cyclotomic polynomials, called the *double-CRT* representation, based on the Chinese Remainder Theorem (CRT). The first CRT layer uses the Residue Number System (RNS) in order to decompose a polynomial into a tuple of polynomials with smaller moduli. The second layer converts each of small polynomials into a vector of modulo integers via the Number Theoretic Transform (NTT). In the double-CRT representation, an arbitrary polynomial is identified with a matrix consisting of small integers, and this enables an efficient polynomial arithmetic by performing componentwise modulo operations. This technique became one of the core optimization techniques used in the implementations of HE schemes [1,25,32].

Cheon et al. [11] recently suggested an HE scheme for arithmetic of approximate numbers, called HEAAN. The main idea of their construction is to consider an RLWE error as a part of an error occurring during approximate computations. Besides homomorphic addition and multiplication, it supports an approximate rounding operation of significant digits on packed ciphertexts. This approximate HE scheme shows remarkable performance in real-world applications that require arithmetic over the real numbers [27,28].

However, the original scheme had one significant problem in the use of the double-CRT representation. The rounding operation of HEAAN can be done by dividing an encrypted plaintext by a ratio of two consecutive ciphertext moduli, so a ciphertext modulus should be chosen as a power of two (or some prime). This parameter choice makes it hard to implement the HE scheme on the RNS representation. Consequently, the previous implementation [10] took a longer time to perform homomorphic operations than other implementations of HE schemes under the same parameter setting.

Our Contribution. In this paper, we present a variant of HEAAN based on the double-CRT representation of cyclotomic polynomial ring elements. The main idea is to exploit a basis consisting of some approximate values of a fixed base as our moduli chain. Every encrypted message in HEAAN contains a small noise from approximate computations. The approximate rounding operation of our scheme yields an additional error from approximation, but it does not destroy the significant digits of an encrypted message as long as the precision of the approximate bases is higher than the precision of the plaintexts. In addition, by selecting approximate bases satisfying some condition for the NTT conversion, we take the advantages of double-CRT representation while maintaining the functionalities of the original scheme.

We also introduce some modulus switching algorithms that can be computed without RNS composition. To be more precise, some homomorphic operations of the original HEAAN scheme (e.g. homomorphic multiplication) require *non-arithmetic* operations such as modulus raising and reduction, which are difficult to perform based on the RNS representation. As a result, the previous implementation required multi-precision arithmetics instead of working on typical word-size integers in hardware architecture (e.g. 64-bit processor). Our new modulus switching techniques can substitute the non-arithmetic operations in the previous scheme. These algorithms are RNS-friendly, that is, they can be represented using only word operations without RNS composition.

We implemented our scheme and compared with the original one to show the performance benefit from a full RNS system. For efficient implementation in the NTT and modulus operations, we adapt harvey's butterfly and barrett modulus reduction techniques. Our full RNS variant improves the performance of basic operations by nearly ten times compared to the original HEAAN [10, 11]. The decryption and homomorphic multiplication timings are reduced from 135 and 1,355 ms down to 7.8 and 164 ms, respectively, when evaluating a circuit of depth 10.

We also present experimental results for homomorphic evaluation of analytic functions and statistic functions. It took 160 ms to compute the multiplicative inverse, exponential function, or sigmoid function with inputs of 32-bit precision on 2^{13} slots, yielding an amortized time of 20 ms per slot. In the case of statistic functions, it took 307 and 518 ms to obtain the mean and variance of 2^{13} real numbers, respectively.

Finally, we implemented a variant of the gradient descent algorithm to show that our HE library can perform complex computations in real-world applications. Our single-core implementation took about 1.8 min to obtain a logistic regression model from homomorphically encrypted dataset consisting of 575 samples each of which has eight features and a binary class information, compared to previous best result of 3.5 min using a machine with four cores [27].

Technical Details. Let N be a power-of-two integer and $R = \mathbb{Z}[X]/(X^N+1)$ be the ring of integers of the $(2N)$-th cyclotomic field. For a fixed base q, we choose an RNS basis $\{q_0, \ldots, q_L\}$ which is a set of coprime integers of approximately the same size as the base q. For an integer $0 \leq \ell \leq L$, a ciphertext at level-ℓ is a pair of polynomials in $R_{Q_\ell} = R/(Q_\ell \cdot R)$ for $Q_\ell = \prod_{i=0}^{\ell} q_i$. The rescaling procedure transforms a level ℓ encryption of m into a level $(\ell - 1)$ encryption of $q_\ell^{-1} \cdot m$, which is an approximation of $q^{-1} \cdot m$ with almost the same precision. The original scheme is more flexible in choice of ciphertext modulus since it can rescale a plaintext by an arbitrary number compared to the fixed base q of our RNS variant. However, our scheme has a significant improvement in performance.

Our scheme can support the NTT representation of RNS decomposed polynomials as the double-CRT representation in the BGV scheme [7,21]. The NTT conversion can be done efficiently when the approximate bases q_ℓ's are prime numbers satisfying $q_\ell \equiv 1 \pmod{2N}$. We give a list of candidate bases to show that there are sufficiently many distinct primes satisfying both conditions for the double-CRT representation.

The homomorphic multiplication algorithm of HEAAN includes modulus switching procedures that convert an element of R_Q into $R_{P \cdot Q}$ for a sufficiently large integer P and switch back to the original modulus Q. These non-arithmetic operations are difficult to perform on the RNS system, so one should recover the coefficient representation of an input polynomial. For an optimization, we adapt an idea of Barjard et al. [3] to suggest approximate modulus switching algorithms with small errors. Instead of exact computation in the original scheme, our approximate modulus raising algorithm finds an element $\tilde{a} \in R_{P \cdot Q}$ satisfying $\tilde{a} \equiv a \pmod{Q}$ and $\|\tilde{a}\| \ll P \cdot Q$ for a given polynomial $a \in R_Q$. Conversely, the approximate modulus reduction algorithm returns an element $b \in R_Q$ such that $P \cdot b \approx \tilde{b}$ for an input polynomial $\tilde{b} \in R_{P \cdot Q}$. These procedures give relaxed conditions on output polynomials, but we can construct algorithms that can be performed on the RNS representation. In addition, we show that the correctness of the HE system is still guaranteed with some small additional error.

Related Works. There have been several studies [5,8,14,15] about homomorphic arithmetic over real or integral numbers besides the HEAAN scheme. However, these approaches do not support the rounding operation which is a core algorithm in approximate computation, and consequently, the required bit-size of a ciphertext modulus grows exponentially on the depth of a circuit to be evaluated.

Many of HE schemes use a polynomial ring structure with large coefficients. Some recent researches accelerated expensive ring operations by exploiting the RNS representation. Bajard et al. [3] proposed a full RNS variant of the BFV scheme [6,18]. Their implementation could avoid the need of conversion between RNS and coefficient representations of an ring element during homomorphic computations. After that, Halevi et al. [24] presented a simplified method with reduced noise growth. Based on this idea, one can implement an HE scheme without any numerical library for big integer arithmetics. This technique has been applied to SEAL [32] after v2.3.1.

Road-Map. In Sect. 2, we review the basics of the HEAAN scheme and introduce fast base conversion. In Sect. 3, we present a method to improve overall homomorphic operations from RNS representation. In Sect. 4, we describe a full RNS variant of HEAAN. Finally, Sect. 5 shows experimental results with optimization techniques.

2   Background

All logarithms are base 2 unless otherwise indicated. We denote vectors in bold, e.g. $\mathbf{a}$, and every vector in this paper will be a column vector. We denote by $\langle \cdot, \cdot \rangle$ the usual dot product of two vectors. For a real number r, $\lfloor r \rceil$ denotes the nearest integer to r, rounding upwards in case of a tie. For an integer q, we identify $\mathbb{Z} \cap (-q/2, q/2]$ as a representative of $\mathbb{Z}_q$ and use $[a]_q$ to denote the reduction of the integer a modulo q into that interval. We use $x \leftarrow D$ to denote the sampling x according to a distribution D and $U(S)$ denotes the uniform distribution over S when S is a finite set. We let λ denote the security parameter throughout the paper: all known valid attacks against the cryptographic scheme under scope should take $\Omega(2^\lambda)$ bit operations. A finite ordered set $\mathcal{B} = \{p_0, p_1, \ldots, p_{k-1}\}$ of integers is called a *basis* if it is pairwise coprime.

2.1   Approximate Homomorphic Encryption

Cheon et al. [11] proposed an HE scheme that supports an approximate arithmetic on encrypted data. The main idea is to consider an error of homomorphic operation (e.g. encryption, multiplication) as part of computational error in approximate computation.

For a power-of-two integer N, we denote by $K = \mathbb{Q}[X]/(X^N + 1)$ the $(2N)$-th cyclotomic field and $R = \mathbb{Z}[X]/(X^N + 1)$ its ring of integers. The residue ring modulo an integer q is denoted by $R_q = R/qR$. The HEAAN scheme uses a fixed *base* integer q and constructs a chain of moduli $Q_\ell = q^\ell$ for $1 \leq \ell \leq L$. For a polynomial $m(X) \in K$, a ciphertext ct is called an encryption of $m(X)$ at level ℓ if $\mathsf{ct} \in R_{Q_\ell}^2$ and $[\langle \mathsf{ct}, \mathsf{sk} \rangle]_{Q_\ell} \approx m(X)$. Homomorphic operations between ciphertexts of HEAAN can be done by the key-switching with special modulus suggested in [21]. For input encryptions of $m_1(X)$ and $m_2(X)$ at a level ℓ, their homomorphic addition and multiplication satisfy $[\langle \mathsf{ct}_{\mathsf{add}}, \mathsf{sk} \rangle]_{Q_\ell} \approx m_1(X) + m_2(X)$ and $[\langle \mathsf{ct}_{\mathsf{mult}}, \mathsf{sk} \rangle]_{Q_\ell} \approx m_1(X) \cdot m_2(X)$, respectively.

The main advantage of this scheme comes from its intrinsic operation called the rescaling procedure. The rescaling algorithm, denoted by $\mathsf{RS}(\cdot)$, transforms a level ℓ encryption of $m(X)$ into an encryption of $q^{-1} \cdot m(X)$ at level $(\ell - 1)$. It can be considered as an approximate rounding operation or an approximate extraction of the most significant bits of the encrypted plaintext. By reducing the size of the plaintext, we can reduce the speed of modulus consumption in the following computation.

For packing of multiple messages, there has been suggested a method to identify an element of a cyclotomic field with a complex vector via a variant of the canonical embedding. Let $\zeta = \exp(-\pi i/N)$ be a $(2N)$-th root of unity in $\mathbb{C}$. Recall that the canonical embedding of K is defined by $a(X) \mapsto (a(\zeta), a(\zeta^3), \ldots, a(\zeta^{2N-1}))$. Note that there is no need to store all entries of $\sigma(a)$ to recover $a(X)$ since $a(\zeta^j) = \overline{a(\zeta^{2N-j})}$. We denote by $\tau : K \to \mathbb{C}^{N/2}$ a variant of the canonical embedding, defined by

$$\tau : a(X) \mapsto (a(\zeta), a(\zeta^5), \ldots, a(\zeta^{2N-3}))_{0 \leq j < N/2},$$

and use it as a decoding function for HEAAN. The inverse of this homomorphism τ is used as the encoding function to pack $(N/2)$ complex numbers in a single polynomial.

This HE scheme can be applied to fixed-point arithmetic on real (complex) numbers. We multiply a scale factor of q to a number z with finite precision and use the value $m = q \cdot z$ for encryption. Then encryption of m will satisfy $[\langle \mathsf{ct}, \mathsf{sk} \rangle]_{Q_\ell} \approx q \cdot z$, which is an approximate and a scaled value of z. A product of two encryptions of $q \cdot z_1$ and $q \cdot z_2$ will return an encryption of $q^2 \cdot z_1 z_2$, which is a scaled value of $z_1 \cdot z_2$ by q^2. We can perform the rescaling procedure to maintain the original scaling factor q.

2.2 The RNS Representation

Let $\mathcal{B} = \{p_0, \ldots, p_{k-1}\}$ be a basis and let $P = \prod_{i=0}^{k-1} p_i$. We denote by $[\cdot]_{\mathcal{B}}$ the map from $\mathbb{Z}_P$ to $\prod_{i=0}^{k-1} \mathbb{Z}_{p_i}$, defined by $a \mapsto [a]_{\mathcal{B}} = ([a]_{p_i})_{0 \le i < k}$. It is a ring isomorphism from the Chinese Remainder Theorem (CRT) and $[a]_{\mathcal{B}}$ is called the *residue number system* (RNS) representation of $a \in \mathbb{Z}_P$. The main advantage of the RNS representation is to perform component-wise arithmetic operations in the small rings $\mathbb{Z}_{p_i}$, which reduces the asymptotic and practical computation cost. This ring isomorphism over the integers can be naturally extended to a ring isomorphism $[\cdot]_{\mathcal{B}} : R_P \to R_{p_0} \times \cdots \times R_{p_{k-1}}$ by applying it coefficient-wise over the cyclotomic rings.

2.3 Fast Basis Conversion

Brakerski [6] introduced a scale-invariant HE scheme based on the LWE problem, and Fan and Vercauteren [18] suggested its ring-based variant called BFV. Barjard et al. [3] proposed a variant of the BFV scheme that maintains the RNS representation of ciphertexts during homomorphic computation. This scheme presents a new algorithm, called the fast basis conversion, to convert the residue of a polynomial into a new basis that is coprime to the original basis.

More precisely, for a basis $\{p_0, \ldots, p_{k-1}, q_0, \ldots, q_{\ell-1}\}$, let $\mathcal{B} = \{p_0, \ldots, p_{k-1}\}$ and $\mathcal{C} = \{q_0, \ldots, q_{\ell-1}\}$ be its subbases. Let us denote their products by $P = \prod_{i=0}^{k-1} p_i$ and $Q = \prod_{j=0}^{\ell-1} q_j$, respectively. Then one can convert the RNS representation $[a]_{\mathcal{C}} = (a^{(0)}, \ldots, a^{(\ell-1)}) \in \mathbb{Z}_{q_0} \times \cdots \times \mathbb{Z}_{q_{\ell-1}}$ of an integer $a \in \mathbb{Z}_Q$ into an element of $\mathbb{Z}_{p_0} \times \cdots \times \mathbb{Z}_{p_{k-1}}$ by computing

$$\mathsf{Conv}_{\mathcal{C} \to \mathcal{B}}([a]_{\mathcal{C}}) = \left(\sum_{j=0}^{\ell-1} [a^{(j)} \cdot \hat{q}_j^{-1}]_{q_j} \cdot \hat{q}_j \pmod{p_i} \right)_{0 \le i < k},$$

where $\hat{q}_j = \prod_{j' \ne j} q_{j'} \in \mathbb{Z}$. We note that $\sum_{j=0}^{\ell-1} [a^{(j)} \cdot \hat{q}_j^{-1}]_{q_j} \cdot \hat{q}_j = a + Q \cdot e$ for some small $e \in \mathbb{Z}$ satisfying $|a + Q \cdot e| \le (\ell/2) \cdot Q$. This implies that $\mathsf{Conv}_{\mathcal{C} \to \mathcal{B}}([a]_{\mathcal{C}}) = [a + Q \cdot e]_{\mathcal{B}}$ can be considered as the RNS representation of the integer $a + Q \cdot e$ with respect to the basis $\mathcal{B}$.

3 Approximate Bases and Full RNS Modulus Switching

The approximate HE scheme of Cheon et al. [11] has its own advantages in arithmetic of approximate numbers. However, a ciphertext modulus could not be chosen as a product of coprime integers, so its implementation [10] requires expensive multi-precision modular arithmetic. In this section, we introduce an idea to avoid the use of a power-of-two base ciphertext modulus and enable the RNS decomposition in the HEAAN scheme. We also propose new algorithms to switch a ciphertext modulus on the RNS components.

3.1 Approximate Basis

The main advantage of HEAAN comes from the rescaling algorithm $RS(\cdot)$. It allows us to perform the rounding of an encrypted plaintext, that is, we can efficiently convert an encryption of m into a ciphertext encrypting the scaled message $q^{-1} \cdot m$. In the case of its application to fixed-point arithmetic, for example, we multiply fixed-point numbers z_i by a common scale factor of q to maintain the precision of plaintexts. After homomorphic multiplication, we obtain an encryption of the product $q^2 \cdot z_1 z_2$ of two numbers $q \cdot z_1$ and $q \cdot z_2$. Then we perform the rescaling algorithm to get an encryption of $q \cdot z_1 z_2$ and maintain the original scale factor q. For this reason, the ciphertext modulus should be chosen as a power of a fixed base $Q_\ell = q^\ell$ to have the same scaling ratio. This point made it difficult to use the RNS representation on HEAAN.

To overcome this obstacle, we propose an idea to use an RNS basis consisting of *approximate* values of a fixed base. In more detail, given the scale factor q and bit precision η, we find a basis $\mathcal{C} = \{q_0, \ldots, q_L\}$ such that $q/q_\ell \in (1-2^{-\eta}, 1+2^{-\eta})$ for $\ell = 1, \ldots, L$. This approximate basis allows us to use the RNS representation of polynomials while keeping the functionality of the HE scheme. We set the level ℓ ciphertext modulus as $Q_\ell = \prod_{i=0}^{\ell} q_i$, so that the ciphertext moduli in the consecutive levels have almost the same ratio $Q_\ell/Q_{\ell-1} = q_\ell \approx q$. The rescaling algorithm with a factor of q_ℓ converts an encryption of m at level ℓ into an encryption of $q_\ell^{-1} \cdot m$ at level $(\ell-1)$. This operation has an additional error from the approximation of q, but we can manage the size of an error not to destroy the significant digits of a plaintext. An approximation error is bounded by

$$\left| q_\ell^{-1} \cdot m - q^{-1} \cdot m \right| = \left| 1 - q_\ell^{-1} \cdot q \right| \cdot \left| q^{-1} \cdot m \right| \leq 2^{-\eta} \cdot \left| q^{-1} \cdot m \right|,$$

so it does not destroy the significant digits of an encrypted plaintext when η is sufficiently larger than the bit precision of an encrypted plaintext.

3.2 Approximate Modulus Switching

The use of an approximate basis enables an implementation of the HEAAN scheme using the RNS representation. However, HEAAN includes some non-arithmetic operations that cannot be directly implemented on the RNS components. Specifically, homomorphic multiplication and rescaling procedure require an exact

---

Algorithm 1. Approximate Modulus Raising

1: **procedure** $\mathtt{ModUp}_{\mathcal{C}\to\mathcal{D}}(a^{(0)}, a^{(1)}, \ldots, a^{(\ell-1)})$
2:     $(\tilde{a}^{(0)}, \ldots, \tilde{a}^{(k-1)}) \leftarrow \mathtt{Conv}_{\mathcal{C}\to\mathcal{B}}([a]_{\mathcal{C}})$.
3:     **return** $(\tilde{a}^{(0)}, \ldots, \tilde{a}^{(k-1)}, a^{(0)}, \ldots, a^{(\ell-1)})$.
4: **end procedure**

---

modulus switching algorithm, and the key-switching technique for rotation and conjugation also contains the same operation (see [9,11] for details).

We remark that the goal of modulus switching algorithms in [11] can be reduced to a problem that finds a ciphertext with a small error while keeping the correctness of the HE scheme. In this section, we propose an idea to approximately perform the modulus switching algorithms on the RNS representation. A full RNS variant of HEAAN will be described in the next section based on this method. Throughout this paper, we will denote by $\mathcal{D} = \{p_0, \ldots, p_{k-1}, q_0, \ldots, q_{\ell-1}\}$, $\mathcal{B} = \{p_0, \ldots, p_{k-1}\}$, and $\mathcal{C} = \{q_0, \ldots, q_{\ell-1}\}$ an RNS basis and its subbases, respectively, with $P = \prod_{i=0}^{k-1} p_i$ and $Q = \prod_{j=0}^{\ell-1} q_j$.

Approximate Modulus Raising. Suppose that we are given the RNS representation $[a]_{\mathcal{C}}$ of an integer $a \in \mathbb{Z}_Q$. The purpose of the approximate modulus raising algorithm, denoted by $\mathtt{ModUp}$, is to find the RNS representation of an integer $\tilde{a} \in \mathbb{Z}_{PQ}$ with respect to the basis $\mathcal{D}$ satisfying two conditions $\tilde{a} \equiv a \pmod{Q}$ and $|\tilde{a}| \ll P \cdot Q$. From the first condition $[\tilde{a}]_{\mathcal{C}} = [a]_{\mathcal{C}}$, we only need to generate the RNS representation of $\tilde{a}$ with the basis $\mathcal{B}$ and it can be done by applying the fast conversion algorithm. See Algorithm 1 for a description of the approximate modulus raising.

As described in Sect. 2.3, the fast conversion algorithm in Algorithm 1 returns $[a + Q \cdot e]_{\mathcal{B}} \in \prod_{i=0}^{k-1} \mathbb{Z}_{p_i}$ for some integer e with $|e| \le \ell/2$. Therefore, the output of $\mathtt{ModUp}$ algorithm is the RNS representation of $\tilde{a} := a + Q \cdot e$ with respect to the basis $\mathcal{D} = \mathcal{B} \cup \mathcal{C}$, as desired.

Approximate Modulus Reduction. Contrary to the modulus raising algorithm, the approximate modulus reduction algorithm, denoted by $\mathtt{ModDown}$, takes an RNS representation $[\tilde{b}]_{\mathcal{D}}$ of an integer $\tilde{b} \in \mathbb{Z}_{P \cdot Q}$ as an input and aims to compute $[b]_{\mathcal{C}}$ for some integer $b \in \mathbb{Z}_Q$ satisfying $b \approx P^{-1} \cdot \tilde{b}$.

We point out that the goal of approximate modulus reduction is reduced to a problem of finding small $\tilde{a} = \tilde{b} - P \cdot b$ satisfying $\tilde{a} \equiv \tilde{b} \pmod{P}$. The RNS representation $[\tilde{b}]_{\mathcal{D}}$ is the concatenation of $[\tilde{b}]_{\mathcal{B}}$ and $[\tilde{b}]_{\mathcal{C}}$. We first take the first component $[\tilde{b}]_{\mathcal{B}} = (\tilde{b}^{(0)}, \ldots, \tilde{b}^{(k-1)})$, which is the same as $[a]_{\mathcal{B}}$ for $a = [\tilde{b}]_P \in \mathbb{Z}_P$. Then we apply the fast conversion algorithm to compute the RNS representation $[\tilde{a}]_{\mathcal{C}}$ of $\tilde{a} = a + P \cdot e$ for some small e. Note that $\tilde{a} \equiv \tilde{b} \pmod{P}$ and $|\tilde{a}| \ll P \cdot Q$ from the property of $\mathtt{Conv}_{\mathcal{B}\to\mathcal{C}}(\cdot)$. Finally, we derive the RNS representation of $b = P^{-1} \cdot (\tilde{b} - \tilde{a})$ with respect to the basis $\mathcal{C}$ by computing $\left(\prod_{i=0}^{k-1} p_i\right)^{-1} \cdot \left([\tilde{b}]_{\mathcal{C}} - [\tilde{a}]_{\mathcal{C}}\right) \in \prod_{j=0}^{\ell-1} \mathbb{Z}_{q_j}$. See Algorithm 2 for a description.

Algorithm 2. Approximate Modulus Reduction

1: **procedure** ModDown$_{\mathcal{D}\rightarrow\mathcal{C}}(\tilde{b}^{(0)}, \tilde{b}^{(1)}, \ldots, \tilde{b}^{(k+\ell-1)})$
2:    $(\tilde{a}^{(0)}, \ldots, \tilde{a}^{(\ell-1)}) \leftarrow \text{Conv}_{\mathcal{B}\rightarrow\mathcal{C}}(\tilde{b}^{(0)}, \ldots, \tilde{b}^{(k-1)})$
3:    **for** $0 \leq j < \ell$ **do**
4:       $b^{(j)} = \left(\prod_{i=0}^{k-1} p_i\right)^{-1} \cdot (\tilde{b}^{(k+j)} - \tilde{a}^{(j)}) \pmod{q_j}.$
5:    **end for**
6: **return** $(b^{(0)}, \ldots, b^{(\ell-1)}).$
7: **end procedure**

Word Operations. In the rest of the paper, the arithmetic operations (e.g. addition and multiplication) modulo a "word-size" integer will be called the *word operations*. Now suppose that p_i's and q_j's are word-size integers. As mentioned before, the fast conversion algorithm $\text{Conv}_{\mathcal{C}\rightarrow\mathcal{B}}([a]_\mathcal{C})$ outputs the tuple $\left(\sum_{j=0}^{\ell-1} [a^{(j)} \cdot \hat{q}_j^{-1}]_{q_j} \cdot \hat{q}_j \pmod{p_i}\right)_{0 \leq i < k}$ for $\hat{q}_j = \prod_{j' \neq j} q_{j'}$. Each component can be computed using the values $[\hat{q}_j^{-1}]_{q_j} = \prod_{j' \neq j} q_{j'}^{-1} \pmod{q_j}$ and $[\hat{q}_j]_{p_i} = \prod_{j' \neq j} q_{j'} \pmod{p_i}$ while avoiding the computation of *big* integers $\hat{q}_j$. In addition, if we pre-compute and store these values, which depend only on the bases $\mathcal{B}$ and $\mathcal{C}$, then the computation cost of $\text{Conv}_{\mathcal{C}\rightarrow\mathcal{B}}(\cdot)$ algorithm can be reduced down to $O(k \cdot \ell)$ word operations.

Complexity of Approximate Modulus Switching. Our modulus switching algorithms have an advantage, in that they can be computed only using word operations. For example, $\text{ModUp}_{\mathcal{C}\rightarrow\mathcal{D}}([a]_\mathcal{C})$ requires exactly the same computation as $\text{Conv}_{\mathcal{C}\rightarrow\mathcal{B}}([a]_\mathcal{C})$, so its total complexity is bounded by $O(k \cdot \ell)$ word operations. The approximate modulus reduction algorithm needs to compute $b^{(j)} = P^{-1} \cdot (\tilde{b}^{(k+j)} - \tilde{a}^{(j)}) \pmod{q_j}$ for $0 \leq j < \ell$ as well as the fast conversion algorithm. The computation of $b^{(j)}$'s can be done in $O(\ell)$ word operations using the pre-computable constants $[P^{-1}]_{q_j} = \left(\prod_{i=0}^{k-1} p_i\right)^{-1} \pmod{q_j}$. Therefore, the total complexity of ModDown is bounded by $O(k \cdot \ell + \ell) = O(k \cdot \ell)$ word operations.

The approximate modulus switching algorithms can be extended to algorithms over the polynomial rings as

$$\text{ModUp}_{\mathcal{C}\rightarrow\mathcal{D}}(\cdot) : \prod_{j=0}^{\ell-1} R_{q_j} \rightarrow \prod_{i=0}^{k-1} R_{p_i} \times \prod_{j=0}^{\ell-1} R_{q_j},$$

$$\text{ModDown}_{\mathcal{D}\rightarrow\mathcal{C}}(\cdot) : \prod_{i=0}^{k-1} R_{p_i} \times \prod_{j=0}^{\ell-1} R_{q_j} \rightarrow \prod_{j=0}^{\ell-1} R_{q_j}$$

by applying them coefficient-wise. These operations require $O(k \cdot \ell \cdot N)$ word operations where N is a degree of a power-of-two cyclotomic ring.

4 A Full RNS Variant of the Approximate HE

In this section, we propose a variant of HEAAN based on the full RNS representation. For simplicity, we choose a power-of-two integer N and consider the $(2N)$-th cyclotomic field $K = \mathbb{Q}[X]/(X^N+1)$ and its ring of integers $R = \mathbb{Z}[X]/(X^N+1)$. An arbitrary element of K is expressed as a polynomial with rational coefficients of degree strictly less than N, and identified with the vector of its coefficients in $\mathbb{Q}^N$. The rounding operation on K and the modulo operation on R will be defined by the coefficient-wise rounding and modulo operations, respectively. In the following, we present a concrete description of a full RNS variant of HEAAN.

Setup$(q, L, \eta; 1^\lambda)$. A base integer q, the number of levels L, and the bit precision η are given as inputs with the security parameter λ.

- Choose a basis $\mathcal{D} = \{p_0, \ldots, p_{k-1}, q_0, q_1, \ldots, q_L\}$ such that $q_j/q \in (1-2^{-\eta}, 1+2^{-\eta})$ for $1 \le j \le L$. We write $\mathcal{B} = \{p_0, \ldots, p_{k-1}\}$, $\mathcal{C}_\ell = \{q_0, \ldots, q_\ell\}$, and $\mathcal{D}_\ell = \mathcal{B} \cup \mathcal{C}_\ell = \{p_0, \ldots, p_{k-1}, q_0, \ldots, q_\ell\}$ for $0 \le \ell \le L$. Let $P = \prod_{i=0}^{k-1} p_i$ and $Q = \prod_{j=0}^{L} q_j$.
- Choose a power-of-two integer N.
- Choose a secret key distribution χ_{key}, an encryption key distribution χ_{enc}, and an error distribution χ_{err} over R.
- Let $\hat{p}_i = \prod_{0 \le i' < k, i' \ne i} p_{i'}$ for $0 \le i < k$. Compute the constants $[\hat{p}_i]_{q_j}$ and $[\hat{p}_i^{-1}]_{p_i}$ for $0 \le i < k$, $0 \le j \le L$.
- Compute the constants $[P^{-1}]_{q_j} = \left(\prod_{i=0}^{k-1} p_i\right)^{-1} \pmod{q_j}$ for $0 \le j \le L$.
- Let $\hat{q}_{\ell,j} = \prod_{0 \le j' \le \ell, j' \ne j} q_{j'}$ for $0 \le j \le \ell \le L$. Compute the constants $[\hat{q}_{\ell,j}]_{p_i}$ and $[\hat{q}_{\ell,j}^{-1}]_{q_j}$ for $0 \le i < k$, $0 \le j \le \ell \le L$.

The constants $[\hat{p}_i]_{q_j}$ and $[\hat{p}_i^{-1}]_{p_i}$ are necessary to compute the conversion $\mathsf{Conv}_{\mathcal{B} \to \mathcal{C}_\ell}(\cdot)$ in the $\mathsf{ModDown}_{\mathcal{D}_\ell \to \mathcal{C}_\ell}(\cdot)$ algorithm. The constants $[P^{-1}]_{q_j}$ are also used in the algorithm. On the other hand, the constants $[\hat{q}_{\ell,j}]_{p_i}$ and $[\hat{q}_{\ell,j}^{-1}]_{q_j}$ are used to compute $\mathsf{Conv}_{\mathcal{C}_\ell \to \mathcal{B}}(\cdot)$ for the $\mathsf{ModUp}_{\mathcal{C}_\ell \to \mathcal{D}_\ell}(\cdot)$ algorithm.

We choose an RNS basis $\mathcal{D}$ consisting of word-size integers, so that every homomorphic arithmetic can be expressed using word operations (e.g. uint64_t). The elements of $\mathcal{B}$ are called the special primes and used in the key-switching procedure. They do not have to be close to q, but their product P should be large enough to get a small key-switching error. The zero-level ciphertext modulus $Q_0 = q_0$ is not necessarily approximate to the base integer q, but it should be larger than the modulus of the encrypted plaintext for the correctness of decryption.

KSGen(s_1, s_2). For given secret polynomials $s_1, s_2 \in R$, sample uniform elements $(a'^{(0)}, \ldots, a'^{(k+L)}) \leftarrow U\left(\prod_{i=0}^{k-1} R_{p_i} \times \prod_{j=0}^{L} R_{q_j}\right)$ and an error $e' \leftarrow \chi_{\mathsf{err}}$. Output the switching key swk as

$$\left(\mathsf{swk}^{(0)} = (b'^{(0)}, a'^{(0)}), \ldots, \mathsf{swk}^{(k+L)} = (b'^{(k+L)}, a'^{(k+L)})\right) \in \prod_{i=0}^{k-1} R_{p_i}^2 \times \prod_{j=0}^{L} R_{q_j}^2$$

where $b'^{(i)} \leftarrow -a'^{(i)} \cdot s_2 + e' \pmod{p_i}$ for $0 \leq i < k$ and $b'^{(k+j)} \leftarrow -a'^{(k+j)} \cdot s_2 + [P]_{q_j} \cdot s_1 + e' \pmod{q_j}$ for $0 \leq j \leq L$.

This procedure generates a switching key to convert a ciphertext with the secret key s_1 into a ciphertext encrypting the same message with the secret key s_2. If a' is the element of $R_{P \cdot Q}$ such that $[a']_{\mathcal{D}} = (a'^{(0)}, \ldots, a'^{(k+L)})$, then the switching key swk can be seen as the RNS representation of $(b', a') \in R_{P \cdot Q}$ in the basis $\mathcal{D}$ for $b' = -a' \cdot s_2 + P \cdot s_1 + e' \pmod{P \cdot Q}$.

KeyGen.

- Sample $s \leftarrow \chi_{\mathsf{key}}$ and set the secret key as $\mathsf{sk} \leftarrow (1, s)$.
- Sample $(a^{(0)}, \ldots, a^{(L)}) \leftarrow U\left(\prod_{j=0}^{L} R_{q_j}\right)$ and $e \leftarrow \chi_{\mathsf{err}}$. Set the public key as

$$\mathsf{pk} \leftarrow \left(\mathsf{pk}^{(j)} = (b^{(j)}, a^{(j)}) \in R_{q_j}^2\right)_{0 \leq j \leq L}$$

  where $b^{(j)} \leftarrow -a^{(j)} \cdot s + e \pmod{q_j}$ for $0 \leq j \leq L$.
- Set the evaluation key as $\mathsf{evk} \leftarrow \mathsf{KSGen}(s^2, s)$.

The encryption key is the RNS representation of an RLWE sample $(b = -a \cdot s + e, a) \in R_{Q_L}^2$ in the basis $\mathcal{C}_L$. The evaluation key evk can be used to perform the relinearization operation during homomorphic multiplication. One can generate additional public keys for more functionalities. For example, we need to publish a rotation key (resp. conjugation key) to compute the permutation (resp. conjugation) on plaintext slots as described in [11].

$\underline{\mathsf{Enc}_{\mathsf{pk}}(m)}$. For $m \in R$, sample $v \leftarrow \chi_{\mathsf{enc}}$ and $e_0, e_1 \leftarrow \chi_{\mathsf{err}}$. Output the ciphertext $\mathsf{ct} = \left(\mathsf{ct}^{(j)}\right)_{0 \leq j \leq L} \in \prod_{j=0}^{L} R_{q_j}^2$ where $\mathsf{ct}^{(j)} \leftarrow v \cdot \mathsf{pk}^{(j)} + (m + e_0, e_1) \pmod{q_j}$ for $0 \leq j \leq L$.

$\underline{\mathsf{Dec}_{\mathsf{sk}}(\mathsf{ct})}$. For $\mathsf{ct} = \left(\mathsf{ct}^{(j)}\right)_{0 \leq j \leq \ell}$, output $\langle \mathsf{ct}^{(0)}, \mathsf{sk} \rangle \pmod{q_0}$.

The encryption algorithm generates the RNS representation of a ciphertext ct satisfying $[\langle \mathsf{ct}, \mathsf{sk} \rangle]_{Q_L} \approx m$. Thus its decryption returns an approximate value of the input plaintext. The encrypted plaintext should satisfy $\|m\|_\infty \leq q_0/2$ in order to recover a correct value.

$\underline{\mathsf{Add}(\mathsf{ct}, \mathsf{ct}')}$. Given two ciphertexts $\mathsf{ct} = \left(\mathsf{ct}^{(0)}, \ldots, \mathsf{ct}^{(\ell)}\right), \mathsf{ct}' = \left(\mathsf{ct}'^{(0)}, \ldots, \mathsf{ct}'^{(\ell)}\right)$ $\in \prod_{j=0}^{\ell} R_{q_j}^2$, output a ciphertext $\mathsf{ct}_{\mathsf{add}} = \left(\mathsf{ct}_{\mathsf{add}}^{(j)}\right)_{0 \leq j \leq \ell}$ where $\mathsf{ct}_{\mathsf{add}}^{(j)} \leftarrow \mathsf{ct}^{(j)} + \mathsf{ct}'^{(j)}$ $\pmod{q_j}$ for $0 \leq j \leq \ell$.

$\underline{\mathsf{Mult}_{\mathsf{evk}}(\mathsf{ct}, \mathsf{ct}')}$. Given two ciphertexts $\mathsf{ct} = \left(\mathsf{ct}^{(j)} = (c_0^{(j)}, c_1^{(j)})\right)_{0 \leq j \leq \ell}$ and $\mathsf{ct}' = \left(\mathsf{ct}'^{(j)} = (c_0'^{(j)}, c_1'^{(j)})\right)_{0 \leq j \leq \ell}$, perform the following procedures and return a ciphertext $\mathsf{ct}_{\mathsf{mult}} \in \prod_{j=0}^{\ell} R_{q_j}^2$.

1. For $0 \leq j \leq \ell$, compute

$$d_0^{(j)} \leftarrow c_0^{(j)} c_0'^{(j)} \pmod{q_j},$$
$$d_1^{(j)} \leftarrow c_0^{(j)} c_1'^{(j)} + c_1^{(j)} c_0'^{(j)} \pmod{q_j},$$
$$d_2^{(j)} \leftarrow c_1^{(j)} c_1'^{(j)} \pmod{q_j}.$$

2. Compute $\mathtt{ModUp}_{\mathcal{C}_\ell \to \mathcal{D}_\ell}(d_2^{(0)}, \ldots, d_2^{(\ell)}) = (\tilde{d}_2^{(0)}, \ldots, \tilde{d}_2^{(k-1)}, d_2^{(0)}, \ldots, d_2^{(\ell)}).$

3. Compute

$$\tilde{\mathsf{ct}} = (\tilde{\mathsf{ct}}^{(0)} = (\tilde{c}_0^{(0)}, \tilde{c}_1^{(0)}), \ldots, \tilde{\mathsf{ct}}^{(k+\ell)} = (\tilde{c}_0^{(k+\ell)}, \tilde{c}_1^{(k+\ell)})) \in \prod_{i=0}^{k-1} R_{p_i}^2 \times \prod_{j=0}^{\ell} R_{q_j}^2$$

where $\tilde{\mathsf{ct}}^{(i)} = \tilde{d}_2^{(i)} \cdot \mathsf{evk}^{(i)} \pmod{p_i}$ and $\tilde{\mathsf{ct}}^{(k+j)} = d_2^{(j)} \cdot \mathsf{evk}^{(k+j)} \pmod{q_j}$ for $0 \leq i < k$, $0 \leq j \leq \ell$.

4. Compute

$$\left(\hat{c}_0^{(0)}, \ldots, \hat{c}_0^{(\ell)}\right) \leftarrow \mathtt{ModDown}_{\mathcal{D}_\ell \to \mathcal{C}_\ell}\left(\tilde{c}_0^{(0)}, \ldots, \tilde{c}_0^{(k+\ell)}\right),$$
$$\left(\hat{c}_1^{(0)}, \ldots, \hat{c}_1^{(\ell)}\right) \leftarrow \mathtt{ModDown}_{\mathcal{D}_\ell \to \mathcal{C}_\ell}\left(\tilde{c}_1^{(0)}, \ldots, \tilde{c}_1^{(k+\ell)}\right).$$

5. Output the ciphertext $\mathsf{ct}_{\mathsf{mult}} = (\mathsf{ct}_{\mathsf{mult}}^{(j)})_{0 \leq j \leq \ell}$ where $\mathsf{ct}_{\mathsf{mult}}^{(j)} \leftarrow (\hat{c}_0^{(j)} + d_0^{(j)}, \hat{c}_1^{(j)} + d_1^{(j)}) \pmod{q_j}$ for $0 \leq j \leq \ell$.

We first generate an extended ciphertext (d_0, d_1, d_2) that decrypts to the product of the input plaintexts under the extended secret key $(1, s, s^2)$. As mentioned before, we use the evaluation key to transform d_2 into a normal ciphertext. Our homomorphic multiplication algorithm is somewhat more complicated compared to the ordinary HEAAN because we switch the ciphertext moduli approximately using our approximate algorithms.

<u>RS(ct).</u> For a level-ℓ ciphertext $\mathsf{ct} = \left(\mathsf{ct}^{(j)} = (c_0^{(j)}, c_1^{(j)})\right)_{0 \leq j \leq \ell} \in \prod_{j=0}^{\ell} R_{q_j}^2$, compute $c_i'^{(j)} \leftarrow q_\ell^{-1} \cdot \left(c_i^{(j)} - c_i^{(\ell)}\right) \pmod{q_j}$ for $i = 0, 1$ and $0 \leq j < \ell$. Output the ciphertext $\mathsf{ct}' \leftarrow \left(\mathsf{ct}'^{(j)} = (c_0'^{(j)}, c_1'^{(j)})\right)_{0 \leq j \leq \ell-1} \in \prod_{j=0}^{\ell-1} R_{q_j}^2$.

For a ciphertext ct encrypting a plaintext m, the rescaling algorithm returns an encryption of $q_\ell^{-1} \cdot m \approx q^{-1} \cdot m$ at level $(\ell-1)$. The output ciphertext contains an additional error from the approximation of q to q_ℓ and the rounding of the input ciphertext. The correctness of our scheme will be shown in Appendix A with noise analysis.

5   Software Implementation

In this section, we provide experimental results with parameter sets. In our implementation, every number is stored as an unsigned 64-bit integer, which is standard on computer system. All homomorphic operations provided in our scheme are expressed as word size operations defined on this standard variable type, so our HE library does not depend on any multi-precision numerical library. Our implementation was performed on a machine with an Intel Core i5 running at 2.9 GHz processor on a single-threaded mode, and its source code is publicly available at https://github.com/HanKyoohyung/FullRNS-HEAAN.

We adapt the discrete Fourier transformation to transform a polynomial represented by its coefficient vector into the vector of evaluations at primitive roots of unity modulo a prime. The modulus switching algorithms require the coefficient representation, but we can manipulate the NTT representation for arithmetic operations. Consequently, the complexity of homomorphic operations mainly depends on this transformation between two representations. We implemented the NTT conversion and its inverse based on the butterfly techniques of Cooley-Tukey [12] and Gentleman-Sande [19], respectively. We also optimized these algorithms using Montgomery modular multiplication and butterfly algorithms [26] and Barrett reduction algorithm [4].

5.1 Parameter Sets and Benchmark

We propose parameter sets for multiplicative depths L from 5 to 15 in Table 1. It also shows experimental results for encryption, decryption, addition, scalar-multiplication, and multiplication (together with the rescaling operation) of the original implementation HEAAN and our RNS variant denoted by HEAAN-RNS.

The smallest ciphertext modulus q_0 should be larger than an encrypted plaintext for the correctness of the decryption circuit. We use $\log q_0 \approx 61$ and $\log q_i \approx 55$ for $i = 1, \ldots, L$. We present a list of primes in Appendix B. For a fair comparison, we choose a power-of-two integer Q_L of the same bit size as the implementation of the original HEAAN. The coefficients of error polynomials are sampled from the discrete Gaussian distribution of standard deviation $\sigma = 3.2$ and a secret key is chosen randomly from the set of signed binary polynomials with the Hamming weight $h = 64$. We used the estimator of Albrecht et al. [2] to guarantee that the proposed parameter sets achieve at least 80-bit security level against the known attacks against the LWE problem.

Our implementation of the RNS variant improved the performance of basic operations by approximately ten times compared to the original HEAAN [10,11]. For example, the encryption, decryption, addition, and multiplication are speedups of 9.1, 17.3, 7.4, and 8.3 times, respectively, when evaluating a circuit of depth $L = 10$.

In Appendix A, we analyze the growth of errors and provide theoretical upper bounds on the growth during homomorphic operations. Figure 1 depicts the bit precisions of an encrypted plaintext during an evaluation of homomorphic multiplications for $L = 10$ with the parameter set in Table 1. We also provide an empirical result on the precision loss.

Our scheme exploits the approximate rounding operation which introduces an additional error. We observed that the precision of an output value is reduced by about three bits compared to the original HEAAN scheme. However, this small

Table 1. Comparison of experimental results of HEAAN and HEAAN-RNS

Variant	L	N	$\log q$	$\lceil \log Q_L \rceil$	Enc (ms)	Dec (ms)	Add (ms)	Cmult (ms)	Mult&RS (ms)
HEAAN	5	2^{15}	55	336	332	106	30	204	740
	10	2^{15}		611	530	135	32	281	1,355
	15	2^{16}		886	1,465	344	70	762	4,169
HEAAN-RNS	5	2^{15}	55	336	31	4.6	2.9	25	85
	10	2^{15}		611	58	7.8	4.3	44	164
	15	2^{16}		886	177	10.0	15.5	125	563

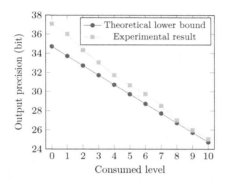

Fig. 1. Bit precision of encrypted plaintext

gap is not an critical issue in most of applications where an approximate result is sufficient for their purposes. In addition, we can easily increase the precision by setting a larger basis while still keeping advantages in the efficiency.

5.2 Homomorphic Evaluation of Statistical and Analytic Functions

The HEAAN scheme can evaluate an arbitrary analytic function by exploiting its polynomial approximation. Table 2 shows a parameter set and evaluation timings for the multiplicative inverse, the exponential function, and the sigmoid function $\sigma(x) = (1 + \exp(-x))^{-1}$. We adapt the approximation method for multiplicative inverse of [11, Algorithm 2] and evaluate the approximate polynomial of degree 15. For the exponential and sigmoid functions, we use the Taylor expansions up to degree 7. The output ciphertexts have at least 32 bits of precision. These computations can be performed over multiple slots simultaneously, yielding a better amortized performance per slot.

We also evaluated mean and variance functions that are the most common quantities in statistical analysis. There have been a few attempts to evaluate these measurements on an HE system. For example, Lauter et al. [30] took about six seconds to obtain the square sum of 100 integers without division by the number of elements.

Table 2. Homomorphic evaluation of analytic functions

Function	Degree	N	$\log q$	$\lceil \log Q_L \rceil$	Total time	Amortized time
x^{-1}	15	2^{14}	55	281	167 ms	21 μs
$\exp x$	7	2^{14}	55	281	164 ms	20 μs
Sigmoid	7	2^{14}	55	281	161 ms	19 μs

For computation of mean and variance of n numbers, we encrypt all the numbers in a single ciphertext and compute their summation by applying the partial sum algorithm [9, Algorithm 2]. It repeats to rotate an encrypted plaintext vector and add it to the original ciphertext. The resulting ciphertext encrypts the mean value in every plaintext slot. The following example describes the partial sum algorithm when $n = 4$.

$$(m_1, m_2, m_3, m_4) \mapsto (m_1, m_2, m_3, m_4) + (m_3, m_4, m_1, m_2)$$
$$= (m_1 + m_3, m_2 + m_4, m_1 + m_3, m_2 + m_4)$$
$$\mapsto \left(\sum_{i=1}^{4} m_i, \sum_{i=1}^{4} m_i, \sum_{i=1}^{4} m_i, \sum_{i=1}^{4} m_i \right)$$

Contrary to previous work, the approximate HE scheme can perform a division by n by multiplying the constant $\lfloor q/n \rceil$ and rescaling by one level. In the case of the variance function, we first square an input ciphertext and apply the same procedure to get a ciphertext encrypting the mean square in its plaintext slots. Then the variance of input data can be computed by subtracting the square of the encrypted mean value. We summarize the parameter and experimental results for homomorphic evaluation of statistical functions on $n = 2^{13}$ numbers in Table 3.

5.3 Homomorphic Training of Logistic Regression Model

The security and privacy issues have arisen on machine learning because the training of a model requires a large database consisting of sensitive information while the prediction phase is based on private information of individuals. The technology of an HE system is a promising solution to address these issues by aggregating encrypted personal data and building a model without information leakage. ML Confidential [23] and CryptoNets [22] are notable examples of leveraging the technology of HE for secure outsourcing of machine learning applications.

In particular, HEAAN [9,11] is a strong candidate for machine learning tasks since most of training and prediction algorithms contain an arithmetic over the real numbers. For example, iDASH Security and Privacy Competition in 2017[1]

---

[1] http://www.humangenomeprivacy.org/2017/.

Table 3. Homomorphic evaluation of statistic functions

	Number of elements (n)	N	$\log q$	$\lceil \log Q_L \rceil$	Total time
Mean	2^{13}	2^{14}	55	171	307 ms
Variance					518 ms

announced a task which aims to build a logistic regression model from homomorphically encrypted genomic data. To be precise, for a given dataset consisting of n samples $(\boldsymbol{x}_i, y_i) \in \mathbb{R}^d \times \{\pm 1\}$ of d features and a binary class, the goal was to find a weight vector $\boldsymbol{\beta} \in \mathbb{R}^{d+1}$ which minimizes the loss function

$$J(\boldsymbol{\beta}) = \sum_{i=1}^{n} \log(1 + \exp(-\boldsymbol{\beta}^T \boldsymbol{z}_i))$$

where $\boldsymbol{z}_i = y_i \cdot (1, \boldsymbol{x}_i)$ for $1 \leq i \leq n$. The best solution [27] adapted the HEAAN library [10] to evaluate Nesterov's accelerated gradient descent method [31].

We implemented the same algorithm based on HEAAN-RNS to show its versatility and efficiency. For a fair comparison, we adapt the previous encoding and evaluation strategies: the whole database is encrypted in a single ciphertext and the sigmoid function of the gradient descent algorithm is approximated to its least squares approximation. Our implementation took about 1.8 min to train a model based on Low Birth Weight Study (lbw) [29] and Umaru Impact Study (uis) [33] datasets using a single core processor, compared to 3.5 min of the previous best solution [27] using four cores, while maintaining the accuracy and area under the ROC curve (AUC) of the resulting classifier (Table 4).

Table 4. Homomorphic training of logistic regression model

Dataset	Num of features	Num of samples	Num of iterations	N	$\log q$	$\lceil \log Q_L \rceil$	Total time	Accuracy	AUC
lbw	9	189	5	2^{16}	40	1061	1.82 min	69.73%	0.62
uis	8	575	5	2^{16}	40	1061	1.83 min	74.43%	0.59

6   Conclusions and Future Work

In this article, we demonstrate a variant of HEAAN based on the RNS representation of polynomials. In the previous implementation, ciphertext moduli were selected as powers of a fixed base for the correctness of rescaling process. We resolve the issue by taking an RNS basis consisting of primes close to the base integer. In addition, we propose variants of modulus switching algorithms which can be computed without any RNS conversion or multi-precision arithmetic.

One disadvantage of our method is that it makes a trade-off between performance and accuracy. Because of the approximation error of an RNS basis, our scheme may have less accuracy compared to the original scheme when using the same parameter. Recently, SEAL version 3.0 [32] has been released. It supports a full RNS variant of HEAAN, which is slightly different from our scheme. The main difference is that a ciphertext of SEAL contains a scaling factor which can be changed during computation. In other words, it continuously tracks the computation and updates the scaling factor information. This method does not have the above accuracy issue, but it is less intuitive and causes new problems related to the management of scaling factors. For example, the addition (resp. multiplication) of ciphertexts of different scaling factors (resp. levels) requires pre (resp. post) processing. It would be an interesting future work to combine the two methods to design a new scheme with enhanced functionality and flexibility.

Acknowledgments. This work was partially supported by the National Research Foundation of Korea (NRF) Grant funded by the Korean Government (MSIT) (No. 2017R1A5A1015626). M. Kim was supported by the National Institute of Health (NIH) under award number U01EB023685 and R01GM118574 as well as Cancer Prevention Research Institute of Texas (CPRIT) grant RR180012.

A   Correctness and Noise Estimation

Our improved HE scheme is based on two main techniques- approximate basis and modulus switching, and both of them induce some additional errors. In this section, we estimate the size of errors and show that they can be managed by choosing a proper HE parameter set. For convenience, we adapt the same notations as in Sect. 4.

A.1   Approximate Modulus Switching

Fast Conversion. Our modulus switching algorithms are based on the fast basis conversion algorithm introduced in [3]. For the RNS representation $[a]_{\mathcal{C}}$ of an integer $a \in \mathbb{Z}_{Q_\ell}$, the fast conversion algorithm $\mathtt{Conv}_{\mathcal{C} \to \mathcal{B}}([a]_{\mathcal{C}})$ computes the RNS representation of $a' = \sum_{j=0}^{\ell-1} [a^{(j)} \cdot \hat{q}_j^{-1}]_{q_j} \cdot \hat{q}_j$ with respect to the basis $\mathcal{B}$. Then there exists an integer $e \in [-\ell/2, \ell/2]$ satisfying $a' = a + Q \cdot e$ since $a' \equiv a \pmod{Q}$ and $|a'| \leq (\ell/2) \cdot Q$.

Approximate Modulus Raising. Let $[a]_{\mathcal{C}}$ be the RNS representation of an integer $a \in \mathbb{Z}_Q$. The approximate modulus raising algorithm $\mathtt{ModUp}_{\mathcal{C} \to \mathcal{D}}([a]_{\mathcal{C}})$ returns the concatenation of $\mathtt{Conv}_{\mathcal{C} \to \mathcal{B}}([a]_{\mathcal{C}})$ and $[a]_{\mathcal{C}}$, which is the RNS representation of $a + Q \cdot e$ for some integer $e \in [-\ell/2, \ell/2]$ from the property of the fast conversion algorithm.

Approximate Modulus Reduction. Let $[\tilde{b}]_{\mathcal{D}} = (\tilde{b}^{(i)})$ for $0 \leq i \leq k+\ell-1$ be the RNS representation of an integer $\tilde{b} \in \mathbb{Z}_{P \cdot Q}$. It satisfies that

$(\tilde{b}^{(0)}, \ldots, \tilde{b}^{(k-1)}) = [a]_{\mathcal{B}}$ for $a = [\tilde{b}]_P$. From the property of the fast conversion algorithm, we have that $(\tilde{a}^{(0)}, \ldots, \tilde{a}^{(\ell-1)}) \leftarrow \text{Conv}_{\mathcal{B} \to \mathcal{C}}([a]_{\mathcal{B}})$ is the RNS representation of $\tilde{a} := a + P \cdot e$ for some integer e such that $|\tilde{a}| \leq (k/2) \cdot P$.

Let $b = P^{-1} \cdot (\tilde{b} - \tilde{a})$. It is an integer from $\tilde{b} \equiv a \equiv \tilde{a} \pmod{P}$. Then the output of $\text{ModDown}_{\mathcal{D} \to \mathcal{C}}([\tilde{b}]_{\mathcal{D}})$ is equal to $[b]_{\mathcal{C}}$ since $b \equiv P^{-1} \cdot (\tilde{b} - \tilde{a}) \equiv \left(\prod_{i=0}^{k-1} p_i\right)^{-1} \cdot (\tilde{b}^{(k+j)} - \tilde{a}^{(j)}) \pmod{q_j}$. Note that the integer $b \in \mathbb{Z}_Q$ satisfies $|b - P^{-1} \cdot \tilde{b}| = P^{-1} \cdot |\tilde{a}| \leq k/2$.

A.2   Homomorphic Operations

In this paragraph, we will focus on homomorphic operations provided in our scheme. We define $\|a\|_\infty$ and $\|a\|_1$ by the relevant norms on the coefficients vector $(a_0, \ldots, a_{N-1})$ of $a(X)$. Let $\zeta = \exp(-\pi i/N)$. Recall that the canonical embedding map on $K = \mathbb{Q}[X]/(X^n + 1)$ is defined by $a(X) \mapsto (a(\zeta), a(\zeta^3), \ldots, a(\zeta^{2N-1}))$. Its ℓ_∞-norm is called the canonical embedding norm, and denoted by $\|a\|_\infty^{\text{can}} = \|\sigma(a)\|_\infty$. Note that $\|a\|_\infty^{\text{can}} = \|\tau(a)\|_\infty$ for the decoding map τ and for any $a \in K$.

We specify the distributions χ_{key}, χ_{err}, and χ_{enc} for noise analysis of our scheme. For an positive integer h, the secret key distribution χ_{key} follows a uniform distribution over the set of signed binary vectors in $\{0, \pm 1\}^N$ whose Hamming weight (the number of nonzero coefficients) is exactly h. The error distribution χ_{err} chooses a polynomial s by sampling its coefficients independently from the discrete Gaussian distribution of variance σ^2 for a real $\sigma > 0$. The encryption key distribution χ_{enc} draws each entry in the vector from $\{0, \pm 1\}$, with probability $1/4$ for each of $+1$ and -1, and probability being zero $1/2$.

We follow the same methodology for noise estimation as in [11,13,21]. Assume that a polynomial $a(X)$ is sampled from one of the distributions used in our HE scheme. Since $a(\zeta)$ is the inner product of coefficient vector of a and the fixed vector $(1, \zeta, \ldots, \zeta^{N-1})$ of Euclidean norm $\sqrt{N}$, the random variable $a(\zeta)$ has variance $V_{\text{err}} = \sigma^2 \cdot N$, where σ^2 is the variance of each coefficient of a. Similarly, $a(\zeta)$ a the variance of $V_q = q^2 N/12$ (resp. $N/2$), when a is sampled from $U(R_q)$ (resp. χ_{enc}). In particular, it has variance h when $a(X)$ is chosen from χ_{key}. Moreover, we can assume that $a(\zeta)$ is distributed similarly to a Gaussian random variable over complex plane since it is a sum of many independent and identically distributed random variables. Every evaluations at root of unity ζ^j share the same variance. Hence, we will use $6 \cdot \sqrt{V}$ as a high-probability bound on the canonical embedding norm of a when each coefficient has a variance V. For a multiplication of two independent random variables close to Gaussian distributions with variances V_1 and V_2, we will use a high-probability bound $16 \cdot \sqrt{V_1 V_2}$.

Encryption. Our encryption algorithm does not use any approximate modulus switching algorithms. Therefore, it has exactly the same noise with the original implementation of HEAAN scheme. For a plaintext $m \in R$, it returns a ciphertext $\text{ct} \in R_{Q_L}^2$ which satisfies $\langle \text{ct}, \text{sk} \rangle \equiv m + e \pmod{Q_L}$ for some $e \in R$ such that $\|e\|_\infty^{\text{can}} \leq B_{\text{enc}} = 8\sqrt{2}\sigma N + 6\sigma\sqrt{N} + 16\sigma\sqrt{hN}$ from Lemma 1 of [11].

Addition. It does not induce any additional error since $\langle \mathsf{ct}_{\mathsf{add}}, \mathsf{sk} \rangle \equiv \langle \mathsf{ct}, \mathsf{sk} \rangle + \langle \mathsf{ct}', \mathsf{sk} \rangle \pmod{Q_\ell}$.

Rescaling. Let $\mathsf{ct} = \left(\mathsf{ct}^{(j)} = (c_0^{(j)}, c_1^{(j)}) \right)_{0 \le j \le \ell} \in \prod_{j=0}^{\ell} R_{q_j}^2$ be an input ciphertext of level ℓ, and $\mathsf{ct}' \leftarrow \left(\mathsf{ct}'^{(j)} = (c_0'^{(j)}, c_1'^{(j)}) \right)_{0 \le j \le \ell-1} \leftarrow \mathsf{RS}(\mathsf{ct})$ be the output ciphertext obtained by $c_i'^{(j)} \leftarrow q_\ell^{-1} \cdot (c_i^{(j)} - c_i^{(\ell)})$ for $i = 0, 1$ and $0 \le j < \ell$.

Let $c_i \in R_{Q_L}$ be the polynomials satisfying $[c_i]_{C_\ell} = (c_i^{(0)}, \dots, c_i^{(\ell)})$ for $i = 0, 1$. Then we have that $[c_i]_{C_{\ell-1}} = (c_i'^{(0)}, \dots, c_i'^{(\ell-1)})$ for $c_i' = q_\ell^{-1} \cdot (c_i - [c_i]_{q_\ell}) = \lfloor q_\ell^{-1} \cdot c_i \rceil$, that is, our rescaling procedure computes the exactly same ciphertext as in the original HEAAN scheme with RNS representation. Therefore, we have $[\langle \mathsf{ct}', \mathsf{sk} \rangle]_{Q_{\ell-1}} = q_\ell^{-1} \cdot [\langle \mathsf{ct}, \mathsf{sk} \rangle]_{Q_\ell} + e_{\mathsf{rs}}$ for some $e_{\mathsf{rs}} \in K$ satisfying $\|e\|_\infty^{\mathsf{can}} \le B_{\mathsf{rs}} = \sqrt{N/3} \cdot (3 + 8\sqrt{h})$ from Lemma 2 of [11].

Multiplication. Suppose that we are given two level-ℓ ciphertext ct and ct'. The output of the first step in the multiplication algorithm is the RNS representation of $(d_0, d_1, d_2) \in R_{Q_\ell}^3$ such that $d_0 + d_1 \cdot s + d_2 \cdot s^2 \equiv \langle \mathsf{ct}, \mathsf{sk} \rangle \cdot \langle \mathsf{ct}', \mathsf{sk} \rangle \pmod{Q_\ell}$. The output of the second step is the RNS representation of $\tilde{d}_2 = d_2 + Q_\ell \cdot e$ with respect to the basis $\mathcal{D}_\ell$ for some $e \in R$ satisfying $\|\tilde{d}_2\|_\infty \le \frac{1}{2}(\ell+1) \cdot Q_\ell$. We may assume that the integral polynomial $\tilde{d}_2$ behaves like the sum of $(\ell+1)$ independent and uniform random variables over R_{Q_ℓ}, so its variance is $V = \frac{1}{2}(\ell+1) \cdot (Q_\ell^2 \cdot N/12)$.

Since the first $(k+\ell+1)$ components of the evaluation key evk can be viewed as an encryption of $P \cdot s^2$ modulo $P \cdot Q_\ell$, the output $\tilde{\mathsf{ct}}$ of the third step is an encryption of $P \cdot \tilde{d}_2 \cdot s^2 \equiv P \cdot d_2 \cdot s^2 \pmod{P \cdot Q_\ell}$. Its error is bounded by $16 \cdot \sqrt{V} \cdot \sqrt{N\sigma^2} = 8\sqrt{(\ell+1)/6} \cdot Q_\ell \cdot \sigma N = \sqrt{(\ell+1)/2} \cdot B_{\mathsf{ks}} \cdot Q_\ell$.

The fourth step reduces the modulus of $\tilde{\mathsf{ct}}$ using the modulus reduction algorithm. It returns a ciphertext $\hat{\mathsf{ct}} \in R_{Q_\ell}^2$ such that $P \cdot \hat{\mathsf{ct}} \approx \tilde{\mathsf{ct}}$. The error $P \cdot \hat{\mathsf{ct}} - \tilde{\mathsf{ct}}$ behaves as if it is a sum of k independent and uniform random variables on R_P, so its variance is $k \cdot V_P = k \cdot P^2 N/12$. Finally, dividing by P, we obtain the error after modulus reduction. Therefore, $\hat{\mathsf{ct}}$ is an encryption of $d_2 \cdot s^2$ with an error bounded by $\sqrt{(\ell+1)/2} \cdot P^{-1} \cdot B_{\mathsf{ks}} \cdot Q_\ell + \sqrt{k} \cdot B_{\mathsf{rs}}$.

B   List of Primes

A ciphertext modulus is chosen to be a product of distinct primes and each of them satisfies the following conditions:

$$\begin{cases} |2^{-\kappa} \cdot q_j - 1| < 2^{-\eta}, \\ q_j \equiv 1 \pmod{2N}, \end{cases}$$

for some integers κ, η, and N. In other words, q_j is an approximation of 2^κ with η-bit precision, and there is a $(2N)$-th primitive root of unity modulo q_j. All primes are expressed using hexadecimal system to show how close they are to powers of two.

There are 22 primes including $q_0 = \texttt{0x20000000000b0001}$ satisfying these conditions for $\kappa = 61$, $\eta = 37$, and $N = 2^{15}$. We have 33 primes when $(\kappa, \eta, N) = (55, 31, 2^{15})$, and 26 prime numbers when $(\kappa, \eta, N) = (49, 25, 2^{15})$. The following is a list of 15 primes (among 33 primes for the second parameter) that were used in the implementation described in Table 1.

$$[80000000080001, 80000000130001, 7ffffffe90001,$$
$$80000000190001, 800000001d0001, 7fffffffbf0001,$$
$$7fffffffbd0001, 80000000440001, 7fffffffba0001,$$
$$80000000490001, 80000000500001, 7fffffffaa0001,$$
$$7fffffffa50001, 800000005e0001, 7fffffff9f0001]$$

References

1. Aguilar-Melchor, C., Barrier, J., Guelton, S., Guinet, A., Killijian, M.-O., Lepoint, T.: NFLLIB: NTT-based fast lattice library. In: Sako, K. (ed.) CT-RSA 2016. LNCS, vol. 9610, pp. 341–356. Springer, Cham (2016). https://doi.org/10.1007/978-3-319-29485-8_20
2. Albrecht, M.R., Player, R., Scott, S.: On the concrete hardness of learning with errors. J. Math. Cryptol. **9**(3), 169–203 (2015)
3. Bajard, J.-C., Eynard, J., Hasan, M.A., Zucca, V.: A full RNS variant of FV like somewhat homomorphic encryption schemes. In: Avanzi, R., Heys, H. (eds.) SAC 2016. LNCS, vol. 10532, pp. 423–442. Springer, Cham (2017). https://doi.org/10.1007/978-3-319-69453-5_23
4. Barrett, P.: Implementing the Rivest Shamir and Adleman public key encryption algorithm on a standard digital signal processor. In: Odlyzko, A.M. (ed.) CRYPTO 1986. LNCS, vol. 263, pp. 311–323. Springer, Heidelberg (1987). https://doi.org/10.1007/3-540-47721-7_24
5. Bonte, C., Bootland, C., Bos, J.W., Castryck, W., Iliashenko, I., Vercauteren, F.: Faster homomorphic function evaluation using non-integral base encoding. In: Fischer, W., Homma, N. (eds.) CHES 2017. LNCS, vol. 10529, pp. 579–600. Springer, Cham (2017). https://doi.org/10.1007/978-3-319-66787-4_28
6. Brakerski, Z.: Fully homomorphic encryption without modulus switching from classical GapSVP. In: Safavi-Naini, R., Canetti, R. (eds.) CRYPTO 2012. LNCS, vol. 7417, pp. 868–886. Springer, Heidelberg (2012). https://doi.org/10.1007/978-3-642-32009-5_50
7. Brakerski, Z., Gentry, C., Vaikuntanathan, V.: (Leveled) fully homomorphic encryption without bootstrapping. In: Proceedings of ITCS, pp. 309–325. ACM (2012)
8. Chen, H., Laine, K., Player, R., Xia, Y.: High-precision arithmetic in homomorphic encryption. In: Smart, N.P. (ed.) CT-RSA 2018. LNCS, vol. 10808, pp. 116–136. Springer, Cham (2018). https://doi.org/10.1007/978-3-319-76953-0_7
9. Cheon, J.H., Han, K., Kim, A., Kim, M., Song, Y.: Bootstrapping for approximate homomorphic encryption. In: Nielsen, J.B., Rijmen, V. (eds.) EUROCRYPT 2018. LNCS, vol. 10820, pp. 360–384. Springer, Cham (2018). https://doi.org/10.1007/978-3-319-78381-9_14
10. Cheon, J.H., Kim, A., Kim, M., Song, Y.: Implementation of HEAAN (2016). https://github.com/kimandrik/HEAAN

11. Cheon, J.H., Kim, A., Kim, M., Song, Y.: Homomorphic encryption for arithmetic of approximate numbers. In: Takagi, T., Peyrin, T. (eds.) ASIACRYPT 2017. LNCS, vol. 10624, pp. 409–437. Springer, Cham (2017). https://doi.org/10.1007/978-3-319-70694-8_15
12. Cooley, J.W., Tukey, J.W.: An algorithm for the machine calculation of complex Fourier series. Math. Comput. **19**(90), 297–301 (1965)
13. Costache, A., Smart, N.P.: Which ring based somewhat homomorphic encryption scheme is best? In: Sako, K. (ed.) CT-RSA 2016. LNCS, vol. 9610, pp. 325–340. Springer, Cham (2016). https://doi.org/10.1007/978-3-319-29485-8_19
14. Costache, A., Smart, N.P., Vivek, S.: Faster homomorphic evaluation of discrete Fourier transforms. In: Kiayias, A. (ed.) FC 2017. LNCS, vol. 10322, pp. 517–529. Springer, Cham (2017). https://doi.org/10.1007/978-3-319-70972-7_29
15. Costache, A., Smart, N.P., Vivek, S., Waller, A.: Fixed-point arithmetic in SHE schemes. In: Avanzi, R., Heys, H. (eds.) SAC 2016. LNCS, vol. 10532, pp. 401–422. Springer, Cham (2017). https://doi.org/10.1007/978-3-319-69453-5_22
16. van Dijk, M., Gentry, C., Halevi, S., Vaikuntanathan, V.: Fully homomorphic encryption over the integers. In: Gilbert, H. (ed.) EUROCRYPT 2010. LNCS, vol. 6110, pp. 24–43. Springer, Heidelberg (2010). https://doi.org/10.1007/978-3-642-13190-5_2
17. Ducas, L., Micciancio, D.: FHEW: bootstrapping homomorphic encryption in less than a second. In: Oswald, E., Fischlin, M. (eds.) EUROCRYPT 2015. LNCS, vol. 9056, pp. 617–640. Springer, Heidelberg (2015). https://doi.org/10.1007/978-3-662-46800-5_24
18. Fan, J., Vercauteren, F.: Somewhat practical fully homomorphic encryption. Cryptology ePrint Archive, Report 2012/144 (2012). https://eprint.iacr.org/2012/144
19. Gentleman, W.M., Sande, G.: Fast Fourier transforms: for fun and profit. In: Proceedings of the Fall Joint Computer Conference, 7–10 November 1966, pp. 563–578. ACM (1966)
20. Gentry, C.: Fully homomorphic encryption using ideal lattices. In: Proceedings of the Forty-First Annual ACM Symposium on Theory of Computing, STOC 2009, pp. 169–178. ACM (2009)
21. Gentry, C., Halevi, S., Smart, N.P.: Homomorphic evaluation of the AES circuit. In: Safavi-Naini, R., Canetti, R. (eds.) CRYPTO 2012. LNCS, vol. 7417, pp. 850–867. Springer, Heidelberg (2012). https://doi.org/10.1007/978-3-642-32009-5_49
22. Gilad-Bachrach, R., Dowlin, N., Laine, K., Lauter, K., Naehrig, M., Wernsing, J.: Cryptonets: applying neural networks to encrypted data with high throughput and accuracy. In: International Conference on Machine Learning, pp. 201–210 (2016)
23. Graepel, T., Lauter, K., Naehrig, M.: ML confidential: machine learning on encrypted data. In: Kwon, T., Lee, M.-K., Kwon, D. (eds.) ICISC 2012. LNCS, vol. 7839, pp. 1–21. Springer, Heidelberg (2013). https://doi.org/10.1007/978-3-642-37682-5_1
24. Halevi, S., Polyakov, Y., Shoup, V.: An improved RNS variant of the BFV homomorphic encryption scheme. Cryptology ePrint Archive, Report 2018/117 (2018). https://eprint.iacr.org/2018/117
25. Halevi, S., Shoup, V.: Design and implementation of a homomorphic-encryption library. IBM Research (Manuscript) (2013)
26. Harvey, D.: Faster arithmetic for number-theoretic transforms. J. Symbolic Comput. **60**, 113–119 (2012)
27. Kim, A., Song, Y., Kim, M., Lee, K., Cheon, J.H.: Logistic regression model training based on the approximate homomorphic encryption. BMC Med. Genomics **11**(4), 83 (2018)

28. Kim, M., Song, Y., Wang, S., Xia, Y., Jiang, X.: Secure logistic regression based on homomorphic encryption: design and evaluation. JMIR Med. Inform. **6**(2), e19 (2018)
29. lbw: Low birth weight study data (2017). https://rdrr.io/rforge/LogisticDx/man/lbw.html
30. Naehrig, M., Lauter, K., Vaikuntanathan, V.: Can homomorphic encryption be practical? In: Proceedings of the 3rd ACM Workshop on Cloud Computing Security Workshop, pp. 113–124. ACM (2011)
31. Nesterov, Y.: A method of solving a convex programming problem with convergence rate $O(1/k^2)$. Soviet Math. Doklady **27**, 372–376 (1983)
32. Simple Encrypted Arithmetic Library (release 3.0.0). Microsoft Research, Redmond, October 2018. http://sealcrypto.org
33. uis: Umaru impact study data (2017). https://rdrr.io/rforge/LogisticDx/man/uis.html

Analysis of Error-Correcting Codes for Lattice-Based Key Exchange

Tim Fritzmann[1]([⊠])[iD], Thomas Pöppelmann[2], and Johanna Sepulveda[1][iD]

[1] Technische Universität München, Munich, Germany
{tim.fritzmann,johanna.sepulveda}@tum.de
[2] Infineon Technologies AG, Munich, Germany
thomas.poeppelmann@infineon.com

Abstract. Lattice problems allow the construction of very efficient key exchange and public-key encryption schemes. When using the Learning with Errors (LWE) or Ring-LWE (RLWE) problem such schemes exhibit an interesting trade-off between decryption error rate and security. The reason is that secret and error distributions with a larger standard deviation lead to better security but also increase the chance of decryption failures. As a consequence, various message/key encoding or reconciliation techniques have been proposed that usually encode one payload bit into several coefficients. In this work, we analyze how error-correcting codes can be used to enhance the error resilience of protocols like NewHope, Frodo, or Kyber. For our case study, we focus on the recently introduced NewHope Simple and propose and analyze four different options for error correction: (i) BCH code; (ii) combination of BCH code and additive threshold encoding; (iii) LDPC code; and (iv) combination of BCH and LDPC code. We show that lattice-based cryptography can profit from classical and modern codes by combining BCH and LDPC codes. This way we achieve quasi-error-free communication and an increase of the estimated post-quantum bit-security level by 20.39% and a decrease of the communication overhead by 12.8%.

Keywords: Post-quantum key exchange · NewHope Simple
Error-correcting codes

1 Introduction

Recently, lattice-based key exchange [3,4,9], public-key encryption (PKE) [11, 22] and signature schemes [6,7,13] have attracted great interest due to their performance, simplicity, and practicality. Aside from NTRU [19] and when focusing on ephemeral key exchange and PKE, the Learning with Errors (LWE) problem and the more structured Ring-LWE (RLWE) problem are the main tools to build state of the art schemes. An interesting property of LWE and RLWE is that the security of the problem depends on the dimension of the underlying lattices but also on the size and shape of the distribution used to generate random secret and

© Springer Nature Switzerland AG 2019
C. Cid and M. J. Jacobson, Jr. (Eds.): SAC 2018, LNCS 11349, pp. 369–390, 2019.
https://doi.org/10.1007/978-3-030-10970-7_17

error elements. When constructing key exchange or PKE schemes this is critical as error elements cannot always be removed by the communicating parties and can lead to differences in the derived key (in key exchange) or differences in the message (in most PKE instances). Thus, small differences in the shared key or decrypted message have to be mitigated by encoding techniques or might finally cause a re-transmission or lead to the inability to decrypt a certain ciphertext.

A reduction of the failure probability by using a better encoding opens up the possibility to (a) increase the LWE/RLWE secret and error terms and thus to strengthen security or (b) to decrease the size of ciphertexts, or in general exchanged data, by removing more information. Moreover, it is important to distinguish between the requirements for ephemeral key exchange and PKE schemes. For ephemeral key exchange, a higher failure probability may be acceptable (e.g., around 2^{-40}) because key agreement errors do not affect the security of the scheme. In the presence of errors, the two parties can just repeat the key exchange process. The issue of decryption errors is more critical when using LWE or RLWE-based schemes to instantiate a PKE scheme. The basic LPR10 [24] scheme is only considered appropriately secured with respect to adaptive chosen plaintext attacks (CPA), which is usually not sufficient in a setting where an adversary has access to a decryption oracle. A commonly used tool for transforming a CPA-secured PKE into a scheme secured against chosen-ciphertext attacks (CCA) is the Fujisaki-Okamoto transformation [15,31]. However, a CCA secured cryptosystem using this transformation requires a decryption/decoding routine with a negligible error rate because an attacker could exploit decryption errors. To increase the resilience against attacks exploiting decryption errors, the failure rate is desired to be lower than 2^{-128}. As in Frodo [2] and Kyber [5], in this work we aim for a failure rate lower than 2^{-140} to have a sufficient margin on the error probability. Note that existing works, such as Hila5 [29] and LAC [23], use an independence assumption to calculate the protocol's failure rate. This assumption is related to the correlation between the coefficients of the error term in LWE/RLWE based schemes. The effect of this correlation on the failure rate is still an open research question and it is not in the scope of this work (see also Sect. 3.2 in [28] for a discussion). However, to decrease decryption errors without decreasing the security of the underlying lattice problem, the reconciliation and en-/decoding techniques are important.

Concurrent to our work, the lattice-based algorithms Hila5 [29], KCL [32], ThreeBears [18] and LAC [23] where developed. They explicitly use forward error correction to achieve better resilience against decryption errors. Except of LAC, which uses a powerful Bose-Chaudhuri-Hocquenghem (BCH) code, all aforementioned schemes apply an error correction that is only capable of correcting a few errors. In this work, we investigate the applicability of more elaborated and modern codes for lattice-based cryptography. Generally, error-correcting codes can be applied in LWE/RLWE schemes when the exchanged key (or message) is chosen by only one of the parties. For example, Frodo, Kyber and NewHope Simple can benefit from the application of powerful error-correcting codes[1]. For our case study, we focus on the RLWE-based NewHope Simple scheme [3], which was submitted with small changes to NIST's call for post-quantum proposals [25]. Compared to an earlier version (called just NewHope) [4], NewHope Simple features a simpler message encoding scheme that uses an additive threshold encoding algorithm and which exhibits a failure rate of less than 2^{-61}. Note that the version of NewHope submitted to the NIST process reaches a failure rate lower than 2^{-140}. This was achieved by reducing the variance of the error distribution. However, in this paper we analyzed different approaches to reduce the failure rate without decreasing the level of security.

Contribution. In this work, we perform an exploration of more powerful error-correcting codes for key exchange mechanisms in order to obtain a quasi-error-free communication and to improve important performance parameters. Our work intensively studies the behaviour of the failure rate when different error-correcting codes and security parameters are applied. For the first time, modern codes, more specifically low-density parity-check (LDPC) codes, are used in this context and compared with the performance of classical BCH codes. In general, the results of the exploration of the design space show that there are several design decisions that make it possible to decrease the failure rate to a value lower than 2^{-140}, increase the security and decrease the communication overhead between the two parties. The selection of a coding option is driven by the requirements of the application. In addition, regarding the protocol's failure rate calculation, we extend the works of [9,10,28], to apply the approach to NewHope Simple. Additionally, we provide first benchmark results. However, we leave the optimization of the implementations with regard to cache and timing attacks to future work as we focus on the exploration of the large design space.

2   NewHope Simple

NewHope Simple, proposed by Alkim, Ducas, Pöppelmann and Schwabe in 2016 [3] as a simplification of NewHope [4], is a lattice-based key exchange, or more specifically a key encapsulation mechanism (KEM), that is built upon the RLWE problem. It allows two entities (Alice and Bob) to agree on a 256-bit shared key μ that is selected by Bob. In the following subsections, the description, security considerations and parameters of NewHope Simple are summarized.

2.1   Notation

Let $\mathcal{R} = \mathbb{Z}_q[x]/(x^n + 1)$ be a ring of integer polynomials. All elements of the ring $\mathcal{R}$ can be written in the form $f(x) = a_0 + a_1 x + a_2 x^2 + \cdots + a_{n-1} x^{n-1}$, where the integer coefficients $a_0, a_1, \ldots, a_{n-1}$ are reduced modulo q. We write $a \xleftarrow{\$} S$ for sampling a value a from the distribution S, where sampling means to take

---

[1] In order to apply error-correcting codes, some changes in the protocol may be necessary, e.g. different parameter selection and/or encoding/decoding functions.

a random value from a set S. Let Ψ_k be a binomial distribution with parameter k. The distribution is determined by $\Psi_k = \sum_{i=0}^{k-1} b_i - b_i'$, where $b_i, b_i' \in \{0,1\}$ are uniform independent bits. The binomial distribution is centered with a zero mean, approximates a discrete Gaussian, has variance $k/2$, and gives a standard deviation of $\psi = \sqrt{k/2}$.

2.2   Protocol

Protocol 1 shows the underlying algorithm of NewHope Simple. Eight steps are highlighted due to the relevance to the present work. For a more detailed description of the algorithm and for details about the application of the CCA transformation, we refer the reader to [3] and [1].

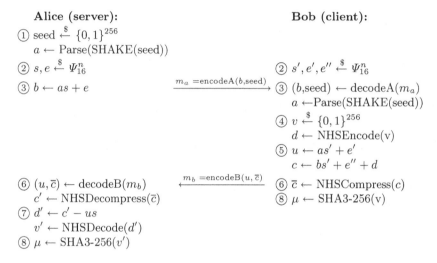

Protocol 1. NewHope Simple protocol. All polynomials are elements of the ring $\mathcal{R} = \mathbb{Z}_q[x]/(x^n + 1)$, where $n = 1024$ and $q = 12289$ [3].

1. Alice samples the seed from a random number generator. The seed is expanded with the SHAKE-128 extendable-output function. The expanded seed is used to generate the public polynomial a.
2. Alice and Bob sample the coefficients of the secret polynomials s and s', and the error polynomials e, e' and e'' according to the error distribution Ψ_k.
3. Alice calculates $b = as + e$ and sends it together with the seed to Bob. Extraction of the secret s from b is hard due to the error term e and because b is exactly an RLWE instance. Similar to Alice, Bob can use the seed to generate the public polynomial a.
4. Bob samples 256 bits from a random number generator and assigns them to the secret key vector v. Then, Bob encodes v into the most significant bit of the coefficients of polynomial $d = \text{NHSEncode}(v)$. The function NHSEncode,

which is given in Appendix A, maps one bit of v into four coefficients of d. This redundancy is used by the NHSDecode function in Step 7 to average out small errors.

5. Bob calculates $u = as' + e'$ and hides the secret key polynomial d in $c = bs' + e'' + d = ass' + es' + e'' + d$. The polynomials u and c are again instances of the RLWE problem.

6. Bob sends to Alice the polynomial u and the compressed polynomial $\bar{c}$. The goal of the compression of polynomial c is the reduction of the communication overhead between Alice and Bob.

7. Alice removes the large noise term ass' from the decompressed polynomial c' by calculating $d' = c' - us \approx bs' + e'' + d - (as' + e')s = ass' + es' + e'' + d - ass' - e's = (es' - e's) + e'' + d$. Alice obtains the term v' after decoding d', using the function NHSDecode, which is also provided in Appendix A.

8. After the decoding, Alice and Bob can use v' and v, respectively, as input for the SHA3-256 function to obtain the shared key.

The functions NHSEncode and NHSDecode of NewHope Simple build an error-correcting code, which is used to remove small errors and to increase the probability that Alice and Bob share a similar key. For the remainder of this paper, this error-correcting code is denoted as *additive threshold encoding algorithm*.

2.3 Security of NewHope Simple

The security level of NewHope Simple depends on three parameters: the dimension n of the ring, the modulus q, and the parameter k that determines the standard deviation of the noise distribution Ψ_k. In this work, the parameters n and q are not modified. Only k is used to improve the security of NewHope Simple as larger noise also leads to a higher security level. For determining the security level, the test script and methodology[2] from [4] can be used.

2.4 Noise Sources of the Protocol

NewHope Simple contains two noise sources: the *difference noise* and the *compression noise*. As noise we define all terms that have an influence on the correctness of the decryption/decoding or reconciliation mechanisms. The reason is that we can model the distortion caused by the convolutions of the secret and error polynomials as noise that is added to encoded data transmitted over a channel.

The *difference noise* emerges from the design of the protocol. Alice is able to remove the strongest noise term ass' from polynomial c (Step 7), but a small noise term remains. This noise term is called difference noise and is equal to $(es' - e's) + e''$. The coefficients of the secret and error polynomials are sampled from the error distribution Ψ_k. When k is increased, the probability of receiving a stronger difference noise increases as well.

---

[2] Script *PQsecurity.py* in https://cryptojedi.org/crypto/#newhope.

The *compression noise* is introduced by the function NHSCompress (Step 6). It compresses the polynomial $c = ass' + es' + e'' + d$ to reduce the communication overhead between Alice and Bob. This is possible as lower-order bits carry a high amount of noise and have low information content. To remove such lower order bits, a coefficient-wise modulus switching between the security parameter q and 2^r is performed, where r is the number of remaining bits. To reduce the number of transmitted bytes between Alice and Bob, the transmitted polynomials b, c and u can be compressed. In the original implementation of NewHope Simple, the compression is only applied on polynomial c, where each coefficient of c is reduced from 14 bits to 3 bits. In this work, we further reduce the communication overhead by compressing polynomial u as in Kyber [10]. To obtain a moderate compression noise, a weaker compression on the coefficients of polynomial u has to be applied. As the uniformly distributed compression noise of u is multiplied with the binomially distributed secret s, the compression noise of u gets magnified.

3   Failure Rate of NewHope Simple

In the original implementation of NewHope Simple, the failure rate is bounded applying the Cramér-Chernoff inequality [3]. This approach provides a probability bound that can be far away from the real failure probability. Previous works, such as Frodo [9], Kyber [10] and Hila5 [28], use probability convolutions to determine the probability distribution of the difference between the keys of Alice and Bob. With the probability distribution of the difference, it is possible to derive the protocol's failure rate. In the following subsections, we shortly explain how to calculate the probability distributions of the two noise terms, *difference noise* and the *compression noise*, mentioned in Subsect. 2.4, and how to calculate the failure rate by a given error distribution.

3.1   Mathematical Operations with Random Variables

In this subsection, the mathematical background for determining the probability distributions of the *difference noise* and the *compression noise* is given.

NewHope Simple uses a binomial distribution for sampling secret and error polynomials. The probability mass function of a binomial random variable (RV) X is $f(i) = Pr(X = i) = \binom{l}{i} p^i (1-p)^{l-i}$ for $i = 0, 1, \ldots, l$. For NewHope Simple, $p = 0.5$ and l is equal to the error distribution parameter k multiplied by two.

Let us define in Theorem 1 the probability distribution of the addition and in Theorem 2 the probability distribution of the multiplication of two independent RVs. Since in NewHope Simple polynomial instead of conventional multiplications are required, we define in Theorem 3 the polynomial product distribution. The proof for Theorem 3 can be found in Appendix B.

Table 1. Calculating distribution of $d = (es' - e's) + e''$

Step	Action	Result
Step 1	Product distribution of two RVs sampled from Ψ_k	Ψ_Z
Step 2	n-fold convolution of the product distribution	es'
Step 3	Convolve distribution of es' with itself	$(es' - e's)$
Step 4	Convolve distribution of $(es' - e's)$ with Ψ_k	$(es' - e's) + e''$

Theorem 1 (Addition of random variables). *Let $\Psi_X(x)$ and $\Psi_Y(y)$ be two probability distributions of the independent random variables X and Y. Then the probability distribution of the sum of both random variables corresponds to the convolution of the individual probability distributions, which can be written as $\Psi_{X+Y} = \Psi_Z(z) = \Psi_X(x) \circledast \Psi_Y(y)$ [17].*

Theorem 2 (Product distribution). *Let $\Psi_X(x)$ and $\Psi_Y(y)$ be two probability distributions of the independent random variables X and Y. Then the product distribution $\Psi_Z(XY = c) = \sum_{x \in X, y \in Y \, s.t. \, xy=c} \Psi_X(x)\Psi_Y(y)$.*

Theorem 3 (Polynomial product distribution). *Let a and b be two polynomials of a ring $\mathcal{R}_q$ with rank n and with independent random coefficients sampled from Ψ_k and let c be the result of the polynomial multiplication of a and b. Then the probability distribution of a random coefficient of c is equal to the n-fold convolution of the product distribution Ψ_Z of two random variables sampled from Ψ_k.*

3.2 Probability Distributions of Difference and Compression Noise

Difference Noise. The partial steps for calculating the probability distribution of the difference term are summarized in Table 1. Note that all calculated probability distributions are related to a single coefficient of a polynomial. The probability distribution of the polynomial product es' can be described as an n-fold convolution of the product distribution of two RVs sampled from Ψ_k. In our case, the probability distributions of an addition and subtraction of two RVs are equal because the RVs are sampled from a symmetrical distribution that is centered at zero. To obtain the probability distribution for $(es' - e's)$, we convolve the probability distribution of es' with itself. Finally, we convolve the distribution of e'' with the result to obtain the probability distribution of $(es' - e's) + e''$.

Compression Noise. The probability distribution of the compression noise can be calculated similar to the probability distribution of the difference noise. The polynomial $c = ass' + es' + e'' + d$ consists of the uniformly distributed public parameter a, some terms sampled from the error distribution and the secret key d. Depending on the respective key bit, the coefficients of polynomial

d are either zero or $\lfloor q/2 \rfloor$. Both values, zero and $\lfloor q/2 \rfloor$, are not affected by the compression. They can be compressed and decompressed without any loss of information. Consequently, the compression noise is only dependent on the term $c_{\text{uncompressed}} = ass' + es' + e''$. The coefficients of the secret and error polynomials are sampled from Ψ_k and the coefficients of a are sampled from a uniform distribution U_q with outcomes between 0 and $q-1$ (after modulus reduction). In Fig. 11 (Appendix C) the probability distribution of the difference and compression noise is plotted.

3.3   From the Noise Distribution to the Failure Rate

Note that the coefficients of the product of two polynomial ring elements are correlated and not independent anymore. This correlation does not influence the validity of Theorem 3 and the calculations done in Subsect. 3.2 because there the calculations are related to a single coefficient. To determine the failure rate, we apply arithmetic operations on correlated coefficients and thus assume that the correlation between the coefficients has a negligible influence to the final result. The experiments discussed in Appendix D have shown that this assumption is valid at least for high failure rates. The independence assumption is also required in Eq. 2 (Subsect. 5.2) to calculate the failure rate of NewHope Simple with a t-bit error-correcting code. To have a safety margin, we aim for a failure rate of 2^{-140} instead of 2^{-128}.

In order to determine the failure rate, given a noise distribution, a closer look at the NHSDecode function must be taken. During the decoding, when one bit is mapped into four coefficients, the absolute values of the four coefficients that are subtracted by $\lfloor q/2 \rfloor$ are summed up. This decoding step is done for all outcomes of the overall error distribution (convolution of difference and compression noise distribution). First, the values of all outcomes are subtracted by $\lfloor q/2 \rfloor$ and the absolute values are built. Let us denote the resulting error distribution as Ψ_{dec}. In the next step, we convolve the distribution of four coefficients $\Psi'_{dec} = \Psi_{dec} \circledast \Psi_{dec} \circledast \Psi_{dec} \circledast \Psi_{dec}$. Note again that for this step we assume that the correlation between the coefficients is negligible. To obtain the bit error rate (BER) of NewHope Simple, all probabilities of the outcomes of Ψ'_{dec} that are lower than or equal to q are summed up. The BER can be multiplied with the secret key length, in our case 256, to compute the union bound. The union bound is the upper bound for the block error rate (BLER) or failure rate of the protocol.

4   Error-Correcting Codes

4.1   Modern and Classical Error-Correcting Codes

Error-correcting codes are an essential technique for realizing reliable data transmissions over a noisy channel. In this work, error-correcting codes are used to mitigate the influence of the difference and compression noise on the failure probability of RLWE based key exchange protocols. Instead of the additive threshold encoding, which is used in the original NewHope Simple scheme, in this work we explore the effect of using more powerful error-correcting codes. The design objectives for the error-correcting code are: (i) good error-correcting capability, to increase the security or decrease the amount of exchanged data; (ii) low failure rate, to avoid repetition of the protocol and to apply CCA transformation; and (iii) reasonable time complexity. The additive threshold encoding has usually a weak error-correcting capability and cannot efficiently achieve low failure rates for certain noise levels. Therefore, more powerful classical[3] and modern[4] codes can be used. The drawback of using powerful error-correcting codes is the increase of computation time.

Modern codes have a strong error-correcting capability and can get close to the channel capacity for long code lengths. The most commonly used error-correcting codes belonging to the class of modern codes are LDPC and Turbo codes. In comparison to Turbo codes, LDPC codes usually have a lower time complexity since they do not require long interleavers and can abort the iteration loop when a correct codeword is found [14]. Moreover, their error floor occurs at lower failure rates [21]. The error floor is a phenomenon of some modern codes that limits the performance for low failure rates. That is, the channel capacity can only be very closely approached for moderate failure rates. Since the goal is to have a low (or even no) error floor and to keep the time complexity low, in this work we select LDPC instead of Turbo codes for obtaining a high error-correcting capability.

The advantages of classical error-correcting codes are the lack of error floor and that the number of correctable errors can be determined during the construction of the code. When the number of correctable errors is known, the performance of the code can be calculated, otherwise, simulations are required. In contrast to classical codes, where the number of correctable errors is known, for modern codes this value is unknown. However, it has been demonstrated by simulation that modern codes achieve a higher error-correcting capability, when compared to the classical approach.

There are a large number of classical error-correcting codes, e.g. Hamming, Reed Muller and BCH codes. Among this alternatives, BCH codes are widely spread in real world applications because of their good performance, the ability to correct multiple errors and their flexibility in terms of code length and code rate. These characteristics motivate us to use BCH codes in the protocol to achieve very low failure rates.

To reach both a high error-correcting performance and a very low failure rate, usually different codes are concatenated. The concatenation of BCH and LDPC codes is a common method, which is used, for example, in the second generation of the digital video broadcast standard for satellite (DVB-S2).

---

[3] Classical codes are described by algebraic coding theory.

[4] Modern codes have a new approach based on probabilistic coding theory.

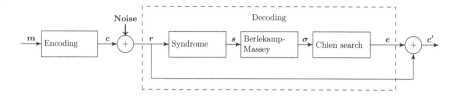

Fig. 1. BCH error correction

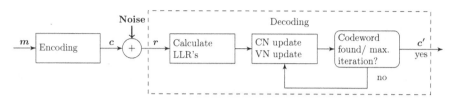

Fig. 2. LDPC error correction with sum-product algorithm

4.2 BCH Codes

BCH codes are a class of powerful classical error-correcting codes that were discovered in 1960. The code length of a BCH code must be $n = q^m - 1$, where $m \in \mathbb{Z}$ is greater or equal to three and q equal to two for the binary BCH codes. There exists a BCH code for any valid code length and any positive integer $t < 2^{m-1}$, where t denotes the number of correctable errors [21]. Figure 1 illustrates the encoding and decoding process of BCH codes. During the encoding, the codeword c is built out of a message m. In the noisy channel, noise is added to the transmitted codeword. At NewHope Simple, this would be the *difference* and *compression noise*. The decoder is used to correct multiple errors in the received codeword r. Generally, the decoding process consists of three parts: determining the syndrome s, error locator polynomial σ and the zeros of σ. Berlekamp's algorithm, which was proposed in 1966, is an efficient method for determining the error locator polynomial [8]. The error polynomial e can be determined by finding the zeros of the error locator polynomial with the Chien search algorithm [12]. The predicted codeword c' is calculated by taking r xor e.

4.3 LDPC Codes

LDPC codes were developed by Gallager in 1962 [16]. They have become attractive since the 90's, when the required computational power has been available. Figure 2 shows a block diagram of an LDPC code. LDPC codes are characterized by its parity check matrix H, which has, in case of LDPC codes, a low density, i.e. a low number of ones. For the encoding, usually, the systematic form of H is computed to derive the generator matrix. With the generator matrix it is possible to calculate the codeword c by a given message m. After transmitting c through the noisy channel, the receiver obtains a noisy codeword r.

The sum-product algorithm is a very efficient soft decision message-passing decoder. It takes as input a parity check matrix, the maximum number of iterations and the log-likelihood ratios (LLR) of the received codeword r. To visualize the decoding process, the Tanner Graph representation of the parity check matrix is used. This representation consists of a bipartite graph with check nodes (CN) and variable nodes (VN), which represent the rows and columns of H, respectively. The sum-product algorithm iteratively sends LLR messages from variable nodes to check nodes and vice versa until a correct codeword is found or the maximum number of iterations is reached. A full description of the algorithm can be found in works like [20] and [26].

4.4   Error-Correcting Codes for NewHope Simple

To meet the requirements mentioned in Subsect. 4.1, we use LDPC codes to maximize the error-correcting capability and BCH codes to achieve very low error rates. In the following paragraphs, we investigate four design options that make use of various combinations of these codes. The respective advantages and disadvantages are summarized in Table 2.

Table 2. Summary of explored coding options

Option	Coding technique	Advantages	Disadvantages
Option 1	BCH	Good error correction	Computationally expensive
Option 2	BCH and additive threshold enc.	Speed up of Option 1 (lower Galois field)	Weaker error correction compared to Option 1
Option 3	LDPC	Closer to channel capacity	Does not achieve very low error rates
Option 4	LDPC and BCH	Lower error rates than Option 3 achievable	Computationally expensive

Option 1. For Option 1, we use a BCH(1023,258) for the error correction. The BCH encoder builds the codeword out of 256 secret key bits, 765 redundancy bits and 2 padding bits. By using the NHSEncode function (Step 4 in Protocol 1), each of the 1023 code bits is mapped to one coefficient of d. Then, in the NHSDecode function (Step 7 in Protocol 1), the coefficients are mapped back to the received codeword with a hard threshold. Finally, the BCH decoder corrects up to 106 bit errors and returns the estimated secret key vector (Fig. 3).

Option 2. For Option 2, we use a BCH(511,259) as outer code and the additive threshold encoding as inner code. In this case, the BCH code uses 252 bits of redundancy in order to correct up to 30 errors. The additive threshold encoding has as input 512 bits (BCH code length with one padding bit). These bits are mapped to 1024 coefficients, resulting into a redundancy of 512 bits. With the

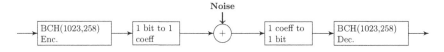

Fig. 3. Option 1, block diagram BCH(1023,258)

additive threshold encoding, it is expected that even more than 30 errors are correctable. In comparison to Option 1, this option is faster because it only requires calculations in $GF(2^9)$. The drawback of this approach is a lower error-correcting capability at the target failure rate (2^{-140}), as shown in Subsect. 5.2 (Fig. 4).

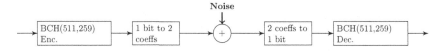

Fig. 4. Option 2, block diagram BCH(511,259) + additive threshold encoding

Option 3. For Option 3, we use an LDPC(1024,256). In this case, all available coefficients are used for the LDPC encoding. Similar to Option 1, one bit is mapped to one coefficient, but within the function NHSDecode, no hard threshold is used. Instead, we apply a transformation on the coefficients in order to allow the usage of the sum-product algorithm (Fig. 5). Each received coefficient d_i' is transformed to

$$d_i'' = \frac{4|d_i' - \lfloor q/2 \rfloor|}{q} - 1. \tag{1}$$

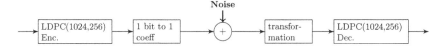

Fig. 5. Option 3, block diagram LDPC(1024,256)

Option 4. For Option 4, we build a concatenation of a BCH(511,259) and an LDPC(1024,512). In this approach, the advantages of BCH and LDPC codes are combined to achieve very low error rates and to get closer to the channel capacity. More specifically, the LDPC(1024,512) is used to remove the strong noise and the BCH(511,259), which can correct up to 30 errors, is applied to remove the remaining errors and thus achieve a very low error rate (Fig. 6).

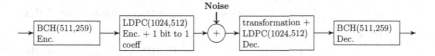

Fig. 6. Option 4, block diagram BCH(511,259) + LDPC(1024,512)

5 Experimental Results

5.1 NewHope Simple Compression Noise

Figure 7 illustrates the influence of the compression noise on the failure rate. The graph shows that the compression has a strong influence on the failure rate for low values of k. For higher values of k, the difference noise dominates. To improve both, security and bandwidth, a balance between difference noise and compression noise has to be found. When applying the error-correcting options described in Subsect. 4.4, we found the optimum at a compression of c from 14 to 3 bits per coefficient and a compression of u from 14 to 10 bits per coefficient. Removing even more bits from the coefficients of c and u leads to a significantly higher compression noise.

The curve with compression of c corresponds to the original implementation of NewHope Simple. For a value of $k = 16$, a failure rate of $2^{-127.88} = 3.20 \cdot 10^{-39}$ is determined, whereas in [3] a failure probability lower than $2^{-61} = 4.34 \cdot 10^{-19}$ is claimed. This difference is not surprising because the Cramér-Chernoff bound is based on an exponential inequality. Due to the exponential behavior, even small changes can entail large differences.

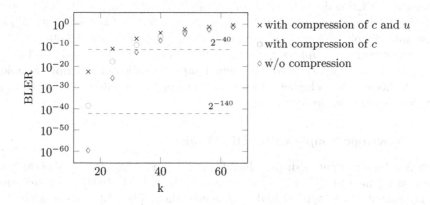

Fig. 7. Influence of compression noise on NewHope Simple's failure rate.

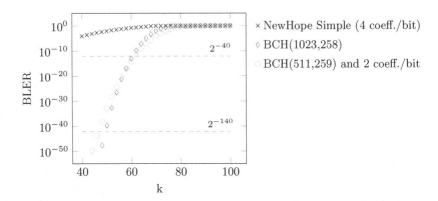

Fig. 8. Improvement of failure rate with Option 1, BCH(1023,258); and Option 2, concatenation of BCH(511,259) and additive threshold encoding. Compression on c and u applied.

5.2 NewHope Simple with BCH Code

When the failure rate of the protocol is known, the improvement using BCH codes can be calculated. The probability that a binary vector of S bits (in our analysis 256) has more than t errors is

$$\text{BLER} = \sum_{i=t+1}^{S} \binom{S}{i} p_b^i (1 - p_b)^{S-i} = 1 - \sum_{i=0}^{t} \binom{S}{i} p_b^i (1 - p_b)^{S-i}, \qquad (2)$$

where p_b denotes the probability of a bit error [30]. Figure 8 shows the improvements with BCH codes. The results show that both BCH variants (Option 1 and Option 2) allow a quasi-error-free communication for k's lower than 46. While NewHope Simple with compression of c and u has a failure rate of $1.69 \cdot 10^{-3}$ for $k = 46$, Option 1 and Option 2 achieve a failure rate of $1.83 \cdot 10^{-57}$ and $2.30 \cdot 10^{-44}$, respectively. In comparison to the original implementation of NewHope Simple, we can choose a much higher k to obtain the same failure rate when BCH codes are used within the protocol.

5.3 NewHope Simple with LDPC Code

When a binary input additive white Gaussian noise channel (BI-AWGNC) is used as channel model and a code length of $n = 1024$ (1023) is chosen, the improvement of the applied LDPC code over the applied BCH code is for the rate 1/2 about 2.8 dB and for the rate 1/4 about 3.8 dB at a BER of 10^{-6}. As a consequence, LDPC codes can get closer to the channel capacity when compared to BCH codes, even with a moderate code length.

Figure 9 compares the original implementation of NewHope Simple (with additional compression of u) with the implementations using an LDPC code (Option 3) and a BCH code (Option 1). The graph shows that LDPC codes

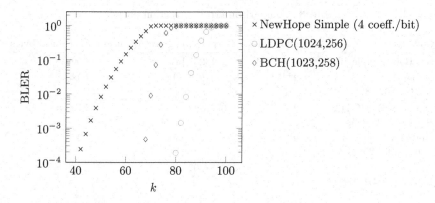

Fig. 9. Improvement of failure rate with Option 3, LDPC(1024,256). Compression on c and u applied.

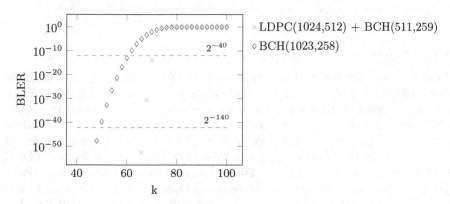

Fig. 10. Improvement of failure rate with Option 4, concatenation of LDPC(1024,512) and BCH(511,259). Compression on c and u applied.

can be used to further improve the error-correcting performance. While the BCH(1023,258) begins to operate in the *waterfall region* for k's smaller than 76, the *waterfall region* for the LDPC(1024,256) begins for k's smaller than 92. However, the error floor is expected to limit the performance of the LDPC code for error rates smaller than about 10^{-10} (see analysis in [27]) so that BCH codes perform better in this region. Interesting is also that the *waterfall region* of the additive threshold encoding is less distinct (lower gradient).

5.4 NewHope Simple with Concatenation of BCH and LDPC Code

To achieve very low error rates and get closer to the channel capacity as a pure BCH implementation, the BCH code is combined with an LDPC code (Option 4). Figure 10 illustrates the performance of the concatenation of the LDPC(1024,512) and the BCH(511,259).

Table 3. Comparison error correction options

Coding option	Failure rate	k	Security classical/quantum	Exchanged bytes
NewHope Simple [3]	$2^{-127.88}$ [a]	16	281/255 bits	4,000
Option 1, BCH(1023,258)	$<2^{-140}$	48	324/294 bits	3,488
Option 2, BCH(511,259) + 1 bit to 2 coeffs.	$<2^{-140}$	46	323/292 bits	3,488
Option 3, LDPC(1024,256)	$<2^{-12}$ [b]	80	348/315 bits	3,488
Option 4, LDPC(1024,512) + BCH(511,259)	$<2^{-140}$	66	338/307 bits	3,488

[a] In the reference, NewHope Simple provides a failure rate of lower than 2^{-61}. This bound was determined using the Cramér-Chernoff inequality. With our approach, we determine a failure rate of $3.20 \cdot 10^{-39} = 2^{-127.88}$.
[b] With Option 3, a failure rate of $\approx 10^{-10} = 2^{-33.22}$ can be efficiently reached.

5.5   Comparison Coding Options

Table 3 summarizes the results of the different coding options. Our analysis shows that NewHope Simple, with the original parameter set, has a much lower failure rate than expected. However, to increase the security and decrease the bandwidth, stronger error-correcting codes have to be applied. To achieve a failure rate of 2^{-140}, parameter k is set for Option 1, Option 2 and Option 4 to 48, 46 and 66, respectively. Since we cannot prove such an error rate for the pure LDPC implementation, we chose a higher failure rate for Option 3. Although Option 1 has a better security strength, we recommend Option 2 because it requires calculations in $GF(2^9)$ instead of $GF(2^{10})$. This reduces the time complexity. For moderate failure rates, Option 3 achieves the best error-correcting capability, but for failure rates lower than about 10^{-10} the error floor limits the performance. Option 4 cannot get as close to the channel capacity as Option 3, but it achieves extremely low error rates. With Option 4, we can realize an error rate of 2^{-140}, an increase of the post quantum security by 20.39% and a decrease of the communication overhead by 12.80%. If k and thus the security level is left unchanged and only the compression on u is increased, the communication overhead can be reduced with Option 4 by 19.20%.

5.6   Benchmark

This section summarizes the run times of the applied algorithms. Table 4 provides an overview of the determined results. All tests were performed on an Intel Core i7-6700HQ (Skylake), which runs at 2.6 GHz (turbo boost disabled). The C-code was compiled with gcc (version 5.4.0) and flags -*O3* -*fomit-frame-pointer* -*march=corei7-avx* -*msse2avx*. In comparison to NewHope Simple, the time complexity increases for Option 1 by 238%; for Option 2 by 40%; for Option 3 by 6462% (when $k = 80$); and for Option 4 by 4455% (when $k = 66$). Option 2 has

Table 4. Benchmark: median and average clock cycles with 1000 test rounds

	Function	Cycles median/average
NewHope Simple:	KeyGen (server)	223952/ 225452
	KeyGen+shared key (client)	353201/ 358821
	Shared key (server)	78216/ 78614
BHC(511,259):	Encoding	104520/ 108738
	Decoding	157704/ 154652
BCH(1023,258):	Encoding	298043/ 302021
	Decoding	1259554/ 1206814
LDPC(1024,512):	Encoding	2069582/ 2073136
	Decoding ($k = 66$)	26862391/ 27464623
LDPC(1024,256):	Encoding	2068959/ 2071198
	Decoding ($k = 80$)	40282855/ 41912347

a relatively small overhead, thus being suitable for applications that require a low time complexity. The costs for the other options are quite high. However, as NewHope Simple is implemented very efficiently and is already very fast, the time overhead can be acceptable. The decoding complexity of LDPC codes depends on the parameter k. To decrease the run time, k can be decreased. Moreover, the min-sum algorithm can be used instead of the sum-product algorithm. Thus, the complexity is reduced at the cost of a lower decoding performance.

6   Conclusion and Future Work

Our analysis has shown that powerful error-correcting codes within lattice-based key exchange protocols can lead to a significant improvement of important performance parameters, such as failure rate, security level and bandwidth. Modern codes, e.g. LDPC codes, can be used to get a high error-correcting capability. However, to obtain very low error rates, classical codes, e.g. BCH codes, should be employed. The concatenation of LDPC and BCH codes combines their advantages to achieve a quasi-error-free key exchange with a high error-correcting capability. With quasi-error-free communication, the CCA transformation can be applied in order to allow protocols, like NewHope Simple, to be also used for encryption. Before LDPC and BCH codes are used in encryption schemes, it is necessary to investigate these codes with respect to the vulnerability to attackers. For instance, constant-time implementations may be challenging. The selection of the encoding technique is driven by the application characteristics. Many applications may not require or may not be able to integrate powerful error-correcting codes. Different application may benefit from the reduction of data transmission by using strong error-correcting codes, even if the computation time increases. Examples are battery-powered wireless devices, where the radio module usually represents a substantial portion of the overall energy consumption.

Acknowledgments. We thank the anonymous reviewers for their valuable comments and suggestions. This work was partly funded by the Fraunhofer High Performance Center for Secure Connected Systems of Munich.

A   NewHope Simple Algorithms NHSEncode and NHSDecode

Algorithm 1: NHSEncode [3]

Input: Randomized vector $v \in \{0,1\}^{256}$
Result: Polynomial $d \in R_q$
for i *from 0 to 255* **do**
$\quad d_i \leftarrow v_i \lfloor q/2 \rfloor$
$\quad d_{i+256} \leftarrow v_i \lfloor q/2 \rfloor$
$\quad d_{i+512} \leftarrow v_i \lfloor q/2 \rfloor$
$\quad d_{i+768} \leftarrow v_i \lfloor q/2 \rfloor$
end

Algorithm 2: NHSDecode [3]

Input: Polynomial $d \in R_q$
Result: Bit vector $v_i \in \{0,1\}^{256}$
for i *from 0 to 255* **do**
$\quad t \leftarrow \sum_{j=0}^{3} |d_{i+256j} - \lfloor q/2 \rfloor|$
$\quad$ **if** $t < q$ **then**
$\quad\quad | \quad v_i \leftarrow 1$
$\quad$ **else**
$\quad\quad | \quad v_i \leftarrow 0$
$\quad$ **end**
end

B   Proof Theorem 3

Proof. Suppose that a and b are polynomials of a ring with coefficients sampled from the probability distribution Ψ_k and let n be the rank of the polynomials. Then the polynomials can be written as $a = a_0 + a_1 x + \cdots + a_{n-1} x^{n-1}$ and $b = b_0 + b_1 x + \cdots + b_{n-1} x^{n-1}$. If we multiply a with b, we can write $c = (a_0 + a_1 x + \cdots + a_{n-1} x^{n-1})(b_0 + b_1 x + \cdots + b_{n-1} x^{n-1})$. By using the distributive law and grouping all terms with the same rank together, it can be obtained $c = (a_0 b_0 + \cdots - a_{n-2} b_2 - a_{n-1} b_1) + (a_0 b_1 + \cdots - a_{n-2} b_3 - a_{n-1} b_2) x + \cdots + (a_0 b_{n-1} + \cdots + a_{n-2} b_1 + a_{n-1} b_0) x^{n-1}$. Where each coefficient of polynomial c is determined by a sum of n products. Since all coefficients of a and b are independently sampled from the probability distribution Ψ_k, the probability distribution of the coefficients of c is an n-fold convolution of the product distribution of two RVs sampled from Ψ_k.

C   Noise Distribution

Figure 11 illustrates the probability distribution of the difference and compression noise for an error distribution parameter of $k = 16$ and a compression from 14 to 3 bits. In the graph, the RV X of a coefficient of a noise polynomial has the outcomes $x = 0, 1, \ldots, q - 1$. The compression noise is uniformly distributed between zero and $q/16$, and between $q - q/16$ and q. All values in between are not affected by the compression.

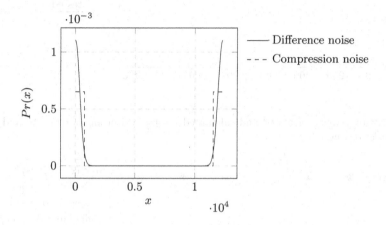

Fig. 11. Noise distributions for $k = 16$ and compression from 14 to 3 bits, where $0 \leq x \leq q - 1$

D   Validation of Failure Rate Analysis

In Fig. 12, the calculated difference and compression noise distribution discussed in Sect. 3 are compared with test measurements. For better visibility, this figure illustrates only values from zero to 1,500. Unlike in Fig. 11, the logarithmic scale is used. For the experiment, the original parameters of NewHope Simple were used. For the tested noise distribution, we used 100,000 test rounds. With $n = 1024$ this leads to 102,400,000 samples. The test measurements match the calculated values. Only for probabilities lower than 10^{-5} the difference noise shows some inaccuracies. With more test samples, the curve is expected to flatten in this region as well.

Figure 13 shows that the independence assumption stated in Subsect. 3.3 can be considered as valid for NewHope Simple with high and moderate failure rates. It illustrates the failure probability with different mapping options within the additive threshold encoding and with varying values of k. Each test value matches with only minor differences the calculated value. For lower values of k the failure

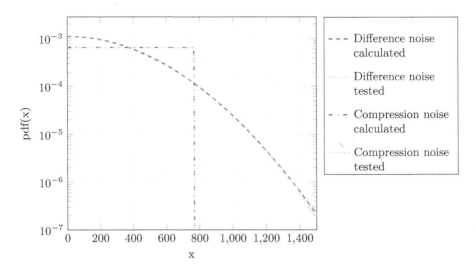

Fig. 12. Comparison of tested and calculated noise distributions for $k = 16$ and compression of c, where $0 \leq x \leq q - 1$

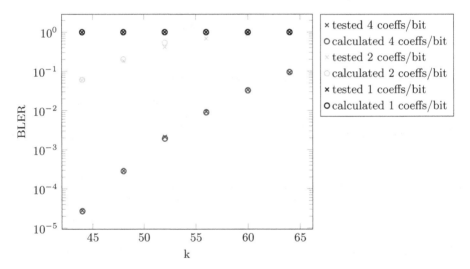

Fig. 13. Comparison of tested and calculated error probability with compression of c

probability is too small in order to find the correct value by testing. Test results have shown that the calculated values for NewHope Simple without compression and NewHope Simple with further compression on polynomial u match the test values as well.

References

1. Alkim, E., et al.: NewHope: Algorithm Specifications and Supporting Documentation (2017). https://newhopecrypto.org/data/NewHope_2017_12_21.pdf
2. Alkim, E., et al.: FrodoKEM - Learning with Errors Key Encapsulation: Algorithm Specifications and Supporting Documentation (2017). https://frodokem.org/files/FrodoKEM-specification-20171130.pdf
3. Alkim, E., Ducas, L., Pöppelmann, T., Schwabe, P.: NewHope without reconciliation. IACR Cryptology ePrint Archive 2016, 1157 (2016)
4. Alkim, E., Ducas, L., Pöppelmann, T., Schwabe, P.: Post-quantum key exchange - a new hope. In: 25th USENIX Security Symposium, USENIX Security 16, 10–12 August 2016, Austin, TX, USA, pp. 327–343 (2016)
5. Avanzi, R., et al.: CRYSTALS-Kyber: Algorithm Specifications and Supporting Documentation (2017). https://www.pq-crystals.org/kyber/data/kyber-specification.pdf
6. Bai, S., Galbraith, S.D.: An improved compression technique for signatures based on learning with errors. In: Benaloh, J. (ed.) CT-RSA 2014. LNCS, vol. 8366, pp. 28–47. Springer, Cham (2014). https://doi.org/10.1007/978-3-319-04852-9_2
7. Barreto, P.S., Longa, P., Naehrig, M., Ricardini, J.E., Zanon, G.: Sharper Ring-LWE signatures. IACR Cryptology ePrint Archive 2016, 1026 (2016)
8. Berlekamp, E.R.: Nonbinary BCH decoding. In: International Symposium on Information Theory, San Remo, Italy (1966)
9. Bos, J.W., et al.: Frodo: take off the ring! practical, quantum-secure key exchange from LWE. In: Proceedings of the 2016 ACM SIGSAC Conference on Computer and Communications Security, 24–28 October 2016, Vienna, Austria, pp. 1006–1018 (2016). https://doi.org/10.1145/2976749.2978425
10. Bos, J.W., et al.: CRYSTALS - Kyber: a CCA-secure module-lattice-based KEM. IACR Cryptology ePrint Archive 2017, 634 (2017)
11. Cheon, J.H., Kim, D., Lee, J., Song, Y.S.: Lizard: Cut off the tail! // practical post-quantum public-key encryption from LWE and LWR. IACR Cryptology ePrint Archive 2016, 1126 (2016)
12. Chien, R.T.: Cyclic decoding procedures for Bose-Chaudhuri-Hocquenghem codes. IEEE Trans. Inf. Theory 10(4), 357–363 (1964). https://doi.org/10.1109/TIT.1964.1053699
13. Ducas, L., Durmus, A., Lepoint, T., Lyubashevsky, V.: Lattice signatures and bimodal Gaussians. Cryptology ePrint Archive, Report 2013/383 (2013)
14. Fan, J.: Constrained Coding and Soft Iterative Decoding. The Springer International Series in Engineering and Computer Science. Springer, Heidelberg (2012)
15. Fujisaki, E., Okamoto, T.: Secure integration of asymmetric and symmetric encryption schemes. In: Wiener, M. (ed.) CRYPTO 1999. LNCS, vol. 1666, pp. 537–554. Springer, Heidelberg (1999). https://doi.org/10.1007/3-540-48405-1_34
16. Gallager, R.G.: Low-density parity-check codes. IRE Trans. Inf. Theory 8(1), 21–28 (1962). https://doi.org/10.1109/TIT.1962.1057683
17. Gitlin, R., Hayes, J., Weinstein, S.: Data Communications Principles. Applications of Communications Theory. Springer, Heidelberg (2012)
18. Hamburg, M.: Supporting documentation: ThreeBears (2017). https://csrc.nist.gov/Projects/Post-Quantum-Cryptography/Round-1-Submissions
19. Hoffstein, J., Pipher, J., Silverman, J.H.: NTRU: a ring-based public key cryptosystem. In: Buhler, J.P. (ed.) ANTS 1998. LNCS, vol. 1423, pp. 267–288. Springer, Heidelberg (1998). https://doi.org/10.1007/BFb0054868

20. Hu, X., Eleftheriou, E., Arnold, D., Dholakia, A.: Efficient implementations of the sum-product algorithm for decoding LDPC codes. In: Proceedings of the Global Telecommunications Conference, GLOBECOM 2001, 25–29 November 2001, San Antonio, TX, USA, p. 1036 (2001). https://doi.org/10.1109/GLOCOM. 2001.965575

21. Lin, S., Costello, D.J.: Error Control Coding, 2nd edn. Prentice-Hall Inc., Upper Saddle River (2004)

22. Lindner, R., Peikert, C.: Better key sizes (and attacks) for LWE-based encryption. In: Kiayias, A. (ed.) CT-RSA 2011. LNCS, vol. 6558, pp. 319–339. Springer, Heidelberg (2011). https://doi.org/10.1007/978-3-642-19074-2_21

23. Lu, X., Liu, Y., Jia, D., Xue, H., He, J., Zhang, Z.: Supporting documentation: LAC (2017). https://csrc.nist.gov/Projects/Post-Quantum-Cryptography/ Round-1-Submissions

24. Lyubashevsky, V., Peikert, C., Regev, O.: On ideal lattices and learning with errors over rings. In: Gilbert, H. (ed.) EUROCRYPT 2010. LNCS, vol. 6110, pp. 1–23. Springer, Heidelberg (2010). https://doi.org/10.1007/978-3-642-13190-5_1

25. National Institute of Standards and Technology: Announcing request for nominations for public-key post-quantum cryptographic algorithms (2016). https://csrc. nist.gov/news/2016/public-key-post-quantum-cryptographic-algorithms

26. Qian, C., Lei, W., Wang, Z.: Low complexity LDPC decoder with modified Sum-Product algorithm. Tsinghua Sci. Technol. **18**(1), 57–61 (2013). https://doi.org/ 10.1109/TST.2013.6449408

27. Richardson, T.: Error floors of LDPC codes. In: Proceedings of the Annual Allerton Conference on Communication Control and Computing, pp. 1426–1435. The University; 1998 (2003)

28. Saarinen, M.O.: HILA5: On reliability, reconciliation, and error correction for Ring-LWE encryption. IACR Cryptology ePrint Archive 2017, 424 (2017)

29. Saarinen, M.J.O.: Supporting documentation: HILA5 (2017). https://csrc.nist. gov/Projects/Post-Quantum-Cryptography/Round-1-Submissions

30. Safak, M.: Digital Communications. Wiley, Hoboken (2017)

31. Targhi, E.E., Unruh, D.: Post-quantum security of the Fujisaki-Okamoto and OAEP transforms. In: Hirt, M., Smith, A. (eds.) TCC 2016. LNCS, vol. 9986, pp. 192–216. Springer, Heidelberg (2016). https://doi.org/10.1007/978-3-662-53644-5_8

32. Zhao, Y., Jin, Z., Gong, B., Sui, G.: Supporting documentation: KCL (2017). https://csrc.nist.gov/Projects/Post-Quantum-Cryptography/Round-1-Submissions

Provably Secure NTRUEncrypt over Any Cyclotomic Field

Yang Wang and Mingqiang Wang[✉]

School of Mathematics, Shandong University, Jinan 250100, Shandong, China
wyang1114@mail.sdu.edu.cn, wangmingqiang@sdu.edu.cn

Abstract. NTRUEncrypt is generally recognized as one of candidate encryption schemes for post quantum cryptography, due to its moderate key sizes, remarkable performance and potential capacity of resistance to quantum computers. However, the previous provably secure NTRUEncrypts are only based on prime-power cyclotomic rings. Whether there are provably secure NTRUEncrypt schemes over more general algebraic number fields is still an open problem. In this paper, we answer this question and present a new provably IND-CPA secure NTRUEncrypt over any cyclotomic field. The security of our scheme is reduced to a variant of learning with errors problem over rings (Ring-LWE). More precisely, the security of our scheme is based on the worst-case approximate shortest independent vectors problem (SIVP$_\gamma$) over ideal lattices. We prove that, once the field is fixed, the bounds of the reduction parameter γ and the modulus q in our scheme are less dependent on the choices of plaintext spaces. This leads to that our scheme provides more flexibility for the choices of plaintext spaces with higher efficiency under stronger security assumption. Furthermore, the probability that the decryption algorithm of our scheme fails to get the correct plaintext is much smaller than that of the previous works.

Keywords: NTRU · Ideal lattices · Canonical embedding
Cyclotomic fields · Ring-LWE

1 Introduction

The NTRU encryption scheme was devised by Hoffstein, Pipher and Silverman in [15]. It is one of the fastest known lattice-based cryptosystems as testified by its inclusion in the IEEE P1363 standard and regarded as an alternative to RSA and ECC due to its potential of resisting attacks by quantum computers. Based on the underlying problem of NTRU, various cryptographic primitives were designed, such as identity-based encryption [8], fully homomorphic encryption [2,20], digital signatures [7,14] and multi-linear maps [11]. Meanwhile, a batch of cryptanalysis works were proposed aiming at NTRU family [1,4,5,9,10,12,16–18].

The security of the first NTRUEncrypt in [15] is heuristic and lacks a solid mathematical proof. This leads to a break-and-repair development history of

© Springer Nature Switzerland AG 2019
C. Cid and M. J. Jacobson, Jr. (Eds.): SAC 2018, LNCS 11349, pp. 391–417, 2019.
https://doi.org/10.1007/978-3-030-10970-7_18

NTRUEncrypt. Stehlé and Steinfeld [29] provided the first provably IND-CPA secure NTRUEncrypt over power of 2 cyclotomic rings. They used the coefficient embedding of polynomial rings and the security of their scheme was based on the corresponding Ring-LWE problem. Although the construction of Stehlé and Steinfeld may be less practical compared with classical NTRUEncrypts [3], their work revealed an important connection between NTRUEncrypt and Ring-LWE, hence between problems over NTRU lattices and worst-case problems (SIVP$_\gamma$) over ideal lattices. An open problem proposed by Stehlé and Steinfeld is whether their construction can be improved to more general rings. Recently, Yu, Xu and Wang [31] modified the scheme in [29] to make it work over cyclotomic rings of the forms $\mathbb{Z}[\zeta_p]$ for prime integer p. The modified scheme in [31] allowed more flexibility choices of cyclotomic rings, but the size requirements for parameters were more limited, making the modified schemes less efficiency. The first NTRUEncrypt scheme using canonical embedding was discussed in [32] which showed that given appropriate parameters, provably secure NTRUEncrypt could work over prime-power cyclotomic rings. The security of the schemes proposed in [31,32] relied on a variant of Ring-LWE problems over cyclotomic rings proposed in [6].

With the calls of post-quantum cryptography by NIST, a better understanding of these problems is necessary and the study of NTRUEncrypt is theoretically valuable as stated in [32]. To our knowledge, till now, provably secure NTRU-Encrypts were all constructed over prime-power cyclotomic rings by using the coefficient embedding. Also, the security parameter γ and the modulus q rely heavily on the choice of plaintext space. That is to say, in order to reach better efficiency in applications, the plaintext space of the existing NTRUEncrypts were all limited to $\{0,1\}^n$-only embed one bit in each coefficient of polynomials in each encrypt process. If we want to embed more bits in each coefficient of polynomials in each encryption process, the lower bounds of γ and q would become pretty bad. These disadvantages restrict the applications of the existing provably secure NTRUEncrypts. Therefore, eliminating the limitation of choices of cyclotomic fields to solve the open problem proposed in [29] and improving the efficiency of the existing provably secure NTRUEncrypts are worth doing. These are also the main motivations of our research.

1.1   Our Contributions

NTRUEncrypt schemes in the standard model by using the canonical embedding over any cyclotomic field. For any fixed cyclotomic field, we design our scheme in the fractional ideal $R^\vee$, i.e. the codifferent ideal of the ring of integers R. In applications, our scheme can also be converted to work in an integral ideal of R.

Once we fix a cyclotomic field, we get an almost uniform bounds for the reduction parameter γ and the modulus q, which are less dependent on the choices of plaintext spaces. Hence, our scheme provides more flexibility for the choices of plaintext spaces and has potential to send more encrypted bits in one encryption process with higher efficiency under stronger hardness assumption.

We use the subgaussian distribution, the decoding basis and the basis-embedding norm to estimate the decryption error. These tools enable us to get tighter lower bounds of q and γ, they also bring us a smaller decryption error. More precisely, our decryption algorithm succeeds in recovering the correct message with an exception of a negligible probability $n^{-\omega(\sqrt{n \log n})}$, much better than the previous $n^{-\omega(1)}$.

We also get a regularity result (a kind of ring-based leftover hash lemma) for all cyclotomic fields, which is useful to design many cryptographic primitives. Set $R_q^\times$ be the set of invertible elements of $R_q = R/(qR)$, the regularity is about how to construct a tuple $(a_1, \cdots, a_m; \sum_{i=1}^m a_i t_i) \approx U((R_q^\times)^m \times R_q))$, where $a_i \hookleftarrow U(R_q^\times)$ are chosen independently and t subjects to some distributions. Our results enrich the choices of the distributions of t.

1.2 Technique Overview

Although the main ideas of our NTRUEncrypt follow Stehlé and Steinfeld's route, many differences exist.

In the previous constructions, analysis of decryption error is the uppermost difficulty which constrains the form of cyclotomic fields. The traditional coefficient embedding decides that this process depends heavily on the form of polynomials f of the corresponding ring $R = \mathbb{Z}[x]/(f(x))$. To overcome this problem, we have a very important observation that the decryption is only relevant to the coefficients corresponding to the basis we choose, and different bases affect the results heavily. The natural choice of coefficient embedding over polynomial rings may mislead us. So we use the decoding basis of $R^\vee$ and define the basis-coefficient embedding to bound the decryption error. These modifications enable us to control the decryption error for all cyclotomic fields in the same way. Then, if we want to enjoy the high computation speed over polynomial rings, it is easy for us to convert our schemes to work in the ring R in theory.

Benefits brought by those tools and our observation are more than these. If we want to reach the highest efficiency, traditional coefficient embedding may limit the number of encrypted bits in each encryption process, i.e. in order to get the highest efficiency, the existing NTRUEncrypts all limited their plaintext space to $\{0, 1\}^n$. This is caused by the coefficient embedding and the perspective that we regard the elements as polynomials in the ring R. If we regard constant polynomials and non-constant polynomials as usual algebraic integers, then the tools we use give us an almost uniform bound for the reduction parameter γ and the modulus q, which is less dependent on the choices of plaintext spaces. Meanwhile, the decryption error is much smaller than that of the existing schemes.

The reason why we design our scheme in $R^\vee$ is that we want to use the hardness results about Ring-LWE showed in [22], other than those proposed in [6]. This is a natural choice when we want to use the canonical embedding and to get rid of the troubles caused by different polynomials. By using the recent hardness results about primal-Ring-LWE (i.e. the secret $s \hookleftarrow U(R_q)$) proved in [28], we can also directly design NTRUEncrypt in R (For more details, see Remark 2). The high level construction outline of our scheme is as follows.

The key generation algorithm is essentially the same as the previous works.

Input: $q \in \mathbb{Z}^+$, $p \in R_q^\times$, $\sigma \in \mathbb{R}^+$.

Output: A key pair $(sk,\ pk) \in R_q^\times \times R_q^\times$.

1. Sample f' from $D_{R,\sigma}$; let $f = p \cdot f' + 1$; if $(f \bmod qR) \notin R_q^\times$, resample.

2. Sample g from $D_{R,\sigma}$; if $(g \bmod qR) \notin R_q^\times$, resample.

3. Return secret key $sk = f$ and public key $pk = h = pg/f \in R_q^\times$.

We use standard method to prove that the algorithm would terminate in expected time. Furthermore, the Gaussian distribution ensures that the secret key is 'short'. Provable security needs the public key to distribute statistically close to uniformity, and the analysis of the public key distribution needs to deal with some kinds of q-ary lattices, in order to bound the corresponding smooth parameters. By an accurate analysis of the relationship between different fractional ideals, we give a lower bound of λ_1 with respect to l_∞ norm of these q-ary lattices. In this section, we consider these problems absolutely in K, hence get a better result compared with [32] in theory.

Our NTRUEncrypt is as following:

Key generation: Use the algorithm, to get $sk = f \in R_q^\times$ with $f = 1 \bmod pR^\vee$, and $pk = h = pg \cdot f^{-1} \in R_q^\times$.

Encryption: Given message $m \in \mathcal{P}$, sample $s, e \hookleftarrow \chi$ and return $c = hs + pe + m \in R_q^\vee$.

Decryption: Given ciphertext c and secret key f, compute $c_1 = fc$. Then return $m = (c_1 \bmod qR^\vee) \bmod pR^\vee$.

Here, χ is the error distribution of the Ring-LWE problem proposed in [22]. The plaintext space of our scheme is $\mathcal{P} = R^\vee/(pR^\vee)$, where p is an invertible element in R_q. By using the decoding basis of $R^\vee$ and the basis-coefficient embedding of elements in $R^\vee$, we get a tight connection between the canonical norms and the basis-coefficient norms. Moreover, by using subgaussian distributions, we also prove that the decryption error is negligible - $n^{-\omega(\sqrt{n \log n})}$, which is better than the existing $n^{-\omega(1)}$. Furthermore, as we remark in Remark 1, we can put all computations and storages in an integral ideal of R and this modification may enjoy the high computation speed over polynomial rings in theory.

Till now, the magnitude of the modulus q is far away from practicality, and this is the common shortcoming of the provably secure NTRUEncrypts. How to reduce the sizes of parameters is an intriguing open problem.

1.3   Organization

In Sect. 2, we introduce some notations and basic results that will be used in our discussion. In Sect. 3, we give a new series of relevant results about some kinds of q-ary lattices. These are important for us to analyze the key generation algorithm of our NTRUEncrypt in Sect. 4. In Sect. 5, we construct the NTRUEncrypt and give a secure reduction from basic lattice problem to the CPA-security of our NTRUEncrypt.

2 Preliminaries

In this section, we introduce some background results and notations.

2.1 Notations

We set $\hat{l} = l$ when l is odd and $\hat{l} = \frac{l}{2}$ when l is even. Functions $\varphi(n)$ and $\mu(n)$ stand for the Euler function and the Möbius function. We use $[n]$ to denote the set $\{1, 2, \cdots, n\}$. For $p = 1, 2, \cdots, \infty$, we use $|| \cdot ||_p$ to represent the l_p norm corresponding to the canonical embedding. When $p = 2$, we usually use $|| \cdot ||$ to represent the l_2 norm. For any matrix $M \in \mathbb{C}^{n \times n}$, we use $\lambda_i(M)$ stand for its eigenvalues and $s_i(M)$ stand for its singular values for $i \in [n]$. We arrange eigenvalues and singular values by their magnitudes, i.e. $\lambda_1(M) \geq \cdots \geq \lambda_n(M)$ and $s_1(M) \geq \cdots \geq s_n(M)$. For two random variables X and Y, $\Delta(X, Y)$ stands for their statistic distance. As usual, $E(X)$ and $Var(X)$ stand for the expectation and the variance of a random variable X. When we write $X \hookleftarrow \xi$, we mean that the random variable X obeys to a distribution ξ. Function rad represents the radical of a positive integer n, i.e. for $n = p_1^{\alpha_1} \cdots p_k^{\alpha_k}$ with different primes p_i, $rad(n) = \prod_{i=1}^{k} p_i$. If S is a finite set, then $|S|$ is its cardinality and $U(S)$ is the uniform distribution over S. Symbols $\mathbb{Z}^+$ and $\mathbb{R}^+$ stand for the sets of positive integers and positive reals. Symbol $\log x$ represents $\log_2 x$ for $x \in \mathbb{R}^+$. For a positive integer a, $\mathbb{Z}_a^\times$ represents the reduced residue system mod a.

2.2 Cyclotomic Fields, Space H and Geometry

Through out this paper, we consider cyclotomic fields. Let $K = \mathbb{Q}(\zeta)$, where $\zeta = \zeta_l$ is a primitive l-th root of unity, which has minimal polynomial $\Phi_l(x) = \prod_{i|l}(x^i - 1)^{\mu(\frac{l}{i})}$ of degree $n = \varphi(l)$. Then $[K : \mathbb{Q}] = n = \varphi(l)$ and $K \cong \mathbb{Q}[x]/\Phi_l(x)$. We set $R = \mathcal{O}_K = \mathbb{Z}[\zeta]$ be the ring of integers of K.

We set $\mathrm{Gal}(K/\mathbb{Q}) = \{\sigma_i : i = 1, \cdots, n\}$ and use the canonical embedding σ on K, who maps $x \in K$ to $(\sigma_1(x), \cdots, \sigma_n(x)) \in H$, where H is a kind of Minkowski space in algebraic number theory. Here we identity $\sigma_i(\zeta) = \zeta^{l_i}$ with l_i the i-th element of $\mathbb{Z}_l^\times$, order the σ_i and define $H = \{(x_1, \cdots, x_n) \in \mathbb{C}^n : x_{n+1-i} = \overline{x_i}, \forall i \in [r]\}$. H is isomorphic to $\mathbb{R}^n$ as an inner product space via the orthonormal basis $h_{i \in [n]}$ defined as follows. Assume $e_j \in \mathbb{C}^n$ be the vector with 1 in its j-th coordinate and 0 elsewhere, i be the imaginary number such that $i^2 = -1$. We then set $h_j = \frac{1}{\sqrt{2}}(e_j + e_{n+1-j})$ and $h_{n+1-j} = \frac{i}{\sqrt{2}}(e_j - e_{n+1-j})$ for $1 \leq j \leq r$.

For any element $x \in K$, we can define the ℓ_p norm of x by $||x||_p = ||\sigma(x)||_p$ for $p < \infty$ and $||x||_\infty = \max_{i \in [n]} |\sigma_i(x)|$. Because multiplication of embedded elements is component-wise, for any $x, y \in K$, we have $||x \cdot y||_p \leq ||x||_\infty \cdot ||y||_p$ for $p \in \{1, \cdots, \infty\}$. The Trace and Norm of $x \in K$ are defined as usual, i.e. $\mathrm{Tr}(x) := \mathrm{Tr}_{K/\mathbb{Q}}(x) = \sum_{i=1}^{n} \sigma_i(x)$ and $\mathrm{N}(x) := \mathrm{N}_{K/\mathbb{Q}}(x) = \prod_{i=1}^{n} \sigma_i(x)$. The discriminant Δ_K of K, the integral and fractional ideals are defined as usual.

Integral ideals can be regarded as special cases of fractional ideals. Recall that, the discriminant of the l-th cyclotomic number field is

$$\Delta_K = (-1)^{\frac{n}{2}} \cdot \left(\frac{l}{\prod_{p|l} p^{\frac{1}{p-1}}} \right)^n \le n^n,$$

where p runs over all prime factors of l.

Let $q \in \mathbb{Z}$ be a prime, then the factorization of the ideal $(q) = qR$ is as follows. Let $d \ge 0$ be the largest integer such that q^d divides l, let $e = \varphi(q^d)$ and let $f \ge 1$ be the multiplicative order of q modulo l/q^d. Then $(q) = \prod_{i=1}^{g} \mathfrak{q}_i^e$, where $\mathfrak{q}_i$ are $g = n/(ef)$ different prime ideals each of norm q^f. In particular, for a prime $q = 1 \bmod l$, we have $e = f = 1$, the ideal (q) splits into n distinct prime ideals as $(q) = \prod_{i \in \mathbb{Z}_l^\times} \mathfrak{q}_i$ with $\mathfrak{q}_i = \langle q, \zeta - \omega^i \rangle$, where ω is a primitive l-th root of unity in $\mathbb{Z}_q^\times$. The norm of $\mathfrak{q}_i$ is q. We have $\Phi_l(x) = \prod_{i \in \mathbb{Z}_l^\times} (x - \omega^i) \bmod q$.

2.3   Lattice and Discretization

We define a lattice as a discrete additive subgroup of H and we only deal with full-rank lattices. The minimum distance $\lambda_1(\Lambda)$ of a lattice is the length of a shortest nonzero lattice vector. We usually use the l_2 norm, i.e. $\lambda_1(\Lambda) = \min_{0 \ne x \in \Lambda} ||x||$. The dual lattice of $\Lambda \subseteq H$ is defined as $\Lambda^\vee = \{ y \in H : \forall\, x \in \Lambda, \langle x, \overline{y} \rangle = \sum_{i=1}^{n} x_i y_i \in \mathbb{Z} \}$. This is actually the complex conjugate of the dual lattice as usually defined in $\mathbb{C}^n$. All of the properties of the dual lattice that we use also hold for the conjugate dual. For any fractional ideal I of K, we can represent I as $\mathbb{Z}\beta_1 + \cdots + \mathbb{Z}\beta_n$ for some $\beta_i \in K$, $i = 1, \cdots, n$. Then $\sigma(I)$ is a lattice of H, and we call $\sigma(I)$ an ideal lattice and identify I with this lattice and associate with I all the usual lattice quantities. We have $|\Delta_K| = \det(\sigma(R))^2$, the squared determinant of the lattice $\sigma(R)$. For any fractional ideal I, we also have $\det(\sigma(I)) = \mathrm{N}(I) \cdot \sqrt{|\Delta_k|}$. The following lemma from [26] gives upper and lower bounds on the minimum distance of an ideal lattice in l_2 norm.

Lemma 1. *For any fractional ideal I in a number field K of degree n,*

$$\sqrt{n} \cdot \mathrm{N}^{\frac{1}{n}}(I) \le \lambda_1(I) \le \sqrt{n} \cdot \mathrm{N}^{\frac{1}{n}}(I) \cdot |\Delta_K|^{\frac{1}{2n}}.$$

For any fractional ideal I in K, its dual is defined as $I^\vee = \{ a \in K : \mathrm{Tr}(aI) \subseteq \mathbb{Z} \}$. It is easy to verify $(I^\vee)^\vee = I$, $I^\vee$ is a fractional ideal and $I^\vee$ embeds under σ as the dual lattice of I as defined before. In fact, an ideal of K and its inverse are related by multiplication with the dual ideal $R^\vee$: $I^\vee = I^{-1} \cdot R^\vee$.

One of the most famous lattice problems is SVP. Given a lattice basis B, try to find a shortest vector in $\Lambda \backslash \{0\}$, where $\Lambda = \mathcal{L}(B)$. The relaxed problem SVP$_\gamma$ is asking for a nonzero lattice vector that is no longer than γ times the length of a solution of SVP. By restricting SVP to the ideal lattice, we obtain Ideal-SVP. No polynomial quantum algorithm is known to solve the worst-case SVP$_\gamma$ problem for $\gamma \le \mathrm{poly}(n)$ and also no algorithm is known to perform non-negligibly better for ideal lattices than classic lattices. The (Ideal-SIVP$_\gamma$) SIVP$_\gamma$

problem is that given a basis of a lattice Λ of dimension n, try to find n linear independent vectors $x_1, \cdots, x_n \in \Lambda$ such that $\max_{1 \leq i \leq n} \|x_i\| \leq \gamma \cdot \lambda_n(\Lambda)$.

We now consider the discretization. We describe the formal definition as in [24], a modified version of [22]. Define $\lceil x \rceil$ to be the smallest integer that is bigger than or equal to x for any $x \in \mathbb{R}$.

Definition 1. *If Bern denotes the Bernoulli distribution, then the univariate Reduction distribution $Red(a) = Bern(\lceil a \rceil - a) - (\lceil a \rceil - a)$ is the discrete probability distribution defined for parameter $a \in \mathbb{R}$ as taking the values*

$-1 + a - \lceil a \rceil$ *with probability* $\lceil a \rceil - a$,
$-a - \lceil a \rceil$ *with probability* $1 - (\lceil a \rceil - a)$.

A random variable $\boldsymbol{R} = (R_1, \cdots, R_n)^T \in \mathbb{R}^n$ has a multivariate Reduction distribution $R \sim Red(\boldsymbol{a})$ on $\mathbb{R}^n$ for parameter $\boldsymbol{a} = (a_1, \cdots, a_n)^T$ if its components $R_j \sim Red(a_j)$ for $j = 1, \cdots, n$ are independent univariate Reduction random variables.

We now describe the coordinate-wise rounding discretisation which is easy to use for our applications.

Definition 2. *Suppose $\Lambda = \mathcal{L}(B)$ is a n-dimensional lattice in space H. For $\boldsymbol{c} \in H$, the coordinate-wise randomized rounding discretisation $\lfloor \boldsymbol{X} \rceil_{\Lambda + c}^B$ of random variable $\boldsymbol{X}$ to the lattice coset $\Lambda + \boldsymbol{c}$ with respect to the basis B is then defined by the conditional random variable*

$$(\lfloor \boldsymbol{X} \rceil_{\Lambda+c}^B | \boldsymbol{X} = \boldsymbol{x}) = \lfloor \boldsymbol{x} \rceil_{\Lambda+c}^B = \boldsymbol{x} + BQ_{\boldsymbol{x},c},$$

where $Q_{\boldsymbol{x},c} \sim Red(B^{-1}(\boldsymbol{c} - \boldsymbol{x}))$.

2.4   Gaussian and Subgaussian Random Variables

For $s > 0$, $\boldsymbol{c} \in H$, define the Gaussian function $\rho_{s,c} : H \to (0, 1]$ as $\rho_{s,c}(\boldsymbol{x}) = e^{-\pi \frac{\|x - c\|^2}{s^2}}$. By normalizing this function, we obtain the continuous Gaussian probability distribution $D_{s,c}$ of parameter s, whose density is given by $s^{-n} \cdot \rho_{s,c}(\boldsymbol{x})$. We usually omit the subscript $\boldsymbol{c}$ when it is 0. Let $\boldsymbol{r} = (r_1, \cdots, r_n) \in (\mathbb{R}^+)^n$ be a vector such that $r_j = r_{n+1-j}$ for $j \in \{1, \cdots, \frac{n}{2}\}$, we can define the elliptical Gaussian distributions in the basis $\{h_i\}_{i \leq n}$ as follows: a sample from D_r is given by $\sum_{i \in [n]} x_i h_i$, where x_i are chosen independently from the Gaussian distribution D_{r_i} over $\mathbb{R}$. Note that, if we define a map $\varphi : H \to \mathbb{R}^n$ by $\varphi(\sum_{i \in [n]} x_i h_i) = (x_1, \cdots, x_n)$, then D_r is also a (elliptical) Gaussian distribution over $\mathbb{R}^n$.

For a lattice $\Lambda \subseteq H$, $\sigma > 0$ and $\boldsymbol{c} \in H$, we define the lattice Gaussian distribution of support Λ, deviation σ and center $\boldsymbol{c}$ by $D_{\Lambda,\sigma,c}(\boldsymbol{x}) = \frac{\rho_{\sigma,c}(\boldsymbol{x})}{\rho_{\sigma,c}(\Lambda)}$ for any $\boldsymbol{x} \in \Lambda$. For $\delta > 0$, we define the smoothing parameter $\eta_\delta(\Lambda)$ as the smallest $\sigma > 0$ such that $\rho_{\frac{1}{\sigma}}(\Lambda^\vee \setminus \mathbf{0}) \leq \delta$. The following theorem comes from [26]. Here we use $\tilde{B}$ to represent the Gram-Schmidt orthogonalization of B and regard the columns of B as a set of vectors. For $B = (b_1, \cdots, b_n)$, define $\|B\| = \max_i \|b_i\|$.

Theorem 1. *There is a probabilistic polynomial time algorithm that, given a basis B of an n-dimensional lattice $\Lambda = \mathcal{L}(B)$, a standard deviation $\sigma \geq ||\tilde{B}|| \cdot \sqrt{\log n}$, and a $\boldsymbol{c} \in H$, outputs a sample whose distribution is $D_{\Lambda,\sigma,\boldsymbol{c}}$.*

We will also use the following lemmas from [23], [25] and [13].

Lemma 2. *For any full-rank lattice Λ and positive real $\varepsilon > 0$, we have $\eta_\varepsilon(\Lambda) \leq \sqrt{\frac{\ln(2n(1+\frac{1}{\varepsilon}))}{\pi}} \cdot \lambda_n(\Lambda)$.*

Lemma 3. *For any full-rank lattice Λ, $\boldsymbol{c} \in H$, $\varepsilon \in (0,1)$ and $\sigma \geq \eta_\varepsilon(\Lambda)$, we have $\Pr_{\boldsymbol{b} \hookleftarrow D_{\Lambda,\sigma,\boldsymbol{c}}}[|| \boldsymbol{b} - \boldsymbol{c}|| \geq \sigma\sqrt{n}] \leq \frac{1+\varepsilon}{1-\varepsilon} \cdot 2^{-n}$.*

Lemma 4. *For any full-rank lattice Λ and any positive real $\varepsilon > 0$, we have $\eta_\varepsilon(\Lambda) \leq \sqrt{\frac{\ln(2n(1+\frac{1}{\varepsilon}))}{\pi}} \cdot \frac{1}{\lambda_1^\infty(\Lambda^\vee)}$.*

Lemma 5. *Let $\Lambda' \subseteq \Lambda$ be full-rank lattices. For any $\boldsymbol{c} \in H$, $\varepsilon \in (0,1/2)$ and $\sigma \geq \eta_\varepsilon(\Lambda')$, we have $\Delta(D_{\Lambda,\sigma,\boldsymbol{c}} \bmod \Lambda', U(\Lambda/\Lambda')) \leq 2\varepsilon$.*

It is convenient for us to use the notion of subguassian random variables in our application. We describe the definitions as in [24].

Definition 3. *For $\delta \geq 0$, a real-valued random variable X is δ-subgaussian with standard parameter $b \geq 0$ if*

$$E(e^{tX}) \leq e^\delta e^{\frac{1}{2}b^2 t^2}, \qquad for\ all\ t \in \mathbb{R}.$$

A real-valued random variable X is δ-subgaussian random variable with scaled parameter $s \geq 0$ if

$$E(e^{2\pi tX}) \leq e^\delta e^{\pi s^2 t^2}, \qquad for\ all\ t \in \mathbb{R}.$$

A real-valued random variable is δ-subgaussian with standard parameter b if and only if it is δ-subgaussian with scaled parameter $\sqrt{2\pi}b$. One can extend the definitions to $\mathbb{R}^n$ or space H.

Definition 4. *For any $\delta \geq 0$, a multivariate random variable $\boldsymbol{X}$ on $\mathbb{R}^n$ is δ-subgaussian with standard parameter $b \geq 0$ if*

$$E(e^{<t,\boldsymbol{X}>}) \leq e^\delta e^{\frac{1}{2}b^2 ||t||^2}, \qquad for\ all\ \boldsymbol{t} \in \mathbb{R}^n.$$

A multivariate random variable $\boldsymbol{Z}$ on H is a δ-subgaussian with standard parameter $b \geq 0$ if

$$E(e^{<t,\boldsymbol{Z}>}) \leq e^\delta e^{\frac{1}{2}b^2 ||t||^2}, \qquad for\ all\ \boldsymbol{t} \in H.$$

This definition is equivalent to say that a random vector $\boldsymbol{X}$ or its distribution is δ-subgaussian with standard parameter b if for all unit vector $\boldsymbol{t}$, the random variable $<\boldsymbol{X}, \boldsymbol{t}>$ is δ-subgaussian with standard parameter b.

Definition 5. *A random variable Z on $\mathbb{R}^n$ (or H) is a noncentral subgaussian random variable with noncentrality parameter $\|E(Z)\| \geq 0$ and deviation parameter $d \geq 0$ if the centered random variable $Z_0 = Z - E(Z)$ is a 0-subgaussian random variable with standard parameter d.*

We regard a central subgaussian random variable as a special case of a noncentral subgaussian random variable. Moreover, we have the following useful lemma which is proposed in [24].

Lemma 6. *Suppose that B is a column basis matrix for a lattice in H with largest singular value $s_1(B)$ and Z is an independent noncentral subgaussian random variable with deviation parameter d_Z. The coordinate-wise randomized rounding discretisation of Z to $\lfloor Z \rceil_{A+c}^B$ is a noncentral subgaussian random variable with noncentrality parameter $\|E(Z)\|$ and deviation parameter $(d_Z^2 + (\frac{1}{2})^2 s_1(B)^2)^{\frac{1}{2}}$.*

2.5 Basis for R and $R^\vee$, Ring-LWE problem

In our application, we hope that the matrices whose columns are consisted of the basis of R or $R^\vee$ have smaller s_1 and larger s_n. So, we introduce the powerful basis and the decoding basis as in [22]. We set τ be the automorphism of K that maps ζ_l to $\zeta_l^{-1} = \zeta_l^{l-1}$, under the canonical embedding it corresponds to complex conjugation $\sigma(\tau(a)) = \overline{\sigma(a)}$.

Definition 6. *The Powerful basis $\overrightarrow{p}$ of $K = \mathbb{Q}(\zeta_l)$ and $R = \mathbb{Z}[\zeta_l]$ is defined as follows:*

- *For a prime power l, define $\overrightarrow{p}$ to be the power basis $(\zeta_l^j)_{(j \in \{0,1,\cdots,n-1\})}$, treated as a vector over $R \subseteq K$.*
- *For l having prime-power factorization $l = \prod l_k = \prod p_k^{\alpha_k}$, define $\overrightarrow{p} = \otimes_k \overrightarrow{p_k}$, the tensor product of the power basis $\overrightarrow{p_k}$ of each $K_k = \mathbb{Q}(\zeta_{l_k})$.*

The Decoding basis of $R^\vee$ is $\overrightarrow{d} = \tau(\overrightarrow{p})^\vee$, the dual of the conjugate of the powerful basis $\overrightarrow{p}$.

Different bases of R (or $R^\vee$) are connected by some unimodular matric, hence the spectral norm (i.e. the s_1) may have different magnitudes. The following lemma comes from [22], which shows the estimates of $s_1(\sigma(\overrightarrow{p}))$ and $s_n(\sigma(\overrightarrow{p}))$.

Lemma 7. *We have $s_1(\sigma(\overrightarrow{p})) = \sqrt{\hat{l}}$, $s_n(\sigma(\overrightarrow{p})) = \sqrt{\frac{l}{rad(l)}}$ and $\|\sigma(\overrightarrow{p})_i\| = \sqrt{n}$ for all $i = 1, \cdots, n$.*

We also need the estimates of $s_1(\sigma(\overrightarrow{d}))$ and $s_n(\sigma(\overrightarrow{d}))$. Assume that $\sigma(\overrightarrow{p}) = T$, Lemma 7 shows that $s_1(T) = \sqrt{\hat{l}}$ and $s_n(T) = \sqrt{\frac{l}{rad(l)}}$. By the definitions of $\overrightarrow{d}$ and the dual ideal, an easy computation shows that $\sigma(\overrightarrow{d}) = (T^*)^{-1}$. Hence we have $s_n(\sigma(\overrightarrow{d})) = \frac{1}{\sqrt{\hat{l}}}$, $s_1(\sigma(\overrightarrow{d})) = \sqrt{\frac{rad(l)}{l}}$. Moreover, one can similarly deduce that $||\sigma(\overrightarrow{d})_i|| \leq \sqrt{\frac{rad(l)}{l}}$ for all $i = 1, 2, \cdots, n$. The following definition is also useful.

Definition 7. *Given a basis B of a fractional ideal J, for any $x \in J$ with $x = x_1 b_1 + \cdots + x_n b_n$, the B-coefficient embedding of x is defined as the vector $(x_1, \cdots, x_n)$ and the B-coefficient embedding norm of x is defined as $||x||_B^c = (\sum_{i=1}^n x_i^2)^{\frac{1}{2}}$.*

If we represent $x \in R$ (or $R^\vee$) with respect to the powerful basis (or decoding basis), we have

$$\sqrt{\frac{l}{rad(l)}} ||x||_{\sigma(\overrightarrow{p})}^c \leq ||\sigma(x)|| \leq \sqrt{\hat{l}} ||x||_{\sigma(\overrightarrow{p})}^c, \qquad for \ x \in R, \tag{1}$$

and

$$\frac{1}{\sqrt{\hat{l}}} ||x||_{\sigma(\overrightarrow{d})}^c \leq ||\sigma(x)|| \leq \sqrt{\frac{rad(l)}{l}} ||x||_{\sigma(\overrightarrow{d})}^c, \qquad for \ x \in R^\vee. \tag{2}$$

We will omit the subscript $\sigma(\overrightarrow{d})$ of $||\cdot||_{\sigma(\overrightarrow{d})}^c$ in the following applications. When we write $x \bmod qR^\vee$, we use the representative element of the coset $x + qR^\vee$ as $\sum_{i=1}^n x_i \overrightarrow{d}_i$ with $x_i \in [-\frac{q}{2}, \frac{q}{2})$. From now on, we only use the decoding basis of $R^\vee$ and the powerful basis of R.

The Ring-LWE distribution and Ring-LWE problem are defined as those in [22]. Define $K_{\mathbb{R}} = K \otimes_{\mathbb{Q}} \mathbb{R}$.

Definition 8. *For a distribution ψ over $K_{\mathbb{R}}$ and a secret $s \hookleftarrow \lfloor \psi \rceil_{R^\vee} \in R_q^\vee$, a sample from Ring-LWE distribution $A_{s,\psi}^\times$ over $R_q^\times \times R_q^\vee$ is generated by choosing $a \hookleftarrow U(R_q^\times)$, $e \hookleftarrow \lfloor \psi \rceil_{R^\vee}$ and outputting $(a, b = a \cdot s + e \bmod qR^\vee)$. The average-case decision version of the Ring-LWE problem, denoted by R-DLWE$_{q,\psi}^\times$, is to distinguish with non-negligible advantage between independent samples from $A_{s,\psi}^\times$, and the same number of uniformly random and independent samples from $R_q^\times \times R_q^\vee$.*

Theorem 2. *Let K be the l-th cyclotomic number field having dimension $n = \varphi(l)$ and $R = \mathcal{O}_K$ be its ring of integers. Let $\alpha = \alpha(n) > 0$, and let $q = q(n) \geq 2$, $q = 1 \bmod l$ be a $poly(n)$-bounded prime such that $\alpha q \geq \omega(\sqrt{\log n})$. Then there is a polynomial-time quantum reduction from $\tilde{O}(\frac{\sqrt{n}}{\alpha})$-approximate SIVP on ideal lattices in K to the problem of solving R-DLWE$_{q,\psi}^\times$ given only k samples, where ψ is the Gaussian distribution $D_{\xi \cdot q}$ with $\xi = \alpha \cdot (\frac{nk}{\log (nk)})^{\frac{1}{4}}$.*

3   Some New Results on q-Ary Lattices

In this section, we shall prove some useful results which will be used in Sect. 4.

3.1 q-Ary Lattices

We know that $R_q = \mathbb{Z}_q[x]/\Phi_l(x)$ and $\mathbb{Z}_q[x]$ is a principal ideal domain, hence R_q is a principal ideal ring. If we set $\phi_i = \omega^{l_i}$, where l_i is the i-th element in $\mathbb{Z}_l^\times$, then $\Phi_l(x) = \prod_{i=1}^n (x - \phi_i) = \prod_{i=1}^n (x - \phi_i^{-1}) \bmod q$. For any proper ideal $I \in R_q$, we can write $I = \langle f(x) \rangle R_q$, where $f(x)$ contains at least one monomials of $x - \phi_i$, i.e. $f(x) = \prod_{i \in S}(x - \phi_i)$ for some non-empty $S \subseteq \{1, 2, \cdots, n\}$. Since any monomials of the form $x - \alpha$ with $\alpha \neq \phi_i$ for $i = 1, 2, \cdots, n$ is an invertible element in R_q, any principal ideal of R_q is of the form described above. We will use I_S to represent the ideal $\prod_{i \in S}(x - \phi_i)R_q$ of R_q.

Let I be a proper ideal of R_q, there is a unique ideal J of R such that $qR \subseteq J \subseteq R$ and $I = J/qR$. In fact, if we set $I = f(x)R_q$, then $J = (f(x), q)R$. Considering the relation $qJ \subseteq qR \subseteq J \subseteq R$, we get $R^\vee \subseteq J^\vee \subseteq (qR)^\vee \subseteq (qJ)^\vee$, which implies $R^\vee \subseteq J^\vee \subseteq \frac{1}{q}(R)^\vee \subseteq \frac{1}{q}(J)^\vee$. Thus we get an R module inclusion relations

$$qR^\vee \subseteq qJ^\vee \subseteq R^\vee \subseteq J^\vee. \tag{3}$$

Moreover, $R^\vee/qJ^\vee$ is an R submodule of $J^\vee/qJ^\vee$. Let $\boldsymbol{a} \in (R_q)^m$, the definitions of the q-ary lattices are as followings:

$$\boldsymbol{a}^\perp(I) = \{(t_1, \cdots, t_m) \in J^m : \sum_{i=1}^m t_i a_i = 0 \bmod qR\},$$

$$L(\boldsymbol{a}, I) = \{(t_1, \cdots, t_m) \in (R^\vee)^m : \exists\, s \in R^\vee,\, \forall i,\, t_i = a_i \cdot s \bmod qJ^\vee\} = R^\vee \cdot \boldsymbol{a} + qJ^\vee.$$

Here, $R^\vee \cdot \boldsymbol{a} = \{t \cdot \boldsymbol{a} = (ta_1, \cdots, ta_m) : t \in R^\vee\}$. We also define $\boldsymbol{a}^\perp$ and $L(\boldsymbol{a})$ as $\boldsymbol{a}^\perp(R_q)$ and $L(\boldsymbol{a}, R_q)$. The dual $M^\vee$ of a lattice $M \subseteq K^m$ is defined as the set of all $\boldsymbol{x} \in K^m$ such that $\mathrm{Tr}(\boldsymbol{x} \cdot \boldsymbol{v}) := \sum_{j=1}^m \mathrm{Tr}(x_j \cdot v_j) \in \mathbb{Z}$ for all $\boldsymbol{v} \in M$. The following lemma shows the dual relations between $\boldsymbol{a}^\perp(I)$ and $L(\boldsymbol{a}, I)$.

Lemma 8. *Let $\boldsymbol{a}^\perp(I)$ and $L(\boldsymbol{a}, I)$ be defined above, then we have $\boldsymbol{a}^\perp(I) = q(L(\boldsymbol{a}, I))^\vee$ and $L(\boldsymbol{a}, I) = q(\boldsymbol{a}^\perp(I))^\vee$.*

Proof. We only need to prove $\boldsymbol{a}^\perp(I) = q(L(\boldsymbol{a}, I))^\vee$, since the other equality can be easily deduced by taking dual in both side of $\boldsymbol{a}^\perp(I) = q(L(\boldsymbol{a}, I))^\vee$.

We start with showing that $\boldsymbol{a}^\perp(I) \subseteq q(L(\boldsymbol{a}, I))^\vee$. For any $\boldsymbol{t} \in \boldsymbol{a}^\perp(I)$ and $\boldsymbol{z} \in L(\boldsymbol{a}, I)$, we only need to show $\sum_{i=1}^m \mathrm{Tr}(t_i \cdot z_i) = 0 \bmod q\mathbb{Z}$. Note that $z_i = a_i \cdot s + q \cdot z_i'$ for some $z_i' \in J^\vee$, we have

$$\sum_{i=1}^m \mathrm{Tr}(t_i \cdot z_i) = \mathrm{Tr}(s \cdot \sum_{i=1}^m t_i \cdot a_i) + q \cdot \sum_{i=1}^m \mathrm{Tr}(t_i \cdot z_i').$$

By the definition, $\sum_{i=1}^{m} t_i \cdot a_i = q \cdot r$ for some $r \in R$. Thus $\sum_{i=1}^{m} \mathrm{Tr}(t_i \cdot z_i) \in q\mathbb{Z}$.

To complete the proof, we will show $q(L(\boldsymbol{a}, I))^\vee \subseteq \boldsymbol{a}^\perp(I)$. For any $\boldsymbol{x} \in (L(\boldsymbol{a}, I))^\vee$, we need to show $q \cdot x_i \in J$ for all $i \in [m]$ and $\sum_{i=1}^{m} qx_i \cdot a_i \in qR$. Note that $q(J^\vee)^m \subseteq L(\boldsymbol{a}, I)$, we can take $\boldsymbol{v}^{(i)}$ be the vectors in $L(\boldsymbol{a}, I)$ such that the i-th coordinate is $q \cdot s'$ with $s' \in J^\vee$ and 0 elsewhere. We have $\mathrm{Tr}(\boldsymbol{x} \cdot \boldsymbol{v}^{(i)}) = \mathrm{Tr}(x_i \cdot q \cdot s') \in \mathbb{Z}$, hence $q \cdot x_i \in J$. Note that $\forall \boldsymbol{t} \in L(\boldsymbol{a}, I)$, $\sum_{i=1}^{m} \mathrm{Tr}(x_i \cdot t_i) \in \mathbb{Z}$. We write t_i as $a_i \cdot s + q \cdot t_i'$ with $t_i' \in J^\vee$, then

$$\sum_{i=1}^{m} \mathrm{Tr}(x_i \cdot t_i) = \mathrm{Tr}(s \cdot \sum_{i=1}^{m} a_i \cdot x_i) + \sum_{i=1}^{m} \mathrm{Tr}(qx_i \cdot t_i'),$$

the latter sum is in $\mathbb{Z}$, hence $\mathrm{Tr}(s \cdot \sum_{i=1}^{m} a_i \cdot x_i) \in \mathbb{Z}$ and we get $\sum_{i=1}^{m} a_i \cdot x_i \in R$. Therefore we have proved $\boldsymbol{a}^\perp(I) = q(L(\boldsymbol{a}, I))^\vee$. We finish the proof.

3.2   Lower Bound of λ_1^∞ in $L(a, I)$

In this section, we shall give an estimate of the lower bound of λ_1^∞ for $L(\boldsymbol{a}, I)$ with $\boldsymbol{a} \hookleftarrow U((R_q^\times)^m)$, where λ_1^∞ is the length of a shortest vector (corresponding to the l_∞ norm) in the lattice $L(\boldsymbol{a}, I)$. The proof mainly follows the thoughts of [29]. Let $I_S = \prod_{i \in S}(x - \phi_i)R_q \subseteq R_q$ and $J_S = (f_S(x), q)R \subseteq R$, where $f_S(x) = \prod_{i \in S}(x - \phi_i)$ for $S \subseteq \{1, 2, \cdots, n\}$. The factorization of ideal $(q)R$ is $\prod_{i=1}^{n} \mathfrak{q}_i$ with $\mathfrak{q}_i = (q, x - \phi_i)R$. Since R is a Dedekind domain, each $\mathfrak{q}_i$ is a maximal ideal, hence $\mathfrak{q}_i$ and $\mathfrak{q}_j$ is coprime for any $i \neq j \in [n]$, $\mathfrak{q}_i \cdot \mathfrak{q}_j = \mathfrak{q}_i \cap \mathfrak{q}_j = (q, (x - \phi_i)(x - \phi_j))R$. Therefore, $J_S = \prod_{i \in S} \mathfrak{q}_i$, $J_S^{-1} = \prod_{i \in S} \mathfrak{q}_i^{-1}$. Further, we have $J_S^\vee = \prod_{i \in S} \mathfrak{q}_i^{-1} R^\vee$.

Lemma 9. *For any $S \subseteq [n]$, $m \geq 2$ and $\varepsilon > 0$, we have $\lambda_1^\infty(L(\boldsymbol{a}, I_S)) \geq B$ with $B = \frac{q^\beta}{n}$, where $\beta = (1 - \frac{1}{m})(1 - \frac{|S|}{n}) - \varepsilon$, except with probability $p \leq 2^{(3m+1)n}q^{-\varepsilon mn}$ over the uniformly random choice of $\boldsymbol{a} \in (R_q^\times)^m$.*

Proof. Let p denote the probability, over the randomness of $\boldsymbol{a}$, that $L(\boldsymbol{a}, I_S)$ contains a non-zero vector $\boldsymbol{t}$ of infinity norm $< B = \frac{q^\beta}{n}$. Recall that, $\boldsymbol{t} \in L(\boldsymbol{a}, I_S)$ if and only if there is an $s \in R^\vee$ such that $t_i = a_i \cdot s \bmod qJ_S^\vee$ for all $i \in [m]$. Meanwhile, for any $s \in R^\vee$, all the elements of the coset $s + qJ_S^\vee$ satisfy the equation $t_i = a_i \cdot s \bmod qJ_S^\vee$ for the same t_i. We give an upper bound of p by the union bound, summing the probabilities $p(\boldsymbol{t}, s) = \mathrm{Pr}_{\boldsymbol{a}}[\, t_i = a_i \cdot s \bmod qJ_S^\vee, \forall i \in [m]]$ over all possible values of $\boldsymbol{t}$ of infinity norm $< B$ and $s \in R^\vee/(qJ_S^\vee)$. Since the $\{a_i\}_{i=1}^{m}$ are independent, we have $p(\boldsymbol{t}, s) = \prod_{i \leq m} p_i(t_i, s)$, where $p_i(t_i, s) = \mathrm{Pr}_{a_i}[t_i = a_i \cdot s \bmod qJ_S^\vee]$. So, we have

$$p \leq \sum_{\substack{\boldsymbol{t} \in (J_S^\vee)^m \\ \forall i,\, 0 < \|t_i\|_\infty < B}} \sum_{s \in R^\vee/qJ_S^\vee} \prod_{i=1}^{m} \mathrm{Pr}_{a_i}[t_i = a_i \cdot s \bmod qJ_S^\vee].$$

Note that $qJ_S^\vee = q \prod_{i \in S} \mathfrak{q}_i^{-1} R^\vee = q \cdot \prod_{i \in S} \mathfrak{q}_i^{-1} \cdot R \cdot R^\vee = \prod_{i \in S'} \mathfrak{q}_i \cdot R^\vee$, where $S' = [n] \setminus S$. We have an isomorphism between $J_S^\vee/qJ_S^\vee$ and $J_S^\vee/(\mathfrak{q}_{i_1} R^\vee) \oplus$

$\cdots \oplus J_S^\vee/(\mathfrak{q}_{i_{|S'|}} R^\vee)$, where $i_j \in S'$ for $j = 1, \cdots, |S'|$. Also we have $R^\vee/qJ_S^\vee \cong R^\vee/(\mathfrak{q}_{i_1} R^\vee) \oplus \cdots \oplus R^\vee/(\mathfrak{q}_{i_{|S'|}} R^\vee)$.

We claim that for the case $p_i(a_i, s) \neq 0$, there must be a set $S'' \subseteq S'$ such that $s, t_i \in \prod_{i \in S''} \mathfrak{q}_i R^\vee$ and $s, t_i \notin \mathfrak{q}_j R^\vee$ for all $j \in S' \setminus S''$. Otherwise, there are some $j \in S'$ such that either $s = 0 \bmod \mathfrak{q}_j R^\vee$ and $t_i \neq 0 \bmod \mathfrak{q}_j R^\vee$, or $s \neq 0 \bmod \mathfrak{q}_j R^\vee$ and $t_i = 0 \bmod \mathfrak{q}_j R^\vee$. In both cases, we have $p_i(a_i, s) = 0$, since $a_i \in R_q^\times$. Then, for $j \in S''$, we have $t_i = a_i \cdot s = 0 \bmod \mathfrak{q}_j R^\vee$, regardless of the value of $a_i \in R_q^\times$. For any $j \in S' \setminus S''$, we have $t_i = a_i \cdot s \neq 0 \bmod \mathfrak{q}_j R^\vee$, the value of a_i is unique, since $s \neq 0 \bmod \mathfrak{q}_j R^\vee$ and $a_i \in R_q^\times$. For $j \in [n] \setminus S'$, the value of a_i can be arbitrary. Hence, overall, if we set $|S''| = d$, we get that there are $(q-1)^{n+d-|S'|}$ different a_i in $R_q^\times$ satisfy $t_i = a_i \cdot s \bmod qJ_S^\vee$, i.e. $p_i(t_i, s) = (q-1)^{d-|S'|}$. Therefore, we can rewrite the sum's conditions by

$$p \leq \sum_{\substack{0 \leq d \leq |S'|}} \sum_{\substack{S'' \subseteq S' \\ |S''| = d \\ \mathfrak{h} := \prod_{i \in S''} \mathfrak{q}_i R^\vee}} \sum_{\substack{s \in R^\vee/(qJ_S^\vee) \\ s \in \mathfrak{h}}} \sum_{\substack{t \in (J_S^\vee)^m \\ \forall i, \, 0 < \|t_i\|_\infty < B \\ t_i \in \mathfrak{h}}} \prod_{i=1}^m (q-1)^{d-|S'|}.$$

Set $\mathfrak{h} = \prod_{i \in S''} \mathfrak{q}_i R^\vee$, where $S'' \subseteq S'$ and $|S''| = d$. Let $N(B, d)$ denote the number of $t \in J_S^\vee$ such that $\|t\|_\infty < B$ and $t \in \mathfrak{h}$. We consider two cases for $N(B, d)$ depending on the magnitudes of d.

Case 1: Suppose that $d \geq \beta \cdot n$. Since $t \in \mathfrak{h} = \prod_{i \in S''} \mathfrak{q}_i R^\vee$, and $\mathfrak{h}$ is a fractional ideal of K, we have $(t) = tR^\vee \subseteq \mathfrak{h}$ and (t) is a full-rank R-submodule of $\mathfrak{h}$. Hence,

$$|N(t)| = N((t)) \geq N(\mathfrak{h}) \geq N(\prod_{i \in S''} \mathfrak{q}_i \cdot R^\vee) = (\prod_{i \in S''} N(\mathfrak{q}_i)) N(R^\vee) = q^d \cdot |\Delta_K|^{-1}.$$

Note that $|\Delta_K| \leq n^n$, we have $|N(t)| \geq \frac{q^d}{n^n}$ and conclude that

$$\|t\|_\infty \geq \frac{1}{\sqrt{n}} \|t\| \geq |N^{\frac{1}{n}}(t)| \geq \frac{q^{\frac{d}{n}}}{n} \geq \frac{q^\beta}{n} = B. \tag{4}$$

Case 2: Suppose now that $d < \beta \cdot n$. Define $\mathfrak{B}(l, c) = \{x \in H : \|x - c\|_\infty < l\}$. Note that $\sigma(\mathfrak{h})$ is a lattice of H, we get $N(B, d)$ is at most the number of points of $\sigma(\mathfrak{h})$ in the region $\mathfrak{B}(B, 0)$. Let $\lambda = \frac{\lambda_1^\infty(\mathfrak{h})}{2}$, then for any two different elements v_1 and $v_2 \in \mathfrak{h}$, we have $\mathfrak{B}(\lambda, v_1) \cap \mathfrak{B}(\lambda, v_2) = \phi$. For any $v \in \mathfrak{B}(B, 0)$, we also have $\mathfrak{B}(\lambda, v) \subseteq \mathfrak{B}(B + \lambda, 0)$. Therefore,

$$N(B, d) \leq \frac{vol(\mathfrak{B}(B + \lambda, 0))}{vol(\mathfrak{B}(\lambda, 0))} = (\frac{B}{\lambda} + 1)^n \leq (2q^{\beta - \frac{d}{n}} + 1)^n \leq 2^{2n} q^{n\beta - d},$$

where we have used the fact that $\lambda_1^\infty(\mathfrak{h}) \geq \frac{q^{\frac{d}{n}}}{n}$ from (4).

We claim that the number of $s \in R^\vee/(qJ_S^\vee)$ and $s \in \mathfrak{h}$ is $q^{|S'|-d}$. In fact, if s satisfies the above conditions, $s \in \mathfrak{h}/(qJ_S^\vee)$. Using a kind of isomorphism relation

(Lemma 2.14 in [21]) which states that for any fractional ideals $\mathfrak{a}, \mathfrak{b}$ and integral ideal $\mathfrak{c}$ with $\mathfrak{b} \subseteq \mathfrak{a}$, $\mathfrak{ac}/\mathfrak{bc} \cong \mathfrak{a}/\mathfrak{b}$, we have

$$\mathfrak{h}/(qJ_S^\vee) = \prod_{i\in S''} \mathfrak{q}_i R^\vee / (\prod_{i\in S'} \mathfrak{q}_i R^\vee) \cong \prod_{i\in S''} \mathfrak{q}_i / (\prod_{i\in S'} \mathfrak{q}_i) \cong R/(\prod_{i\in (S'\backslash S'')} \mathfrak{q}_i).$$

Hence, we have $|\mathfrak{h}/(qJ_S^\vee)| = |R/(\prod_{i\in(S'\backslash S'')} \mathfrak{q}_i)| = q^{|S'|-d}$. Using the above $N(B,d)$-bounds and the fact that the number of subsets of S' of cardinality d is $\leq 2^d$, setting $\mathfrak{P} = \prod_{i=1}^m (q-1)^{d-|S'|}$, we can rewrite the inequality of p as

$$p \leq \left(\sum_{0\leq d < \beta\cdot n} + \sum_{\beta\cdot n \leq d \leq |S'|} \right) \sum_{\substack{S'' \subseteq S' \\ |S''| = d \\ \mathfrak{h} = \prod_{i\in S''} \mathfrak{q}_i R^\vee}} \sum_{\substack{s \in R^\vee/(qJ_S^\vee) \\ s\in\mathfrak{h}}} \sum_{\substack{t \in (J_S^\vee)^m \\ \forall i,\, 0 < ||t_i||_\infty < B \\ t_i \in \mathfrak{h}}} \mathfrak{P}$$

$$\leq \sum_{0\leq d < \beta\cdot n} \sum_{\substack{S'' \subseteq S' \\ |S''| = d \\ \mathfrak{h} = \prod_{i\in S''} \mathfrak{q}_i R^\vee}} \sum_{\substack{s \in R^\vee/(qJ_S^\vee) \\ s\in\mathfrak{h}}} \sum_{\substack{t \in (J_S^\vee)^m \\ \forall i,\, 0 < ||t_i||_\infty < B \\ t_i \in \mathfrak{h}}} \mathfrak{P}$$

$$\leq 2^{|S'|} \max_{d < \beta\cdot n} \frac{q^{|S'|-d} N^m(B,d)}{(q-1)^{m(|S'|-d)}}$$

$$\leq 2^{n(1+3m)} \cdot q^{-\varepsilon m n}.$$

We finish the proof.

Remark: The estimate of $N(B,d)$ in the case $d < \beta \cdot n$ is originally inspired by [32], it may be standard. This lemma and the following regularity theorem can be regarded as a special case of Lemma 5.2 and Theorem 5.3 in [28].

3.3 Improved Results on Regularity

In this subsection, we discuss the regularity results of any cyclotomic ring. The following result is a direct consequence of Lemmata 4, 5, 8 and 9. By Lemmas 9 and 8, we have $\lambda_1^\infty((\mathfrak{a}^\perp(I_S))^\vee) = \frac{1}{q}\lambda_1^\infty(L(\mathfrak{a}, I_S)) \geq \frac{1}{q} q^{\frac{|S|}{mn} - \frac{|S|}{n} - \frac{1}{m} - \varepsilon}$, except with a fraction of $2^{(3m+1)n} q^{-\varepsilon m n}$ of $\mathfrak{a} \in (R_q^\times)^m$ for $S \subseteq [n]$ and $m \geq 2$. Then Lemma 4 tells us that $\eta_\delta((\mathfrak{a}^\perp(I_S))^\vee) \leq n\sqrt{\frac{\ln(2mn(1+\frac{1}{\delta}))}{\pi}} \cdot q^{\frac{|S|}{n} + \frac{1}{m} - \frac{|S|}{mn} + \varepsilon}$ for any $\delta > 0$. Therefore, Lemma 5 gives us the following lemma.

Lemma 10. *Let $q = 1 \bmod l$ be a prime, $K = \mathbb{Q}(\zeta_l)$, $R = \mathcal{O}_K$, $m \geq 2$, $\delta \in (0, \frac{1}{2})$, $\varepsilon > 0$, $S \subseteq [n]$, $\mathbf{c} \in R^m$ and $\mathbf{t} \hookleftarrow D_{R^m, \sigma, \mathbf{c}}$, where $\sigma \geq n\sqrt{\frac{\ln(2mn(1+\frac{1}{\delta}))}{\pi}} \cdot q^{\frac{|S|}{n} + \frac{1}{m} - \frac{|S|}{mn} + \varepsilon}$. Then for all except a fraction of $2^{(3m+1)n} q^{-\varepsilon m n}$ of $\mathfrak{a} \in (R_q^\times)^m$, we have*

$$\Delta\left(\mathbf{t} \bmod \mathbf{a}^\perp(I_S); U(R^m/\mathbf{a}^\perp(I_S))\right) \leq 2\delta.$$

Let $\mathbb{D}_\chi$ be the distribution of such tuple $(a_1, \cdots, a_m, \sum_{i=1}^{m} t_i a_i) \in (R_q^\times)^m \times R_q$, where $a_i \hookleftarrow U(R_q^\times)$ are chosen independently and $t \hookleftarrow D_{R^m,\sigma}$. The regularity of the generalized knapsack function $(t_1, \cdots, t_m) \to \sum_{i=1}^{m} t_i a_i$ is the statistical distance between $\mathbb{D}_\chi$ and $U((R_q^\times)^m \times R_q)$. Note that for each $\boldsymbol{a} \hookleftarrow U((R_q^\times)^m)$, the map $\boldsymbol{t} \mapsto \sum_{i=1}^{m} a_i t_i$ induces an isomorphism from the quotient $R^m/\boldsymbol{a}^\perp$ to its range. The latter is R_q, thanks to the invertibility of a_i's. By taking $S = \phi$ and $c = 0$ in Lemma 10, we deduce the following result.

Theorem 3. Let $q = 1 \bmod l$ be a prime, $K = \mathbb{Q}(\zeta_l)$, $R = \mathcal{O}_K$, $m \geq 2$, $\delta \in (0, \frac{1}{2})$, $\varepsilon > 0$ and $a_i \hookleftarrow U(R_q^\times)$ for all $i \in [m]$. Assume $t \hookleftarrow D_{R^m,\sigma}$, where $\sigma \geq n\sqrt{\frac{\ln(2mn(1+\frac{1}{\delta}))}{\pi}} \cdot q^{\frac{1}{m}+\varepsilon}$. Then we have

$$\Delta\left((a_1, \cdots, a_m, \sum_{i=1}^{m} t_i a_i);\ U((R_q^\times)^m \times R_q)\right) \leq 2\delta + 2^{(3m+1)n} q^{-\varepsilon mn}.$$

4 Analysis of Key Generation Algorithm

With the results in Sect. 3, we can derive a key generation algorithm for NTRU-Encrypt as in [29]. Further, by choosing appropriate parameters, we can show that the key generation algorithm terminates in expected time and the public key distribution is very closed to the uniform distribution.

The key generation algorithm is as follows:

Input: $q \in \mathbb{Z}^+$, $p \in R_q^\times$, $\sigma \in \mathbb{R}^+$.

Output: A key pair $(sk, pk) \in R_q^\times \times R_q^\times$.

1. Sample f' from $D_{R,\sigma}$; let $f = p \cdot f' + 1$; if $(f \bmod qR) \notin R_q^\times$, resample.
2. Sample g from $D_{R,\sigma}$; if $(g \bmod qR) \notin R_q^\times$, resample.
3. Return secret key $sk = f$ and public key $pk = h = pg/f \in R_q^\times$.

Notice that for powerful basis $\overrightarrow{p}$ of R, we have $\|\overrightarrow{p}\| = \sqrt{n}$. Hence, as long as $\sigma \geq \sqrt{n} \cdot \sqrt{\log n}$, we can sample an element in polynomial time to obey the distribution $D_{R,\sigma}$ by using Theorem 1. The following lemma shows that the key generation algorithm can terminate with high probability by executing only several times. Proofs in this section are standard and are put in Appendix A.

Lemma 11. Let l be a positive integer, $n = \varphi(l)$ and q be a prime such that $q = 1 \bmod l$. Assume $\sigma \geq n \cdot \sqrt{\frac{\ln(2n(1+\frac{1}{\varepsilon}))}{\pi}} \cdot q^{\frac{1}{n}}$, for an arbitrary $\varepsilon \in (0, \frac{1}{2})$. Let $a \in R$ and $p \in R_q^\times$. Then

$$\Pr_{f' \hookleftarrow D_{R,\sigma}}[(p \cdot f' + a \bmod qR) \notin R_q^\times] \leq n(\frac{1}{q} + 2\varepsilon).$$

Next, we show that the generated secret key by the key generation algorithm is short. This lemma is very useful for us to analyze the decryption error in Sect. 5.

Lemma 12. *Let $n \geq 5$, $q \geq 8n$, $q = 1 \bmod l$ be a prime and $\sigma \geq \sqrt{\frac{2\ln(6n)}{\pi}}$. $n \cdot q^{\frac{1}{n}}$. Then with probability at least $1 - 2^{3-n}$, the secret key f, g satisfy $||f|| \leq 2\sqrt{n}\sigma||p||_{\infty}$ and $||g|| \leq \sqrt{n}\sigma$.*

The last lemma of this section estimates the statistic distance between the distribution of public key and the uniform distribution over $R_q^{\times}$. The proof is essentially the same as Theorem 3 in [29]. We denote by $D_{\sigma,z}^{\times}$ the discrete Gaussian $D_{R,\sigma}$ restricted to $R_q^{\times} + z$.

Lemma 13. *Let $\varepsilon > 0$, $n \geq 5$, $q \geq 8n$ and $\sigma \geq n^{\frac{3}{2}}\sqrt{\ln(8nq)} \cdot q^{\frac{1}{2}+2\varepsilon}$. Let $p \in R_q^{\times}$, $y_i \in R_q$ and $z_i = -y_i p^{-1} \bmod qR$ for $i \in \{1, 2\}$. Then*

$$\Delta \left[\frac{y_1 + p \cdot D_{\sigma,z_1}^{\times}}{y_2 + p \cdot D_{\sigma,z_2}^{\times}} \bmod qR, \ U(R_q^{\times}) \right] \leq \frac{2^{9n}}{q^{\lfloor \varepsilon n \rfloor}}.$$

5   NTRUEncrypt Scheme and Security Analysis

In this section, we give our modified NTRUEncrypt. Meanwhile, we shall analyze the decryption error and give an elementary reduction from R-DLWE$_{q,D_{q\xi}}^{\times}$ to the CPA-security of our scheme.

The plaintext space of our scheme is $\mathcal{P} = R^{\vee}/pR^{\vee}$ with $p \in R_q^{\times}$. Denote $\chi = \lfloor D_{\xi \cdot q} \rceil_{R^{\vee}}$ with $\xi = \alpha \cdot (\frac{nk}{\log(nk)})^{\frac{1}{4}}$, where $k = O(1)$ is a positive integer. We will use the decoding basis for element $x \in R \subseteq R^{\vee}$. One should note that $f = 1 \bmod pR$ implies $f = 1 \bmod pR^{\vee}$.

> **Key generation:** *Use the algorithm described in Section 4, return $sk = f$ $\in R_q^{\times}$ with $f = 1 \bmod pR^{\vee}$, and $pk = h = pg \cdot f^{-1} \in R_q^{\times}$.*
>
> **Encryption:** *Given message $m \in \mathcal{P}$, sample $s, e \hookleftarrow \chi$ and return $c = hs$ $+ pe + m \in R_q^{\vee}$.*
>
> **Decryption:** *Given ciphertext c and secret key f, compute $c_1 = fc$. Then return $m = (c_1 \bmod qR^{\vee}) \bmod pR^{\vee}$.*

We first give an accurate estimate of the infinite norm of elements sampled from the discretisation of a Gaussian distribution.

Lemma 14. *Assume that $\xi = \alpha \left(\frac{nk}{\log(nk)} \right)^{\frac{1}{4}}$, $\chi = \lfloor D_{\xi \cdot q} \rceil_{R^{\vee}}$, $\alpha \cdot q \geq \omega(\sqrt{\log n})$ and $k = O(1)$. Set $\delta = \omega(\sqrt{n \log n} \cdot \alpha^2 \cdot q^2)$ and B the decoding basis of $R^{\vee}$, then for any $t \in H$, we have $\Pr_{x \hookleftarrow \chi}(| < t, x > | > \delta ||t||^2) \leq n^{-\omega(\sqrt{n \log n}) \cdot ||t||^2}$.*

Proof. Note that a gaussian random variable $x \hookleftarrow D_{q \cdot \xi}$ has mean 0 and deviation $\frac{q \cdot \xi}{\sqrt{2\pi}}$, the discretisation $\lfloor x \rceil$ is a noncentral subgaussian random variable with noncentrality parameter 0 and deviation parameter $(\frac{q^2 \xi^2}{2\pi} + \frac{1}{4}s_1(B)^2)^{\frac{1}{2}}$, by Lemma 6. Therefore, by the Definition 5, we have

$$E(e^{<t,\lfloor x \rceil>}) \leq e^{\frac{1}{2} \left(\frac{q^2 \xi^2}{2\pi} + \frac{1}{4}s_1(B)^2 \right) \cdot ||t||^2}.$$

For any $x \hookleftarrow D_{q \cdot \xi}$, by taking the Chernoff bound, we get

$$\Pr(| < t, \lfloor x \rceil > | > \delta \cdot ||t||^2) = \Pr(e^{|<t, \lfloor x \rceil>|} > e^{\delta \cdot ||t||^2})$$
$$\leq 2 \cdot e^{\frac{1}{2} \cdot \left(\frac{q^2 \xi^2}{2\pi} + \frac{1}{4} s_1^2(B) \right) \cdot ||t||^2 - \delta \cdot ||t||^2}.$$

Now, we estimate the value of $\frac{1}{2} \cdot \left(\frac{q^2 \xi^2}{2\pi} + \frac{1}{4} s_1^2(B) \right) \cdot ||t||^2$. Since $s_1(B) = \sqrt{\frac{rad(l)}{l}} \leq 1$, we have $\frac{1}{2} \cdot \left(\frac{q^2 \xi^2}{2\pi} + \frac{1}{4} s_1^2(B) \right) \cdot ||t||^2 = \Omega(\alpha^2 \cdot q^2 \cdot \sqrt{n} \log^{-\frac{1}{2}} n \cdot ||t||^2)$. Therefore,

$$\Pr(| < t, \lfloor x \rceil > | > \delta \cdot ||t||^2) \leq n^{-\omega(\sqrt{n \log n}) \cdot ||t||^2}.$$

We finish the proof.

By using Lemma 14, we can get an estimate for $||x||_\infty$ with $x \hookleftarrow \chi = \lfloor D_{q \cdot \xi} \rceil$. Choosing $t = (\frac{1}{\sqrt{2}}, 0, \cdots, 0, \frac{1}{\sqrt{2}})$ and $t = (\frac{i}{\sqrt{2}}, 0, \cdots, 0, -\frac{i}{\sqrt{2}})$, where i is the imaginary number such that $i^2 = -1$, we get

$$\Pr_{x \hookleftarrow \chi}(|\mathrm{Re}(\sigma_1(x))| > \frac{1}{\sqrt{2}} \omega(\sqrt{n \log n} \cdot \alpha^2 \cdot q^2) \leq n^{-\omega(\sqrt{n \log n})}$$

and

$$\Pr_{x \hookleftarrow \chi}(|\mathrm{Im}(\sigma_1(x))| > \frac{1}{\sqrt{2}} \omega(\sqrt{n \log n} \cdot \alpha^2 \cdot q^2) \leq n^{-\omega(\sqrt{n \log n})}.$$

Hence, we have $\Pr_{x \hookleftarrow \chi}(|\sigma_1(x)| > \omega(\sqrt{n \log n} \alpha^2 q^2)) \leq 2n^{-\omega(\sqrt{n \log n})}$. Similarly, one can also prove that $\Pr_{x \hookleftarrow \chi}(|\sigma_k(x)| > \omega(\sqrt{n \log n} \alpha^2 q^2)) \leq 2n^{-\omega(\sqrt{n \log n})}$ for any $k = 1, 2 \cdots, \frac{n}{2}$. Therefore, we conclude that

$$\Pr_{x \hookleftarrow \chi}(||\sigma(x)||_\infty > \omega(\sqrt{n \log n} \cdot \alpha^2 \cdot q^2)) \leq n \cdot n^{-\omega(\sqrt{n \log n})} \leq n^{-\omega'(\sqrt{n \log n})}. \tag{5}$$

In order to show that the decryption algorithm succeeds in recovering the correct message with high probability, we need the parameters C_1 and C_2 such that $C_1 ||x||^c \leq ||x|| \leq C_2 ||x||^c$.

Lemma 15. *Let $n \geq 5$, $q \geq 8n$, $q = 1 \bmod l$, $\sigma \geq \sqrt{\frac{2 \ln (6n)}{\pi}} \cdot n \cdot q^{\frac{1}{n}}$, $C_1 = \sqrt{l}$ and $C_2 = \sqrt{\frac{rad(l)}{l}}$. If $\omega(n^{\frac{3}{2}} \sqrt{\log n \log \log n}) \cdot \alpha^2 \cdot q^2 \cdot \sigma \cdot ||p||_\infty^2 < \frac{q}{2}$, then with probability $1 - n^{-\omega(\sqrt{n \log n})}$, the decryption algorithm of NTRUEncrtpt recovers m.*

Proof. Notice that $f \cdot h \cdot s = p \cdot g \cdot s \bmod qR^\vee$, we have $fc = pgs + pfe + fm \bmod qR^\vee \in R^\vee$. If $||pgs + pfe + fm||_\infty^c < \frac{q}{2}$, then we have fc has the representation of the form $pgs + pfe + fm$ in $R_q^\vee$. Hence, we have $m = (fc \bmod qR^\vee) \bmod pR^\vee$. It thus suffices to give an upper bound on the probability that $||pgs + pfe + fm||_\infty^c \geq \frac{q}{2}$.

Note that $||fc||_\infty^c \leq ||fc||^c \leq C_1 ||fc|| = C_1 ||pgs + pfe + fm|| \leq C_1(||pgs|| + ||pfe|| + ||fm||)$. By the choice of σ and Lemma 12, with probability greater than

$1 - 2^{3-n}$, $||f|| \leq 2\sqrt{n}\sigma||p||_\infty$ and $||g|| \leq \sqrt{n}\sigma$. Hence, combining with (5), we get

$$||pfe|| + ||pgs|| \leq 2\sqrt{n}\sigma||p||_\infty^2 \cdot ||e||_\infty + \sqrt{n}\sigma||p||_\infty \cdot ||s||_\infty$$
$$\leq \omega(n\sqrt{\log n} \cdot \alpha^2 \cdot q^2)\sigma||p||_\infty^2$$

with probability $1-n^{-\omega(\sqrt{n\log n})}$. Since $m \in R^\vee/(pR^\vee) \subseteq K$, by reducing modulo the $p\sigma(\overrightarrow{d})_i$'s, we can write m into $\sum_{i=1}^n \varepsilon_i p\sigma(\overrightarrow{d})_i$ with $\varepsilon_i \in (-\frac{1}{2}, \frac{1}{2}]$. We have

$$||m|| = ||\sum_{i=1}^n \varepsilon_i p\sigma(\overrightarrow{d})_i|| \leq ||p||_\infty||\sum_{i=1}^n \varepsilon_i\sigma(\overrightarrow{d})_i|| \leq \frac{\sqrt{n}}{2}||p||_\infty C_2,$$

where we have used that

$$||\sum_{i=1}^n \varepsilon_i\sigma(\overrightarrow{d}_i)|| \leq C_2 \cdot ||\sum_{i=1}^n \varepsilon_i\sigma(\overrightarrow{d}_i)||^c \leq C_2 \cdot \frac{\sqrt{n}}{2}.$$

So, we have $||fm|| \leq ||f|| \cdot ||m|| \leq n\sigma||p||_\infty^2 C_2$ with probability $\geq 1 - 2^{3-n}$. Therefore, putting these results together, we have

$$||fc||_\infty^c \leq C_1(\omega(n\sqrt{\log n} \cdot \alpha^2 \cdot q^2) \cdot \sigma \cdot ||p||_\infty^2 + n \cdot \sigma \cdot ||p||_\infty^2 \cdot C_2)$$
$$\leq \omega(n^{\frac{3}{2}}\sqrt{\log n \log\log n} \cdot \alpha^2 \cdot q^2) \cdot \sigma \cdot ||p||_\infty^2$$

with probability $1 - n^{-\omega(\sqrt{sn\log n})}$, where we have used the fact that $C_2 \leq 1$ and $C_1 = O(\sqrt{n\log\log n})$. We conclude the results we need.

Remark 1. We remark that we can put all computations in an integral ideal $I = \hat{l} \cdot R^\vee \subseteq R$ by multiplying an integer $\hat{l}$(in this case, the corresponding q is $\hat{l}$ times bigger than the q in Lemma 15). We use symbol $\hat{a}$ to represent the corresponding element of $a \in R^\vee$, i.e. $\hat{a} = \hat{l} \cdot a$. Note that $f = 1 \mod pR^\vee$, we have $\hat{l} \cdot f = \hat{l} \mod pI$. Therefore, $\hat{m} = \hat{l}^{-1}(\hat{l}((f \cdot \hat{c} \mod qI) \mod pI) \mod pI)$ with $\hat{m} \in I/(pI)$ and $gcd(p, \hat{l}) = 1$. Since the corresponding 'decoding basis' of I is connected with the usual power basis of R by an invertible matrix $M \in \mathbb{Z}^{n \times n}$, this modification may enjoy the high computation speed over polynomial rings.

Remark 2. By using the recent hardness results about primal-Ring-LWE (i.e. the secret $s \hookleftarrow U(R_q)$) proved in [28], we can directly design NTRUEncrypt in R. If we set $\mathcal{P} = R/pR$ and choose $s, e \hookleftarrow \lfloor D_{\xi \cdot q} \rceil_R$ (techniques used in [22, Lemma 2.23] can be modified to R), then the same encryption and decryption process also work. In this case, we use the powerful basis of R. Correspondingly, if we set $\alpha \cdot q = \omega(\sqrt{\log n})$, magnitudes of $||s||_\infty$ and $||e||_\infty$ are $\tilde{O}(n)$. Then, we can estimate that $q = \tilde{O}(\sqrt{\frac{rad(l)}{l}} \cdot n^{\frac{3}{2}} \cdot \sigma)$ is sufficient to decrypt correctly with probability greater than $1 - n^{-\tilde{O}(n)}$. Therefore, we have $q = \tilde{O}(n^6 \cdot \sqrt{\frac{rad(l)}{l}}) \in (\tilde{O}(n^5), \tilde{O}(n^6)]$. But, the reduction parameter $\gamma \leq \tilde{O}(n^{12.5})$, due to the reduction loss of primal-Ring-LWE problem, see [28]. In this situation, we can have high efficiency with weaker hardness guarantee, so, an assessment from the view of actual attacks need be done as in [8].

Remark 3. The reason why we constrain our NTRUEncrypt schemes in cyclotomic fields is that we want to use the decoding basis of $R^\vee$. If a general number field has such a good basis, we can also design NTRUEncrypt over general fields by using our techniques, together with the hardness results showed in [27]. More details are discussed in [30].

Remark 4. By using similar techniques, we can also give a module version of NTRUEncrypt. The security reduction of this modified version of NTRUEncrypt can be reduced to the corresponding Module-LWE problems. More details are put in Appendix B.

The security of our scheme follows by an elementary reduction from R-$\text{DLWE}^\times_{q,D_{q\xi}}$, exploiting the uniformity of the public key in $R_q^\times$ and the invertibility of $p \in R_q$. We put the proof in Appendix C.

Lemma 16. *Let $n \geq 5$, $q \geq 8n$, $q = 1 \bmod l$, $\sigma \geq \sqrt{\ln(8nq)} \cdot n^{\frac{3}{2}} \cdot q^{\frac{1}{2}+\varepsilon}$, $\delta > 0$ and $\varepsilon \in (0, \frac{1}{2})$. If there exists an IND-CPA attack against NTRUEncrypt that runs in time T with advantage δ, then there exists an algorithm solving* R-$\text{DLWE}^\times$ *with parameters q and $q\xi$ that runs in time $T' = T + O(n)$ with advantage $\delta' = \delta - q^{-\Omega(n)}$.*

In a summary, we have the following result.

Theorem 4. *Let l be a positive integer, $n = \varphi(l) \geq 5$, $q \geq 8n$, $q = 1 \bmod l$ be a prime of size $poly(n)$ and $K = \mathbb{Q}(\zeta_l)$. Assume that $\alpha \in (0,1)$ satisfies $\alpha q \geq \omega(\sqrt{\log n})$. Let $\xi = \alpha \cdot (\frac{nk}{\log(nk)})^{\frac{1}{4}}$ with $k = O(1)$, $\varepsilon \in (0, \frac{1}{2})$ and $p \in R_q^\times$. Moreover, let $\sigma \geq n^{\frac{3}{2}} \cdot \sqrt{\ln(8nq)} \cdot q^{\frac{1}{2}+\varepsilon}$ and $\omega(n^{\frac{3}{2}}\sqrt{\log n \log\log n} \cdot \alpha^2 \cdot q^2) \cdot \sigma \cdot ||p||_\infty^2 < q$. Then if there exists an IND-CPA attack against NTRUEncrypt(n,q,p,σ,ξ) that runs in time $poly(n)$ with advantage $\frac{1}{poly(n)}$, there exists a $poly(n)$-time algorithm solving Ideal-SIVP$_\gamma$ on any ideal lattice of K with $\gamma = \tilde{O}(\frac{\sqrt{n}}{\alpha})$. Moreover, the decryption algorithm succeeds in regaining the correct message with probability $1 - n^{-\omega(\sqrt{n\log n})}$ over the choice of the encryption randomness.*

To sum up, though the magnitude of q is little far away from practicality, the biggest advantage of our scheme is that it is less dependent on the choice of p and is not limited by the cyclotomic fields it bases on. Hence, our schemes provide more flexibility for the choices of plaintext spaces and get rid of the dependence of the cyclotomic fields, so that our NTRUEncrypt has potentialities to send more encrypted bits in each encrypt process with higher efficiency and stronger security. Further, our decryption algorithm succeeds in recovering the correct message with a probability of $1 - n^{-\omega(\sqrt{n\log n})}$, while the previous works were $1 - n^{-\omega(1)}$. Therefore, we believe, our scheme may have more advantages in theory.

Acknowledgement. We would like to express our gratitude to Bin Guan and Yang Yu for helpful discussions. We also thank the anonymous SAC'18 reviewers for their valuable comments and suggestions. The authors are supported by National Cryptography Development Fund (Grant No. MMJJ20180210), NSFC Grant 61832012, NSFC Grant 61672019 and the Fundamental Research Funds of Shandong University (Grant No. 2016JC029).

A   Missing Proofs in Sect. 4

Proof of Lemma 11: Thanks to the Chinese Remainder Theorem, we only need to bound the probability that $p \cdot f' + a \in \mathfrak{q}_i$ is no more than $\frac{1}{q} + 2\varepsilon$, for any $i \leq n$. By Lemma 1 and the properties of cyclotomic ring, we have $\lambda_1(\mathfrak{q}_i) = \lambda_n(\mathfrak{q}_i) \leq \sqrt{n} N(\mathfrak{q}_i)^{\frac{1}{n}} (\sqrt{|\Delta_K|})^{\frac{1}{n}} \leq n q^{\frac{1}{n}}$. By Lemmas 2 and 5, we know that $f' \bmod \mathfrak{q}_i$ is within distance 2ε to uniformity on $R/\mathfrak{q}_i$, so we have $f' = -a/p \bmod \mathfrak{q}_i$ with probability less than $\frac{1}{q} + 2\varepsilon$ as we need.

Proof of Lemma 12: Set $\varepsilon = \frac{1}{3n-1}$. Note that $\lambda_n(R) = \lambda_1(R) \leq \sqrt{n} \cdot (\sqrt{|\Delta_K|})^{\frac{1}{n}} \leq n$. By Lemma 2, we have $\eta_\varepsilon(R) \leq \sqrt{\frac{2 \ln (6n)}{\pi}} \cdot n$. Hence, $\Pr_{x \hookleftarrow D_{R,\sigma,c}}(\|x\| \geq \sqrt{n}\sigma) \leq \frac{3n}{3n-2} 2^{-n}$. Meanwhile, σ satisfies the condition in Lemma 11, so we get

$$\begin{aligned} \Pr_{g \hookleftarrow D_{R,\sigma}}(\|g\| \geq \sqrt{n}\sigma \mid g \in R_q^\times) &= \frac{\Pr_{g \hookleftarrow D_{R,\sigma}}(\|g\| \geq \sqrt{n}\sigma \text{ and } g \in R_q^\times)}{\Pr_{g \hookleftarrow D_{R,\sigma}}(g \in R_q^\times)} \\ &\leq \frac{\Pr_{g \hookleftarrow D_{R,\sigma}}(\|g\| \geq \sqrt{n}\sigma)}{\Pr_{g \hookleftarrow D_{R,\sigma}}(g \in R_q^\times)} \\ &\leq \frac{3n}{3n-2} \cdot 2^{-n} \cdot \frac{1}{1 - n(\frac{1}{q} + 2\varepsilon)} \leq 2^{3-n}. \end{aligned}$$

Therefore, we have $\|f'\|, \|g\| \leq \sqrt{n}\sigma$ with probability no less than $1 - 2^{3-n}$. Moreover we can estimate $\|f\| \leq 1 + \|p\|_\infty \cdot \|f'\| \leq 2\sqrt{n}\sigma \|p\|_\infty$.

Proof of Lemma 13: For $a \in R_q^\times$, we define $\Pr_a = \Pr_{f_1,f_2}[(y_1 + pf_1)/(y_2 + pf_2) = a]$, where $f_i \hookleftarrow D_{\sigma,z_i}^\times$. It is suffice to show that $|\Pr_a - (q-1)^{-n}| \leq 2^{2n+5} q^{-\lfloor \varepsilon n \rfloor} \cdot (q-1)^{-n} =: \varepsilon'$ except a fraction $\leq 2^{8n} q^{-2n\varepsilon}$ of $a \in R_q^\times$. Note that $a_1 f_1 + a_2 f_2 = a_1 z_1 + a_2 z_2$ is equivalent to $(y_1 + pf_1)/(y_2 + pf_2) = -a_2/a_1$ in $R_q^\times$ and $-a_2/a_1 \hookleftarrow U(R_q^\times)$ when $\mathbf{a} \hookleftarrow U(R_q^\times)^2$, we get $\Pr_a := \Pr_{f_1,f_2}[a_1 f_1 + a_2 f_2 = a_1 z_1 + a_2 z_2] = \Pr_{-a_2/a_1}$ for $\mathbf{a} \in (R_q^\times)^2$.

The set of solutions $(f_1, f_2) \in R^2$, $f_i \hookleftarrow D_{\sigma,z_i}^\times$, to the equation $a_1 f_1 + a_2 f_2 = a_1 z_1 + a_2 z_2 \bmod qR$ is $\mathbf{z} + \mathbf{a}^{\perp \times}$, where $\mathbf{z} = (z_1, z_2)$ and $\mathbf{a}^{\perp \times} = \mathbf{a}^\perp \cap (R_q^\times + qR)^2$. Therefore

$$\Pr_a = \frac{D_{R^2,\sigma}(\mathbf{z} + \mathbf{a}^{\perp \times})}{D_{R,\sigma}(z_1 + R_q^\times + qR) \cdot D_{R,\sigma}(z_2 + R_q^\times + qR)}.$$

Note that $\mathbf{a} \in (R_q^\times)^2$, we know for any $\mathbf{t} \in \mathbf{a}^\perp$, $t_2 = -t_1 \frac{a_1}{a_2}$, so t_1 and t_2 are in the same ideal I of R_q. It follows that $\mathbf{a}^{\perp \times} = \mathbf{a}^\perp \setminus (\cup_{I \subsetneq R_q} \mathbf{a}^\perp(I)) =$

$a^\perp \setminus (\cup_{S \subseteq [n], S \neq \phi} a^\perp(I_S))$. Similarly, we have $R_q^\times + qR = R \setminus (\cup_{S \subseteq [n], S \neq \phi}(I_S + qR))$. Using the inclusion-exclusion principal, we get

$$D_{R^2, \sigma}(z + a^{\perp \times}) = \sum_{S \subseteq [n]} (-1)^{|S|} \cdot D_{R^2, \sigma}(z + a^\perp(I_S)), \qquad (6)$$

$$D_{R, \sigma}(z_i + R_q^\times + qR) = \sum_{S \subseteq [n]} (-1)^{|S|} \cdot D_{R, \sigma}(z_i + I_S + qR), \quad \forall\, i \in \{1, 2\}. \quad (7)$$

In the rest of the proof, we show that, except for a fraction $\leq 2^{8n} q^{-2n\varepsilon}$ of $a \in (R_q^\times)^2$:

$$D_{R^2, \sigma}(z + a^{\perp \times}) = (1 + \delta_0) \cdot \frac{(q-1)^n}{q^{2n}},$$

$$D_{R, \sigma}(z_i + R_q^\times + qR) = (1 + \delta_i) \cdot \frac{(q-1)^n}{q^n}, \quad \forall\, i \in \{1, 2\},$$

where $|\delta_i| \leq 2^{2n+2} q^{-\lfloor \varepsilon n \rfloor}$ for $i \in \{0, 1, 2\}$. These imply that $|Pr_a - (q-1)^{-n}| \leq \varepsilon'$.

Handling (6): When $|S| \leq \varepsilon n$, we apply Lemma 10 with $m = 2$ and $\delta = q^{-n-\lfloor \varepsilon n \rfloor}$. Note that $qR^2 \subseteq a^\perp(I_S) \subseteq R^2$, we have $|R^2/a^\perp(I_S)| = \frac{|R^2/(qR^2)|}{|a^\perp(I_S)/(qR^2)|}$. Meanwhile, $|R^2/(qR^2)| = q^{2n}$ and $|a^\perp(I_S)/(qR^2)| = |I_S| = q^{n-|S|}$, since $|R_q|/|I_S| = |R_q/I_S| = q^{|S|}$. Therefore for all except a fraction $\leq \frac{2^{7n}}{q^{2n\varepsilon}}$ of $a \in (R_q^\times)^2$,

$$\left| D_{R^2, \sigma}(z + a^\perp(I_S)) - q^{-n-|S|} \right| = |D_{R^2, \sigma, -z}(a^\perp(I_S)) - q^{-n-|S|}| \leq 2\delta.$$

When $|S| > \varepsilon n$, we can choose $S' \subseteq S$ with $|S'| = \lfloor \varepsilon n \rfloor$. Then we have $a^\perp(I_S) \subseteq a^\perp(I_{S'})$ and hence $D_{R^2, \sigma, -z}(a^\perp(I_S)) \leq D_{R^2, \sigma, -z}(a^\perp(I_{S'}))$. Using the result proven above, we conclude that $D_{R^2, \sigma, -z}(a^\perp(I_S)) \leq 2\delta + q^{-n-\lfloor \varepsilon n \rfloor}$. Overall, we get

$$\left| D_{R^2, \sigma}(z + a^{\perp \times}) - \frac{(q-1)^n}{q^{2n}} \right| = \left| D_{R^2, \sigma}(z + a^{\perp \times}) - \sum_{k=0}^{n} (-1)^k \binom{n}{k} q^{-n-k} \right|$$

$$\leq 2^{n+1} \delta + 2 \sum_{k=\lceil \varepsilon n \rceil}^{n} \binom{n}{k} q^{-n-\lfloor \varepsilon n \rfloor}$$

$$\leq 2^{n+1} (\delta + q^{-n-\lfloor \varepsilon n \rfloor})$$

for all except a fraction $\leq \frac{2^{8n}}{q^{2n\varepsilon}}$ of $a \in (R_q^\times)^2$, since the are 2^n choices of S. The δ_0 satisfies $|\delta_0| \leq \frac{q^{2n}}{(q-1)^n} 2^{n+1} (\delta + q^{-n-\lfloor \varepsilon n \rfloor}) = (\frac{q}{q-1})^n \cdot 2^{n+2} \cdot q^{-\lfloor \varepsilon n \rfloor} \leq 2^{2n+2} q^{-\lfloor \varepsilon n \rfloor}$, as required.

Handling (7): Note that for any $S \in [n]$, $\det(I_S + qR) = |R/J_S| \cdot \sqrt{|\Delta_K|} = q^{|S|} \cdot \sqrt{|\Delta_K|}$, where J_S is the ideal of R such that $J_S/(qR) = I_S$. By Minkowski's Theorem, we have $\lambda_1(I_S + qR) = \lambda_n(I_S + qR) \leq n \cdot q^{\frac{|S|}{n}}$. Lemma 2 implies that

$\sigma > \eta_\delta(I_S + qR)$ for any $|S| \leq \frac{n}{2}$ with $\delta = q^{-\frac{n}{2}}$. Therefore, Lemma 5 shows that $|D_{R,\sigma,-z_i}(I_S + qR) - q^{-|S|}| \leq 2\delta$. For the case $|S| > \frac{n}{2}$, we can choose $S' \subseteq S$ with $|S'| \leq \frac{n}{2}$. Using the same argument above, we get $D_{R,\sigma,-z_i}(I'_S + qR) \leq D_{R,\sigma,-z_i}(I_S + qR) \leq 2\delta + q^{-\frac{n}{2}}$. Therefore,

$$\left| D_{R,\sigma}(z_i + R_q^\times + qR) - \frac{(q-1)^n}{q^n} \right| = \left| D_{R,\sigma}(z_i + R_q^\times + qR) - \sum_{k=0}^{n}(-1)^k \binom{n}{k} q^{-k} \right|$$

$$\leq 2^{n+1}\delta + 2\sum_{k=\frac{n}{2}}^{n} \binom{n}{k} q^{-k}$$

$$\leq 2^{n+1}(\delta + q^{-\frac{n}{2}}),$$

which leads to the desired bound on δ_i for $i = 1,\, 2$.

B   Module NTRUEncrypt

The hardness assumption of Ring-LWE may be possible weaker than the classic LWE: classic LWE is known to be as hard as the standard worst-case problems on Euclidean lattices, whereas Ring-LWE is only known to be as hard as their restrictions to special classes of ideal lattices which are a subset of Euclidean lattices. To 'overcome' this shortcoming, Langlois and Stehlé gave some worst-case to average-case reducitons for module lattices in 2015. In this section, we give a modified version of NTRUEncrypt over modules and a reduction from Module-LWE to the Module-NTRUEncrypt.

B.1   Basic Hard Problems

We first introduce some basic definitions and corresponding results about Module-LWE (MLWE). A subset $M \subseteq K^d$ is an R-module if it is closed under addition and under multiplication by elements of R. It is a finitely generated module if there exists a finite family $\{b_k\}$ of vectors in K^d such that $M = \sum_k R \cdot b_k$. When K is a cyclotomic field as we required, there exists a so-called pseudo-bases for M as stated in [19]: For every module M, there exist $I_{k1 \leq k \leq d}$ with I_k nonzero ideal of R and $\{b\}_{1 \leq k \leq d}$ linearly independent vectors of K^d such that $M = \sum_{1 \leq k \leq d} I_k \cdot b_k$. We call $[\{I_k\}, \{b_k\}]$ a pseudo-basis of M. We remark that we only deal with the full-rank modules, i.e. the number of ideals and vectors is equal to d.

The canonical embedding can be extend to K^d in the usual way. For any $x \in K^d$ with $x = (x_1, \cdots, x_d)$, we define the map σ by $\sigma(x) = (\sigma(x_1), \cdots, \sigma(x_n))$. Therefore, $\sigma(K^d) \subseteq H^d \cong \mathbb{R}^{nd}$ and any module of K^d is a full-rank lattice in H^d, we regard a module M as a module lattice.

The definitions of Module-LWE distribution and Module-LWE problem are as followings. We define $T_{R^\vee} = K \otimes_\mathbb{Q} \mathbb{R}/R^\vee$.

Definition 9. *Let ψ be some distribution on $T_{R^\vee}$ and $s \in (R_q^\vee)^d$ be a vector. The Module-LWE distribution $A_{s,\psi}^{(M)}$ is a distribution on $(R_q)^d \times T_{R^\vee}$ obtained by choosing a vector $a \in (R_q)^d$ uniformly at random, and $e \hookleftarrow \psi \in T_{R^\vee}$, and returning $(a, \frac{1}{q}\sum_{i=1}^d a_i \cdot s_i + e)$.*

Let $q \geq 2$ and Ψ be a family of distributions on $T_{R^\vee}$.

- The search version of the Module-LWE denoted by MSLWE$_{q,\Psi}$ is as follows: Let $s \in (R_q^\vee)^d$ be a secret and $\psi \in \Psi$; Given arbitrarily many samples from $A_{s,\psi}^{(M)}$, the goal is to find s.
- The decision version of the Module-LWE denoted by MDLWE$_{q,\Psi}$ is as follows: Let $s \in (R_q^\vee)^d$ be uniformly random and $\psi \in \Psi$; The goal is to distinguish between arbitrarily many independent samples from $A_{s,\psi}^{(M)}$ and the same number of independent samples from $U((R_q)^d \times T_{R^\vee})$.

In [19], an elementary reduction from Module-SIVP to Module-LWE is given.

Theorem 5. *Let $M \subseteq K^d$, $\varepsilon(N) = N^{-\omega(1)}$ with $N = nd$, $\alpha \in (0,1)$ and $q \geq 2$ be a prime, with $q \leq \text{poly}(N)$ and $q = 1 \bmod l$ such that $\alpha q \geq 2\sqrt{d} \cdot \omega(\sqrt{\log(n)})$. There is a quantum reduction from solving M-SIVP$_{\tilde{\omega}(\frac{\sqrt{N}d}{\alpha})}$ to solving MDLWE$_{q,D_\xi}$, given only k samples, in polynomial time with non-negligible advantage with $\xi = \alpha(\frac{nk}{\log(nk)})$.*

As in the case of Ring-LWE, we can also modify the distribution of $A_{s,\psi}^{(M)}$ to $(R_q^\times)^d \times R_q^\vee$. We scale the b component by a factor of q, so that it is an element of $K_\mathbb{R}/(qR^\vee)$. The corresponding error distribution is $D_{q\xi}$ with $\xi = \alpha \cdot (\frac{nk}{\log(nk)})$ and k the number of samples. Then we discretize the error, by taking $e \hookleftarrow \lfloor D_{q\xi} \rceil$. The decision version of MLWE becomes to distinguish between the modified distribution of $A_{s,\lfloor D_{q\xi}\rceil}^{(M)}$ and the uniform samples from $(R_q)^d \times R_q^\vee$. Notice that by using the same method proposed in [24, Lemma 2.24], we can change the secret s to obey the distribution of the errors, i.e. $s = (s_1, \cdots, s_d)$ with $s_i \hookleftarrow \lfloor D_{q\xi}\rceil$. At last, if we restrict $a \in (R_q^\times)^d$, the difficult of this problem does not decrease. We still use symbol $A_{s,D_{q\xi}}^{(M)}$ to denote the distribution of (a,b) obtained by choosing $a \hookleftarrow U((R_q^\times)^d)$, $s \hookleftarrow (\lfloor D_{q\xi}\rceil)^d$, $e \hookleftarrow \lfloor D_{q\xi}\rceil$ and $b = \sum_{i=1}^d a_i \cdot s_i + e$. We will use the symbol MDLWE$_{q,D_{q\xi}}^\times$ to denote the problem of distinguish the samples from $A_{s,D_{q\xi}}^{(M)}$ and $U((R_q^\times)^d \times R_q^\vee)$.

B.2   Modified Module NTRUEncrypt

In this subsection, we give a modified version of NTRUEncrypt whose security rely on the corresponding MDLWE problem. The key generation algorithm is as follows:

Input: $n, q \in \mathbb{Z}^+$, $p \in R_q^\times$, $\sigma \in \mathbb{R}^+$.

Output: *A key pair* $(sk, pk) \in R_q^\times \times (R_q^\times)^d$.

1. *Sample* f' *from* $D_{R,\sigma}$; *let* $f = p \cdot f' + 1$; *if* $(f \bmod qR) \notin R_q^\times$, *resample.*
2. *For* $i = 1, \cdots, d$, *sample* g_i *from* $D_{R,\sigma}$; *if* $(g_i \bmod qR) \notin R_q^\times$, *resample.*
3. *Return* $sk = f$ *and* $pk = (h_1, \cdots, h_d) = (pg_1/f, \cdots, pg_d/f) \in (R_q^\times)^d$.

By the results of Sect. 4, the statistical distance of the distribution of pk and $U((R_q^\times)^d)$ is less than $d \cdot \frac{9n}{q^{\lfloor \varepsilon n \rfloor}}$. Then algorithm can terminate in expected time and for all $i = 1, \cdots, d$, the l_2 norm of f_i and g_i is small with overwhelming probabilities.

We also set the plaintext message space $\mathcal{P} = R^\vee / pR^\vee$, denote $\chi = \lfloor D_{\xi \cdot q} \rceil_{R^\vee}$ with $\xi = \alpha \cdot (\frac{nk}{\log(nk)})^{\frac{1}{4}}$, where $k = O(1)$ is a positive integer and use decoding basis for element $x \in R \subseteq R^\vee$. The Module-NTRUEncrypt is as follows:

Key generation: *Use the algorithm describe above, return* $sk = f \in R_q^\times$ *with*
$$f = 1 \bmod pR^\vee, \text{ and } pk = \boldsymbol{h} \in (R_q^\times)^d.$$

Encryption: *Given message* $m \in \mathcal{P}$, *set* $\boldsymbol{s} \hookleftarrow \chi^d$, $e \hookleftarrow \chi$ *and return the cipher*
$$c = \sum_{i=1}^d h_i \cdot s_i + pe + m \in R_q^\vee.$$

Decryption: *Given ciphertext* c *and secret key* f, *compute* $c_1 = fc$. *Then*
$$return \ m = (c_1 \bmod qR^\vee) \bmod pR^\vee.$$

Notice that $c_1 = f \cdot c = p \sum_{i=1}^d g_i \cdot s_i + pfe + fm \bmod qR^\vee$, hence under the decoding basis, we have $||c_1||_\infty^c \leq \omega(d \cdot n^{\frac{3}{2}} \cdot \sqrt{\log n \log \log n} \cdot \alpha^2 \cdot q^2) \cdot \sigma \cdot ||p||_\infty^2$ with probability $1 - n^{-\omega(\sqrt{n \log n})}$. Therefore, we get the following lemma.

Lemma 17. *Let* $n \geq 5$, $q \geq 8n$, $q = 1 \bmod l$, $\sigma \geq \sqrt{\frac{2\ln(6n)}{\pi}} \cdot n \cdot q^{\frac{1}{n}}$, $C = \sqrt{l}$ *and* $C_2 = \sqrt{\frac{rad(l)}{l}}$. *If* $\omega(d \cdot n^{\frac{3}{2}} \sqrt{\log n \log \log n}) \cdot \alpha^2 \cdot q^2 \cdot \sigma \cdot ||p||_\infty^2 < q$, *then with probability* $1 - n^{-\omega(\sqrt{n \log n})}$, *the decryption algorithm of* Module-NTRUEncrtpt *recovers* $\hat{m}$.

The security of the scheme follows by an elementary reduction from $\text{MDLWE}_{q,D_{q\xi}}^\times$, exploiting the uniformity of the public key in $(R_q^\times)^d$ and the invertibility of $p \in R_q$. It's proof is similar to Lemma 16.

Lemma 18. *Let* $n \geq 5$, $q \geq 8n$, $q = 1 \bmod l$, $\sigma \geq \sqrt{\ln(8nq)} \cdot n^{\frac{3}{2}} \cdot q^{\frac{1}{2}+\varepsilon}$, $\delta > 0$ *and* $\varepsilon \in (0, \frac{1}{2})$. *If there exists an IND-CPA attack against* Module-NTRUEncrypt *that runs in time* T *with advantage* δ, *then there exists an algorithm solving* $\text{MDLWE}^\times$ *with parameters* q *and* $q\xi$ *that runs in time* $T' = T + O(n)$ *with advantage* $\delta' = \delta - q^{-\Omega(n)}$.

In a summary, we have the following results.

Theorem 6. *Let l be a positive integer, $n = \varphi(l) \geq 5$, $q \geq 8n$, $q = 1 \bmod l$ be a prime of size $poly(n)$, $K = \mathbb{Q}(\zeta_l)$, $R = \mathcal{O}_k$, $M \subseteq K^d$ with d a positive integer and $N = nd$. Assume that $\alpha \in (0,1)$ satisfies $\alpha q \geq 2\sqrt{d} \cdot \omega(\sqrt{\log n})$. Let $\xi = \alpha \cdot (\frac{nk}{\log(nk)})^{\frac{1}{4}}$ with $k = O(1)$, $\varepsilon \in (0, \frac{1}{2})$ and $p \in R_q^\times$. Moreover, let $\sigma \geq n^{\frac{3}{2}} \cdot \sqrt{\ln(8nq)} \cdot q^{\frac{1}{2}+\varepsilon}$ and $\omega(d \cdot n^{\frac{3}{2}}\sqrt{\log n \log\log n} \cdot \alpha^2 \cdot q^2) \cdot \sigma \cdot \|p\|_\infty^2 < q$. Then, if there exists an IND-CPA attack against* Module-NTRUEncrypt(n, q, p, σ, ξ) *that runs in time $poly(n)$ and has success probability $\frac{1}{2} + \frac{1}{poly(n)}$, there exists a $poly(n)$-time algorithm solving γ-Module-SIVP with $\gamma = \tilde{\omega}(\frac{\sqrt{N}d}{\alpha})$. Moreover, the decryption algorithm succeeds with probability $1 - n^{-\omega(\sqrt{n\log n})}$ over the choice of the encryption randomness.*

C Proof of Lemma 16

Let $\mathfrak{A}$ be the given *IND-CPA* attack algorithm, we construct an algorithm $\mathfrak{B}$ against R-DLWE$_{q, D_{q\xi}}^\times$ as follows. Given oracle $\mathfrak{O}$ that samples from either $U(R_q^\times \times R_q^\vee)$ or $A_{s, D_{q\xi}}^\times$ for some $s \hookleftarrow \chi$, $\mathfrak{B}$ calls $\mathfrak{O}$ to get a sample (h', c') from $R_q^\times \times R_q^\vee$, then runs $\mathfrak{A}$ with public key $h = p \cdot h' \in R_q^\times$. When $\mathfrak{A}$ outputs challenge messages $m_0, m_1 \in \mathcal{P}$, $\mathfrak{B}$ picks $b \hookleftarrow U(0,1)$, computes $c = p \cdot c' + m_b \in R_q^\vee$ and give it to $\mathfrak{A}$. When $\mathfrak{A}$ returns its guess b', $\mathfrak{B}$ returns 1 when $b' = b$ and 0 otherwise.

Note that h' is uniformly random in $R_q^\times$, so is the public key h given to $\mathfrak{A}$. Thus, it is within statistical distance $q^{-\Omega(n)}$ of the public key distribution in the attack. Moreover, when $c' = hs + e$ with $s, e \hookleftarrow \chi$, the ciphertext c given to $\mathfrak{A}$ has the right distribution as in the *IND-CPA* attack. Therefore, if $\mathfrak{O}$ outputs samples from $A_{s, D_{q\xi}}^\times$, $\mathfrak{A}$ succeeds and $\mathfrak{B}$ returns 1 with probability $\geq \frac{1}{2} + \delta - q^{-\Omega(n)}$.

Now, if $\mathfrak{O}$ outputs samples from $U(R_q^\times \times R_q^\vee)$, then c is uniformly random in R_q and independent of b. Hence, $\mathfrak{B}$ outputs 1 with probability $\frac{1}{2}$. The claimed advantage of $\mathfrak{B}$ follows.

References

1. Albrecht, M., Bai, S., Ducas, L.: A subfield lattice attack on overstretched NTRU assumptions. In: Robshaw, M., Katz, J. (eds.) CRYPTO 2016. LNCS, vol. 9814, pp. 153–178. Springer, Heidelberg (2016). https://doi.org/10.1007/978-3-662-53018-4_6
2. Bos, J.W., Lauter, K., Loftus, J., Naehrig, M.: Improved security for a ring-based fully homomorphic encryption scheme. In: Stam, M. (ed.) IMACC 2013. LNCS, vol. 8308, pp. 45–64. Springer, Heidelberg (2013). https://doi.org/10.1007/978-3-642-45239-0_4
3. Cabarcas, D., Weiden, P., Buchmann, J.: On the efficiency of provably secure NTRU. In: Mosca, M. (ed.) PQCrypto 2014. LNCS, vol. 8772, pp. 22–39. Springer, Cham (2014). https://doi.org/10.1007/978-3-319-11659-4_2
4. Cheon, J.H., Jeong, J., Lee, C.: An algorithm for NTRU problems and cryptanalysis of the GGH multilinear map without a low-level encoding of zero. LMS J. Comput. Math. 19(A), 255–266 (2016). https://doi.org/10.1112/S1461157016000371

5. Coppersmith, D., Shamir, A.: Lattice attacks on NTRU. In: Fumy, W. (ed.) EURO-CRYPT 1997. LNCS, vol. 1233, pp. 52–61. Springer, Heidelberg (1997). https://doi.org/10.1007/3-540-69053-0_5

6. Ducas, L., Durmus, A.: Ring-LWE in polynomial rings. In: Fischlin, M., Buchmann, J., Manulis, M. (eds.) PKC 2012. LNCS, vol. 7293, pp. 34–51. Springer, Heidelberg (2012). https://doi.org/10.1007/978-3-642-30057-8_3

7. Ducas, L., Durmus, A., Lepoint, T., Lyubashevsky, V.: Lattice signatures and bimodal Gaussians. In: Canetti, R., Garay, J.A. (eds.) CRYPTO 2013. LNCS, vol. 8042, pp. 40–56. Springer, Heidelberg (2013). https://doi.org/10.1007/978-3-642-40041-4_3

8. Ducas, L., Lyubashevsky, V., Prest, T.: Efficient identity-based encryption over NTRU lattices. In: Sarkar, P., Iwata, T. (eds.) ASIACRYPT 2014. LNCS, vol. 8874, pp. 22–41. Springer, Heidelberg (2014). https://doi.org/10.1007/978-3-662-45608-8_2

9. Ducas, L., Nguyen, P.Q.: Learning a zonotope and more: cryptanalysis of NTRUSign countermeasures. In: Wang, X., Sako, K. (eds.) ASIACRYPT 2012. LNCS, vol. 7658, pp. 433–450. Springer, Heidelberg (2012). https://doi.org/10.1007/978-3-642-34961-4_27

10. Gama, N., Nguyen, P.Q.: New chosen-ciphertext attacks on NTRU. In: Okamoto, T., Wang, X. (eds.) PKC 2007. LNCS, vol. 4450, pp. 89–106. Springer, Heidelberg (2007). https://doi.org/10.1007/978-3-540-71677-8_7

11. Garg, S., Gentry, C., Halevi, S.: Candidate multilinear maps from ideal lattices. In: Johansson, T., Nguyen, P.Q. (eds.) EUROCRYPT 2013. LNCS, vol. 7881, pp. 1–17. Springer, Heidelberg (2013). https://doi.org/10.1007/978-3-642-38348-9_1

12. Gentry, C.: Key recovery and message attacks on NTRU-composite. In: Pfitzmann, B. (ed.) EUROCRYPT 2001. LNCS, vol. 2045, pp. 182–194. Springer, Heidelberg (2001). https://doi.org/10.1007/3-540-44987-6_12

13. Gentry, C., Peikert, C., Vaikuntanathan, V.: Trapdoors for hard lattices and new cryptographic constructions. In: Proceedings of the Fortieth Annual ACM Symposium on Theory of Computing, STOC 2008, pp. 197–206, ACM, New York (2008). https://doi.org/10.1145/1374376.1374407

14. Hoffstein, J., Howgrave-Graham, N., Pipher, J., Silverman, J.H., Whyte, W.: NTRUSign: digital signatures using the NTRU lattice. In: Joye, M. (ed.) CT-RSA 2003. LNCS, vol. 2612, pp. 122–140. Springer, Heidelberg (2003). https://doi.org/10.1007/3-540-36563-X_9

15. Hoffstein, J., Pipher, J., Silverman, J.H.: NTRU: a ring-based public key cryptosystem. In: Buhler, J.P. (ed.) ANTS 1998. LNCS, vol. 1423, pp. 267–288. Springer, Heidelberg (1998). https://doi.org/10.1007/BFb0054868

16. Howgrave-Graham, N.: A hybrid lattice-reduction and meet-in-the-middle attack against NTRU. In: Menezes, A. (ed.) CRYPTO 2007. LNCS, vol. 4622, pp. 150–169. Springer, Heidelberg (2007). https://doi.org/10.1007/978-3-540-74143-5_9

17. Jaulmes, É., Joux, A.: A chosen-ciphertext attack against NTRU. In: Bellare, M. (ed.) CRYPTO 2000. LNCS, vol. 1880, pp. 20–35. Springer, Heidelberg (2000). https://doi.org/10.1007/3-540-44598-6_2

18. Kirchner, P., Fouque, P.-A.: Revisiting lattice attacks on overstretched NTRU parameters. In: Coron, J.-S., Nielsen, J.B. (eds.) EUROCRYPT 2017. LNCS, vol. 10210, pp. 3–26. Springer, Cham (2017). https://doi.org/10.1007/978-3-319-56620-7_1

19. Langlois, A., Stehlé, D.: Worst-case to average-case reductions for module lattices. Des. Codes Cryptogr. **75**(3), 565–599 (2015). https://doi.org/10.1007/s10623-014-9938-4

20. López-Alt, A., Tromer, E., Vaikuntanathan, V.: On-the-fly multiparty computation on the cloud via multikey fully homomorphic encryption. In: Proceedings of the Forty-Fourth Annual ACM Symposium on Theory of Computing, STOC 2012, pp. 1219–1234. ACM, New York (2012). https://doi.org/10.1145/2213977.2214086

21. Lyubashevsky, V., Peikert, C., Regev, O.: On ideal lattices and learning with errors over rings. In: Gilbert, H. (ed.) EUROCRYPT 2010. LNCS, vol. 6110, pp. 1–23. Springer, Heidelberg (2010). https://doi.org/10.1007/978-3-642-13190-5_1

22. Lyubashevsky, V., Peikert, C., Regev, O.: A toolkit for ring-LWE cryptography. In: Johansson, T., Nguyen, P.Q. (eds.) EUROCRYPT 2013. LNCS, vol. 7881, pp. 35–54. Springer, Heidelberg (2013). https://doi.org/10.1007/978-3-642-38348-9_3

23. Micciancio, D., Regev, O.: Worst-case to average-case reductions based on Gaussian measures. SIAM J. Comput. **37**(1), 267–302 (2007). https://doi.org/10.1137/S0097539705447360

24. Murphy, S., Player, R.: Noise distributions in homomorphic ring-LWE. Cryptology ePrint Archive, Report 2017/698 (2017). https://eprint.iacr.org/2017/698

25. Peikert, C.: Limits on the hardness of lattice problems in ℓ_p norms. In: Proceedings of the Twenty-Second Annual IEEE Conference on Computational Complexity, CCC 2007, pp. 333–346. IEEE Computer Society, Washington (2007). https://doi.org/10.1109/CCC.2007.12

26. Peikert, C.: An efficient and parallel Gaussian sampler for lattices. In: Rabin, T. (ed.) CRYPTO 2010. LNCS, vol. 6223, pp. 80–97. Springer, Heidelberg (2010). https://doi.org/10.1007/978-3-642-14623-7_5

27. Peikert, C., Regev, O., Stephens-Davidowitz, N.: Pseudorandomness of ring-LWE for any ring and modulus. In: Proceedings of the 49th Annual ACM SIGACT Symposium on Theory of Computing, STOC 2017, pp. 461–473. ACM, New York (2017). https://doi.org/10.1145/3055399.3055489

28. Rosca, M., Stehlé, D., Wallet, A.: On the ring-LWE and polynomial-LWE problems. Cryptology ePrint Archive, Report 2018/170 (2018). https://eprint.iacr.org/2018/170

29. Stehlé, D., Steinfeld, R.: Making NTRU as secure as worst-case problems over ideal lattices. In: Paterson, K.G. (ed.) EUROCRYPT 2011. LNCS, vol. 6632, pp. 27–47. Springer, Heidelberg (2011). https://doi.org/10.1007/978-3-642-20465-4_4

30. Wang, Y., Wang, M.: CRPSF and NTRU signatures over cyclotomic fields. Cryptology ePrint Archive, Report 2018/445 (2018). https://eprint.iacr.org/2018/445

31. Yu, Y., Xu, G., Wang, X.: Provably secure NTRU instances over prime cyclotomic rings. In: Fehr, S. (ed.) PKC 2017. LNCS, vol. 10174, pp. 409–434. Springer, Heidelberg (2017). https://doi.org/10.1007/978-3-662-54365-8_17

32. Yu, Y., Xu, G., Wang, X.: Provably secure NTRUEncrypt over more general cyclotomic rings. Cryptology ePrint Archive, Report 2017/304 (2017). https://refeprint.iacr.org/2017/304

Classical Public Key Cryptography

A Generalized Attack on Some Variants of the RSA Cryptosystem

Abderrahmane Nitaj[1(✉)], Yanbin Pan[2], and Joseph Tonien[3]

[1] Laboratoire de Mathématiques Nicolas Oresme, Université de Caen Normandie,
Caen, France
abderrahmane.nitaj@unicaen.fr
[2] Key Laboratory of Mathematics Mechanization, NCMIS,
Academy of Mathematics and Systems Science,
Chinese Academy of Sciences, Beijing, China
panyanbin@amss.ac.cn
[3] School of Computing and Information Technology, University of Wollongong,
Wollongong, Australia
joseph.tonien@uow.edu.au

Abstract. Let $N = pq$ be an RSA modulus with unknown factorization. The RSA cryptosystem can be attacked by using the key equation $ed - k(p-1)(q-1) = 1$. Similarly, some variants of RSA, such as RSA combined with singular elliptic curves, LUC and RSA with Gaussian primes can be attacked by using the key equation $ed - k\left(p^2 - 1\right)\left(q^2 - 1\right) = 1$. In this paper, we consider the more general equation $eu - \left(p^2 - 1\right)\left(q^2 - 1\right)v = w$ and present a new attack that finds the prime factors p and q in the case that u, v and w satisfy some specific conditions. The attack is based on Coppersmith's technique and improves the former attacks.

Keywords: RSA variants · Coppersmith's technique
Lattice reduction

1 Introduction

In 1978, Rivest, Shamir and Adleman [19] invented the RSA cryptosystem. Nowadays, it is the most widely used public key cryptosystem and serves for encryption and signature. The security of RSA is based on the difficulty of factoring specific large integers, called RSA moduli. An RSA modulus is in the form $N = pq$ where p and q are large prime numbers of the same size. The public exponent in RSA is an integer e satisfying $\gcd(e, (p-1)(q-1)) = 1$ while the private exponent is the integer d satisfying $ed \equiv 1 \pmod{(p-1)(q-1)}$. Since its invention, the RSA cryptosystem has been intensively studied for vulnerabilities. Many attacks on RSA exploit the RSA key equation $ed - k(p-1)(q-1) = 1$.

Y. Pan was supported by the NNSF of China (No. 61572490 and No. 11471314), and by the National Center for Mathematics and Interdisciplinary Sciences, CAS.

© Springer Nature Switzerland AG 2019
C. Cid and M. J. Jacobson, Jr. (Eds.): SAC 2018, LNCS 11349, pp. 421–433, 2019.
https://doi.org/10.1007/978-3-030-10970-7_19

A few attacks are based on the continued fraction algorithm such as Wiener's attack [22] and most of the attacks are based on lattice reduction techniques, introduced by Coppersmith [8] (see [2,3,10,15]). Combining both techniques, Blömer and May [1] presented an attack using the generalized key equation $ex + y = k(p - 1)(q - 1)$ for suitably small integers x, k and y.

Many variants of RSA have been proposed for improving the security or reducing the encryption or the decryption time (see [4,18,21]). The variants of RSA in [7,9,13,20] make use of a public exponent e and a private exponent d satisfying the equation

$$ed - k\left(p^2 - 1\right)\left(q^2 - 1\right) = 1. \tag{1}$$

In [5], Bunder et al. proposed an attack on these variants by using the continued fraction algorithm approach. Setting $e = N^\beta$, they showed that one can solve the Eq. 1 and find the prime factors p and q if $d = N^\delta$ and $\delta < \frac{1}{2}(3 - \beta)$. This was recently improved to $\delta < 2 - \sqrt{\beta}$ by Peng et al. [17] and by Zheng et al. [23] by using lattice reduction techniques and Coppersmith's method.

In this paper we consider the generalized equation

$$eu - \left(p^2 - 1\right)\left(q^2 - 1\right)v = w. \tag{2}$$

This equation can be transformed into the modular equation

$$v(p + q)^2 - (N + 1)^2 v - w \equiv 0 \pmod{e}. \tag{3}$$

We set $e = N^\beta$, $u = N^\delta$, $w = N^\gamma$ and using lattice reduction techniques and Coppersmith's method, we show that one can solve the Eq. (3) and find the prime factors p and q under the condition

$$\delta < \frac{7}{3} - \gamma - \frac{2}{3}\sqrt{1 + 3\beta - 3\gamma} - \varepsilon, \tag{4}$$

where ε is a small positive constant. Observe that the key Eq. (1) is a special case of the Eq. (3) where $w = 1$ and $\gamma = 0$. In this special case, the condition (4) becomes

$$\delta < \frac{7}{3} - \frac{2}{3}\sqrt{1 + 3\beta} - \varepsilon,$$

which is slightly worst than the condition $\delta < 2 - \sqrt{\beta}$ derived by the method of Peng et al. [17]. Apart this special case, our method supersedes the method of Peng et al. since their method works only for $w = 1$ while our method works for any $w = N^\gamma$ under the condition (4).

In [6], Bunder et al. studied the Eq. (2) using a combination of the continued fraction algorithm and Coppersmith's method. They showed that this equation can be solved whenever

$$uv < 2N - 4\sqrt{2}N^{\frac{3}{4}} \quad \text{and} \quad |w| < (p - q)N^{\frac{1}{4}}v.$$

The first condition implies the following one

$$\delta < \frac{3 - \beta}{2},$$

which is worst than our condition with $\gamma = 0$. As a consequence, our new method can be seen as an extension of the method of Bunder et al. [6].

The rest of the paper is organized as follows. In Sect. 2, we briefly describe the RSA variants that use exponents satisfying $ed \equiv 1 \pmod{(p^2 - 1)(q^2 - 1)}$. We also recall some facts on Coppersmith's method and lattice basis reduction. In Sect. 3, we present our attack. In Sect. 4, we present a comparison with existing attacks. We conclude the paper in Sect. 5.

2 Preliminaries

In this section, we briefly present some variants of the RSA cryptosystem that use the key equation $ed \equiv 1 \pmod{(p^2 - 1)(q^2 - 1)}$. We also present Coppersmith's method and lattice basis reduction.

2.1 LUC Cryptosystem

LUC cryptosystem, introduced by Smith and Lennon [20] in 1993 is based on Lucas functions. A related cryptosystem was propose by Castagnos [7] in 2007. Both cryptosystems use an RSA modulus $N = pq$, a public exponent e, and a private exponent satisfying a key equation $ed - k(p^2 - 1)(q^2 - 1) = 1$ which can be generalized by the equation $eu - (p^2 - 1)(q^2 - 1)v = w$.

2.2 RSA Type Schemes Based on Singular Cubic Curves

In 1995, Kuwakado, Koyama, and Tsuruoka [13] proposed a new cryptosystem based on the singular cubic with equation

$$y^2 = x^3 + bx^2 \mod N.$$

where $N = pq$ is an RSA modulus. In this cryptosystem, the encryption and the decryption keys satisfy an equation of the form $ed - k(p^2 - 1)(q^2 - 1) = 1$. A generalization of this equation is $eu - (p^2 - 1)(q^2 - 1)v = w$.

2.3 RSA with Gaussian Primes

A variant of RSA was introduced in 2002 by Elkamchouchi, Elshenawy and Shaban [9]. It is an extension of the RSA cryptosystem to the domain of Guassian integers. Gaussian integers are complex numbers of the form $z = a + bi$ where a and b are integers and $i^2 = -1$. The norm of a Gaussian integer is $|a + bi| = \sqrt{a^2 + b^2}$. In the RSA variant with Gaussian integers, the modulus is $N = PQ$, a product of two Gaussian integer primes P and Q and the public and private exponents satisfy $ed - k(|P|^2 - 1)(|Q|^2 - 1) = 1$. If $P = p$ and $Q = q$ are integer primes, then $ed - k(p^2 - 1)(q^2 - 1) = 1$. This can be generalized as $eu - (p^2 - 1)(q^2 - 1)v = w$.

2.4   Coppersmith's Method

In 1996, Coppersmith [8] proposed two methods related to finding small modular roots of univariate polynomials and small integer roots of bivariate polynomials. Since then, many techniques have been proposed for more variables (see [16]). Let

$$h(x, y, z) = \sum_{i,j,k} a_{i,j,k} x^i y^j z^k \in \mathbb{Z}[x, y, z],$$

be a polynomial with ω monomials. Its Euclidean norm is

$$\|h(x, y, z)\| = \sqrt{\sum_{i,j,k} a_{i,j,k}^2}.$$

The following result was proposed by Howgrave-Graham [11] to find the small modular roots of a polynomial.

Theorem 1. *Let e be a positive integer and $h(x, y, z) \in \mathbb{Z}[x, y, z]$ be a polynomial with at most ω monomials. Suppose that*

$$\|h(xX, yY, zZ)\| < \frac{e^m}{\sqrt{\omega}} \quad and \quad h(x_0, y_0, z_0) \equiv 0 \pmod{e^m},$$

where $|x_0| < X$, $|y_0| < Y$, $|z_0| < Z$. Then $h(x_0, y_0, z_0) = 0$ holds over the integers.

Coppersmith's method enables to find several polynomials that can be used in Howgrave-Graham's Theorem 1. This is possible by applying a lattice reduction technique such as the LLL algorithm [14] to a lattice with a given basis. In general, the LLL algorithm produces a reduced basis with relatively small norms such as in the following result (see [15]).

Theorem 2 (LLL). *Let $\mathcal{L}$ be a lattice spanned by a basis $(u_1, \ldots, u_\omega)$. Then the LLL algorithm outputs a new basis $(b_1, \ldots, b_\omega)$ satisfying*

$$\|b_1\| \leq \ldots \leq \|b_i\| \leq 2^{\frac{\omega(\omega-1)}{4(\omega+1-i)}} \det(\mathcal{L})^{\frac{1}{\omega+1-i}}, \quad i = 1, \ldots, \omega - 1,$$

where $\det(\mathcal{L})$ is the determinant of the lattice.

We assume that if $h_1, h_2, h_3 \in \mathbb{Z}[x, y, z]$ are three polynomials produced by Coppersmith's method, then the ideal generated by the polynomial equations $h_1(x, y, z) = 0$, $h_2(x, y, z) = 0$, $h_3(x, y, z) = 0$ has dimension zero. Then, a system of polynomials sharing the root can be solved by using Gröbner basis computation or resultant techniques.

3   The Attack

Theorem 3. *Let $N = pq$ be an RSA modulus and $e = N^\beta$ be a public exponent. Suppose that e satisfies the equation $eu - (p^2 - 1)(q^2 - 1)v = w$ with $u < N^\delta$ and $|w| < N^\gamma$. If*

$$\delta < \frac{7}{3} - \gamma - \frac{2}{3}\sqrt{1 + 3\beta - 3\gamma} - \varepsilon,$$

then one can factor N in polynomial time.

Proof. Let $N = pq$ be an RSA modulus. Let e be a public exponent satisfying $eu - (p^2 - 1)(q^2 - 1)v = w$ with $|w| < eu$. Suppose that $e = N^\beta$, $u < N^\delta$ and $|w| < N^\gamma$. Then

$$v = \frac{eu - w}{(p^2 - 1)(q^2 - 1)} < \frac{eu + |w|}{(p^2 - 1)(q^2 - 1)} < 2N^{\beta + \delta - 2},$$

where we used $(p^2 - 1)(q^2 - 1) \approx N^2$. It follows that the solution (u, v, w) of the equation $eu - (p^2 - 1)(q^2 - 1)v = w$ satisfies $u < N^\delta$, $v < 2N^{\beta + \delta - 2}$ and $|w| < N^\gamma$. We set

$$X = 2N^{\beta + \delta - 2}, Y = 3N^{\frac{1}{2}}, Z = N^\gamma. \tag{5}$$

This means that the solution (u, v, w) satisfies $u < N^\delta$, $v < X$ and $|w| < Z$. Moreover, since p and q are of the same size, then we have $p + q < 3N^{\frac{1}{2}} = Y$.

Transforming the equation $eu - (p^2 - 1)(q^2 - 1)v = w$, we get a modular one, namely $-v((N + 1)^2 - (p + q)^2) - w \equiv 0 \pmod{e}$. This can be rewritten as

$$v(p + q)^2 - (N + 1)^2 v - w \equiv 0 \pmod{e}.$$

Consider the polynomial

$$f(x, y, z) = xy^2 + a_1 x + z,$$

where $a_1 = -(N + 1)^2$. Then $(x, y, z) = (v, p + q, -w)$ is a solution of the polynomial modular equation $f(x, y, z) \equiv 0 \pmod{e}$. To find the small solutions of the equation $f(x, y, z) \equiv 0 \pmod{e}$, we apply Coppersmith's method combined with the extended strategy of Jochemsz and May [12] for finding small modular roots.

Let m and t be positive integers to be specified later. For $0 \le k \le m$, define the set

$$M_k = \bigcup_{0 \le j \le t} \{x^{i_1} y^{2i_2 + j} z^{i_3} \mid x^{i_1} y^{2i_2} z^{i_3} \text{ is a monomial of } f^m(x, y, z)$$

$$\text{and } \frac{x^{i_1} y^{2i_2} z^{i_3}}{(xy^2)^k} \text{ is a monomial of } f^{m-k}\}.$$

A straightforward calculation shows that $f^m(x, y, z)$ is

$$f^m(x, y, z) = \sum_{i_1=0}^{m} \sum_{i_2=0}^{i_1} \binom{m}{i_1}\binom{i_1}{i_2} a_1^{i_1-i_2} x^{i_1} y^{2i_2} z^{m-i_1}.$$

Hence, $x^{i_1} y^{2i_2} z^{i_3}$ is a monomial of $f^m(x, y, z)$ if

$$i_1 = 0, \ldots, m, \quad i_2 = 0, \ldots, i_1, \quad i_3 = m - i_1.$$

Similarly, $x^{i_1} y^{2i_2} z^{i_3}$ is a monomial of $f^{m-k}(x, y, z)$ if

$$i_1 = 0, \ldots, m - k, \quad i_2 = 0, \ldots, i_1, \quad i_3 = m - k - i_1.$$

From this, we deduce that for $0 \le k \le m$, if $x^{i_1} y^{2i_2} z^{i_3}$ is a monomial of $f^m(x, y, z)$, then $\frac{x^{i_1} y^{2i_2} z^{i_3}}{(xy^2)^k}$ is a monomial of $f^{m-k}(x, y, z)$ if

$$i_1 = k, \ldots, m, \quad i_2 = k, \ldots, i_1, \quad i_3 = m - i_1.$$

This leads to a characterization of the set M_k. For $0 \le k \le m$, we obtain

$$x^{i_1} y^{i_2} z^{i_3} \in M_k \text{ if } i_1 = k, \ldots, m, \ i_2 = 2k, \ldots, 2i_1 + t, \ i_3 = m - i_1.$$

Replacing k by $k + 1$, we get

$$x^{i_1} y^{i_2} z^{i_3} \in M_{k+1} \text{ if }$$
$$i_1 = k + 1, \ldots, m, \ i_2 = 2k + 2, \ldots, 2i_1 + t, \ i_3 = m - i_1.$$

For $0 \le k \le m$, define the polynomials

$$g_{k,i_1,i_2,i_3}(x, y, z) = \frac{x^{i_1} y^{i_2} z^{i_3}}{(xy^2)^k} f(x, y, y)^k e^{m-k} \quad \text{with} \quad x^{i_1} y^{i_2} z^{i_3} \in M_k \backslash M_{k+1}.$$

Since for $t \ge 1$, we have

$$x^{i_1} y^{i_2} z^{i_3} \in M_k \backslash M_{k+1}$$
$$\text{if } i_1 = k, \ldots, m, \ i_2 = 2k, 2k + 1, \ i_3 = m - i_1,$$
$$\text{or } i_1 = k, \ i_2 = 2k + 2, \ldots, 2i_1 + t, \ i_3 = m - i_1,$$

then the polynomials $g_{k,i_1,i_2,i_3}(x,y,z)$ reduce to the polynomials $G_{k,i_1,i_2,i_3}(x,y,z)$ and $H_{k,i_1,i_2,i_3}(x,y,z)$ where

$$G_{k,i_1,i_2,i_3}(x,y,z) = x^{i_1-k}y^{i_2-2k}z^{i_3}f(x,y,z)^k e^{m-k},$$
$$\text{for} \quad k = 0,\ldots m, \ i_1 = k,\ldots,m, \ i_2 = 2k, 2k+1, \ i_3 = m - i_1,$$
$$H_{k,i_1,i_2,i_3}(x,y,z) = y^{i_2-2k}z^{i_3}f(x,y,z)^k e^{m-k},$$
$$\text{for} \quad k = 0,\ldots m, \ i_1 = k, \ i_2 = 2k+2,\ldots,2i_1+t, \ i_3 = m - i_1.$$

Observe that for the target solution $(x,y,z) = (v, p+q, -w)$, the former polynomials satisfy

$$G_{k,i_1,i_2,i_3}(x,y,z) \equiv H_{k,i_1,i_2,i_3}(x,y,z) \equiv 0 \pmod{e^m}.$$

Let $\mathcal{L}$ denote the lattice spanned by the coefficient vectors of the polynomials $G_{k,i_1,i_2,i_3}(xX, yY, zZ)$ and $H_{k,i_1,i_2,i_3}(xX, yY, zZ)$ where X, Y and Z are positive integers to be defined later. The ordering of rows is such that any polynomial $G_{k,i_1,i_2,i_3}(xX, yY, zZ)$ is prior to any polynomial $H_{k,i_1,i_2,i_3}(xX, yY, zZ)$. Inside each type of polynomial, the ordering of the tuples (k, i_1, i_2, i_3) follows rule

$$(k, i_1, i_2, i_3) \prec (k', i_1', i_2', i_3') \text{ if } \begin{cases} k < k', \\ k = k', i_1 < i_1' \\ k = k', i_1 = i_1', i_2 < i_2', \\ k = k', i_1 = i_1', i_2 = i_2', i_3 < i_3'. \end{cases}$$

Similarly, the monomials $x^{i_1}y^{i_1}z^{i_1}$ in the columns are ordered following the rule

$$x^{i_1}y^{i_1}z^{i_1} \prec x^{i_1'}y^{i_2'}z^{i_3'} \text{ if } \begin{cases} i_1 < i_1' \\ i_1 = i_1', i_2 < i_2', \\ i_1 = i_1', i_2 = i_2', i_3 < i_3'. \end{cases}$$

This leads to a left triangular matrix. As an example, for $m = 2$ and $t = 3$, the matrix is presented in the following triangular table where the non-zero terms are denoted $*$.

Polynomial	z^2	yz^2	xz	xyz	x^2	x^2y	xy^2z	xy^3z	x^2y^2	x^2y^3	x^2y^4	x^2y^5	y^2z^2	y^3z^2	xy^4z	xy^5z	x^2y^6	x^2y^7
$G_{0,0,0,2}$	Z^2e^2	0	0	0	0	0	0	0	0	0	0	0	0	0	0	0	0	0
$G_{0,0,1,2}$	0	YZ^2e^2	0	0	0	0	0	0	0	0	0	0	0	0	0	0	0	0
$G_{0,1,0,1}$	0	0	XZe^2	0	0	0	0	0	0	0	0	0	0	0	0	0	0	0
$G_{0,1,1,1}$	0	0	0	$XYZe^2$	0	0	0	0	0	0	0	0	0	0	0	0	0	0
$G_{0,2,0,0}$	0	0	0	0	X^2e^2	0	0	0	0	0	0	0	0	0	0	0	0	0
$G_{0,2,1,0}$	*	0	0	0	0	X^2Ye^2	0	0	0	0	0	0	0	0	0	0	0	0
$G_{1,1,2,1}$	0	0	0	0	0	0	ZXY^2e	0	0	0	0	0	0	0	0	0	0	0
$G_{1,1,3,1}$	0	0	0	0	0	0	0	Y^3ZXe	0	0	0	0	0	0	0	0	0	0
$G_{1,2,2,0}$	0	0	*	0	*	0	*	0	X^2Y^2e	0	0	0	0	0	0	0	0	0
$G_{1,2,3,0}$	*	0	0	0	0	*	0	*	0	X^2Y^3e	0	0	0	0	0	0	0	0
$G_{2,2,4,0}$	0	0	0	0	0	0	*	0	*	0	X^2Y^4	0	0	0	0	0	0	0
$G_{2,2,5,0}$	*	0	0	0	*	0	0	*	0	*	0	X^2Y^5	0	0	0	0	0	0
$H_{0,0,2,2}$	0	0	0	0	0	0	0	0	0	0	0	0	$Y^2Z^2e^2$	0	0	0	0	0
$H_{0,0,3,2}$	0	0	0	0	0	0	0	0	0	0	0	0	*	$Y^3Z^2e^2$	0	0	0	0
$H_{1,1,4,1}$	0	0	0	0	0	0	*	0	0	0	0	0	0	*	Y^4ZXe	0	0	0
$H_{1,1,5,1}$	0	0	0	0	0	0	0	*	0	0	0	0	*	0	0	Y^5ZXe	0	0
$H_{2,2,6,0}$	0	0	0	0	0	0	0	0	0	*	0	0	0	*	0	0	Y^6X^2	0
$H_{2,2,7,0}$	0	0	0	0	0	0	0	0	0	0	*	0	0	0	*	0	0	Y^7X^2

Since the matrix is triangular, then only the diagonal terms contribute to the determinant. On the other hand, only e, X, Y and Z contribute to the determinant and we get the form

$$\det(\mathcal{L}) = e^{n_e} X^{n_X} Y^{n_Y} Z^{n_Z}. \tag{6}$$

Using the construction of the polynomials $G_{k,i_1,i_2,i_3}(x, y, z)$ and $H_{k,i_1,i_2,i_3}(x, y, z)$, the exponents n_e, n_X, n_Y, n_Z, and the dimension ω of the lattice are as follows

$$n_e = \sum_{k=0}^{m} \sum_{i_1=k}^{m} \sum_{i_2=2k}^{2k+1} \sum_{i_3=m-i_1}^{m-i_1} (m-k) + \sum_{k=0}^{m} \sum_{i_1=k}^{k} \sum_{i_2=2k+2}^{2i_1+t} \sum_{i_3=m-i_1}^{m-i_1} (m-k)$$
$$= \frac{1}{6}m(m+1)(4m+3t+5),$$

$$n_X = \sum_{k=0}^{m} \sum_{i_1=k}^{m} \sum_{i_2=2k}^{2k+1} \sum_{i_3=m-i_1}^{m-i_1} i_1 + \sum_{k=0}^{m} \sum_{i_1=k}^{k} \sum_{i_2=2k+2}^{2i_1+t} \sum_{i_3=m-i_1}^{m-i_1} i_1$$
$$= \frac{1}{6}m(m+1)(4m+3t+5),$$

$$n_Y = \sum_{k=0}^{m} \sum_{i_1=k}^{m} \sum_{i_2=2k}^{2k+1} \sum_{i_3=m-i_1}^{m-i_1} i_2 + \sum_{k=0}^{m} \sum_{i_1=k}^{k} \sum_{i_2=2k+2}^{2i_1+t} \sum_{i_3=m-i_1}^{m-i_1} i_2 \tag{7}$$
$$= \frac{1}{6}(m+1)\left(4m^2 + 6mt + 3t^2 + 5m + 3t\right),$$

$$n_Z = \sum_{k=0}^{m} \sum_{i_1=k}^{m} \sum_{i_2=2k}^{2k+1} \sum_{i_3=m-i_1}^{m-i_1} i_3 + \sum_{k=0}^{m} \sum_{i_1=k}^{k} \sum_{i_2=2k+2}^{2i_1+t} \sum_{i_3=m-i_1}^{m-i_1} i_3$$
$$= \frac{1}{6}m(m+1)(2m+3t+1).$$

$$\omega = \sum_{k=0}^{m} \sum_{i_1=k}^{m} \sum_{i_2=2k}^{2k+1} \sum_{i_3=m-i_1}^{m-i_1} 1 + \sum_{k=0}^{m} \sum_{i_1=k}^{k} \sum_{i_2=2k+2}^{2i_1+t} \sum_{i_3=m-i_1}^{m-i_1} 1$$
$$= (m+1)(m+t+1).$$

For $t = \tau m$ and sufficiently large m, we can approximate the exponents n_e, n_X, n_Y, n_Z by their leading term and get

$$n_e = \frac{1}{6}(3\tau + 4)m^3 + o(m^3),$$

$$n_X = \frac{1}{6}(3\tau + 4)m^3 + o(m^3),$$

$$n_Y = \frac{1}{6}(3\tau^2 + 6\tau + 4)m^3 + o(m^3), \tag{8}$$

$$n_Z = \frac{1}{6}(3\tau + 2)m^3 + o(m^3),$$

$$\omega = (\tau + 1)m^2 + o(m^2).$$

Applying the LLL algorithm to the lattice $\mathcal{L}$, we get a reduced basis where the three first vectors $h_i(Xx, Yy, Zz)$, $i = 1, 2, 3$ satisfy the conditions $\|h_1(Xx, Yy, Zz)\| \leq \|h_2(Xx, Yy, Zz)\| \leq \|h_3(Xx, Yy, Zz)\|$, and

$$\|h_3(Xx, Yy, Zz)\| \leq 2^{\frac{\omega(\omega-1)}{4(\omega-2)}} \det(\mathcal{L})^{\frac{1}{\omega-2}}.$$

For comparison, Theorem 1 can be applied if

$$\|h_3(Xx, Yy, Zz)\| < \frac{e^m}{\sqrt{\omega}}.$$

To this end, we set

$$2^{\frac{\omega(\omega-1)}{4(\omega-2)}} \det(\mathcal{L})^{\frac{1}{\omega-2}} < \frac{e^m}{\sqrt{\omega}},$$

or equivalently

$$\det(\mathcal{L}) < \frac{2^{-\frac{\omega(\omega-1)}{4}}}{(\sqrt{\omega})^{\omega-2}} e^{m(\omega-2)}.$$

Hence, using (6), we get

$$e^{n_e - m\omega} X^{n_X} Y^{n_Y} Z^{n_Z} < \frac{2^{-\frac{\omega(\omega-1)}{4}}}{(\sqrt{\omega})^{\omega-2}} e^{-2m}, \tag{9}$$

where the right side term is a small constant depending only on e and m. Plugging the values of n_e, n_X, n_Y, n_Z and ω from (8) as well as the values $e = N^\beta$, $X = 2N^{\beta+\delta-2}$, $Y = 3N^{\frac{1}{2}}$, $Z = N^\gamma$ in each term of (9), we get

$$e^{n_e - m\omega} = N^{\left(-\frac{1}{2}\tau - \frac{1}{3}\right)\beta m^3 + o(m^3)},$$

$$X^{n_X} = N^{\left(\frac{1}{2}\tau + \frac{2}{3}\right)(\beta+\delta-2)m^3 + o(m^3)} \cdot 2^{\left(\frac{1}{2}\tau + \frac{2}{3}\right)m^3 + o(m^3)}$$

$$= N^{\left(\frac{1}{2}\tau + \frac{2}{3}\right)(\beta+\delta-2)m^3 + o(m^3) + \varepsilon_1},$$

$$Y^{n_Y} = N^{\frac{1}{2}\left(\frac{1}{2}\tau^2 + \tau + \frac{2}{3}\right)m^3 + o(m^3)} \cdot 3^{\left(\frac{1}{2}\tau^2 + \frac{1}{2}\tau + \frac{1}{6}\right)m^3 + o(m^3)}$$

$$= N^{\frac{1}{2}\left(\frac{1}{2}\tau^2 + \tau + \frac{2}{3}\right)m^3 + o(m^3) + \varepsilon_2},$$

$$Z^{n_Z} = N^{\left(\frac{1}{2}\tau + \frac{1}{3}\right)\gamma m^3 + o(m^3)},$$

$$\frac{2^{-\frac{\omega(\omega-1)}{4}}}{(\sqrt{\omega})^{\omega-2}} e^{-2m} = N^{-2\beta m - \varepsilon_3},$$

where $\varepsilon_1, \varepsilon_2$ and ε_3 are small positive constants depending on m, and N. It follows that the inequality (9) can be rewritten in terms of the exponents as

$$\left(-\frac{1}{2}\tau - \frac{1}{3}\right)\beta + \left(\frac{1}{2}\tau + \frac{2}{3}\right)(\beta + \delta - 2)$$

$$+ \frac{1}{2}\left(\frac{1}{2}\tau^2 + \tau + \frac{2}{3}\right) + \left(\frac{1}{2}\tau + \frac{1}{3}\right)\gamma < \frac{-2\beta m - \varepsilon_3 - \varepsilon_1 - \varepsilon_2}{m^3}.$$

Setting $\frac{-2\beta m - \varepsilon_3 - \varepsilon_3 - \varepsilon_1\varepsilon_2}{m^3} = -\varepsilon_4$ and rearranging, we get

$$3\tau^2 + 6(\delta + \gamma - 1)\tau + 4\beta + 8\delta + 4\gamma - 12 < -12\varepsilon_4. \tag{10}$$

The left side of (10) is optimal for $\tau_0 = 1 - \delta - \gamma$. Plugging τ_0 in (10), we get

$$-3\delta^2 + (14 - 6\gamma)\delta - \gamma^2 + 4\beta + 10\gamma - 15 < -12\varepsilon_4.$$

This inequality is valid if

$$\delta < \frac{7}{3} - \gamma - \frac{2}{3}\sqrt{1 + 3\beta - 3\gamma} - \varepsilon, \tag{11}$$

where ε is a small positive constant depending on m and N. This terminates the proof.   □

4   Comparison with Existing Results

In [6], Bunder et al. combined the continued fraction algorithm and Copper-smith's method to study the equation $eu - (p^2 - 1)(q^2 - 1)v = w$. They showed that it is possible to solve it if

$$uv < 2N - 4\sqrt{2}N^{\frac{3}{4}} \quad \text{and} \quad |w| < (p - q)N^{\frac{1}{4}}v.$$

In terms of $e = N^\beta$, $u = N^\delta$ and $|w| = N^\gamma$, the first condition implies the following one

$$\delta < \frac{3 - \beta}{2}.$$

For $\gamma = 0$, that is $w = 1$, the bound of Theorem 3 becomes

$$\delta < \frac{7}{3} - \frac{2}{3}\sqrt{1 + 3\beta} - \varepsilon.$$

Neglecting the ε term, the difference between the former bound and the bound of [6] is

$$\delta_1 = \frac{7}{3} - \frac{2}{3}\sqrt{1 + 3\beta} - \frac{3 - \beta}{2} = \frac{5}{6} + \frac{b}{2} - \frac{2}{3}\sqrt{1 + 3\beta}.$$

A straightforward calculation shows that $\delta_1 \geq 0$. This shows that the bound of Theorem 3 is better than the bound of [6].

In [17], Peng et al. proposed a lattice based method to solve the equation $ed - k(p^2 - 1)(q^2 - 1) = 1$ under the condition $\delta < 2 - \sqrt{\beta}$ and $\beta > 1$. This is a special case of the general equation $eu - (p^2 - 1)(q^2 - 1)v = w$. In this special case, we have $w = N^\gamma = 1$ and $\gamma = 0$, and the difference between the bound of Theorem 3 and the bound of [17] is

$$\delta_2 = 2 - \sqrt{\beta} - \left(\frac{7}{3} - \frac{2}{3}\sqrt{1 + 3\beta}\right) = \frac{2}{3}\sqrt{1 + 3\beta} - \frac{1}{3} - \sqrt{\beta}.$$

Again, a straightforward calculation shows that $\delta_2 \geq 0$. This means that the condition of Theorem 3 is not better than Peng al.'s bound. Nevertheless, our method is more general and can solve a variety of equations with $w \neq 1$.

5   Conclusion

In this paper, we have studied the equation $eu - (p^2 - 1)(q^2 - 1)v = w$ which is a generalization of the equation $ed - k(p^2 - 1)(q^2 - 1) = 1$. The latter equation is the key equation of some variants of the RSA cryptosystem with modulus $N = pq$, public exponent e and private key d. We have showed that, under some conditions, it is possible to solve the equation $eu - (p^2 - 1)(q^2 - 1)v = w$ and break the cryptosystem. The attack is based on applying Coppersmith's method to a multivariate modular equation and can be seen as an extension of former attacks on such cryptosystems.

References

1. Blömer, J., May, A.: A generalized Wiener attack on RSA. In: Bao, F., Deng, R., Zhou, J. (eds.) PKC 2004. LNCS, vol. 2947, pp. 1–13. Springer, Heidelberg (2004). https://doi.org/10.1007/978-3-540-24632-9_1
2. Boneh, D., Durfee, G.: Cryptanalysis of RSA with private key d less than $N^{0.292}$. In: Stern, J. (ed.) EUROCRYPT 1999. LNCS, vol. 1592, pp. 1–11. Springer, Heidelberg (1999). https://doi.org/10.1007/3-540-48910-X_1
3. Boneh, D.: Twenty years of attacks on the RSA cryptosystem. Notices Am. Math. Soc. **46**(2), 203–213 (1999)
4. Boneh, D., Shacham, H.: Fast variants of RSA. CryptoBytes **5**(1), 1–9 (2002)
5. Bunder, M., Nitaj, A., Susilo, W., Tonien, J.: A new attack on three variants of the RSA cryptosystem. In: Liu, J.K., Steinfeld, R. (eds.) ACISP 2016. LNCS, vol. 9723, pp. 258–268. Springer, Cham (2016). https://doi.org/10.1007/978-3-319-40367-0_16
6. Bunder, M., Nitaj, A., Susilo, W., Tonien, J.: A generalized attack on RSA type cryptosystems. Theor. Comput. Sci. **704**, 74–81 (2017)
7. Castagnos, G.: An efficient probabilistic public-key cryptosystem over quadratic field quotients. Finite Fields Appl. **13**(3–13), 563–576 (2007)
8. Coppersmith, D.: Small solutions to polynomial equations, and low exponent RSA vulnerabilities. J. Cryptol. **10**(4), 233–260 (1997)
9. Elkamchouchi, H., Elshenawy, K., Shaban, H., Extended RSA cryptosystem and digital signature schemes in the domain of Gaussian integers. In: Proceedings of the 8th International Conference on Communication Systems, pp. 91–95 (2002)
10. Hinek, M.J.: Cryptanalysis of RSA and its Variants. Chapman & Hall/CRC Cryptography and Network Security. CRC Press, Boca Raton (2010)
11. Howgrave-Graham, N.: Finding small roots of univariate modular equations revisited. In: Darnell, M. (ed.) Cryptography and Coding 1997. LNCS, vol. 1355, pp. 131–142. Springer, Heidelberg (1997). https://doi.org/10.1007/BFb0024458
12. Jochemsz, E., May, A.: A strategy for finding roots of multivariate polynomials with new applications in attacking RSA variants. In: Lai, X., Chen, K. (eds.) ASIACRYPT 2006. LNCS, vol. 4284, pp. 267–282. Springer, Heidelberg (2006). https://doi.org/10.1007/11935230_18
13. Kuwakado, H., Koyama, K., Tsuruoka, Y.: A new RSA-type scheme based on singular cubic curves $y^2 = x^3 + bx^2 \pmod{n}$. IEICE Trans. Fundam. **E78–A**, 27–33 (1995)
14. Lenstra, A.K., Lenstra, H.W., Lovász, L.: Factoring polynomials with rational coefficients. Math. Ann. **261**, 513–534 (1982)

15. May, A.: New RSA vulnerabilities using lattice reduction methods. Ph.D. thesis. University of Paderborn (2003). http://www.cits.rub.de/imperia/md/content/may/paper/bp.ps
16. May, A.: Using LLL-reduction for solving RSA and factorization problems. In: Nguyen, P., Vallée, B. (eds.) The LLL Algorithm. Information Security and Cryptography, pp. 315–348. Springer, Heidelberg (2007). https://doi.org/10.1007/978-3-642-02295-1_10
17. Peng, L., Hu, L., Lu, Y., Wei, H.: An improved analysis on three variants of the RSA cryptosystem. In: Chen, K., Lin, D., Yung, M. (eds.) Inscrypt 2016. LNCS, vol. 10143, pp. 140–149. Springer, Cham (2017). https://doi.org/10.1007/978-3-319-54705-3_9
18. Quisquater, J.J., Couvreur, C.: Fast decipherment algorithm for RSA public-key cryptosystem. Electron. Lett. **18**(21), 905–907 (1982)
19. Rivest, R., Shamir, A., Adleman, L.: A method for obtaining digital signatures and public-key cryptosystems. Commun. ACM **21**(2), 120–126 (1978)
20. Smith, P.J., Lennon, G.J.J.: LUC: a new public key cryptosystem. In: Ninth IFIP Symposium on Computer Science Security, pp. 103–117. Elseviver Science Publishers (1993)
21. Takagi, T.: Fast RSA-type cryptosystem modulo $p^k q$. In: Krawczyk, H. (ed.) CRYPTO 1998. LNCS, vol. 1462, pp. 318–326. Springer, Heidelberg (1998). https://doi.org/10.1007/BFb0055738
22. Wiener, M.: Cryptanalysis of short RSA secret exponents. IEEE Trans. Inf. Theory **36**, 553–558 (1990)
23. Zheng, M., Kunihiro, N., Hu, H.: Cryptanalysis of RSA variants with modified Euler quotient. In: Joux, A., Nitaj, A., Rachidi, T. (eds.) AFRICACRYPT 2018. LNCS, vol. 10831, pp. 266–281. Springer, Cham (2018). https://doi.org/10.1007/978-3-319-89339-6_15

Injective Encodings to Binary Ordinary Elliptic Curves

Mojtaba Fadavi[1], Reza Rezaeian Farashahi[1,2(✉)], and Soheila Sabbaghian[1]

[1] Department of Mathematical Sciences, Isfahan University of Technology,
84156-83111 Isfahan, Iran
{mojtaba.fadavi,s.sabbaghian}@math.iut.ac.ir, farashahi@cc.iut.ac.ir
[2] School of Mathematics, Institute for Research in Fundamental Sciences (IPM),
P.O. Box 19395-5746, Tehran, Iran

Abstract. Representing points of elliptic curves in a way that no pattern can be detected by sensors in the transmitted data is a crucial problem in elliptic curve cryptography. One of the methods that we can represent points of the elliptic curves in a way to be indistinguishable from random bit strings is using injective encoding function. So far, several injective encodings to elliptic curves have been presented, but the previous encoding functions have not supported the binary elliptic curves. More precisely, the only injective encoding to binary elliptic curves was given for Hessian curves, the family of elliptic curves with a point of order 3. In this paper, we propose approaches for constructing injective encoding algorithms to the ordinary binary elliptic curves $y^2 + xy = x^3 + ax^2 + b$ with $\mathrm{Tr}(a) = 1$ as well as those with $\mathrm{Tr}(a + 1) = 0$.

Keywords: Elliptic curve · Cryptography · Injective encoding

2010 Mathematics Subject Classification: 11G05 · 11T06 · 14H52.

1 Introduction

The problem of finding encoding functions from a finite field $\mathbb{F}_q$ into the $\mathbb{F}_q$-rational points of the given curve was stated by Schoof in 1985 [16]. Such an encoding function is a crucial requirement in the curve-based cryptosystems. For instance, the public key for identity $id \in \{0,1\}^*$ in the *IBE* scheme, is a $\mathbb{F}_q$-rational point $Q_{id} = \mathfrak{H}(id)$, where $\mathfrak{H}$ is the desired encoding function. This function is also a requirement for PAKE (Password Authenticated Key Exchange) [5] such as SPEKE (Simple Password Exponential Key Exchange) [13], and PSI (Private Set Intersection) protocols [15].

Bernstein et al. in [3] explained that the traditional methods for encoding to elliptic curves do not disguise the points properly so the encoded points are distinguishable from uniform random bit strings, and consequently, censors can recognize patterns in the transmitted data. To avoid this important drawback, they suggested using a bijection between bit strings and about half of all $\mathbb{F}_q$-rational points of an elliptic curve E (of j-invariant not equal to 1728) over $\mathbb{F}_q$

© Springer Nature Switzerland AG 2019
C. Cid and M. J. Jacobson, Jr. (Eds.): SAC 2018, LNCS 11349, pp. 434–449, 2019.
https://doi.org/10.1007/978-3-030-10970-7_20

with odd q, where E has a $\mathbb{F}_q$-rational point of order 2. In the other word, they suggested using injective encoding function, which allows to correspond the set of bit strings $\{0,1\}^n$ to a subset of $E(\mathbb{F}_q)$. When we use injective encoding function $f : \{0,1\}^n \to E(\mathbb{F}_q)$, instead of transferring a point $P \in f(\{0,1\}^n) \subset E(\mathbb{F}_q)$ we easily transfer the corresponding bit string of P. So far, injective encoding functions are presented for ordinary elliptic curves with non-trivial 3 torsion point by Farashahi [8], non-trivial 4 torsion point by Fouque et al. [9] and non-trivial 2 torsion point by Bernstein et al. [3]. However, for binary ordinary elliptic curves, up to now the only injective encoding function is proposed for binary Hessian elliptic curves [8]. After that, Aranha et al. in [2] using λ-affine coordinate and some computational tricks improved the algorithm in [4] [Appendix E]. But, they did not propose any injective encoding function to binary elliptic curves. To the best of our knowledge, no injective encoding function to ordinary binary elliptic curves has been presented, and this is the main contribution of this paper.

The motivation of this paper is constructing injective encoding function for all ordinary binary elliptic curves, because the previous injective encoding is restricted to binary elliptic curves with a point of order 3 [8]. Two approaches will be proposed in this paper, the first one is applicable to ordinary binary elliptic curves $y^2 + xy = x^3 + ax^2 + b$ where $\mathrm{Tr}(a) = 1$. And, the second is for ordinary binary elliptic curves $y^2 + xy = x^3 + ax^2 + b$ where $\mathrm{Tr}(a+1) = 0$. In fact, ordinary elliptic curves $y^2 + xy = x^3 + ax^2 + b$ over prime extensions of $\mathbb{F}_2$ with $\mathrm{Tr}(a) = 1$, which are of paramount importance in binary elliptic curve cryptography, belong to both classes at the same time. For instant, all of the five recommended binary elliptic curves by *NIST* have cofactor 2 i.e. the recommended curves are of the form $y^2 + xy = x^3 + ax^2 + b$ with $\mathrm{Tr}(a) = 1$. The proposed algorithms can be applied in protocols which require admissible encoding function to binary elliptic curves. Moreover, since the encoding is injective it behaves as the same as Elligator 2 in [3].

This paper is organized as follows: In Sect. 2, we talk about some different encoding methods from $\mathbb{F}_{2^n}$ to binary elliptic curves and explain what kind of encoding functions are appropriate for cryptography. Besides, we explain an injective function from bit strings to binary finite fields. In Sect. 3, we briefly review the injective encoding function to the binary Hessian curves, and then we will explain our approaches for finding injective encodings to binary elliptic curves.

Throughout the paper, the cardinality of a finite set $\mathcal{S}$ is shown by $\#\mathcal{S}$ and $\|$ denote the concatenation. Also, $\mathcal{H}$ shows a standard hash function.

2   Background

2.1   Elliptic Curves

An elliptic curve is a smooth projective genus 1 curve over a field $\mathbb{F}$, with a given $\mathbb{F}$-rational point. Traditionally, an elliptic curve E over a field $\mathbb{F}$ is presented by the Weierstrass equation

$$E: \quad y^2 + a_1 xy + a_3 y = x^3 + a_2 x^2 + a_4 x + a_6, \tag{1}$$

where the coefficients $a_1, a_2, a_3, a_4, a_6 \in \mathbb{F}$. Using suitable change of variables, Eq. (1) can be written in the following forms

$$E : y^2 = x^3 + ax^2 + bx + c, \qquad \text{char}(\mathbb{F}) \neq 2,$$

$$E : y^2 + xy = x^3 + ax^2 + b, \qquad \text{char}(\mathbb{F}) = 2, \; j(E) \neq 0, \; \Delta(E) = b, \qquad (2)$$

$$E : y^2 + cy = x^3 + ax + b, \qquad \text{char}(\mathbb{F}) = 2, \; j(E) = 0, \; \Delta(E) = a^3, \qquad (3)$$

where $\Delta(E)$ and $j(E)$ are the discriminant and the j-invariant of the elliptic curve. Elliptic curves in Eqs. (2) and (3) are called ordinary and supersingular binary elliptic curves, respectively. Elliptic curves can be represented by several other models such as Edwards, Hessian, Montgomery, Jacobi intersection, and Jacobi quartic ([1, Chap. 13], [19, Chap. 2], [17]).

The *trace* function $\text{Tr} : \mathbb{F}_{2^n} \to \mathbb{F}_2$ is a linear transformation that is defined as follows:

$$\text{Tr}(a) = \sum_{k=0}^{n-1} a^{2^k}.$$

In addition, for n odd, the *half trace* is the function $\text{HTr} : \mathbb{F}_{2^n} \to \mathbb{F}_{2^n}$, where

$$\text{HTr}(c) = \sum_{k=0}^{\frac{n-1}{2}} c^{2^{2k}}.$$

An ordinary binary elliptic curve such as E can be transformed into the equation $z^2 + z = g(x)$, where $z = \frac{y}{x}$ and $g(x) = \frac{x^3 + ax^2 + b}{x^2}$. Consequently, points on E can be found using the solutions of equation $z^2 + z = g(x)$. Also, It is well-known that the equation $z^2 + z = c$ over $\mathbb{F}_{2^n}$ has solution if and only if $\text{Tr}(c) = 0$. And, if z_0 is a solution of this equation then $z_0 + 1$ is one other. More precisely, let $y \in \mathbb{F}_{2^n}$ be an element of trace 1. If $\text{Tr}(c) = 0$ then the solution of equation $z^2 + z = c$ is as follows:

$$Z(c) = \begin{cases} \sum_{k=0}^{\frac{n-3}{2}} c^{2^{2k+1}} & \text{if } n \text{ is odd}, \\ \sum_{k=0}^{n-1} (\sum_{j=0}^{k} c^{2^j}) y^{2^k} & \text{if } n \text{ is even}, \end{cases} \qquad (4)$$

Hence, having an element $y \in \mathbb{F}_{2^n}$, where $\text{Tr}(y) = 1$, we can deterministically compute the roots of quadratic equation $z^2 + z = c$ where $\text{Tr}(c) = 0$.

It is well-known that two ordinary elliptic curves

$$E_1 : y^2 + xy = x^3 + ax^2 + b, \qquad \Delta(E_1) = b,$$

$$E_2 : y^2 + xy = x^3 + \bar{a}x^2 + \bar{b}, \qquad \Delta(E_2) = \bar{b},$$

over $\mathbb{F}_{2^n}$ are isomorphic over $\mathbb{F}_{2^n}$ if and only if $b = \bar{b}$ and $\text{Tr}(a) = \text{Tr}(\bar{a})$ [11]. As a result, we conclude that the number of isomorphism classes of ordinary elliptic curves over $\mathbb{F}_{2^n}$ is $2^{n+1} - 2$. More precisely, fix two elements $\mu, \gamma \in \mathbb{F}_{2^n}$ with $\text{Tr}(\mu) = 0$ and $\text{Tr}(\gamma) = 1$. The set of representative of the isomorphism classes is

$$\mathcal{I} = \mathcal{I}_\mu \cup \mathcal{I}_\gamma,$$

where $\mathcal{I}_a = \{y^2 + xy = x^3 + ax^2 + b \mid b \in \mathbb{F}_{2^n}^*\}$ for $a \in \mathbb{F}_{2^n}$.

Also, any given elliptic curve $E \in \mathcal{I}$ has a non-trivial 2 torsion group, and $\#E(\mathbb{F}_{2^n})$ is divisible by 4 if and only if $E \in \mathcal{I}_\mu$.

Now, we partition $\mathcal{I}$ in terms of n as follows, because we want to investigate different cases separately, according to the requirement of Algorithms 4 and 6.

1. n is odd. Since, $\mathrm{Tr}(1) = 1$ we have

$$\mathcal{I} = \mathcal{I}_1 \cup \mathcal{I}_0,$$

2. n is even. Since, $\mathrm{Tr}(1) = 0$ we have

$$\mathcal{I} = \mathcal{I}_\gamma \cup \mathcal{I}_1,$$

where $\mathrm{Tr}(\gamma) = 1$.

In Sect. 3, we provide injective encoding algorithms for all ordinary binary elliptic curves $y^2 + xy = x^3 + ax^2 + b$ over $\mathbb{F}_{2^n}$ with $\mathrm{Tr}(a) = 1$ as well as those with $\mathrm{Tr}(a + 1) = 0$.

2.2   Encoding into Elliptic Curves

Encoding into Elliptic Curves: Boneh, et al. in [7] proposed the try-and-increment method. This method is probabilistic, hence it does not run in a constant time so is vulnerable to timing attacks. A variant form of the try-and-increment for elliptic curves over $\mathbb{F}_{2^n}$ with n odd is as follows:

Algorithm 1 Try-and-Increment Algorithm for Ordinary Binary Elliptic Curves

---

Input: $M \in \{0,1\}^*$, **a random oracle** $\mathcal{H} := \{0,1\}^* \to \mathbb{F}_{2^n} \times \{0,1\}$, n **odd,**
$E/\mathbb{F}_{2^n} : y^2 + xy = x^3 + ax^2 + b,$ **and** $k \in \mathbb{N}.$
Output: $(x, y) \in E(\mathbb{F}_{2^n})$ or $\perp$.

1: $i = 0$;
2: **while** $i < k$ **do**
3:     $(x, v) = \mathcal{H}(M \| i);$ $\triangleright$ v is the least significant bit of $\mathcal{H}(M \| i)$
4:     $g(x) = \frac{x^3 + ax^2 + b}{x^2};$
5:     **if** $\mathrm{Tr}(g(x)) = 0$ **then return** $(x, x(\mathrm{HTr}(g(x)) + v));$
6:     **end if**
7:     $i = i + 1$;
8: **end while**
     return $\perp$.

The probability of success for any arbitrary $M \in \{0,1\}^*$ is close to $\frac{1}{2}$. Hence, the probability of failure after up to k rounds is about 2^{-k}, and by taking $k \approx 128$ we are sure that the algorithm will be successful unless in a very rare situations.

Boneh and Franklin in [6] suggested a deterministic method of encoding into elliptic curves, but their method was restricted to the supersingular elliptic curves of the form $E : y^2 = x^3 + b$ over $\mathbb{F}_q$, where $q \equiv 2 \pmod{3}$. Although their method was efficient, the MOV attack [14] can be used for transforming the elliptic curve discrete logarithm problem to the finite field version, which has subexponential complexity. They also introduced the following notion of admissible encoding.

Definition 1. *([6]) A function $f : S \to R$, where S and R are two finite sets, is an admissible encoding if it satisfies the following conditions:*
1: Computable: f is computable in deterministic polynomial time.
2: l to 1: for any $r \in R$, $\#f^{-1}(r) = l$.
3: Samplable: there exists a probabilistic polynomial time algorithm that for any $r \in R$, it returns a random element $s \in f^{-1}(r)$.

Let $f : \mathbb{F}_{2^n} \to f(\mathbb{F}_{2^n}) \subset E(\mathbb{F}_{2^n})$ be an admissible encoding and $\mathfrak{h} : \{0,1\}^* \to \mathbb{F}_{2^n}$ be a random oracle. Brier et al. in [4] proved that $\mathbf{h} : \{0,1\}^* \to E(\mathbb{F}_{2^n})$, where $\mathbf{h}(m) = f(\mathfrak{h}(m))$, is indifferentiable from a random oracle to $f(\mathbb{F}_{2^n}) \subseteq E(\mathbb{F}_{2^n})$. So, having an admissible encoding function $f : \mathbb{F}_{2^n} \to f(\mathbb{F}_{2^n}) \subset E(\mathbb{F}_{2^n})$ is required for having a random oracle into $E(\mathbb{F}_{2^n})$. Also, the inverse function of injective encoding functions can be used in representing points of binary elliptic curves in a way that the preimage is indistinguishable from a uniform bit-string. Now, we review important encoding methods and state their drawbacks.

Icart's Method: As it was stated above, the Boneh et al.'s encoding function can just be applied to supersingular elliptic curves $y^2 = x^3 + b$. Icart in [12] extended their method and proposed an explicit encoding to elliptic curves $y^2 = x^3 + ax + b$ over $\mathbb{F}_q$ where $q \equiv 2 \pmod{3}$, and ordinary elliptic curves $E : y^2 + xy = x^3 + ax^2 + b$ over $\mathbb{F}_{2^n}$ where n is odd. His binary encoding function is as follows:

$$f_{a,b} : \mathbb{F}_{2^n} \to E(\mathbb{F}_{2^n})$$
$$u \to (x, ux + v^2),$$

where $v = a + u + u^2$ and $x = (v^4 + v^3 + b)^{\frac{1}{3}} + v$. In addition, $\#f_{a,b}^{-1}(P) \leq 4$ so it is not $l : 1$, for some small positive integer, and as a result it is not an admissible encoding.

SW Method: Another completely different method that was given before Icart's method is the Shallue-Woestijne's method. Their method covers all isomorphism classes of elliptic curves over all finite fields [18], but it was at most 8:1 and as a result it is not an admissible encoding. Here, we recall the binary case of the SW method for ordinary elliptic curves.

Let E be an ordinary binary elliptic curve $y^2 + xy = x^3 + ax^2 + b$, $g(x) = (x^3 + ax^2 + b)/x^2$ and

$$X_1(t,w) = \frac{t(a+w+w^2)}{1+t+t^2}, \qquad X_2(t,w) = t.X_1(t,w) + (a + w + w^2),$$
$$X_3(t,w) = \frac{X_1(t,w).X_2(t,w)}{X_1(t,w)+X_2(t,w)}. \tag{5}$$

Then

$$g(X_1(t, w)) + g(X_2(t, w)) + g(X_3(t, w)) \in h(\mathbb{F}_{2^n}),$$

where $h : \mathbb{F}_{2^n} \to \mathbb{F}_{2^n}$ and $h(z) = z^2 + z$. Since the trace function is a linear transformation on $\mathbb{F}_{2^n}$, then either one or all of $g(X_i) \in h(\mathbb{F}_{2^n})$. In the other words, we have $\text{Tr}(g(X_i)) = 0$ for either one of $i \in \{1, 2, 3\}$ or all of them. Given such X_i, the solutions of equation $z^2 + z = g(X_i)$ is computable, so the Algorithm 2 always returns 2 or 6 solutions.

Remark 1. *We use* $Z(c)$ *for computing the root of the equation* $z^2 + z = c$ *in all of the following algorithm. However, for* n *odd* $Z(c)$ *is exactly the same as the* $\text{HTr}(c)$.

Algorithm 2 is the Shallue-Woestijne algorithm for binary elliptic curves over $\mathbb{F}_{2^n}$, where n is odd and w is fixed.

Algorithm 2   Binary SW Algorithm

Input: $a, b, t \in \mathbb{F}_{2^n}$, $c = a + w + w^2 \neq 0$, n odd, **and** $E : y^2 + xy = x^3 + ax^2 + b$.
Output: $(x, y) \in E(\mathbb{F}_{2^n})$.

1: **if** $t^2 + t + 1 = 0$ **then return** $\mathcal{O}$;
2: **end if**
3: $X_1(t) = \frac{tc}{1+t+t^2}$; $X_2(t) = tX_1(t) + c$; $X_3(t) = \frac{X_1(t)X_2(t)}{X_1(t)+X_2(t)}$;
4: **for** $i = 1$ to 3 **do**
5:     $g(X_i) = \frac{X_i^3 + aX_i^2 + b}{X_i^2}$;
6:     **if** $\text{Tr}(g(X_i)) = 0$ **then return** $(X_i, X_i\text{HTr}(g(X_i)))$;
7:     **end if**
8: **end for**

The equation of elliptic curve $y^2 + xy = x^3 + ax^2 + b$ in the λ−affine coordinate is of the form $(\lambda^2 + \lambda + a)x^2 = x^4 + b$, where $\lambda = x + \frac{y}{x}$. Aranha et al. in [2] improved Algorithm 2 using the λ−affine coordinate of elliptic curves. More precisely, they fixed t and considered w as the variable parameter and showed that the number of inversions in the Eq. 5 can be decreased to one inversion by using the pre-computed values $\frac{t}{t^2+t+1}$, $\frac{t+1}{t^2+t+1}$, $\frac{t^2+t}{t^2+t+1}$. They also used the λ−affine coordinates as a computational trick to eliminate computing inversion to have more efficient binary elliptic curve arithmetic.

2.3   Injective Functions from Bit Strings to $\mathbb{F}_{2^n}$

To construct an injective encoding function to binary elliptic curves we require an injective encoding from $\{0, 1\}^{n-1}$ to a determined subset S of $\mathbb{F}_{2^n}$. Here, we explain function κ^l for $l \in \{0, 1\}$ and we use it in Sect. 3.

Let $\Lambda = \{\lambda_1, \ldots, \lambda_n\}$ be an arbitrary basis for $\mathbb{F}_{2^n}$. Then every element $b \in \mathbb{F}_{2^n}$ is uniquely represented by a bit string $b_1, b_2, \cdots, b_n$ with $b = \sum_{j=1}^{n} b_j \lambda_j$.

In particular, $1 = \sum_{j=1}^{n} c_j \lambda_j$, with $c_i \neq 0$ for a fixed i. Let $\mathcal{B}^l$, for $l \in \{0, 1\}$, be the subset of $\mathbb{F}_{2^n}$ given by

$$\mathcal{B}^l = \{b : b \in \mathbb{F}_{2^n} \mid b_i = l\}.$$

Now, we can define the function $\kappa^l : \{0, 1\}^{n-1} \to \mathcal{B}^l \subset \mathbb{F}_{2^n}$, where

$$\kappa^l(b_1, \cdots, b_{i-1}, b_{i+1}, \cdots, b_n) = \sum_{j=1}^{n} b_j \lambda_j$$

and $b_i = l$. Clearly, function κ^l is a bijective function and none of the elements of $b = (b_1, \cdots, b_{i-1}, b_{i+1}, \cdots, b_n) \in \{0, 1\}^{n-1}$ is sent to $\sum_{j=1}^{n}(b_j + c_j)\lambda_j = 1 + \sum_{j=1}^{n} b_j \lambda_j$. As a result, for any $w \in \mathbb{F}_{2^n}$ one and only one of w or $w + 1$ belongs to $\mathcal{B}^l$.

For example, if $\Lambda = \{1, \alpha, \alpha^2, \cdots, \alpha^{n-1}\}$ is the polynomial basis of $\mathbb{F}_{2^n}$, then $\mathcal{B}^l$ is the set of elements in $\mathbb{F}_{2^n}$ with the least significant bit l, where $l \in \{0, 1\}$. Now, we can define the bijective functions $\kappa^l : \{0, 1\}^{n-1} \to \mathcal{B}^l \subset \mathbb{F}_{2^n}$ for $l \in \{0, 1\}$, where

$$\kappa^l(b_2, b_3, \cdots, b_n) = l + \sum_{j=2}^{n} b_j \alpha^{j-1}.$$

Hereafter, we let $\kappa = \kappa^0$.

3   Injective Encoding to Binary Elliptic Curves

In this section, we first recall the injective encoding function to binary elliptic curves with a point of order 3 [8]. Then, we present two Algorithms which bring about injective encoding for all ordinary binary elliptic curves $E : y^2 + xy = x^3 + ax^2 + b$ with $\text{Tr}(a) = 1$ or $\text{Tr}(a + 1) = 0$, respectively.

3.1   Encoding into Hessian Curves

Up to now, the only injective encoding to binary elliptic curves has been given for the Hessian form of elliptic curves over $\mathbb{F}_{2^n}$ with n odd [8]. A binary Hessian elliptic curve has a point of order 3, therefore that injective encoding is applicable only to the family of binary elliptic curves with a point of order 3. More precisely, let

$$H_d : x^3 + y^3 + 1 = dxy,$$

where $d \in \mathbb{F}_{2^n}$ and $d^3 \neq 1$, be an Hessian curve over a finite field $\mathbb{F}_{2^n}$ with n odd [10]. It is shown in [8] that there is an injective function $\texttt{elt} : \{0, 1\}^{n-1} \to \mathbb{F}_{2^n}$ in which $\text{Tr}(d^3(\texttt{elt}(b)^2 + \texttt{elt}(b))) = 0$, for all $b \in \{0, 1\}^{n-1}$. Therefore, the following map is well defined and injective.

$$\texttt{i}_d : \{0, 1\}^{n-1} \longrightarrow H_d(\mathbb{F}_{2^n})$$

where $\mathtt{i}_d(b) = (x, y)$ if $\mathtt{elt}(b) \neq 0$, and $x = duv$, $y = d(u + v)$ with

$$u = \frac{1}{d}\left(\frac{w}{\mathtt{elt}(b)}\right)^{1/3}, \qquad v = \frac{1}{d}\left(\frac{w+1}{\mathtt{elt}(b)}\right)^{1/3},$$
$$w^2 + w = d^3(\mathtt{elt}(b)^2 + \mathtt{elt}(b)),$$

and $\mathtt{i}_d(b) = (1, 0)$ if $\mathtt{elt}(b) = 0$.

3.2   Injective Encoding to Binary Elliptic Curves with $\mathrm{Tr}(a) = 1$.

Let E be the following ordinary binary elliptic curve

$$E : y^2 + xy = x^3 + ax^2 + b, \qquad \mathrm{Tr}(a) = 1. \tag{6}$$

Here, we explain our first approach for finding injective encoding function from $\{0, 1\}^{n-1}$ to elliptic curves with Eq. (6).

As we recall, Eq. (5) is a two variables function in w and t. The main idea for finding a new injective encoding from $\{0, 1\}^{n-1}$ to the ordinary binary elliptic curves, is fixing t and going through all $w \in \mathbb{F}_{2^n}$. However, to achieve such injective encoding we require to have $\mathrm{Tr}(a) = 1$, and binary elliptic curves which are used in elliptic curve cryptography are exactly ordinary binary elliptic curves with $\mathrm{Tr}(a) = 1$. SW algorithm for binary elliptic curves $y^2 + xy = x^3 + ax^2 + b$, when we fix $t \in \mathbb{F}_{2^n}$ and consider $w \in \mathbb{F}_{2^n}$ as a variable, is the following algorithm and we use the notation $\mathfrak{f}$ for Algorithm 3 to call it in Algorithm 4.

Algorithm 3   Encoding to Binary Elliptic Curves $y^2 + xy = x^3 + ax^2 + b$.

Input: $w, a, b \in \mathbb{F}_{2^n}$, where $t(t+1)(t^2+t+1) \neq 0$, $s = \frac{t}{t^2+t+1}$, $r = \frac{t+1}{t}$,
         and $E/\mathbb{F}_{2^n} : y^2 + xy = x^3 + ax^2 + b$.
Output: $(x, y) \in E(\mathbb{F}_{2^n})$.

1: $c = a + w + w^2$;
2: **if** $c(1 + c) = 0$ **then**, Return $\mathcal{O}$;
3: **end if**
4: $X_1 = sc$; $X_2 = rX_1$; $X_3 = trX_1$;

5: **for** $i = 1$ to 3 **do**
6:    $g(X_i) = \frac{X_i^3 + aX_i^2 + b}{X_i^2}$;
7:       **if** $\mathrm{Tr}(g(X_i)) = 0$ **then** $x = X_i$; $y = X_i.Z(g(X_i))$;
8:       **end if**
9: **end for**
    return (x, y).

Remark 2. *Clearly, $\mathfrak{f}(w) = P$ if and only if $\mathfrak{f}(w+1) = P$, so for a given point $P \in \mathfrak{f}(\mathbb{F}_{2^n})$, $\mathcal{V}_P = \{w_1, w_1+1, w_2, w_2+1, w_3, w_3+1\} \subset \mathbb{F}_{2^n}$ is the largest possible preimage set of P and by considering the set $\mathcal{W}_P = \{w_1, w_2, w_3\} \subset \mathcal{V}_P$ as the preimage set of P, we do not lose information about the preimages of P.*

The following proposition shows that there is an interesting feature in Algorithm 3 which can be used for providing a 2:1 encoding from $\mathbb{F}_{2^n}$ to binary elliptic curves of Eq. (6).

Proposition 3. *If* $\mathrm{Tr}(a) = 1$, *then Algorithm 3 is at most* $4 : 1$.

Proof. Since $\mathrm{Tr}(a) = 1$ we conclude that at most two elements of $\mathbb{F}_{2^n}$ are sent to $\mathcal{O}$. Now, suppose that we are given a point $P = (x_0, y_0) \in \mathfrak{f}(\mathbb{F}_{2^n}) \subset E(\mathbb{F}_{2^n})$. We consider two possibilities.

1. If $X_1 = \frac{t}{t^2+t+1}(a + w + w^2) \neq x_0$ for all $w \in \mathbb{F}_{2^n}$. Clearly, $\#\mathfrak{f}^{-1}(P) \leq 4$ because $\deg(X_2) = \deg(X_3) = 2$, and we are done.
2. If $X_1(w_1) = X_1(w_1 + 1) = \frac{t}{t^2+t+1}(a + w_1 + w_1^2) = x_0$. In this case, it is impossible that we have $X_2(w_2) = \frac{t+1}{t^2+t+1}(w_2^2 + w_2 + a) = x_0$ and $X_3(w_3) = \frac{t(t+1)}{t^2+t+1}(w_3^2 + w_3 + a) = x_0$ simultaneously.
   Because, if this happens then $\mathfrak{f}^{-1}(P) = \{w_1, w_1 + 1, w_2, w_2 + 1, w_3, w_3 + 1\} \subset \mathbb{F}_{2^n}$ and we have

$$x_0 = \frac{t}{t^2+t+1}(w_1^2 + w_1 + a) = \frac{t+1}{t^2+t+1}(w_2^2 + w_2 + a), \qquad (7)$$

$$x_0 = \frac{t}{t^2+t+1}(w_1^2 + w_1 + a) = \frac{t(t+1)}{t^2+t+1}(w_3^2 + w_3 + a), \qquad (8)$$

or equivalently

$$w_2^2 + w_2 + \left(\frac{t(w_1^2 + w_1) + a}{t+1}\right) = 0, \qquad (9)$$

$$w_3^2 + w_3 + \left(\frac{w_1^2 + w_1 + ta}{t+1}\right) = 0. \qquad (10)$$

Now, if we let $A = \frac{t(w_1^2 + w_1) + a}{t+1}$ and $B = \frac{w_1^2 + w_1 + ta}{t+1}$, we see that $A + B = w_1^2 + w_1 + a$ and $\mathrm{Tr}(A + B) = \mathrm{Tr}(w_1^2 + w_1 + a) = \mathrm{Tr}(a) = 1$. Therefore, we conclude that one and only one of the Eqs. (9) or (10) has solution. Hence, one of the Eqs. (7) or (8) is held and $\#\mathfrak{f}^{-1}(P) = 4$. ☐

Now, let we are given a point $P \in \mathfrak{f}(\mathbb{F}_{2^n})$. Since P has at most four preimages, we have to first modify Algorithm 3 to have a 2:1 encoding function then using the bijective function in Sect. 2.3 we can construct our desired injective encoding from $\{0,1\}^{n-1}$ to $E(\mathbb{F}_{2^n})$.

Theorem 4. *Let E be the elliptic curve of Eq. (6). There is a function* $\mathfrak{g} :$ $\mathbb{F}_{2^n} \to E(\mathbb{F}_{2^n})$ *which is 2:1. In addition,* $\mathfrak{g}^{-1}$ *is computable.*

Proof. Since $\mathrm{Tr}(a) = 1$, by Proposition 3 we conclude that Algorithm 3 is at most $4 : 1$. So, we have two main possibilities for the preimage set of any point $P = (x_0, y_0) \in \mathfrak{f}(\mathbb{F}_{2^n})$.

1. $\#\mathfrak{f}^{-1}(P) = 4$. Let $\mathcal{W}_P = \{w, \lambda\}$ be the preimage set of $\mathfrak{f}^{-1}(P)$, then $\mathcal{W}_P = \{w_i, w_j\}$, where $i, j \in \{1, 2, 3\}$ and $i \neq j$. Also, the index i of w_i refers to the index of X_i which produces point P. So, we have the following three cases.

(a) $\{w, \lambda\} = \{w_1, w_2\}$. Let R_1 and R_{31} be the sets of roots of equations

$$x^2 + x + \left(\frac{t(w^2 + w) + a}{t + 1}\right) = 0,$$

$$x^2 + x + \frac{(t + 1)(w^2 + w) + a}{t} = 0.$$

These two equations are related to each other, in the sense that if we simplify the first equation regarding to w and swap x with w, we get the other equation and vice versa. Now, there are two cases

i. If $w = w_1$ and $\lambda = w_2$ then $R_1 \subseteq \{\lambda, \lambda + 1\}$ and $R_{31} \subseteq \{\zeta, \zeta + 1\}$.

ii. If $w = w_2$ and $\lambda = w_1$ then $R_1 \subseteq \{\zeta, \zeta + 1\}$ and $R_{31} \subseteq \{w, w + 1\}$, where $\{\zeta, \zeta + 1\} \cap f^{-1}(P) = \emptyset$. In each case, we can use the function f of Algorithm 3 to investigate which set is the suitable set. For the first case, we define $g(w) = f(w)$, and for the second case we define $g(w) = -f(w)$.

(b) $\{w, \lambda\} = \{w_1, w_3\}$. Let R_2 and R_{32} be the sets of roots of equations

$$x^2 + x + \frac{w^2 + w + ta}{t + 1} = 0,$$

$$x^2 + x + (t + 1)(w^2 + w) + ta = 0.$$

i. If $w = w_1$ and $\lambda = w_3$ then $R_2 \subseteq \{\lambda, \lambda + 1\}$ and $R_{32} \subseteq \{\zeta, \zeta + 1\}$.

ii. If $w = w_3$ and $\lambda = w_1$ then $R_2 \subseteq \{\zeta, \zeta + 1\}$ and $R_{32} \subseteq \{w, w + 1\}$, where again $\{\zeta, \zeta + 1\} \cap f^{-1}(P) = \emptyset$. Similar to the first case, we use the function f to investigate which set is the suitable set. For the first case, we define $g(w) = f(w)$, and for the second case we define $g(w) = -f(w)$.

(c) $\{w, \lambda\} = \{w_2, w_3\}$. Let R_4 and R_5 be the sets of roots of equations

$$x^2 + x + \frac{w^2 + w + (t + 1)a}{t} = 0,$$

$$x^2 + x + t(w^2 + w) + (t + 1)a = 0.$$

i. If $w = w_2$ and $\lambda = w_3$ then $R_4 \subseteq \{\lambda, \lambda + 1\}$ and $R_5 \subseteq \{\zeta, \zeta + 1\}$.

ii. If $w = w_3$ and $\lambda = w_2$ then $R_4 \subseteq \{\zeta, \zeta + 1\}$ and $R_5 \subseteq \{w, w + 1\}$.

Like the previous cases, for the first case we define $g(w) = f(w)$, and for the second case we define $g(w) = -f(w)$.

2. $\#f^{-1}(P) = 2$. Let $\mathcal{W}_P = \{w\}$.

In this case, none of the sets of $R_1, R_2, R_{31}, R_{32}, R_4$ and R_5 are allowed to output. So, we just define $g(w) = f(w)$.

For computing $g^{-1}(P)$, where $P = (x, y) \in g(\mathbb{F}_{2^n})$, we consider the list

$$L = [a + x(t + 1 + s_1), a + x(t + s_2), a + x(1 + s_1 s_2)],$$

where $s_1 = \frac{1}{t}, s_2 = \frac{1}{t+1}$ and since t is fixed we only use the precomputed value of s_1 and s_2. The preimage of P is the element $l \in L$ which satisfies the necessary property $\mathrm{Tr}(l) = 0$. For such $l \in L$, we accept the solution $w_0 = Z(l)$ of the equation $w^2 + w + l = 0$ as the desired preimage if $g(w_0) = P$.   □

Algorithm 4 Encoding to Binary Elliptic curves $y^2 + xy = x^3 + ax^2 + b$ with $\text{Tr}(a) = 1$.

Input: $w, a, b \in \mathbb{F}_{2^n}$, where $\text{Tr}(a) = 1$, $t(t+1)(t^2+t+1) \neq 0$, $s = \frac{t}{t^2+t+1}$, $r = \frac{t+1}{t}$, $s_1 = \frac{1}{t}$, $s_2 = \frac{1}{t+1}$ and $E : y^2 + xy = x^3 + ax^2 + b$.
Output: $(x, y) \in E(\mathbb{F}_{2^n})$.

1: **if** $(1 + a + w + w^2) = 0$ **then, return** $\mathcal{O}$;
2: **end if**
3: $L = [(t(w^2 + w) + a)s_2, ((t+1)(w^2 + w) + a)s_1, (w^2 + w + ta)s_2, (t+1)(w^2 + w) + ta, (w^2 + w + (t+1)a)s_1, t(w^2 + w) + (t+1)a];$
4: **for** $i = 1$ to 6 **do**
5:     **if** $\text{Tr}(L[i]) = 0$ **then** $z = Z(L[i]));$
6:         **if** $\mathfrak{f}(w) = \mathfrak{f}(z)$ **then return** $(-1)^{i+1}\mathfrak{f}(w);$
7:     **end if**
8:     **end if**
9: **end for**
    return $\mathfrak{f}(w);$

Algorithm 5 explains details of computing the preimage of a point $P \in \mathfrak{g}(\mathbb{F}_{2^n})$. It should be mentioned that we don't require the exact preimage of a point P. In fact, since $\mathfrak{g}^{-1}(P) = \{w, w+1\}$ we are able to find another preimage of P using output of Algorithm 5 and that is sufficient for constructing our injective encoding function.

The following proposition describes how we can extract an injective encoding function by composing the functions $\mathfrak{g}$ and $\kappa : \{0,1\}^{n-1} \to \mathbb{F}_{2^n}$.

Proposition 5. *Function* $\mathfrak{g} \circ \kappa : \{0,1\}^{n-1} \to E(\mathbb{F}_{2^n})$ *is an injective encoding function.*

Proof. Function $\mathfrak{g}$ is 2:1 with this property that, for all $w \in \mathbb{F}_{2^n}$, $\mathfrak{g}(w) = \mathfrak{g}(w+1)$. On the other hand, the injective function $\kappa : \{0,1\}^{n-1} \to \mathbb{F}_{2^n}$ covers one and only one of the elements w or $w+1$. Therefore, function $\mathfrak{g} \circ \kappa : \{0,1\}^{n-1} \to E(\mathbb{F}_{2^n})$ will be the desired injective encoding function. $\square$

Algorithm 5 Computing the preimage of $P \in \mathfrak{g}(\mathbb{F}_{2^n})$.

Input: $E : y^2 + xy = x^3 + ax^2 + b$, where $a, b \in \mathbb{F}_{2^n}$, $\text{Tr}(a) = 1$, $t(t+1)(t^2+t+1) \neq 0$, $s_1 = \frac{1}{t}$, $s_2 = \frac{1}{t+1}$ and $P \in E(\mathbb{F}_{2^n})$.
Output: $w \in \mathbb{F}_{2^n}$, where $\mathfrak{g}(w) = P$, or $\emptyset$.

1: **if** $P = \mathcal{O}$ **then, return** $w = Z(a+1);$
2: **end if**
3: $L = [a + x(t+1+s_1), a + x(t+s_2), a + x(1+s_1 s_2)];$
4: **for** $i = 1$ to 3 **do**
5:     **if** $\text{Tr}(L[i]) = 0$ **then** $w = Z(L[i]));$
6:         **if** $\mathfrak{g}(w) = P$ **then return** $w;$
7:     **end if**
8:     **end if**
9: **end for**
    return $\emptyset$

3.3   Injective Encoding to Binary Elliptic Curves with $\text{Tr}(a+1) = 0$

Here, we describe our second simple approach for finding an injective encoding to the family of binary elliptic curves

$$E : y^2 + xy = x^3 + ax^2 + b, \qquad \text{Tr}(a+1) = 0. \tag{11}$$

We remark that, this method can be seen as simplified SW algorithm.

Proposition 6. *Let E be an elliptic curve over $\mathbb{F}_{2^n}$ given by the Eq. (11). Then, for every $t \in \mathbb{F}_{2^n}$ there exits a point P on E with $x(P)$ equals t, $t+1$ or $t^2 + t$.*

Proof. For t where $t^2 + t = 0$ we have the point $P = (0, \sqrt{b})$. Now, let $g(x) = x + a + \frac{b}{x^2}$. Then, for all $t \in \mathbb{F}_{2^n}^*$, we have $g(t) + g(t+1) + g(t^2 + t) = t^2 + t + a + 1$. Using the linearity of Trace function, we have

$$\text{Tr}(g(t)) + \text{Tr}(g(t+1)) + \text{Tr}(g(t^2 + t)) = \text{Tr}(t^2 + t + a + 1) = 0.$$

So, there exist a point P on E where $x(P) \in \{t, t+1, t^2 + t\}$ and $y(P) = x(P)Z(g(x(P)))$ (see Eq. 4). □

The trivial solution to correspond an element $t \in \mathbb{F}_{2^n}$ to a point on binary elliptic curve E, is to check whether there is a point with x-coordinate equals t or $t+1$. But, what about the case if it fails? For the family of elliptic curves E of the form (11), Proposition 6 shows there is a point on E with x-coordinate equals $t^2 + t$ if there is no points with x-coordinate equal to t and $t+1$. To make this encoding uniform 2:1, the first step is to find a point on E for the value $t^2 + t$ and if it fails the second step is for the values t and $t+1$. Here the output of encoding is the same for input values t and $t+1$. Also, the main technical point is using the negation map on E to make a distinction between these two steps.

Now, we present Algorithm 6 which is 2:1 from $\mathbb{F}_{2^n}$ to $E(\mathbb{F}_{2^n})$, where E is the elliptic curve with Eq. (11). Using Algorithm 6 we can construct our desired injective encoding from $\{0,1\}^{n-1}$ to $E(\mathbb{F}_{2^n})$.

Algorithm 6 Encoding to Binary Elliptic Curves $y^2 + xy = x^3 + ax^2 + b$ with $\text{Tr}(a+1) = 0$.

Input: $t, a, b \in \mathbb{F}_{2^n}$, $E : y^2 + xy = x^3 + ax^2 + b$, with $\text{Tr}(a+1) = 0$.
Output: $(x, y) \in E(\mathbb{F}_{2^n})$.

1: $X_1 = t$; $X_2 = t+1$; $X_3 = t^2 + t$;
2: **if** $X_3 = 0$ **then return** $(0, \sqrt{b})$;
3: **end if**
4: $g_1 = X_1 + a + \frac{b}{X_1^2}$; $g_2 = X_2 + a + \frac{b}{X_2^2}$; $g_3 = X_3 + a + \frac{b}{X_3^2}$;
5: **if** $\text{Tr}(g_3) = 0$ **then return** $(X_3, X_3(Z(g_3) + 1))$;
6: **else**
7:    **if** $\text{Tr}(g_1) = 0$ **then return** $(X_1, X_1 Z(g_1))$;
8:    **else return** $(X_2, X_2 Z(g_2))$;
9:    **end if**
10: **end if**

Theorem 7. *Function* $\mathfrak{e} : \mathbb{F}_{2^n} \to E(\mathbb{F}_{2^n})$ *given by Algorithm 6 is 2:1. Furthermore,* $\mathfrak{e}^{-1}(P)$ *is computable.*

Proof. Let $P = (u, v)$ be an affine point of $E(\mathbb{F}_{2^n})$. Clearly $P = (u, v) \in \mathfrak{e}(\mathbb{F}_{2^n})$ only if there exists some $t \in \mathbb{F}_{2^n}$ such that u equals $t, t + 1$ or $t^2 + t$. In other words $\mathfrak{e}^{-1}(P) \subset \{u, u + 1, w, w + 1\}$, where $w \in \mathbb{F}_{2^n}$ and $w^2 + w = u$. Obviously, to compute $\mathfrak{e}^{-1}(P)$, we find elements $t \in \{u, u + 1, w, w + 1\}$ that is mapped to P by $\mathfrak{e}$.

If $u = 0$ then $P = (0, \sqrt{b})$. Clearly, $\mathfrak{e}^{-1}(P) = \{0, 1\}$. From now on, we assume $u \neq 0$. For $x \in \mathbb{F}_{2^n}^*$, let $g(x) = x + a + b/x^2$ and $T(x) = \text{Tr}(g(x))$. Clearly there exists a point $P = (u, v)$ on E if and only if $T(u) = 0$. Then we have, $v = uZ(g(u))$ or $v = u(Z(g(u)) + 1)$ (see Sect. 2.1). For the point $P = (u, v)$ on E with $u \neq 0$, let $c(P) = v/u + Z(g(u))$. Then, $P = (u, u(Z(g(u)) + c(P)))$. Clearly the compression of point P or $-P$ is given by $x(P) = x(-P) = u$ and the bit $c(P)$ or $1 + c(P)$ respectively.

We consider the following cases for u.

1. Let u be such that $\text{Tr}(u) = 0$, then let fix $w \in \mathbb{F}_{2^n}$ such that $w^2 + w = u$. Then

$$\mathfrak{e}^{-1}(P) \subset \{u, u + 1, w, w + 1\}.$$

For the point P with $u = 1$, we have $\mathfrak{e}^{-1}(P) = \{w, w + 1\}$ if $c(P) = 1$ and $\mathfrak{e}^{-1}(P) = \emptyset$ otherwise. Now, we assume $u \neq 1$. From Proposition 6, we have

$$T(u) + T(u + 1) + T(u^2 + u) = 0, \qquad T(w) + T(w + 1) + T(u) = 0.$$

Since $T(u) = 0$, there are 4 possibilities for the values $T(w), T(w+1), T(u+1)$ and $T(u^2 + u)$.

From Algorithm 6, we check the output of $\mathfrak{e}$ for following cases of the input t.
   - For all $t \in \{u, u + 1\}$, if $T(u^2 + u) = 0$ we have $x(\mathfrak{e}(t)) = u^2 + u \neq u$, since $u \neq 0$, so $\mathfrak{e}(t) \neq \pm P$. Also, if $T(u^2 + u) = 1$, we have $\mathfrak{e}(t) = P$ if $c(P) = 0$ and $\mathfrak{e}(t) = -P$ if $c(P) = 1$.
   - For all $t \in \{w, w + 1\}$, we have $x(\mathfrak{e}(t)) = w^2 + w = u$. Then $y(\mathfrak{e}(t)) = u(Z(g(u)) + 1) \neq y(P)$ if $c(P) = 0$, and $y(\mathfrak{e}(t)) = u(Z(g(u)) + 1) = y(P)$ if $c(P) = 1$. In other words, for all $t \in \{w, w + 1\}$, we have $\mathfrak{e}(t) = -P$ if $c(P) = 0$ and $\mathfrak{e}(t) = P$ if $c(P) = 1$.

   Then, we compute $\mathfrak{e}^{-1}(P)$ for all possible cases of $c(P)$ and $T(u^2 + u)$.
   - If $c(P) = 0$ and $T(u^2 + u) = 0$, then we have $\mathfrak{e}^{-1}(P) = \emptyset$.
   - If $c(P) = 0$ and $T(u^2 + u) = 1$, then we see that $\mathfrak{e}^{-1}(P) = \{u, u + 1\}$.
   - If $c(P) = 1$ then $\mathfrak{e}^{-1}(P) = \{w, w + 1\}$.

2. If $\text{Tr}(u) = 1$, then there is no element $w \in \mathbb{F}_{2^n}$ such that $w^2 + w = u$. So,

$$\mathfrak{e}^{-1}(P) \subset \{u, u + 1\}.$$

Similar to the previous case, we have $T(u) + T(u + 1) + T(u^2 + u) = 0$. Also, $\mathfrak{e}^{-1}(P) = \{u, u + 1\}$ if $T(u^2 + u) = 1$ and $c(P) = 0$ and $\mathfrak{e}^{-1}(P) = \emptyset$ otherwise.

Briefly, for all $P = (u, v) \in \mathfrak{e}(\mathbb{F}_{2^n})$, we have

$$\mathfrak{e}^{-1}(P) = \begin{cases} \{u, u+1\} & \text{if } u = 0, \\ \{u, u+1\} & \text{if } c(P) = 0 \text{ and } \mathrm{Tr}(a + \frac{b}{u^4 + u^2}) = 1, \ u \neq 0, 1, \\ \{w, w+1\} & \text{if } c(P) = 1 \text{ and } \mathrm{Tr}(u) = 0, \\ \emptyset & \text{otherwise.} \end{cases}$$

Hence, the function $\mathfrak{e}$ is 2:1. $\qquad\square$

Algorithm 7 describes computing preimage of a point $P \in \mathfrak{e}(\mathbb{F}_{2^n})$.

Proposition 8. *Function* $\mathfrak{e} \circ \kappa : \{0, 1\}^{n-1} \to E(\mathbb{F}_{2^n})$, $l \in \{0, 1\}$, *is an injective encoding function.*

Proof. The proof line is the same as Proposition 5. $\qquad\square$

Note that Algorithm 7 for a given point $P \in E(\mathbb{F}_{2^n})$ outputs an element t in $\mathbb{F}_{2^n}$ or gives nothing. Notice, t is represented by a bit string of length n. For computing the preimage of P by $\mathfrak{e} \circ \kappa$, the required output is a bit sting of length $n - 1$, where simply is obtained by removing a single bit of t in the suitable fixed position. More precisely, for the basis $\Lambda = \{\lambda_1, \ldots, \lambda_n\}$ of $\mathbb{F}_{2^n}$, let fix i such that $1 = \sum_{j=1}^{n} c_j \lambda_j$, with $c_i \neq 0$. From Sect. 2.3, we recall the injective functions $\kappa^l : \{0, 1\}^{n-1} \to \mathbb{F}_{2^n}$, for $l = 0, 1$. The preimage of $t = \sum_{j=1}^{n} t_j \lambda_j$ by one of these functions is the required output bit string $(t_1, \cdots, t_{i-1}, t_{i+1}, \cdots, t_n)$ in $\{0, 1\}^{n-1}$.

Algorithm 7 Computing the preimage of $P \in E(\mathbb{F}_{2^n})$.

Input: $E : y^2 + xy = x^3 + ax^2 + b$, **where** $a, b \in \mathbb{F}_{2^n}$, $\mathrm{Tr}(a + 1) = 0$, **and** $P = (x, y) \in E(\mathbb{F}_{2^n})$.
Output: $t \in \mathbb{F}_{2^n}$, **where** $\mathfrak{e}(t) = P$, **or** $\emptyset$.

1: $u = x(P)$;
2: **if** $u = 0$ **then return** 0;
3: **end if**
4: $v = y(P)$;
5: $T_u = \mathrm{Tr}(u)$;
6: **if** $T_u = 0$ **then,** $w = Z(u)$;
7: **end if**
8: **if** $\frac{v}{u} = Z(u + a + b/u^2)$ **then** $c_P = 0$;
9: **else** $c_P = 1$;
10: **end if**
11: **if** $u = 1$ **then** $T = 0$;
12: **else** $T = \mathrm{Tr}(a + \frac{b}{(u^4 + u^2)})$;
13: **end if**
14: **if** $c_P = 0$ **and** $T = 1$ **then return** u;
15: **end if**
16: **if** $c_P = 1$ **and** $T_u = 0$ **then return** w;
17: **end if**
    return $\emptyset$

---

4   Concluding Remarks

It is well-known that the encoding functions from $\mathbb{F}_{2^n}$ to the binary elliptic curves are non-uniform. In fact, the SW-method and the Icart's method, are at most 6:1 and 4:1, respectively. But, we require to have uniform encoding function, because the transmitted data have to be indistinguishable from the uniform bit strings. In this regard, we can use the injective encoding function to the binary elliptic curves as an admissible encoding. So far, the only injective encoding function to binary elliptic curves is given for those with a point of order 3. In this paper, we studied the general case of binary elliptic curves, and we proposed encoding algorithms which provide us injective encoding functions into binary elliptic curves. Algorithms 4 and 6 covers elliptic curves with equation $y^2 + xy = x^3 + ax^2 + b$ with $\mathrm{Tr}(a) = 1$ and $\mathrm{Tr}(a+1) = 0$, respectively. These algorithms are both 2:1 and the preimage of a point P in the image of functions is $\{w, w+1\}$, for some $w \in \mathbb{F}_{2^n}$. So using a suitable injective function $\kappa : \{0,1\}^{n-1} \to \mathbb{F}_{2^n}$, which covers one and only one of the elements of the set $\{w, w+1\}$, we construct injective encoding function from $\{0,1\}^{n-1}$ to the given elliptic curves.

Acknowledgment. The authors thank Diego Aranha and Anonymous reviewers for the useful comments of this work. This research was in part supported by a grant from IPM (No. 96050416).

References

1. Avanzi, R., et al.: Handbook of Elliptic and Hyperelliptic Curve Cryptography. CRC Press, Boca Raton (2005)
2. Aranha, D.F., Fouque, P.-A., Qian, C., Tibouchi, M., Zapalowicz, J.-C.: Binary elligator squared. In: Joux, A., Youssef, A. (eds.) SAC 2014. LNCS, vol. 8781, pp. 20–37. Springer, Cham (2014). https://doi.org/10.1007/978-3-319-13051-4_2
3. Bernstein, D.J., Hamburg, M., Krasnova, A., Lange, T.: Elligator: elliptic-curve points indistinguishable from uniform random strings. In: Sadeghi, A.R., Gligor, V.D., Yung, M. (eds.) ACM Conference on Computer and Communications Security, pp. 967–980. ACM (2013)
4. Brier, E., Coron, J.-S., Icart, T., Madore, D., Randriam, H., Tibouchi, M.: Efficient indifferentiable hashing into ordinary elliptic curves. In: Rabin, T. (ed.) CRYPTO 2010. LNCS, vol. 6223, pp. 237–254. Springer, Heidelberg (2010). https://doi.org/10.1007/978-3-642-14623-7_13
5. Boyko, V., MacKenzie, P., Patel, S.: Provably secure password-authenticated key exchange using Diffie-Hellman. In: Preneel, B. (ed.) EUROCRYPT 2000. LNCS, vol. 1807, pp. 156–171. Springer, Heidelberg (2000). https://doi.org/10.1007/3-540-45539-6_12
6. Boneh, D., Franklin, M.: Identity-based encryption from the weil pairing. In: Kilian, J. (ed.) CRYPTO 2001. LNCS, vol. 2139, pp. 213–229. Springer, Heidelberg (2001). https://doi.org/10.1007/3-540-44647-8_13
7. Boneh, D., Lynn, B., Shacham, H.: Short signatures from the weil pairing. In: Boyd, C. (ed.) ASIACRYPT 2001. LNCS, vol. 2248, pp. 514–532. Springer, Heidelberg (2001). https://doi.org/10.1007/3-540-45682-1_30

8. Farashahi, R.R.: Hashing into Hessian curves. In: Nitaj, A., Pointcheval, D. (eds.) AFRICACRYPT 2011. LNCS, vol. 6737, pp. 278–289. Springer, Heidelberg (2011). https://doi.org/10.1007/978-3-642-21969-6_17

9. Fouque, P.-A., Joux, A., Tibouchi, M.: Injective encodings to elliptic curves. In: Boyd, C., Simpson, L. (eds.) ACISP 2013. LNCS, vol. 7959, pp. 203–218. Springer, Heidelberg (2013). https://doi.org/10.1007/978-3-642-39059-3_14

10. Hesse, O.: Über die Elimination der Variabeln aus drei algebraischen Gleichungen vom zweiten Grade mit zwei Variabeln. J. Reine Angew. Math. **10**, 68–96 (1844)

11. Hankerson, D., Menezes, A.J., Vanstone, S.: Guide to Elliptic Curve Cryptography, 1st edn. Springer, New York (2004). https://doi.org/10.1007/b97644

12. Icart, T.: How to hash into elliptic curves. In: Halevi, S. (ed.) CRYPTO 2009. LNCS, vol. 5677, pp. 303–316. Springer, Heidelberg (2009). https://doi.org/10.1007/978-3-642-03356-8_18

13. Jablon, D.P.: Strong password-only authenticated key exchange. SIGCOMM Comput. Commun. **26**(5), 5–26 (1996)

14. Menezes, A., Okamoto, T., Vanstone, S.A.: Reducing elliptic curve logarithms to logarithms in a finite field, pp. 1639–1647. IEEE (1993)

15. Resende, A.C.D., Aranha, D.F.: Faster unbalanced private set intersection. J. Internet Serv. Appl. **9**(1), 1–18 (2018)

16. Schoof, R.: Elliptic curves over finite fields and the computation of square roots mod p. Math. Comput. **44**(170), 483–494 (1985)

17. Silverman, J.H.: The Arithmetic of Elliptic Curves. Springer, Berlin (1995)

18. Shallue, A., van de Woestijne, C.E.: Construction of rational points on elliptic curves over finite fields. In: Hess, F., Pauli, S., Pohst, M. (eds.) ANTS 2006. LNCS, vol. 4076, pp. 510–524. Springer, Heidelberg (2006). https://doi.org/10.1007/11792086_36

19. Washington, L.C.: Elliptic Curves: Number Theory and Cryptography, 2nd edn. CRC Press, Boca Raton (2008)

Machine Learning and Cryptography

Unsupervised Machine Learning
on Encrypted Data

Angela Jäschke[(✉)] and Frederik Armknecht

University of Mannheim, Mannheim, Germany
{jaeschke,armknecht}@uni-mannheim.de

Abstract. In the context of Fully Homomorphic Encryption, which allows computations on encrypted data, Machine Learning has been one of the most popular applications in the recent past. All of these works, however, have focused on supervised learning, where there is a labeled training set that is used to configure the model. In this work, we take the first step into the realm of unsupervised learning, which is an important area in Machine Learning and has many real-world applications, by addressing the clustering problem. To this end, we show how to implement the K-Means-Algorithm. This algorithm poses several challenges in the FHE context, including a division, which we tackle by using a natural encoding that allows division and may be of independent interest. While this theoretically solves the problem, performance in practice is not optimal, so we then propose some changes to the clustering algorithm to make it executable under more conventional encodings. We show that our new algorithm achieves a clustering accuracy comparable to the original K-Means-Algorithm, but has less than 5% of its runtime.

Keywords: Machine Learning · Clustering
Fully Homomorphic Encryption

1 Introduction

1.1 Motivation

Fully Homomorphic Encryption (FHE) schemes can in theory perform arbitrary computations on encrypted data. Since the discovery of FHE, many applications have been proposed, ranging from medical over financial to advertising scenarios. The underlying idea is mostly the same: Suppose Alice has some confidential data X which she would like to utilize, and Bob has an algorithm $\mathcal{A}$ which he could apply to Alice's data for money. However, conventionally, either Alice would have to give her confidential data to Bob, or run the algorithm herself, for which she may not have the know-how or computational power. FHE allows Alice to encrypt her data to $C := \mathrm{Enc}(X)$ and send it to Bob. Bob

A. Jäschke was financed by the Baden-Wurttemberg Stiftung as a part of the PAL SAaaS project.

can convert his algorithm $\mathcal{A}$ into a function $\mathcal{A}'$ over the ciphertext space and apply it to the encrypted data, resulting in $R := \mathcal{A}'(C)$. He can then send this result back to Alice, who can decrypt it with her secret key. FHE promises that indeed $\text{Dec}(R) = \text{Dec}(\mathcal{A}'(\text{Enc}(X))) = \mathcal{A}(X)$. Since Alice's data was encrypted the whole time, Bob learns nothing about the data entries. Note that the functionality where Bob's algorithm is also kept secret from Alice is not traditionally guaranteed by FHE, but can in practice be achieved via a property called *circuit privacy*, in the sense that Alice learns nothing except the result $\mathcal{A}(X)$.

One of the most popular applications of FHE has been Machine Learning, with many works focusing on Neural Networks and different variants of regression. To our knowledge, all works in this line are concerned with *supervised* learning. This means that there is a training set with known outcomes, and the algorithm tries to build a model that matches the desired outputs to the inputs as well as possible. When the training phase is done, the algorithm can be applied to new instances to predict unknown outcomes. However, there is a second branch in Machine Learning that has not been touched by FHE research: *Unsupervised* learning. For these kinds of algorithms, there are no labeled training examples, there is simply a dataset on which some kind of analysis shall be performed. An example of this is clustering, where the aim is to group data entries that are similar in some way. The number of clusters might be a parameter that the user enters, or it may be automatically selected by the algorithm. Clustering has numerous applications like genome sequence analysis, market research, medical imaging or social network analysis, to name a few, some of which inherently involve sensitive data – making a privacy-preserving evaluation with FHE even more interesting.

1.2 Contribution

In this work, we approach this unexplored branch of Machine Learning and show how to implement the K-Means-Algorithm, an important clustering algorithm, on encrypted data. We discuss the problems that arise when trying to evaluate the K-Means-Algorithm on encrypted data, and show how to solve them. To this end, we first present a natural encoding that allows the execution of the algorithm as it is (including the usually challenging division by an encrypted value), but is not optimal in terms of performance. We then present a modification to the K-Means-Algorithm that performs comparably in terms of clustering accuracy, but is much more FHE-friendly in that it avoids division by an encrypted value. We include another modification that trades accuracy for efficiency in the involved comparison operation, and compare the runtimes of these approaches.

2   Related Work

Encryption schemes that allow one type of operation on ciphertexts have been around for some time and have a comprehensive security characterization [3]. Fully Homomorphic Encryption however, which allows both unlimited additions and multiplications, was only first solved in [19]. Since then, many other schemes have been developed, for example [8,12–15,18,20,37], to name just a few. An overview can be found in [2]. There are several libraries offering FHE implementations, like [11,16,23], and the one we use, [38].

Machine Learning as an application of FHE was first proposed in [35], and subsequently there have been numerous works on the subject, to our knowledge all concerned with supervised learning. The most popular of these applications seem to be (Deep) Neural Networks (see [7,10,21,26,36]) and (Linear) Regression (e.g., [4,17,32] or [22]), though there is also some work on other algorithm classes like decision trees and random forests [41], or logistic regression ([5,6,29,30]). In contrast, our work is concerned with the clustering problem from unsupervised Machine Learning.

The K-Means-Algorithm has been a subject of interest in the context of privacy-preserving computations for some time, but to our knowledge all previous works like [9,24,25,31,42] require interaction between several parties, e.g. via Multiparty Computation (MPC). For a more comprehensive overview of the K-Means-Algorithm in the context of MPC, we refer the reader to [34]. While this interactivity may certainly be a feasible requirement in many situations, and indeed MPC is likely to be faster than FHE in these cases, we feel that there are several reasons why a non-interactive solution as we present it is an important contribution.

1. **Client Economics:** In MPC, the computation is split between different parties, each performing computations every round and combining the results. In FHE computations, the entire computation is performed by the service provider. Even if this computation on encrypted data is more expensive than the total MPC computation, the client reduces his effort to zero this way, making this solution attractive to him and thus generating a demand for it.
2. **Function Privacy:** Imagine the K-Means-Algorithm in this paper as a placeholder for a more complex proprietary algorithm that the service provider executes on the client's data as a service. This algorithm could utilize building blocks from the K-Means-Algorithm that we present in this paper, or involve the K-Means-Algorithm as a whole in the context of pipelining several algorithms together, or be something completely new. Here, the service provider would want to prevent the user from learning the details of this algorithm, as it is his business secret. While FHE per se does not guarantee this functionality, all schemes today fulfill the requirement of *circuit privacy* needed to achieve it. Thus for this case, FHE would be the preferred solution.
3. **Future Efficiency Gain:** MPC is much older than FHE, and efficiency for the latter has increased by a factor of 10^4 in the last six years alone. To argue that MPC is faster and thus FHE solutions are superfluous seems premature at this point, and our contributions are not specific to any implementation, but work on all FHE schemes that support a $\{0,1\}$ plaintext space.

Also, many of these interactive solutions rely on a vertical (in [40]) or horizontal (in [28]) partitioning of the data for security. In contrast, FHE allows a non-interactive setting with a single database owner who wishes to outsource the computation.

3   Preliminaries

In this section, we cover underlying concepts like the K-Means-Algorithm, encoding issues, our choice of implementation library, and the datasets we use.

3.1   The K-Means Algorithm

The K-Means-Algorithm is one of the most well-known clustering algorithms in unsupervised learning. Published in [33], it is considered an important benchmark algorithm and is frequently the subject of current research to this day. It takes as input the data $X = \{x_1, \ldots, x_m\}$ and a number K of clusters to be used, and begins by choosing K randomly chosen data entries as so-called **cluster centroids** c_k. Then, in a step called `Cluster Assignment`, it computes for each data entry x_i which cluster centroid c_k is nearest, and assigns the data entry to that centroid. When this has been done for all data entries, the second step begins: During the `Move Centroids` step, the cluster centroids are moved by setting each centroid as the average of all data entries that were assigned to it in the previous step. These two steps are repeated for a set number of times T or until the centroids do not change anymore. We use the first method.

The output of the algorithm is the values of the centroids, or the cluster assignment for the data entries (which can easily be computed from the former). We opt for the first approach. The pseudocode for the algorithm as we use it can be found in Appendix A, along with a visualization. Accuracy can either be measured in terms of correctly classified data entries, which assumes that the correct classification is known (there might not even exist a unique best solution), or via the so-called cost function, which measures the (average) distance of the data entries to their assigned cluster centroids. We opt for the first approach because our datasets are benchmarking sets for which the labels are indeed provided, and it allows better comparability between the different algorithms.

3.2   Encoding

FHE schemes generally have finite fields as a plaintext space, and any rational numbers (which can be scaled to integers) must be embedded into this plaintext space. There are two main approaches in literature, which we quickly compare side by side in Table 1. Note that for absolute value computation and comparison, we need to use the digitwise encoding.

3.3   FHE Library Choice

In [27], it was shown that among all bases p for digitwise p-adic encoding in FHE computations, the choice $p = 2$ is best in terms of the number of additions and multiplications to be performed on the ciphertexts. Hence, we use an FHE scheme with a plaintext space of $\{0, 1\}$. The currently fastest FHE implementation for this plaintext space, TFHE [38], states that *"an optimal circuit for*

Table 1. Two mainstream encoding approaches.

	Digitwise	Embedded
Description	For a base p, display the number in p-adic[a] representation and encrypt each digit separately	Choose the plaintext space large enough to accommodate all computations
Supports comparison?	✓	✗
Supports absolute value?	✓	✗
Supports division?	✗	✗
Efficiency	Slower	Faster
Flexibility	Full	The function that is being computed must be known (at least a bound) at setup, as computations fail if the result gets too big. This is actually *Somewhat* Homomorphic Encryption, not *Fully* Homomorphic Encryption

[a] This can be extended to plaintext spaces $GF(p^k)$ if the scheme supports them.

TFHE is most likely a circuit with the smallest possible number of gates" – thus, this library is a perfect choice for us, and we will use the binary encoding for signed integers and tweaks presented in [26] for maximum efficiency.

3.4 Datasets

To evaluate performance, we use four datasets from the FCPS dataset [39]:

- The Hepta dataset consists of 212 data points of 3 dimensions. There are 7 clearly defined clusters.
- The Lsun dataset is 2-dimensional with 400 entries and 3 classes. The clusters have different variances and sizes.
- The Tetra dataset is comprised of 400 entries in 3 dimensions. There are 4 clusters, which almost touch.
- The Wingnut dataset has only 2 clusters, which are side-by-side rectangles in 2-dimensional space. There are 1016 entries.

For accuracy measurements, each version of the algorithm was run 1000 times (with varying starting centroids) for number of iterations $T = 5, 10, ..., 45, 50$ on each dataset. For runtimes on encrypted data, we used the Lsun dataset.

4   Approach 1: Implementing the Exact K-Means-Algorithm

We now show a method of implementing the K-Means algorithm largely as it is. To this end, we first discuss challenges that arise in the context of FHE computation of this algorithm. We then address these challenges by changing the distance metric, and then present an encoding that supports the division required in computing the average in the MoveCentroid-step. As this method is in no way restricted to the K-Means-Algorithm, the result is of independent interest. As it turns out, there are some issues with this approach, which we will also discuss.

4.1   FHE Challenges

Fully homomorphic encryption schemes can easily compute additions and multiplications on the underlying plaintext space, and most also offer subtraction. Using these operations as building blocks, more complex functionalities can be obtained. However, there are three elements in the K-Means-Algorithm that pose challenges, as it is not immediately clear how to obtain them from these building blocks. We list these (with the line numbers referring to the pseudocode on page 20 in Appendix A.2) and quickly explain how we solve them.

- The distance metric (Line 9, $\Delta(x,y) = ||x-y||_2 := \sqrt{\sum_i (x_i - y_i)^2}$): To our knowledge, taking the square root of encrypted data has not been implemented yet. In Sect. 4.2, we will argue that the Euclidean norm is an arbitrary choice in this context and solve this problem by using the L_1-distance $\Delta(x,y) = ||x-y||_1 := \sum_i (|x_i - y_i|)$ instead of the Euclidean distance.
- Comparison (Line 10, $\tilde{\Delta} < \Delta$) in finding the centroid with the smallest distance to the data entry: This has been constructed from bit multiplications and additions in [26] for bitwise encoding, so we view this issue as solved. A detailed explanation can be found in the extended version of this paper.
- Division (Line 25, $c_k = c_k/d_k$) in computing the new centroid value as the average of the assigned data points: In FHE computations, division by an encrypted value is usually not possible (whereas division by an unencrypted value is no problem). We present a way of implementing the division with a new encoding in Sect. 4.3, and propose a modified version of the Algorithm in Sect. 5 that only needs division by a constant.

4.2   The Distance Metric

Traditionally, the distance measure used with the K-Means Algorithm is the Euclidean Distance $\Delta(x,y) = ||x-y||_2 := \sqrt{\sum_i (x_i - y_i)^2}$, also known as the L_2-Norm, as it is analytically smooth and thus reasonably well-behaved. However, in the context of K-Means Clustering, smoothness is irrelevant, and we may look to other distance metrics. Concretely, we consider the L_1-Norm[1] (also known as the Manhattan-Metric) $\Delta(x, y) := \sum_i(|x_i - y_i|)$. This has a considerable advantage over the Euclidean distance: Firstly, we do not need to take a square root, which to our knowledge has not yet been achieved on encrypted data. Secondly, of course one could apply the standard trick and not take the root, working instead with the sum of squared distances. However, this would mean a considerable efficiency loss due to numerous multiplications and the greatly increased bitlengths of their results. These long numbers are then summed up, and the result is input into the algorithm that finds the minimum (Algorithm 2 on page 12). These two steps already constitute bottlenecks in the entire computation when working with short numbers in the L_1 norm, so an increase in the bitlengths would greatly increase computation time.

Taking the absolute value can easily be achieved through a digit-wise encoding like the binary encoding which we use: We can use the MSB as the conditional (it is 1 if the number is negative and 0 if it is positive) and use a multiplexer[2] gate applied to the value and its negative. The concrete algorithm can be seen in the extended version of this paper. Thus, using the L_1-Norm is not only justified by the arbitrariness of the Euclidean Norm, but is also much more efficient. We compare the clustering accuracy in Fig. 1.

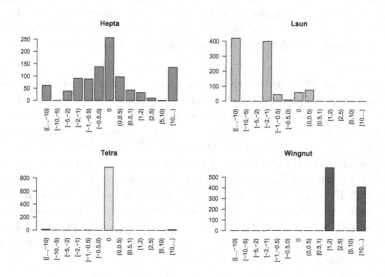

Fig. 1. Difference in percent of data points mislabeled for L_1-norm compared to the L_2-norm $((\% \text{ mislabeled } L_1) - (\% \text{ mislabeled } L_2))$.

---

[1] [1] in fact argues that for high-dimensional spaces, the L_1-Norm is more meaningful than the Euclidean Norm.

[2] $\text{MUX}(c, a, b) = \begin{cases} a, & c = 1 \\ b, & c = 0 \end{cases}$.

For both versions of the distance metric, we calculated the percentage of wrongly labeled data points for 1000 runs, which we can do because the datasets we use come with the correct labels. We plotted histograms of the difference (in percent mislabeled) between the L_1-norm and the L_2-norm for each run. Thus, a value of 0.5 means that the L_1 norm version misclassified 0.5% more data entries than the L_2-version, and -2 means that the L_1 version misclassified 2% less entries than the L_2-version. Each subplot corresponds to one of the four datasets. We see that indeed, it is impossible to say which metric is better – for the Hepta dataset, the performance is very balanced, for the Lsun dataset, the L_1-norm performs much better, for the Tetra dataset, they nearly always perform exactly the same, and for the Wingnut dataset, the L_2-norm is consistently better.

4.3   Fractional Encoding

Suppose we have routines to perform addition, multiplication and comparison on bitwise encoded numbers. The idea is to express the number we wish to encode as a fraction and encode the numerator and denominator separately. Concretely, we choose the denominator a_d randomly in a certain range (like $a_d \in [2^k, 2^{k+1})$ for some k) and compute the nominator a_n as $a_n = \lfloor a \cdot a_d \rceil$. We then encode both separately, so we have $a = (a_n, a_d)$. If we then want to perform computations (including division) on values encoded in this way, we can express the operations using the subroutines from the binary encoding through the regular computation rules for fractions. The details can be seen in Appendix B.

Controlling the Bitlength. Every single one of these operations requires a multiplication of some sort, which means that the bitlengths of the nominators and denominators double with each operation, as there is no cancellation when the data is encrypted. However, in bitwise encoding, deleting the last k least significant bits corresponds to dividing by 2^k and truncating. Doing this for both nominator and denominator yields roughly the same result as before, but with lower bitlengths. As an example, suppose that we have encoded our integers with 15 bits, and after multiplication we thus have 30 bits in nominator and denominator, e.g. $651049779/1053588274 \approx 0.617936$. Then dividing both nominator and denominator by 2^{15} and truncating yields $19868/32152$, which evaluates to $0.617939 \approx 0.617936$. The accuracy can be set through the original encoding bitlength (15 here).

4.4   Evaluation

While this new encoding theoretically allows us to perform the K-Means-Algorithm and solves the division problem in FHE, we now discuss the practical performance in terms of accuracy and runtime.

Accuracy. To see how the exact algorithm performs, we use the four datasets from Sect. 3.4. We ran the exact algorithm 1000 times for number of iterations

$T = 5, 10, ..., 45, 50$, and for sake of completeness we include both distance metrics. The results in this section were obtained by running the algorithms in unencrypted form. We first examine the effect of T on the exact version of the algorithm by looking at the average (over the 1000 runs) misclassification rate for both metrics. The result can be seen in Fig. 2 – we see that the rate levels off after about 15 rounds in all cases, so there is no reason to iterate further.

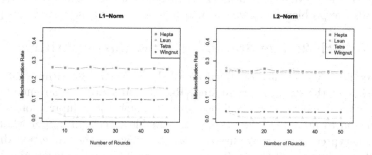

Fig. 2. Misclassification rate with increasing rounds for exact algorithms.

In practice, however, our Fractional Encoding does have some problems: The first issue is the procedure to shorten the bitlengths from Subsect. 4.3. While it works reasonably well for short computations, we found it nearly impossible to set the number of bits to delete such that the entire algorithm ran correctly. The reason is simple: If not enough bits are cut off, the bitlength grows, propagating with each operation and resulting in an overflow when the number becomes too large for the allocated bitlength. If too many bits are cut off, one loses too much accuracy or may even end with a 0 in the denominator. Both these cases result in completely arbitrary and unusable results. The reason why it is so hard to set the shortening parameter properly is that generally, nominator and denominator will not require the same number of bits. Also, because the data is encrypted, we cannot see the actual size of the underlying data, so the shortening parameter cannot be set dynamically – in fact, if this were possible, it would imply that the FHE scheme is insecure. Even setting the parameter roughly requires extensive knowledge about the encrypted data, which the data owner may not want to share with the computing party.

Runtime. The second issue with this encoding is the runtime. Even though TFHE is the most efficient FHE library with which many computational tasks approach practically feasible runtimes, the fact that this encoding requires several multiplications on binary numbers for each elementary operation slows it down considerably. We compare the runtimes of all our algorithms in Sect. 7, and as we will see, running the K-Means-Algorithm on a real-world dataset with this Fractional Encoding would take almost 1.5 years on our computer.

4.5   Conclusion

In conclusion, this encoding is theoretically possible, but we would not recommend it for practical use due to its inefficiency and hardness of setting the shortening parameter (or even higher inefficiency if little to no shortening is done). However, for very flat computations (in the sense that there are not many successive operations performed), this encoding that allows division may still be of interest. For the K-Means-Algorithm, we instead change the algorithm in a way that avoids the problematic division, which we present in the rest of this paper.

5   Approach 2: The Stabilized K-Means-Algorithm

In this section, we present a modification of the K-Means algorithm that avoids the division in the MoveCentroid-step. Recall that conventional encodings in FHE, like the binary one we will use, do not allow the computation of c_1/c_2 where c_1 and c_2 are ciphertexts, but it is possible to compute c_1/a where a is some unencrypted number. We use this fact to exchange the ciphertext division in Line 25 of Algorithm 3 (page 20) for a constant division, resulting in a variant that can be computed with more established and efficient encodings than the one from Sect. 4.3. We present this new algorithm in Sect. 5.2, and compare the accuracy of the results to the original K-Means-Algorithm in Sect. 5.3.

5.1   Encoding

The dataset we use to evaluate our algorithms consists of rational numbers. To encode these so that we can encrypt them bit by bit, we scaled them with a factor of 2^{20} and truncated to obtain an integer. We then used Two's Complement encoding to accommodate signed numbers, and switched to Sign-Magnitude Encoding for multiplication. Note that deleting the last 20 bits corresponds to dividing the number by 2^{20} and truncating, so the scaling factor can remain constant even after multiplication, where it would normally square.

5.2   The Algorithm

Recall that in the original K-Means-Algorithm, the MoveCentroid-step consists of computing each centroid as the average of all data entries that have been assigned to it. More specifically, suppose that we have a $(m \times K)$-dimensional cluster assignment matrix A, where

$$A_{ik} = \begin{cases} 1, & \text{Data entry } x_i \text{ is assigned to centroid } c_k \\ 0 & \text{else.} \end{cases}$$

Then computing the new centroid value c_k consists of multiplying the data entries x_i with the corresponding entry A_{ik} and summing up the results before dividing by the sum over the respective column k of A:

$$c_k = \sum_{i=1}^{m} x_i \cdot A_{ik} \Big/ \sum_{i=1}^{m} A_{ik}.$$

Algorithm 1. The Stabilized K-Means-Algorithm

Input: Data set $X = \{x_1, \ldots, x_m\}$ // $x_i \in \mathbb{R}^\ell$ for some ℓ
Input: Number of clusters K
Input: Number of iterations T
// Initialization
1 Randomly reorder X;
2 Set centroids $c_k = x_k$ for $k = 1$ to K;
// Keep track of centroid assignments
3 Generate $(m \times K)$-dimensional boolean matrix A set to 0;
4 **for** $j = 1$ to T **do**
        // Cluster Assignment
5   **for** $i = 1$ to m **do**
6           $\Delta = \infty$;
7   **for** $k = 1$ to K **do**
                    // Compute distances to all centroids
8               $\Delta_k := \|x_i - c_k\|_1$;
9   **end**
            // The ith row of A has all 0's except at the column corresponding to the
              centroid with the minimum distance
10          $A[i, \cdot] \leftarrow \texttt{FindMin}(\Delta_1, \ldots, \Delta_K)$;
11   **end**
        // Move Centroids
12   **for** $k = 1$ to K **do**
            // Keep old centroid value
13          $\bar{c}_k = c_k$;
14          $c_k = 0$;
15   **for** $i = 1$ to m **do**
                    // If $A_{ik} == 1$, add x_i to c_k, otherwise add $\bar{c}_k$ to c_k
16              $c_k = c_k + \texttt{MUX}(A_{ik}, x_i, \bar{c}_k)$;
17   **end**
            // Divide by number of terms m
18          $c_k = c_k/m$
19   **end**
20 **end**
Output: $\{c_1, \ldots, c_K\}$

Our modification now replaces this procedure with the following idea: To compute the new centroid c_k, add the corresponding data entry x_i to the running sum if $A_{ik} = 1$, otherwise add the old centroid value $\bar{c}_k$ if $A_{ik} = 0$. This can be easily done with a multiplexer gate (or more specifically, by abuse of notation, a multiplexer gate applied to each bit of the two inputs) with the entry A_{ik} as the conditional boolean variable:

$$c_k = \sum_{i=1}^{m} \texttt{MUX}(A_{ik}, x_i, \bar{c}_k)/m.$$

The sum now always consists of m terms, so we can divide by the unencrypted constant m. It is also now obvious why we call it the *stabilized* K-Means-Algorithm: We expect the centroids to move much more slowly, because the old centroid values stabilize the value in the computation. The details of this new algorithm can be found in Algorithm 1, with the changes compared to the original K-Means-Algorithm shaded.

Computing the Minimum. As the reader may have noticed in Line 10, we have replaced the comparison step in finding the nearest centroid for a data entry with a new function $\texttt{FindMin}(\Delta_1, \ldots, \Delta_K)$ due the change in data structure of A (from an integer vector to a boolean matrix). This new function outputs

$$A[i, \cdot] \leftarrow \texttt{FindMin}(\Delta_1, \ldots, \Delta_K)$$

such that the ith row of A, $A[i, \cdot]$, has all 0's except at the column corresponding to the centroid with the minimum distance to x_i. The idea is to run the $\texttt{Compare}$ circuit to obtain a Boolean value: $\texttt{Compare}(x, y) = 1$ if $x < y$, and 0 otherwise.

We start by comparing the first two distances Δ_1 and Δ_2 and setting the Boolean value as $C := \texttt{Compare}(\Delta_1, \Delta_2)$. Then we can write $A[i, 1] = C$ and $A[i, 2] = \neg C$ and keep track of the current minimum through $\texttt{minval} := \texttt{MUX}(C, \Delta_1, \Delta_2)$. We then compare $\texttt{minval}$ to Δ_3 etc. until we have reached Δ_K. Note that we need to modify all entries $A[i, k]$ with k smaller than the current index by multiplying them with the current Boolean value, preserving the indices if the minimum doesn't change through the comparison, and setting them to 0 if it does. The exact workings can be found in Algorithm 2, and an example of how the algorithm works can be found in the extended version of this paper.

If the encryption scheme is one where multiplicative depth is important, it is easy to modify $\texttt{FindMin}$ to be depth-optimal: Instead of comparing Δ_1 and Δ_2, then comparing the result to Δ_3, then comparing that result to Δ_4 etc., we could instead compare Δ_1 to Δ_2 and Δ_3 to Δ_4 and then compare those two results etc., reducing the multiplicative depth from linear in the number of clusters K to logarithmic. Since depth is not important for our implementation choice TFHE, we implemented the function as described in Algorithm 2.

Algorithm 2. $\texttt{FindMin}(\Delta_1, \ldots, \Delta_K)$

Input: Distances $\Delta_1, \ldots, \Delta_K$ of current data entry i to all centroids $c_1 \ldots, c_K$
Input: Row i of Cluster Assignment matrix A, denoted $A[i, \cdot]$
// Set all entries 0 except the first
1 Set $A[i, \cdot] = [1, 0, \ldots, 0]$;
   // Set the minimum to Δ_1
2 Set $\texttt{minval} = \Delta_1$;
3 **for** $k = 2$ *to* K **do**
      // C is a Boolean value, $C = 1$ iff $\texttt{minval} \leq \Delta_k$
4     $C = \texttt{Compare}(\texttt{minval}, \Delta_k)$;
5   **for** $r = 1$ *to* $k - 1$ **do**
         // Set all previous values to 0 if new min is Δ_k, don't change if new min is old min
6        $A[i, r] = A[i, r] \cdot C$;
7   **end**
         // Set $A[i, k]$ to 1 if Δ_k is new min, 0 otherwise
8     $A[i, k] = \neg C$;
9   **if** $k \neq K$ **then**
            // Update the minval variable unless we're done
10       $\texttt{minval} = \texttt{MUX}(C, \texttt{minval}, \Delta_k)$;
11   **end**
12 **end**
   Output: $A[i, \cdot]$

5.3 Evaluation

In this section, we will investigate the performance of our Stabilized K-Means-Algorithm compared to the traditional K-Means-Algorithm.

Accuracy. The results in this section were obtained by running the algorithms in unencrypted form. As we are interested in relative rather than absolute performance, we merely care about the difference in the output of the modified and exact algorithms on the same input (i.e., datasets and starting centroids), not so much about the output itself. Recall that we obtained $T = 15$ as a good choice for number of rounds for the exact algorithm – however, as we have already explained above, the cluster centroids converge more slowly in the stabilized version, so we will likely need more iterations here. We now compare the performance of the stabilized version to the exact version. We perform this comparison by examining the average (over the 1000 iterations) difference in the misclassification rate. Thus, a value of 2 means that the stabilized version mislabeled 2% more instances than the exact version, and a difference of -1 means that the stabilized version misclassified 1% less data points than the exact version.

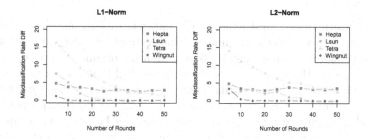

Fig. 3. Average difference in misclassification rate between the stabilized and the exact algorithm ((average % mislabeled stabilized) − (average % mislabeled exact)).

The results for both distance metrics can be seen in Fig. 3. We see that while behavior varies slightly depending on the dataset, $T = 40$ iterations is a reasonable choice since the algorithms do not generally seem to converge further with more rounds. We will fix this parameter from here on, as it also exceeds the required amount of iterations for the exact version to converge.

While the values in Fig. 3 do converge, they do not generally reach a difference of 0, which would imply similar performance. However, this is not surprising - we significantly modified the original algorithm, not with the intention of improving clustering accuracy, but rather to make it executable under an FHE scheme at all. This added functionality comes as a tradeoff, and we will now examine the magnitude of the loss in accuracy in Fig. 4. The corresponding histogram for the L_2-norm can be found in the extended version of this paper.

We can see that in the vast majority of instances, the stabilized version performs exactly the same as the the original K-Means-Algorithm. We also see that

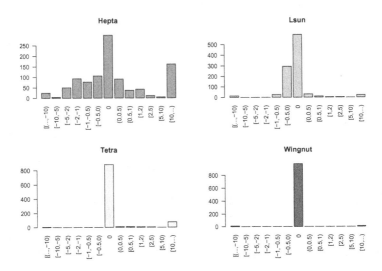

Fig. 4. Distribution of the difference in misclassification rate for stabilized vs. exact K-Means-Algorithm $\big((\% \text{ mislabeled stabilized}) - (\% \text{ mislabeled exact})\big)$, L_1-norm.

concrete performance does depend on the dataset. In some cases, the modified version even outperforms the original one: Interestingly, for the Lsun dataset, the stabilized version is actually slightly better than the original algorithm in about 30% of the cases. However, most of the time, we feel that there will be a slight performance decrease. The fact that there are some outliers where performance is drastically worse can easily be solved by running the algorithm several times in parallel, and only keeping the best run. This can be done under homomorphic encryption much like computing the minimum in Sect. 5.2, but will not be implemented in this paper.

Runtime. While we will have a more detailed discussion of the runtime of all our algorithms in Sect. 7, we would like to already present the performance gain at this point: Recall that we estimated that running the exact algorithm from Sect. 4 would take almost 1.5 years. In contrast, our Stabilized Algorithm can be run in 25.93 days, or less than a month. This is less than 5% of the runtime of the exact version.

Conclusion. In conclusion to this section, we feel that by modifying the K-Means-Algorithm, we have traded a very small amount of accuracy for the ability to perform clustering on encrypted data in a more reasonable amount of time, which is a functionality that has not been achieved previously. The next section will deal with an idea to improve runtimes even more.

6   Approach 3: The Approximate Version

We now present another modification which trades in a bit of accuracy for improved runtime. Due to space constraints, the details have been moved to Appendix C and we give only a high-level sketch at this point: Since the Compare function is linear in its inputs lengths, speeding up this building block would make the entire computation more efficient. First recall that we encode our numbers bitwise after having scaled them to integers. This means that we have access to the individual bits and can delete the S least significant bits, which corresponds to dividing the number by 2^S and truncating. Let $\tilde{X}$ denote this truncated version of a number X, and $\tilde{Y}$ that of a number Y. Then $\text{Compare}(\tilde{X}, \tilde{Y}) = \text{Compare}(X, Y)$ if $|X - Y| \geq 2^S$, and may or may not return the correct result if $|X - Y| < 2^S$. However, correspondingly, if the result is wrong, the centroid that is wrongly assigned to the data entry is no more than 2^S further from the data entry than the correct one. We propose to pick an initial S and decrease it over the course of the algorithm, so that accuracy increases as we near the end. We call this variant of the (stabilized) algorithm the *approximate* version.

In our experiments with $S = 5$, we saw that accuracy is comparable to the stabilized version, and the gain is around 210.7 min for the entire algorithm. Unfortunately, this is swallowed by the magnitude of the total computation time, as the main bottlenecks lie elsewhere. However, running just the comparison and approximate comparison functions with the same parameters as in our implementation of the K-Means-Algorithm (35 bits, 5 bits deleted for approximate comparison) yielded a drop in average runtime from 3.24 to 1.51 s. We see that this does make a big difference and may be of independent interest for computations involving many comparisons, which is why we choose to present the modification even though the effect was outweighed by other bottlenecks in the K-Means-Algorithm computation.

7   Implementation Results

We now present runtimes for the stabilized and approximate versions of the K-Means-Algorithm, and the times for the exact version using Fractional Encoding. Computations were done in a virtual machine with 20 GB of RAM and 4 cores, running an Intel i7-3770 processor with 3.4 GHz. We used the TFHE library [38] without the SPQLIOS_FMA-option, as our processor did not support this.

The dataset we used was the Lsun dataset from [39], which consists of 400 rational data entries of 2 dimensions, and $K = 3$ clusters. We encoded the binary numbers with 35 bits and scaled to integers using 2^{20}. The timings measured were for one round, and the approximate version used a deletion parameter of $S = 5$. For the Fractional Encoding, the data was encoded with nominator in $[2^{11}, 2^{12})$ and denominator in roughly the same range. We allotted 35 bits total for nominator and denominator each to allow a growth in required bitlength, and set the shortening parameter to 12, but shortened by 11 every once in a while (we derived this approach experimentally, see the discussion of the shortcoming of this approach in Sect. 4.4). The Fractional exact version was so slow that we ran it only on the first 10 data entries of the dataset - we will extrapolate the runtimes in Sect. 7.1.

7.1   Runtimes for the Entire Algorithm on a Single Core

We now present the runtimes for the entire K-Means-Algorithm on encrypted data on our specific machine with single-thread computation. There is some extrapolation involved, as the measured runtimes were for one round (so we multiplied by the round number, which differs between the exact version and the other two), and in the Fractional (exact) case, only for 10 data entries, so we multiplied that time by 40. Note that these times (which are with no parallelization) can be found in Table 2. We see that even though the stabilized version needs more rounds than the exact version, the latter is still significantly slower due to the Fractional Encoding. The approximate version (always with $S = 5$ deleted bits in the comparison) would save about 210.7 min.

Table 2. Single-thread runtimes (extrapolated) on our machine.

	Exact (fractional)	Stabilized	Approximate
Runtime per round	873.46 h (36.39 days)	15.56 h	15.47 h
Rounds required	15	40	40
Total runtime	545.91 days ≈ 17.95 months	25.93 days ≈ 0.85 months	25.79 days ≈ 0.85 months

7.2   Further Speedup

We would now like to address the subject of parallelism. At the moment (last accessed April 24th 2018), the TFHE library only supplies single-thread computations - i.e., there is no parallelism. However, version 1.5 is expected soon, and this will allegedly support multithreading. We first explain the huge difference this would make for the runtime, and then quantify the involved timings.

Parallelism. It is easy to see that all our versions of the K-Means-Algorithm are highly parallelizable: The `Cluster Assignment` step trivially so over the data entries (without any time needed for recombination), and the `Move Centroids` similarly over the cluster centroids (also over the data entries with very small recombination effort). Since both steps are linear in the number K of centroids, the number m of data entries, and the number T of round iterations, we present our runtimes in this subsection as *per centroid, per data entry, per round, per core*. This allows a flexible estimate for when multithreading is supported.

Round Runtimes. We now present the runtime results for each of the three variants on encrypted data per centroid, per data entry, per round, per core in Table 3. We do not include runtimes for encoding/encryption and decryption/decoding, as these would be performed on the user side, whereas the computation would be outsourced (encoding/encryption is ca. 1.5 s, and decoding/decryption is around 5 ms). We see that the Fractional Encoding is extremely slow, which motivated the Stabilized Algorithm in the first place.

Table 3. Runtimes per centroid, per data entry, per round, per core.

	Exact (fractional)	Stabilized	Approximate
`Cluster Assignment`	$1650.91\,\text{s} \approx 27.5\,\text{min}$	$35.59\,\text{s}$	$35.39\,\text{s}$
`Move Centroids`	$969.47\,\text{s} \approx 16.2\,\text{min}$	$11.09\,\text{s}$	$11.03\,\text{s}$
Total	$2620.38\,\text{s} \approx 43.7\,\text{min}$	$46.68\,\text{s}$	$46.42\,\text{s}$

A   Supplementary Material for the K-Means-Algorithm

This appendix contains some supplemental material for the K-Means-Algorithm.

A.1   Visualization of the K-Means-Algorithm

We first present a visualization of the K-Means-Algorithm in Fig. 5.

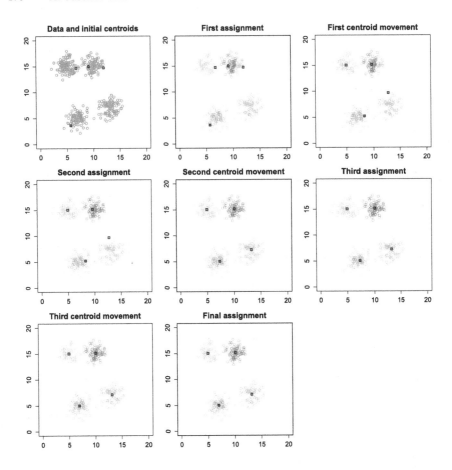

Fig. 5. An illustration of the K-Means-Algorithm.

A.2   Pseudocode

We now present the exact workings of the K-Means-Algorithm in Algorithm 3, where operations like addition and division are performed component-wise if applied to vectors.

Algorithm 3. The K-Means-Algorithm

Input: Data set $X = \{x_1, \ldots, x_m\}$ // $x_i \in \mathbb{R}^\ell$ for some ℓ
Input: Number of clusters K
Input: Number of iterations T
// Initialization
1 Randomly reorder X;
2 Set centroids $c_k = x_k$ for $k = 1$ to K;
  // Keep track of centroid assignments
3 Generate m-dimensional vector A;
  // Keep track of denominators in average computation
4 Generate K-dimensional vector $d = (d_1, \ldots, d_K)$;
5 **for** $j = 1$ *to* T **do**
        // Cluster Assignment
6   **for** $i = 1$ *to* m **do**
7         $\Delta = \infty$;
8   **for** $k = 1$ *to* K **do**
9            $\tilde{\Delta} := \|x_i - c_k\|_2$;
          // Check if current cluster is closer than previous closest
10   **if** $\tilde{\Delta} < \Delta$ **then**
              // If so, update Δ and assign data entry to current cluster
11              $\Delta = \tilde{\Delta}$;
12              $A_i = k$;
13   **end**
14   **end**
15   **end**
        // Move Centroids
16   **for** $k = 1$ *to* K **do**
17        $c_k = 0$;
18        $d_k = 0$;
19   **end**
20   **for** $i = 1$ *to* m **do**
        // Add the data entry to its assigned centroid
21        $c_{A_i} = c_{A_i} + x_i$;
        // Increase the appropriate denominator
22        $d_{A_i} = d_{A_i} 1$
23   **end**
24   **for** $k = 1$ *to* K **do**
        // Divide centroid by number of assigned data entries to get average
25        $c_k = c_k / d_k$;
26   **end**
27 **end**
   Output: $\{c_1, \ldots, c_K\}$

B   Operations for Fractional Encoding

This section presents how to build the elementary operations for Fractional Encoding from routines to perform addition, multiplication and comparison on numbers that are encoded in binary fashion. We denote these routines with $\mathtt{Add}(a, b), \mathtt{Mult}(a, b)$ and $\mathtt{Comp}(a, b)$, where the latter returns 1 (encrypted) if $a < b$ and 0 otherwise. Then if we want to operate on values encoded in this way, we can express the operations using the subroutines from the binary encoding as follows:

- $a + b$: $\texttt{FracAdd}((a_n, a_d), (b_n, b_d))$
  $= \big(\texttt{Add}(\texttt{Mult}(a_n, b_d), \texttt{Mult}(a_d, b_n)), \texttt{Mult}(a_d, b_d)\big)$
- $a \cdot b$: $\texttt{FracMult}((a_n, a_d), (b_n, b_d)) = \big(\texttt{Mult}(a_n, b_n), \texttt{Mult}(a_d, b_d)\big)$
- a/b : $\texttt{FracDiv}((a_n, a_d), (b_n, b_d)) = \big(\texttt{Mult}(a_n, b_d), \texttt{Mult}(a_d, b_n)\big)$
- $a \leq b$: $\texttt{FracComp}((a_n, a_d), (b_n, b_d))$:
  This is slightly more involved. Note that the MSB determines the sign of the number (1 if it is negative and 0 otherwise). Let

$$c := \text{Sign}(a_d) \oplus \text{Sign}(b_d),$$

and let

$$\texttt{MUX}(c, a, b) = \begin{cases} a, & c = 1 \\ b, & c = 0 \end{cases}$$

be the multiplexer gate.
Then we set

$$d := \texttt{MUX}(c, \texttt{Mult}(a_n, b_d), \texttt{Mult}(a_d, b_n))$$

and

$$e := \texttt{MUX}(c, \texttt{Mult}(a_d, b_n), \texttt{Mult}(a_n, b_d))$$

and output the result as $\texttt{Comp}(e, d)$.
A more detailed explanation can be found in the extended version of this paper.

C   Details of the Approximate Algorithm

In this section, we present the details of the approximate version of our algorithm.

C.1   The Algorithm

Recall the main idea: Since the $\texttt{Compare}$ function is linear in the length of its inputs, speeding up this building block would make the entire computation more efficient. To do this, first recall that we encode our numbers in a bitwise fashion after having scaled them to integers. This means that we have access to the individual bits and can, for example, delete the S least significant bits, which corresponds to dividing the number by 2^S and truncating. Let $\tilde{X}$ denote this truncated version of a number X, and $\tilde{Y}$ that of a number Y. Then $\texttt{Compare}(\tilde{X}, \tilde{Y}) = \texttt{Compare}(X, Y)$ if $|X - Y| \geq 2^S$, and may or may not return the correct result if $|X - Y| < 2^S$. However, correspondingly, if the result is wrong, the centroid that is wrongly assigned to the data entry is no more than 2^S further from the data entry than the correct one. We propose to pick an initial S and decrease it over the course of the algorithm, so that accuracy increases as we near the end. The exact workings of this approximate comparison, denoted $\texttt{ApproxCompare}$, can be seen in Algorithm 4.

Algorithm 4. `ApproxCompare`(X, Y, S)

Input: The two arguments X, Y, encoded bitwise
Input: The accuracy factor S
   // Corresponds to $\tilde{X} = \lfloor X/2^S \rfloor$
1   Remove last S bits from X, denote $\tilde{X}$;
   // Corresponds to $\tilde{Y} = \lfloor Y/2^S \rfloor$
2   Remove last S bits from Y, denote $\tilde{Y}$;
   // Regular comparison function, $C \in \{0, 1\}$
3   $C = $ `Compare`$(\tilde{X}, \tilde{Y})$;
Output: C

C.2   Evaluation

In this section, we compare the performance of the stabilized K-Means-Algorithm using this approximate comparison, denoted simply by "Approximate Version", to the original and stabilized K-Means-Algorithm on our data sets.

Accuracy. Recall from Sect. 5.1 that we scaled the data with the factor 2^{20} and truncated to obtain the input data. This means that for $S = 5$, a wrongly assigned centroid would be at most 2^5 further from the data entry than the correct centroid on the scaled data - or no more than 2^{-15} on the original data scale. We set $S = \min\{7, (T/5) - 1\}$ where T is the number of iterations, and

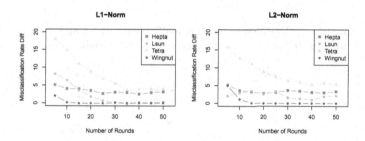

Fig. 6. Average difference in misclassification rate for approximate vs. stabilized algorithm ((average % mislabeled approximate) − (average % mislabeled stabilized)).

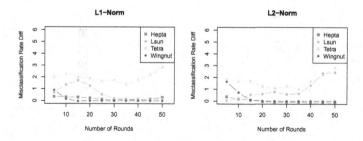

Fig. 7. Average difference in misclassification rate for approximate vs. stabilized algorithm ((average % mislabeled approximate) − (average % mislabeled stabilized)).

reduce S by one every 5 rounds. We again examine the average (over 1000 iterations) difference in the misclassification rate to both the exact algorithm and the stabilized algorithm.

The results for both distance metrics can be seen in Figs. 6 and 7. We see that again, $T = 40$ iterations is a reasonable choice because the algorithms do not seem to converge further with more rounds. We now again look at

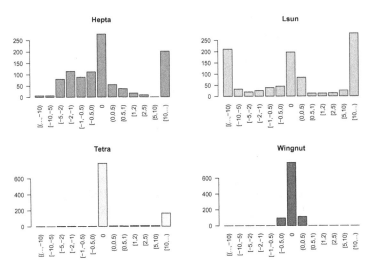

Fig. 8. Distribution of the difference in misclassification rate for approximate vs. exact K-Means-Algorithm $\big((\%\text{ mislabeled approximate}) - (\%\text{ mislabeled exact})\big)$, L_1-norm.

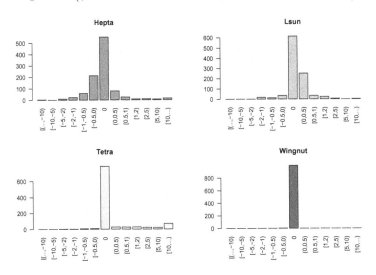

Fig. 9. Distribution of the difference in misclassification rate for approximate vs. stabilized K-Means-Algorithm $\big((\%\text{ mislabeled approx.}) - (\%\text{ mislabeled stab.})\big)$, L_1-norm.

the distribution of the ratios in Fig. 8 (for the approximate versus the exact K-Means-Algorithm) and Fig. 9 (for the approximate versus the stabilized K-Means-Algorithm). Figures for the L_2-norm can be found in the extended version of this paper.

We see that usually, the approximate version performs only slightly worse than the stabilized version. There is still the effect in the Lsun dataset that the approximate version outperforms the original K-Means-Algorithm in a significant amount of cases (though this effect mostly occurs for the L_1-norm), but it rarely does better than the stabilized version. This is not surprising, as it is in essence the stabilized version but with an opportunity for errors.

Runtime. We now examine how much gain in terms of runtime we have from this modification. Recall that it took about 1.5 years to run the exact algorithm, and 25.93 days to run the stabilized version. The approximate version runs in 25.79 days, which means a difference of about 210.7 min.

Obviously, the effect of the approximate comparison is not as big as anticipated. This is due to the bottleneck actually being the computation of the L_1-norm rather than the FindMin-procedure. Thus, for this specific application, the approximate version may not be the best choice - however, for an algorithm that has a high number of comparisons relative to other operations, there can still be huge performance gains in terms of runtime. To see this, we ran just the comparison and approximate comparison functions with the same parameters as in our implementation of the K-Means-Algorithm (35 bits, 5 bits deleted for approximate comparison). The average (over 1000 runs each) runtime was 3.24 s for the regular comparison and 1.51 s for the approximate comparison. We see that this does make a big difference, which is why we choose to present the modification even though the effect was outweighed by other bottlenecks in the K-Means-Algorithm computation.

Conclusion. In conclusion, the approximate comparison provides the user with an easy method of trading in accuracy for faster computation, and most importantly, this loss of accuracy can be decreased as computations near the end. However, for the specific application of the K-Means-Algorithm, these gains were unfortunately swallowed by the rest of the computation.

References

1. Aggarwal, C.C., Hinneburg, A., Keim, D.A.: On the surprising behavior of distance metrics in high dimensional space. In: Van den Bussche, J., Vianu, V. (eds.) ICDT 2001. LNCS, vol. 1973, pp. 420–434. Springer, Heidelberg (2001). https://doi.org/10.1007/3-540-44503-X_27
2. Armknecht, F., et al.: A guide to fully homomorphic encryption. IACR Cryptology ePrint Archive (2015/1192)
3. Armknecht, F., Katzenbeisser, S., Peter, A.: Group homomorphic encryption: characterizations, impossibility results, and applications. DCC **67**, 209–232 (2013)

4. Barnett, A., et al.: Image classification using non-linear support vector machines on encrypted data. IACR Cryptology ePrint Archive (2017/857)
5. Bonte, C., Vercauteren, F.: Privacy-preserving logistic regression training. IACR Cryptology ePrint Archive 233 (2018)
6. Bos, J.W., Lauter, K.E., Naehrig, M.: Private predictive analysis on encrypted medical data. J. Biomed. Inform. **50**, 234–243 (2014)
7. Bost, R., Popa, R.A., Tu, S., Goldwasser, S.: Machine learning classification over encrypted data. In: NDSS (2015)
8. Brakerski, Z., Gentry, C., Vaikuntanathan, V.: Fully homomorphic encryption without bootstrapping. In: ECCC, vol. 18 (2011)
9. Bunn, P., Ostrovsky, R.: Secure two-party k-means clustering. In: CCS (2007)
10. Chabanne, H., de Wargny, A., Milgram, J., Morel, C., Prouff, E.: Privacy-preserving classification on deep neural network. IACR Cryptology ePrint Archive (2017/035)
11. Chen, H., Laine, K., Player, R.: Simple encrypted arithmetic library - SEAL v2.1. IACR Cryptology ePrint Archive 2017, 224 (2017)
12. Chillotti, I., Gama, N., Georgieva, M., Izabachène, M.: Faster fully homomorphic encryption: bootstrapping in less than 0.1 seconds. In: Cheon, J.H., Takagi, T. (eds.) ASIACRYPT 2016. LNCS, vol. 10031, pp. 3–33. Springer, Heidelberg (2016). https://doi.org/10.1007/978-3-662-53887-6_1
13. Coron, J.-S., Lepoint, T., Tibouchi, M.: Scale-invariant fully homomorphic encryption over the integers. In: Krawczyk, H. (ed.) PKC 2014. LNCS, vol. 8383, pp. 311–328. Springer, Heidelberg (2014). https://doi.org/10.1007/978-3-642-54631-0_18
14. Coron, J.-S., Naccache, D., Tibouchi, M.: Public key compression and modulus switching for fully homomorphic encryption over the integers. In: Pointcheval, D., Johansson, T. (eds.) EUROCRYPT 2012. LNCS, vol. 7237, pp. 446–464. Springer, Heidelberg (2012). https://doi.org/10.1007/978-3-642-29011-4_27
15. van Dijk, M., Gentry, C., Halevi, S., Vaikuntanathan, V.: Fully homomorphic encryption over the integers. In: Gilbert, H. (ed.) EUROCRYPT 2010. LNCS, vol. 6110, pp. 24–43. Springer, Heidelberg (2010). https://doi.org/10.1007/978-3-642-13190-5_2
16. Ducas, L., Micciancio, D.: FHEW: bootstrapping homomorphic encryption in less than a second. In: Oswald, E., Fischlin, M. (eds.) EUROCRYPT 2015. LNCS, vol. 9056, pp. 617–640. Springer, Heidelberg (2015). https://doi.org/10.1007/978-3-662-46800-5_24
17. Esperança, P.M., Aslett, L.J.M., Holmes, C.C.: Encrypted accelerated least squares regression. In: Singh, A., Zhu, X.J. (eds.) AISTATS (2017)
18. Fan, J., Vercauteren, F.: Somewhat practical fully homomorphic encryption. IACR Cryptology ePrint Archive (2012/144)
19. Gentry, C.: A fully homomorphic encryption scheme. Ph.D. thesis, Stanford University (2009)
20. Gentry, C., Sahai, A., Waters, B.: Homomorphic encryption from learning with errors: conceptually-simpler, asymptotically-faster, attribute-based. In: Canetti, R., Garay, J.A. (eds.) CRYPTO 2013. LNCS, vol. 8042, pp. 75–92. Springer, Heidelberg (2013). https://doi.org/10.1007/978-3-642-40041-4_5
21. Gilad-Bachrach, R., Dowlin, N., Laine, K., Lauter, K.E., Naehrig, M., Wernsing, J.: CryptoNets: applying neural networks to encrypted data with high throughput and accuracy. In: ICML (2016)

22. Graepel, T., Lauter, K., Naehrig, M.: ML confidential: machine learning on encrypted data. In: Kwon, T., Lee, M.-K., Kwon, D. (eds.) ICISC 2012. LNCS, vol. 7839, pp. 1–21. Springer, Heidelberg (2013). https://doi.org/10.1007/978-3-642-37682-5_1

23. Halevi, S., Shoup, V.: Algorithms in HElib. In: Garay, J.A., Gennaro, R. (eds.) CRYPTO 2014. LNCS, vol. 8616, pp. 554–571. Springer, Heidelberg (2014). https://doi.org/10.1007/978-3-662-44371-2_31

24. Jagannathan, G., Pillaipakkamnatt, K., Wright, R.N., Umano, D.: Communication-efficient privacy-preserving clustering. Trans. Data Priv. **3**, 1–25 (2010)

25. Jagannathan, G., Wright, R.N.: Privacy-preserving distributed k-means clustering over arbitrarily partitioned data. In: SIGKDD (2005)

26. Jäschke, A., Armknecht, F.: Accelerating homomorphic computations on rational numbers. In: Manulis, M., Sadeghi, A.-R., Schneider, S. (eds.) ACNS 2016. LNCS, vol. 9696, pp. 405–423. Springer, Cham (2016). https://doi.org/10.1007/978-3-319-39555-5_22

27. Jäschke, A., Armknecht, F.: (Finite) field work: choosing the best encoding of numbers for FHE computation. In: Capkun, S., Chow, S. (eds.) Cryptology and Network Security. CANS 2017, vol. 11261, pp. 482–492. Springer, Cham (2017). https://doi.org/10.1007/978-3-030-02641-7_23

28. Jha, S., Kruger, L., McDaniel, P.: Privacy preserving clustering. In: di Vimercati, S.C., Syverson, P., Gollmann, D. (eds.) ESORICS 2005. LNCS, vol. 3679, pp. 397–417. Springer, Heidelberg (2005). https://doi.org/10.1007/11555827_23

29. Kim, A., Song, Y., Kim, M., Lee, K., Cheon, J.H.: Logistic regression model training based on the approximate homomorphic encryption. IACR Cryptology ePrint Archive (254) (2018)

30. Kim, M., Song, Y., Wang, S., Xia, Y., Jiang, X.: Secure logistic regression based on homomorphic encryption. IACR Cryptology ePrint Archive (074) (2018)

31. Liu, X., et al.: Outsourcing two-party privacy preserving k-means clustering protocol in wireless sensor networks. In: MSN (2015)

32. Lu, W., Kawasaki, S., Sakuma, J.: Using fully homomorphic encryption for statistical analysis of categorical, ordinal and numerical data. IACR Cryptology ePrint Archive (2016/1163)

33. MacQueen, J., et al.: Some methods for classification and analysis of multivariate observations. In: Proceedings of the Fifth Berkeley Symposium on Mathematical Statistics and Probability (1967)

34. Meskine, F., Bahloul, S.N.: Privacy preserving k-means clustering: a survey research. Int. Arab J. Inf. Technol. **9**, 194–200 (2012)

35. Naehrig, M., Lauter, K.E., Vaikuntanathan, V.: Can homomorphic encryption be practical? In: CCSW (2011)

36. Phong, L.T., Aono, Y., Hayashi, T., Wang, L., Moriai, S.: Privacy-preserving deep learning via additively homomorphic encryption. IACR Cryptology ePrint Archive (2017/715)

37. Smart, N.P., Vercauteren, F.: Fully homomorphic encryption with relatively small key and ciphertext sizes. In: Nguyen, P.Q., Pointcheval, D. (eds.) PKC 2010. LNCS, vol. 6056, pp. 420–443. Springer, Heidelberg (2010). https://doi.org/10.1007/978-3-642-13013-7_25

38. TFHE Library. https://tfhe.github.io/tfhe

39. Ultsch, A.: Clustering with SOM: U* c. In: Proceedings of Workshop on Self-Organizing Maps (2005)

40. Vaidya, J., Clifton, C.: Privacy-preserving k-means clustering over vertically partitioned data. In: SIGKDD (2003)
41. Wu, D.J., Feng, T., Naehrig, M., Lauter, K.E.: Privately evaluating decision trees and random forests. PoPETs, (4) (2016)
42. Xing, K., Hu, C., Yu, J., Cheng, X., Zhang, F.: Mutual privacy preserving k-means clustering in social participatory sensing. IEEE Trans. Ind. Inform. **13**, 2066–2076 (2017)

Profiled Power Analysis Attacks Using Convolutional Neural Networks with Domain Knowledge

Benjamin Hettwer[1]([✉]) [ID], Stefan Gehrer[1] [ID], and Tim Güneysu[2] [ID]

[1] Robert Bosch GmbH, Corporate Sector Research, Stuttgart, Germany
{benjamin.hettwer,stefan.gehrer}@de.bosch.com
[2] Horst Görtz Institute for IT-Security, Ruhr University Bochum, Bochum, Germany
tim.gueneysu@rub.de

Abstract. Evaluation of cryptographic implementations against profiled side-channel attacks plays a fundamental role in security testing nowadays. Recently, deep neural networks and especially Convolutional Neural Networks have been introduced as a new tool for that purpose. Although having several practical advantages over common Gaussian templates such as intrinsic feature extraction, the deep-learning-based profiling techniques proposed in literature still require a suitable leakage model for the implementation under test. Since this is a crucial task, we are introducing domain knowledge to exploit the full power of approximating very complex functions with neural networks. By doing so, we are able to attack the secret key directly without any assumption about the leakage behavior. Our experiments confirmed that our method is much more efficient than state-of-the-art profiling approaches when targeting an unprotected hardware and a protected software implementation of the AES.

Keywords: Side-channel attacks · Deep learning
Convolutional Neural Networks

1 Introduction

Power-based Side-Channel Attacks (SCAs) are a well-known and powerful class of threats for security enabled devices, for example in context of the Internet of Things. They exploit information leakages gained from the power consumption or electromagnetic emanations of a device to extract secret information such as cryptographic keys, even though the employed algorithms are mathematically sound. This is caused by the correlation of power consumption and processed data. Since the advent of power-based SCAs by Kocher et al. in 1999 [14], numerous papers have been published on this topic. Most of them fit into one of the following categories:

© Springer Nature Switzerland AG 2019
C. Cid and M. J. Jacobson, Jr. (Eds.): SAC 2018, LNCS 11349, pp. 479–498, 2019.
https://doi.org/10.1007/978-3-030-10970-7_22

Non-profiled SCAs techniques aim to recover the secret key by performing statistical calculations on power measurements of the device under attack regarding a hypothesis of the device's leakage. Typical examples are Differential Power Analysis [15], Correlation Power Analysis [5], and Mutual Information Analysis [9].

Profiled SCAs assume a stronger adversary who is in possession of a profiling device. It is an open copy of the attacked device which the adversary can manipulate to characterize the leakages very precisely in a first step. Once this has been done, the built model can be used to attack the actual target device in the key extraction phase. Template Attacks (TAs) [7], Stochastic attacks [26] and machine-learning-based attacks [3,12,16] are common approaches in this area.

In the same manner, researchers and industry developed methods to counteract SCAs. *Masking*, for instance, aims for randomizing intermediate values that are internally processed by the cryptographic device in order to break the connection between the secret (respectively some intermediate value that depends on the secret) and its power footprint [19]. The concept of *Hiding* countermeasures are different from masking in a sense that their goal is to change the power characteristics directly. This can be achieved, for example, by making every operation consume the same amount of energy. However, it has been shown that protected implementations can be broken as well, whereby particularly profiled SCAs are a reasonable choice [10,23].

There is a recent line of work that deals with the application of *Deep Learning (DL)* techniques for profiled side-channel analysis. A common factor that motivates the usage of DL models in general is that they intrinsically incorporate a feature extraction mechanism. That is, unlike most standard Machine Learning (ML) classifiers, DL models can learn from the raw input data set as they are able to identify the most informative data themselves without human engineering. Within the SCA community, Maghrebi et al. [18] showed in a series of experiments that DL can outperform TAs and standard ML techniques like support vector machines when targeting hard- and software implementations of AES. One year later, Cagli et al. [6] investigated *Convolutional Neural Networks (CNNs)* combined with data augmentation to defeat cryptographic implementations which are protected with different jitter-based countermeasures. Again, better results were reported for the DL network compared to TAs with manual trace realignment. Summarizing the insights of the two studies, it becomes evident that DL techniques and in particular CNNs gives two major advantages that make them interesting for profiled SCAs:

- They are able to automatically extract the areas in the side-channel traces which contains the most information. When using standard SCA techniques, the selection of the so-called Points of Interests (POIs) is often done manually as preprocessing step ahead of the actual attack. This is not only tedious, but also error prone as proper POI selection has shown to have a significant impact on the attack efficiency [35].

– CNNs are invariant to small input modifications such as noise (also artificially generated). Furthermore, they integrate time samples from the complete traces efficiently (meaning they require fewer parameters which needs to be optimized during training) for their decision. This property enables them to perform a higher-order SCA and defeat masking countermeasures.

All studies on deep-learning-based SCAs assumed that the attacker has some implicit knowledge about the leakage behavior of the attacked implementation. However, the choice of an adequate leakage model (i.e., an approximation of the physical signal that is generated by the device when computing some sensitive intermediate value) is usually crucial for the success of SCAs [8] and heavily depends on how much information about the target architecture is available to the adversary. Since this is may be difficult to determine upfront, we present a black-box approach for evaluating cryptographic implementations without a leakage model by using CNNs with Domain Knowledge (DK) neurons.

1.1 Contributions

The contributions of this paper are twofold:

1. We introduce a novel CNN architecture for profiled SCAs which allows to encode domain specific information. By doing so, it is possible to feed the plaintext or ciphertext as an additional source of information into the network (apart from the power measurements). The CNN with DK is dedicated to autonomously learn the leakage of the device with regard to the secret key.
2. We perform practical experiments with an unprotected hardware and a protected software implementation of AES. The results confirm that our method reduces the search space for breaking the secret key in the attack phase by at least three orders of magnitude for the hardware implementation, and more than ten orders in case of the protected software implementation.

The rest of this paper is structured as follows: In Sect. 2, background on profiled SCAs, Neural Networks (NNs) and DL is provided. Section 3 introduces CNNs and our architectural extension with domain neurons. In Sect. 4, the results of our experiments are presented and discussed. The last Section summarizes the paper and gives insights on possible future work.

2 Preliminaries

This section serves as entry point to profiled SCAs, NNs and DL. We refer the reader to [19] for a more profound introduction into power-based SCAs, and [11] for a comprehensive summary on NN and DL.

2.1 Profiled Side-Channel Analysis

Profiled SCAs are considered as the most powerful type of SCAs and are divided in two phases. In a first step, the adversary takes advantage of a profiling device on which he can fully control input and secret key parameters of the cryptographic algorithm. He uses that to acquire a set of N_P profiling side-channel traces $X \in \mathbb{R}^D$, where D denotes the number of sample points in the measurements. Let $V = g(t, k)$ be a random variable representing the result of an intermediate operation of the target cipher which depends partly on public information t (plaintext or ciphertext chunk) and secret key $k \in \mathcal{K}$, where $\mathcal{K}$ is the set of possible key values. V is assumed to have an influence on the deterministic part of the side-channel measurements. The ultimate goal of the attacker during the profiling phase is then to estimate the probability:

$$\Pr[X|V = v] \tag{1}$$

for every possible value $v \in V$ from the profiling base $\{X_i, v_i\}_{i=1,\ldots,N_P}$. In TAs for example, the Probability Density Function (PDF) of (1) is assumed to be multivariate Gaussian and can described by the parameter pairs (μ_v, Σ_v), depicting the mean values and covariance matrices for the corresponding values of v [7].

During the attack phase, the adversary generates a new set of N_A attack traces from the actual target device (which is structurally identical to the profiling device) whereby the secret key k is fixed and unknown. In order to retrieve it, estimations for all possible key candidates $k^* \in \mathcal{K}$ are made and combined following the *maximum likelihood* strategy such that:

$$k = \underset{k^* \in \mathcal{K}}{\operatorname{argmax}} \prod_{i=1}^{N_A} \Pr[V = v_i | X_i] \tag{2}$$

where the probabilities on the right are retrieved with the help of the built profile and public information t which is also available for the attack traces. In order to avoid numerical instabilities, it is common to process the logarithms of the likelihoods.

Although the Gaussian model assumption is often fairly realistic in practice [19], arbitrary functions of the side-channel leakage cannot be captured with templates. In settings where the PDF of the leakage is not known upfront, ML-based profiling methods are more promising. Another issue that comes with TAs is the necessity to find a small number of POIs in the high-dimensional side-channel measurements. This is due to size restriction of the covariance matrices Σ_v, which are $(N_S \times N_S)$ large when N_S is the number of POIs. In order to discover the POIs, dimensionality reduction techniques such as (PCA) can be employed. PCA captures the data with the largest variance and thus helps to reduce the amount of noise in the traces. That is why PCA is a heavily used technique in side-channel analysis, not only for TAs, but also in settings where the profiling is done with ML techniques [16,25,33]. However, in general one can say that ML-based attacks are more suitable when it is difficult to restrict the number of POIs effectively [17].

2.2   Neural Networks and Deep Learning

NNs were partly inspired by biological learning systems (e.g. the human brain) and date back at least into the 1960s. They are nowadays the privileged choice for supervised classification tasks. For these, the learning system is fed with training examples from a data set consisting of input data vectors (=features) and associated outcome measurement (=label) and the goal is to find a suitable relationship in order to map new inputs to the correct label. Note that in the context of profiled SCA, the first set is equal to the profiling base N_P and the second one corresponds to the attack set N_A.

NNs are composed of densely interconnected units called neurons, which take a number of real-valued inputs and produce a single real-value output [20]. The simplest type of a NN is the *perceptron*. As illustrated in Fig. 1, it receives a vector of input features $X = (x_1, \ldots, x_D)$ and performs a linear combination with the weight values $w_1, \ldots, w_D$ of its input connections and a bias value w_0. The result is passed through an Activation (ACT) function f, e.g., the Rectified Linear Unit (ReLU) [21] in order to calculate the output value $\tilde{y}$. For learning the perceptron, the weights are adjusted according to the training data set.

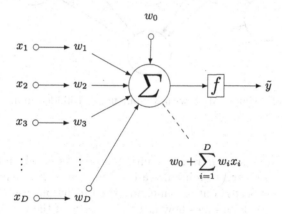

Fig. 1. Perceptron

Single-layer perceptrons are only able to represent functions whose underlying data set is linearly separable such as the Boolean AND function. To overcome this limitation and represent more complex mappings, many perceptrons can be stacked together to form a whole network which are generally referred to as *Multi-Layer Perceptrons (MLPs)*. An MLP consists of three types of units, typically arranged in layers as shown in Fig. 2. The input layer is just a representation of the raw input features. All neurons of the input layer are connected to each neuron of the following hidden layer. The number of hidden layers in an MLP and the number of units per hidden varies, depending on the required model capacity to fit the training data. In general, too many units in the hidden layer may lead to overfitting, while underestimating the number of neurons has an negative effect on the classification performance of the MLP [11]. The units in the output layer, finally, directly correspond to the predictions of the classification problem to solve.

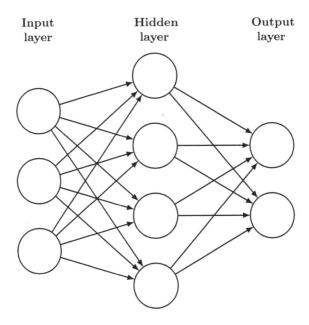

Fig. 2. Example of a simple MLP with 3 input units, 4 hidden units, 2 output units (bias units omitted).

Training the MLP is an iterative, multi-step process by which the weights of the network are optimized to minimize a loss function, which depicts the difference between the expected output label and the prediction result. The learning rate hyperparameter determines how fast the weights of the network are driven towards the optimal solution. In practice, optimizer algorithms such as Stochastic Gradient Descent (SGD) or ADAM are employed for that purpose [11].

In recent years there has been a growing interest in NN models with multiple hidden layers stacked upon each other, which are commonly specified under the term *deep learning*. It is a particular powerful type of ML techniques that are able to represent the learning task as nested hierarchy of concepts, where more abstract concept representations are built from simpler ones. The usage of deep NNs is motivated by the fact that they have and outperformed classical ML approaches in solving central problems in artificial intelligence such as speech recognition and image classification. These tasks usually deal with high-dimensional data which makes it exponentially more difficult to learn a classifier that generalizes well on unseen examples, a challenge that is also known as the curse of dimensionality [11]. Since this applies in exactly the same manner for the SCA domain as discussed before, deep NNs and especially CNNs seem like a promising choice as tool for profiled SCAs.

3 Convolutional Neural Networks

In this section, we first describe the primary building blocks of CNNs until we present our architectural extension with DK neurons.

3.1 Core Constructions

CNNs tackle the problem of large input data dimensions by including task-specific mechanisms into their architecture that allow to reduce the number of parameters of the model, while keeping or even increasing the accuracy of the network [22]. CNNs are primarily used in the field of pattern recognition within images, however they can also be used to process 1-D time-series data (as it is the case for side-channel traces). Additional to the Fully-Connected (FC) layers used in classical MLPs, CNNs include two other types of layers, namely Convolutional (CONV) layers and Pooling (POOL) layers:

CONV layers determine the output of neurons which are connected to small spatial regions of the input by calculating the scalar product with a set of so-called kernels or filters as illustrated in Fig. 3. The movement policy of the filters can be modified by the strides parameter. The weight parameters of the kernels are learned to activate when they detect a specific feature or pattern at a certain position in the input. In order to perceive enough information, different filters are used yielding several outputs which increases the depth of the network. CONV layers are to some extent, shift, scale, and distortion invariant. This property has shown to be very useful against de-synchronized side-channel traces [6].

POOL layers perform downsampling of their given input in order to reduce the number of parameters and the computational complexity of the network, by considering the max (=max-pooling) or average (=average-pooling) of a certain spatial extent as the output. They are important for getting low-dimensional abstract feature representations and compressing the information that is extracted in the CONV layers.

Apart from the CONV and POOL layers which are specific for CNNs, there are two additional techniques that can be found in common architectures of CNNs. These are dropout and batch normalization:

Dropout is a regularization technique that helps the network to increase generalization and reduce the phenomena of overfitting [28]. The key idea of dropout is to randomly drop units (along with their connections) from the NN during training. The probability to drop a unit can be controlled by the probability coefficient $P_{Drop} \in [0, 1)$. Because of that, dropout can be seen as an ensemble method that combines a exponential number of different "thinned" NN architectures efficiently during training. At test time, a single network with downscaled weights is used for predictions.

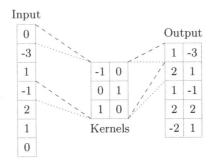

Fig. 3. Example of a 1-D convolution operation with 2 kernels of length 3 and stride of 1. The output is formed by applying the kernel to each part of the input (as with a sliding window).

Batch Normalization was introduced by Ioffe et al. [13] to establish a stable distribution of activation values throughout the whole layered structure of a network. A stable distribution makes the network more robust to parameter tuning since the input of one layer depends on the output of the previous layer. Therefore, normalization is incorporated into the network architecture by applying it to each mini-batch of training examples. This eventually allows the usage of higher learning rates.

3.2 Principal Architecture

Following the Input (IN) layer, CNN architectures typically consist of repetitive blocks of CONV and POOL layers. The basic concepts of sparse, local connectivity, weight sharing and subsampling enable the network to extract more abstract representations of given inputs, until spatial output dimensions are small enough to be connected to subsequent FC layers. Additionally, the use of non-linear ACT functions such as ReLU or sigmoid right after each CONV and FC layer enables the network to learn more complex functions. In a classification setting (as for example the one we describe in Sect. 2.1), the neurons of the last layer in the network output probabilities over discrete classes. These are calculated by means of the *Softmax (SOFT)* function.

To sum up, the architecture of a typical CNN consists of two major parts. A feature extractor and a feature combinator. The feature extractor consists of alternating CONV and POOL layers. It yields low dimensional representations of the input (in our case a side-channel trace), giving crucial information to the subsequent layers for solving the classification task. FC layers act as feature combinators and connect information to the desired output. A current CNN can therefore characterized by the following construction:

$$\text{IN} \circ [\text{CONV} \circ \text{ACT} \circ \text{POOL}]^{n_1} \circ [\text{FC} \circ \text{ACT}]^{n_2} \circ \text{FC} \circ \text{SOFT}$$

where n_1 and n_2 denote the number of feature extractor blocks, respectively the number of FC layers used.

3.3   CNNs with Domain Knowledge Neurons

In our approach, we study the effect of additional DK neurons in the CNN architecture for profiled SCAs. Their addition is motivated by the fact that merging domain specific information with extracted features of the CONV layers enables the network to go into different statistics at decision level [32]. In that sense, we propose a multimodal CNN with late information fusion strategy where additional public data is fed to the network in order to increase the efficiency of the attack. Since we have targeted the first byte of the AES key in our experiments as described in Sect. 4, we decided to use the corresponding plaintext byte as input for the DK neurons. However, it is also conceivable to exploit other related data that is available to the attacker (e.g. the ciphertext or information about the internal structure of the attacked implementation). Introducing a chunk of the plaintext into the network as second input brings two major advantages that motivates our approach:

– We do not have to stick to a certain leakage model. Instead of assuming that the attacked implementation leaks information regarding a certain operation for which we do the profiling (for example the output of the AES S-Box, respectively the hamming weight of the S-Box), we directly use the secret key k as a label. By doing so, we give the network the ability to autonomously learn the most meaningful representation of the leakage which is needed to classify the used secret key.
– The second advantage is a direct consequence of our generic leakage model. In the attack phase, we do not make a key guess on all possible candidates and combine the estimations on it via maximum-likelihood as seen in Sect. 2.1. Instead, the network gives us a direct key estimation in form of the probabilities:

$$\Pr[k|X, t] \tag{3}$$

for every attack trace X and associated plaintext t. This leads to a faster convergence of the key rank as we will see later in the experiments section.

Our developed CNN architecture is illustrated in Fig. 4. A detailed description is given in Table 2 in the Appendix. In summary, the feature extractor part of the model consists of three CONV layers and two POOL layers. All CONV layers use the same kernel size of eight, but the number of filters is increased from eight, to 16, up to 32. Dimensionality reduction of the features is reached by max-pooling across two data points after the first two CONV layers. After flattening the spatial depth of the feature extractors into a single dimension, it is concatenated with the input of the DK neurons. Since we merge one byte of the plaintext one-hot encoded into network, the DK layer contains 256 neurons (one for each possible value). One-hot encoding represent the plaintext byte as vector of 256 binary variables where only the correct value is set to one. The information from the feature extractor part and the DK neurons is combined by a following FC layer. The Output (OUT) layer consists of 256 neurons as we make a classification for one key byte. In order to avoid overfitting, four dropout layers are included into the network architecture with a consistent dropout rate $P_{Drop} = 0.2$. Furthermore, batch normalization is employed after all CONV and FC layers. Throughout the network, ReLU is used as an activation function.

We stress that our CNN design is not the product of some architectural optimization technique. It was rather obtained by following best practices for developing deep NN architectures [27], and examination of related work [6,18].

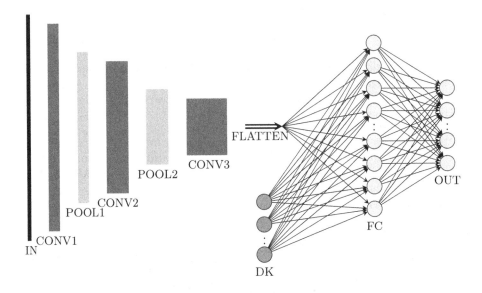

Fig. 4. Simplified visualization of CNN with domain input neurons.

4   Experiments

In the following section, we present our experimental results. After explaining the general attack setup, we compare our CNN with DK approach against four different profiling attacks from literature regarding attack efficiency when targeting an unprotected hardware and a protected software implementation of AES.

4.1   Baseline

For our experiments, we have implemented three deep NNs which were proposed in literature as baseline for our CNN with DK neurons. An overview of the evaluated models and associated target operations compared to our approach is given in Table 1. The numbers in the first column represent the number of layers with trainable weights. We chose these networks as reference, since the proposing authors applied them to break the same or very similar targets (unprotected hardware and protected software implementations of AES). Additionally, we performed a classical TA for both attacked data sets. In all experiments we aim to recover the first byte of the AES key. However, we stress that if one is able to retrieve one byte of the key successfully, the remaining bytes can be attacked likewise.

Table 1. Overview of implemented attacks

Type	Profiling target (Label)	Source
2-layer MLP	$V = \text{S-box}(t[0], k[0])$	[18]
3-layer CNN	$V = \text{S-box}(t[0], k[0])$	[18]
5-layer CNN	$V = \text{S-box}(t[0], k[0])$	[24]
TA	$V = \text{S-box}(t[0], k[0])$	[7]
5-layer CNN w/ DK	$k[0]$	This paper

Not all baseline models are described in the same level of detail in the according papers. For example, the activation functions for the MLP in [18] are not given. Therefore, we performed a so-called *grid search* for estimating the missing hyperparameters that are needed to rebuild and train the networks. It works as follows: First, an interval or set of possible values has to be selected for each parameter that should be optimized. Grid search is then just a simple strategy that tries all possible parameter combinations over the predefined ranges. We list the optimized parameters and associated search intervals for each of the evaluated models in Table 3 in the Appendix due space restrictions. The applied methodology, however, has been the same for all attacks and is described in the following section.

4.2 Methodology

Data Sets. For the conducted experiments, we have considered data sets of $N_P = 200\,000$ profiling traces with random plaintext and keys. The number of attack traces N_A with random plaintext but fixed secret key k varies for the attacks. We have used four sets each having $10\,000$ attack traces for the unprotected hardware implementation, and two sets each containing $10\,000$ attack traces for the protected software implementation. All attack sets were acquired with a different key in order to prevent any bias in the results due to overfitting to a certain key value.

Evaluation Metric. A single, well-know metric from the SCA domain has been used to evaluate the performance of the attacks: The Key Guessing Entropy (KGE) or key rank function. It is a technique which quantifies the difficulty to retrieve the correct value of the key regarding the required number of attack traces [29]. In principle, the KGE is calculated by summing up the log-likelihoods obtained in Eq. (2) over all key guesses $k^* \in \mathcal{K}$ (respectively the log-likelihoods of (3)) and do a ranking of the result. This ranking is updated after each attack trace. The KGE has the advantage of taking the full information on the probability distributions that are given in (2) or (3) into account, whereas the standard accuracy metric from the DL domain only considers the label with the highest confidence.

Attack Scenario. In order to have a fair comparison, we have applied the following strategy for all attacks:

1. We have done a grid search hyperparameter optimization for all models according to the values in Table 3, meaning we trained each model for all possible parameter combinations with the full profiling set N_P and validated its performance with 2000 attack traces from N_A. The model variants that yield the lowest KGE were considered for further analysis.
2. Next, we performed 20 (10 for the software implementation) independent attacks using the models obtained in the first step and calculated the mean KGE, whereas each attack was conducted with an independent set of 2000 traces from N_A.

The experiments last around three weeks on a single Nvidia GTX 1080 Ti graphics card. All implemented attacks are based on the Keras [1] and scikit-learn [2] frameworks.

4.3   Results for Unprotected Hardware Implementation

Our first series of experiments have been based on the public data set of the *DPA Contest v2* [30]. These side-channel traces were acquired from an unprotected AES design running on an FPGA platform. The used AES module performs one round per clock cycle. Each trace contains 3253 sample points and covers a complete encryption operation.

As a preprocessing step, we transformed all traces to have zero mean and unit variance (sometimes referred to as data standardization). We also investigated the effect of normalizing the traces into a range of [0, 1] or having no preprocessing at all, but got the best results with standardization.

We have not reduced the dimension of the traces, except for TA. TAs requires the attacker to determine a small number of sample points which contain the most discriminative information. Otherwise they can become computationally intractable as laid out in Sect. 2.1. We employed a PCA for that purpose with the number of components to keep as hyperparameter. The exact parameter configurations for the networks can be found in the Appendix.

Figure 5 shows the mean key ranks according to the number of traces for each implemented attack. From that, we can make the following observations:

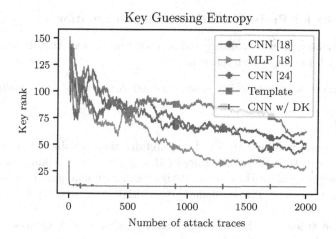

Fig. 5. Mean ranks when targeting the first key byte of an unprotected AES hardware implementation.

- Our CNN with domain neurons outperforms all other approaches, meaning it has the lowest mean KGE after 2000 attack traces (8 vs. 30 when comparing it with the MLP-based attack).
- None of the attacks reaches a stable key rank of zero. We indeed examined that a larger number of traces is necessary to recover the key with a success rate of 100% (approximately 5000 with the MLP). This is not completely in line with the good results obtained in [18] and could be a direct consequence of our hyperparameter optimization process and the assumptions we had to make when reimplementing the networks. Additionally, targeting an S-box that is not followed by a register may not be the optimal choice in a hardware setting since the leakage of combinatorial logic is typically lower than register leakage.
- Even though our developed CNN is not able to converge to a key rank of zero (also not with more than 2000 attack traces), it stabilizes under the top ten with less than ten attempts. The CNN with DK converges so much faster due to higher probabilities for the top ranked key estimations. For example, the top five probabilities obtained after the SOFT layer account for approximately 95% of the complete probability distribution, an effect that is not visible for the baseline models with such an intensity. This makes our attack especially interesting for settings where the number of attack traces is restricted to a few of tens or even less.

4.4   Results for Protected Software Implementation

The second platform we have targeted is a software-based AES implementation equipped with two SCA countermeasures:

- A first-order secure masking scheme called *Rotating Sbox Masking (RSM)*, and
- Shuffling.

In RSM, the mask values are fixed to carefully chosen values, but rotated for every execution. It is therefore considered a lightweight masking scheme. The employed shuffling algorithm in the design randomly changes the order of execution of the S-boxes. The implementation originates from DPA Contest v4.2 [4].

Since the traces which were provided within the DPA Contest v4.2 were generated with a single fixed key and we are required to have random keys for the profiling, we self-acquired the data sets N_P and N_A on a ChipWhisperer-Lite board for the second series of experiments. The board was running with a clock frequency of 7.37 MHz. Each trace is composed of 10 000 sample points representing approximately the first one and a half rounds of an encryption operation. As an example, we have plotted three measurements in Fig. 6.

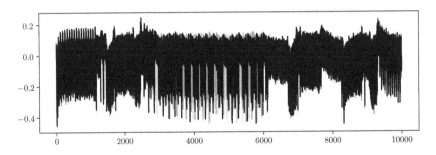

Fig. 6. Three example traces of the protected software AES implementation. The shuffling of the S-boxes is clearly visible in the range between the time samples 2500 and 6500.

We have applied the same data standardization preprocessing as for the hardware target also to the traces of the software implementation. Additionally, a separate hyperparameter optimization for the software data set has been conducted. The results of the attacks are illustrated in Fig. 7. One can notice that:

- The CNN with DK performs very well on the software implementation. Indeed, it takes roughly 20 traces to get to key rank zero for the first time and stabilizes after roughly 600 attack traces. This demonstrates that our developed method is also able to defeat cryptographic implementations which are secured with several countermeasures.

Key Guessing Entropy

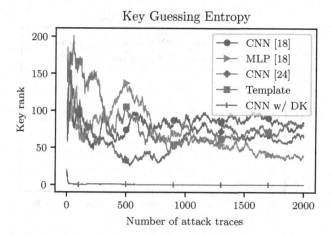

Fig. 7. Mean ranks when targeting the first key byte of an protected AES software implementation.

– Compared to the results for the unprotected hardware implementation, all approaches except ours perform worse for the software implementation. This indicates that the employed masking and shuffling countermeasures effectively decrease the leakage of the targeted S-box. We have also tested the effect of using a whole attack data set with fixed key (10 000 traces) but were not able to reach a constant KGE of zero with the TA and the networks from related work.

Examining the Effect of Domain Knowledge Neurons. In order to assess the effect of DK on the attack success, we have trained our developed CNN architecture from scratch under the exact same conditions but without the additional input of the plaintext. Afterwards, we have computed the mean KGE for the CNN without domain neurons in the same manner as we have done for the other implemented attacks. The results are shown in Fig. 8.

From the plots, it can be concluded that the information provided by the domain neurons in fact improve the performance of the network. Both CNNs (with and without DK) are able to reach a key rank below five after less than 20 traces, which indicates that our generic architecture by itself leads to a significant performance boost. However, only the network which is equipped with the domain input converges to zero. This demonstrates our assumption that additional knowledge, that is present to the attacker anyway, can be used more efficiently as it was done in state-of-the-art approaches. Maghrebi et al., e.g., used the plaintext only to generate the labels for training/profiling and was therefore not given to the networks in the attack phase to classify unseen traces [18].

Key Guessing Entropy

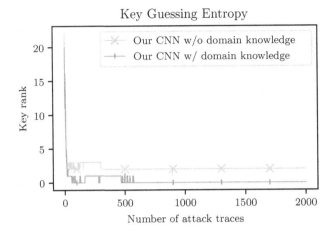

Fig. 8. Mean KGE when targeting a protected AES software implementation with, and without DK.

Hybrid learning systems (as our approach can be considered) have shown remarkably result on several real-world problems [31,34]. Our developed strategy adapts the idea to the SCA domain. The experiments presented in this section clearly illustrate that combining different types of information (e.g. side-channel traces and the plaintext) into one DL classifier can boost the performance of profiled SCAs up to several orders of magnitude (compared to state-of-the art attack methods). Furthermore, we stress that our approach may also be beneficial to evaluate other kinds of cryptographic implementations apart from AES as we make not us of any internal algorithmic structures or implementation details.

5   Conclusion

In this paper we have introduced CNNs with DK neurons as a tool for profiled SCAs. The addition of domain neurons supplies the network with extra information such as the plaintext. We showed that this feature gives a great practical advantage compared to state-of-the-art profiling attacks [18,24], which require to manually choose a certain operation of the attacked implementation for which the profiling is done. Instead, we have demonstrated by experiments with two different data sets that our proposed CNN with DK effectively manages to autonomously capture the function with the highest leakage for breaking the secret key directly. Our method can thus be seen as a novel and generic tool to assess the side-channel resistance of cryptographic implementations in a real black-box manner (i.e. assuming an attacker with no knowledge about internal implementation structures).

Future work might explore other kinds of DK than the plaintext. For instance, one could try to attack the AES subkey in the last round and feed the corresponding ciphertext into the network. An alternative path of future work could be to study the effect of domain neurons in combination with other deep NN architectures (e.g. Recurrent Neural Networks).

Acknowledgment. The authors would like to thank the reviewers for their comments. This work is supported in parts by the German Federal Ministry of Education and Research (BMBF) under grant agreement number 16KIS0606K (SecRec).

A   Network Parameters

Table 2. Network configuration of CNN with domain neurons.

Layer type	Hyperparameters
Trace input	-
Convolution 1D	Filters = 8, filter length = 8
Max-pooling	Pool length = 2
Dropout	$P_{Drop} = 0.2$
Convolution 1D	Filters = 16, filter length = 8
Batch normalization	-
Max-pooling	Pool length = 2
Dropout	$P_{Drop} = 0.2$
Convolution 1D	Filters = 32, filter length = 8
Batch normalization	-
Dropout	$P_{Drop} = 0.2$
Flatten	-
Domain input	Neurons = 256
Concatenate	-
Fully-connected	Neurons = 400
Batch normalization	-
Dropout	$P_{Drop} = 0.2$
Output	Neurons = 256

Table 3. Results of grid search hyperparameter optimization for all implemented attacks. Chosen values for the hardware attack are marked in bold letters, chosen values for the software attack are marked by underlining.

Type	Hyperparameter
2-layer MLP	Batch size: [**50**, <u>100</u>]
	Epochs: [100, **<u>200</u>**]
	Optimizers: [SGD, **RMSprop**, <u>Adam</u>, Nadam]
	Activation: [ReLU, **sigmoid**, tanh]
	Learn rate: [0.001, <u>0.0001</u>, **0.00001**]
3-layer CNN	Batch size: [**<u>50</u>**, 100]
	Epochs: [**<u>100</u>**, 200]
	Optimizers: [SGD, **RMSprop**, <u>Adam</u>, Nadam]
	Learn rate: [0.001, **<u>0.0001</u>**, 0.00001]
	P_{Drop}: [**<u>0.2</u>**, 0.3, 0.4, 0.5]
5-layer CNN	Batch size: [**<u>50</u>**, 100]
	Epochs: [**<u>100</u>**, 200]
	Optimizers: [<u>SGD</u>, RMSprop, Adam, **Nadam**]
	Learn rate: [**<u>0.001</u>**, 0.0001, 0.00001]
TA	PCA components: [1, ..., **5**, <u>6</u>, ..., 100]
5-layer CNN w/DK	Batch size: [**<u>50</u>**, 100]
	Epochs: [**<u>100</u>**, 200]
	Optimizers: [<u>SGD</u>, **RMSprop**, Adam, Nadam]
	Activation: [**<u>ReLU</u>**, sigmoid, tanh]
	Learn rate: [**<u>0.001</u>**, 0.0001, 0.00001]
	P_{Drop}: [<u>0.2</u>, **0.3**, 0.4, 0.5]

References

1. Keras Documentation. https://keras.io/
2. Scikit-learn: machine learning in Python. http://scikit-learn.org/stable/
3. Bartkewitz, T., Lemke-Rust, K.: Efficient template attacks based on probabilistic multi-class support vector machines. In: Mangard, S. (ed.) CARDIS 2012. LNCS, vol. 7771, pp. 263–276. Springer, Heidelberg (2013). https://doi.org/10.1007/978-3-642-37288-9_18
4. Bhasin, S., Bruneau, N., Danger, J.-L., Guilley, S., Najm, Z.: Analysis and improvements of the DPA contest v4 implementation. In: Chakraborty, R.S., Matyas, V., Schaumont, P. (eds.) SPACE 2014. LNCS, vol. 8804, pp. 201–218. Springer, Cham (2014). https://doi.org/10.1007/978-3-319-12060-7_14
5. Brier, E., Clavier, C., Olivier, F.: Correlation power analysis with a leakage model. In: Joye, M., Quisquater, J.-J. (eds.) CHES 2004. LNCS, vol. 3156, pp. 16–29. Springer, Heidelberg (2004). https://doi.org/10.1007/978-3-540-28632-5_2
6. Cagli, E., Dumas, C., Prouff, E.: Convolutional neural networks with data augmentation against jitter-based countermeasures. In: Fischer, W., Homma, N. (eds.)

CHES 2017. LNCS, vol. 10529, pp. 45–68. Springer, Cham (2017). https://doi.org/ 10.1007/978-3-319-66787-4_3

7. Chari, S., Rao, J.R., Rohatgi, P.: Template attacks. In: Kaliski, B.S., Koç, K., Paar, C. (eds.) CHES 2002. LNCS, vol. 2523, pp. 13–28. Springer, Heidelberg (2003). https://doi.org/10.1007/3-540-36400-5_3

8. Doget, J., Prouff, E., Rivain, M., Standaert, F.X.: Univariate side channel attacks and leakage modeling. J. Cryptogr. Eng. 1(2), 123 (2011). https://doi.org/10.1007/ s13389-011-0010-2

9. Gierlichs, B., Batina, L., Tuyls, P., Preneel, B.: Mutual information analysis. In: Oswald, E., Rohatgi, P. (eds.) CHES 2008. LNCS, vol. 5154, pp. 426–442. Springer, Heidelberg (2008). https://doi.org/10.1007/978-3-540-85053-3_27

10. Gilmore, R., Hanley, N., O'Neill, M.: Neural network based attack on a masked implementation of AES. In: 2015 IEEE International Symposium on Hardware Oriented Security and Trust, HOST, pp. 106–111, May 2015. https://doi.org/10. 1109/HST.2015.7140247

11. Goodfellow, I., Bengio, Y., Courville, A.: Deep Learning. MIT Press, Cambridge (2016). http://www.deeplearningbook.org

12. Hospodar, G., Gierlichs, B., De Mulder, E., Verbauwhede, I., Vandewalle, J.: Machine learning in side-channel analysis: a first study. J. Cryptogr. Eng. 1(4), 293 (2011). https://doi.org/10.1007/s13389-011-0023-x

13. Ioffe, S., Szegedy, C.: Batch normalization: accelerating deep network training by reducing internal covariate shift. CoRR abs/1502.03167 (2015). http://arxiv.org/ abs/1502.03167

14. Kocher, P., Jaffe, J., Jun, B.: Differential power analysis. In: Wiener, M. (ed.) CRYPTO 1999. LNCS, vol. 1666, pp. 388–397. Springer, Heidelberg (1999). https://doi.org/10.1007/3-540-48405-1_25

15. Kocher, P., Jaffe, J., Jun, B., Rohatgi, P.: Introduction to differential power analysis. J. Cryptogr. Eng. 1(1), 5–27 (2011). https://doi.org/10.1007/s13389-011-0006-y

16. Lerman, L., Bontempi, G., Markowitch, O.: Side channel attack: an approach based on machine learning. In: Second International Workshop on Constructive Side-Channel Analysis and Secure Design, COSADE 2011 (2011)

17. Lerman, L., Poussier, R., Bontempi, G., Markowitch, O., Standaert, F.-X.: Template attacks vs. machine learning revisited (and the curse of dimensionality in side-channel analysis). In: Mangard, S., Poschmann, A.Y. (eds.) COSADE 2014. LNCS, vol. 9064, pp. 20–33. Springer, Cham (2015). https://doi.org/10.1007/978-3-319-21476-4_2

18. Maghrebi, H., Portigliatti, T., Prouff, E.: Breaking cryptographic implementations using deep learning techniques. In: Carlet, C., Hasan, M.A., Saraswat, V. (eds.) SPACE 2016. LNCS, vol. 10076, pp. 3–26. Springer, Cham (2016). https://doi. org/10.1007/978-3-319-49445-6_1

19. Mangard, S., Oswald, E., Popp, T.: Power Analysis Attacks. Revealing the Secrets of Smart Cards, 1st edn. Springer, Boston (2007). https://doi.org/10.1007/978-0-387-38162-6

20. Mitchell, T.M.: Machine Learning, 1st edn. McGraw-Hill Inc., New York (1997)

21. Nair, V., Hinton, G.E.: Rectified linear units improve restricted Boltzmann machines. In: Proceedings of the 27th International Conference on International Conference on Machine Learning, ICML 2010, pp. 807–814. Omnipress, USA (2010). http://dl.acm.org/citation.cfm?id=3104322.3104425

22. O'Shea, K., Nash, R.: An introduction to convolutional neural networks. CoRR abs/1511.08458 (2015)

23. Oswald, E., Mangard, S.: Template attacks on masking—resistance is futile. In: Abe, M. (ed.) CT-RSA 2007. LNCS, vol. 4377, pp. 243–256. Springer, Heidelberg (2006). https://doi.org/10.1007/11967668_16

24. Picek, S., Samiotis, I.P., Heuser, A., Kim, J., Bhasin, S., Legay, A.: On the performance of deep learning for side-channel analysis. Cryptology ePrint Archive, Report 2018/004 (2018). https://eprint.iacr.org/2018/004

25. Saravanan, P., Kalpana, P., Preethisri, V., Sneha, V.: Power analysis attack using neural networks with wavelet transform as pre-processor. In: 18th International Symposium on VLSI Design and Test, pp. 1–6, July 2014. https://doi.org/10.1109/ISVDAT.2014.6881059

26. Schindler, W., Lemke, K., Paar, C.: A stochastic model for differential side channel cryptanalysis. In: Rao, J.R., Sunar, B. (eds.) CHES 2005. LNCS, vol. 3659, pp. 30–46. Springer, Heidelberg (2005). https://doi.org/10.1007/11545262_3

27. Smith, L.N., Topin, N.: Deep convolutional neural network design patterns. CoRR abs/1611.00847 (2016). http://arxiv.org/abs/1611.00847

28. Srivastava, N., Hinton, G., Krizhevsky, A., Sutskever, I., Salakhutdinov, R.: Dropout: a simple way to prevent neural networks from overfitting. J. Mach. Learn. Res. **15**, 1929–1958 (2014)

29. Standaert, F.-X., Malkin, T.G., Yung, M.: A unified framework for the analysis of side-channel key recovery attacks. In: Joux, A. (ed.) EUROCRYPT 2009. LNCS, vol. 5479, pp. 443–461. Springer, Heidelberg (2009). https://doi.org/10.1007/978-3-642-01001-9_26

30. TELECOM ParisTech SEN research group: DPA Contest v2. http://www.dpacontest.org/v2/

31. Towell, G.G., Shavlik, J.W.: Knowledge-based artificial neural networks. Artif. Intell. **70**(1–2), 119–165 (1994)

32. Wang, D., Mao, K., Ng, G.W.: Convolutional neural networks and multimodal fusion for text aided image classification. In: 2017 20th International Conference on Information Fusion, Fusion, pp. 1–7, July 2017. https://doi.org/10.23919/ICIF.2017.8009768

33. Whitnall, C., Oswald, E.: Robust profiling for DPA-style attacks. In: Güneysu, T., Handschuh, H. (eds.) CHES 2015. LNCS, vol. 9293, pp. 3–21. Springer, Heidelberg (2015). https://doi.org/10.1007/978-3-662-48324-4_1

34. Xie, G.S., Zhang, X.Y., Yan, S., Liu, C.L.: Hybrid CNN and dictionary-based models for scene recognition and domain adaptation. ArXiv e-prints, January 2016

35. Zheng, Y., Zhou, Y., Yu, Z., Hu, C., Zhang, H.: How to compare selections of points of interest for side-channel distinguishers in practice? In: Hui, L.C.K., Qing, S.H., Shi, E., Yiu, S.M. (eds.) ICICS 2014. LNCS, vol. 8958, pp. 200–214. Springer, Cham (2015). https://doi.org/10.1007/978-3-319-21966-0_15

Author Index

Printed in the United States
By Bookmasters